Essential Environmental Studies

S.P. Misra

&

S.N. Pandey

Ane Books Pvt. Ltd.

New Delhi ♦ Chennai ♦ Mumbai

Bengaluru ♦ Kolkata ♦ Thiruvananthapuram ♦ Lucknow

Essential Environmental Studies

S.P. Misra & S.N. Pandey

First Edition : 2008
Reprint : 2009 (Paperback Edition)
Second Edition : **2010**

Published by

Ane Books Pvt. Ltd.
4821, Parwana Bhawan, 1st Floor, 24 Ansari Road,
Darya Ganj, **New Delhi** - 110 002, India
Tel.: +91(011) 23276843-44, Fax: +91(011) 23276863
e-mail: kapoor@anebooks.com, Website: www.anebooks.com

Branches

- Avantika Niwas, 1st Floor, 19 Doraiswamy Road, T. Nagar, **Chennai** - 600 017, Tel.: +91(044) 28141554, 28141209 e-mail: anebooks_tn@airtelmail.in
- Plot No. 59, Sector-1, Shirwane, Nerul, **Navi Mumbai** - 400 706, Tel.: +91(022) 27720842, 27720851 e-mail: anebooksmum@mtnl.net.in
- 38/1, 1st Floor, Model House, First Street, Opp. Shamanna Park, Basavannagudi, **Bengaluru** - 560 004, Tel.: +91(080) 41681432, 26620045 e-mail: anebang@airtelmail.in
- Flat No. 16A, 220 Vivekananda Road, Maniktalla, **Kolkata** - 700 006, Tel.: +91(033) 23547119, 23523639 e-mail: anekol@vsnl.net
- # 6, TC 25/2710, Kohinoor Flats, Lukes Lane, Ambujavilasam Road, **Thiruvananthapuram** - 01, Kerala, Tel.: +91(0471) 4068777, 4068333 e-mail: anebookstvm@airtelmail.in

Representative Office

- C-26, Sector-A, Mahanagar, **Lucknow** - 226 006 Mobile - +91 93352 29971

ISBN : 978-93-8015-657-6

Printed at : Gopaljee Enterprises, Delhi

Preface to the Second Edition

The first edition and subsequent reprint of the book has been well received and accepted by a wide circle of students and teachers. Considering the urgent requirement of the book in the market, some quick changes to update the text and improvement in certain illustrations have been made. The number of objective questions has been further increased. We hope that the present edition shall continue to satisfy the needs of the readers. We further record our appreciation to the publishers for their sincerity and effeciency.

Authors

Preface to the First Edition

Environment today has become a matter of global concern. A tremendous rise in human population integrated with technological advancement all around the world, has caused ruthless exploitation of scarce natural resources and infused several environment problems. The world is heading towards an environmental disaster. As environment belongs to all of us, these problems have become cancerous to human welfare. It is now essential that every citizen be educated and made aware of various environmental issues, their mitigations and social aspects.

The book is made on 'Core Module Syllabus' of University Grants Commission for environmental studies for undergraduate course, of all branches of higher education in the wake of the landmark judgement of Supreme Court of India (2003). A short extract of the judgement is as :

> "..... for more than a century, there was a growing realization that mankind has to live in tune with nature, if life was to be peaceful, happy and satisfied. In the name of scientific development, man started distancing himself from nature and even developed an urge to conquer nature. Our ancestors had known that nature was not subduable and, therefore, had made it an obligation for man to surrender to nature and live in tune with it."

The textbook, in toto is based on the new mandatory qualifying course on Environmental Studies for undergraduates of Indian Universities, irrespective of any discipline and even for those with no background in the science. It will cater not only to the needs of the students, but will also prove quite useful for the future of their family, community, humanity and the earth. This is infact more than a book: it is about reader's life, what is in store for him and future of this fragile planet. Owing to this fact it has been entitled :Essential Environmental Studies", a must book for every responsible citizen of the country.

The text matter, following exactly the Core Module Syllabus of UGC, is spread over 8 units and 58 chapters. The book is aimed at creating awareness amongst readers about various environmental issues and their solutions. Every effort has been made to create increased sensitivity among readers about environmental problems like dwindling natural resources-freshwater crisis, deforestation, desertification, energy crisis, hunger and poverty; imperilled ecosystems; depleting biodiversity-endangered species and wildlife trade; escalating pollution-global warming, acid rains, ozone layer depletion and nuclear holocausts; exponential population growth-urbanisation; and many social issues. The last unit-Field Work, will inculcate inspiring ideas and guidance for meaningful project work on various environmental problems of the area.

Written in student friendly manner no effort has been spared to make subject matter up-to-date and concise supported with relevant case studies to exemplify problems, solutions, successes and failures. The book has been supplemented with a large number of illustrations and informative figures with chapters independent and self-contained. Chapter summary, review and objective questions are provided at the end of each chapter. Glossary of terms and some valuable appendices are included penultimately.

The authors acknowledge a deep sense of gratitude to authors and publishers of various books, magazines and newspapers and various TV channels for elucidating their information in endeavouring this work. Also, they would remiss if thanks to Mr. Sunil Saxena and Mr. J.R. Kapoor of Ane Books India, for their cooperation and favour are not recorded. Suggestions from its users and experts to improve the book are welcome.

Authors

Contents

9. Role of an Individual in Conservation of Natural Resources 179

10. Equitable Use of Resources for Sustainable Life Styles 186

UNIT III: Ecosystems

11. Concept of an Ecosystem 197

12. Structure and Function of Ecosystem 205

UNIT VIII: Field Work

List of Boxes

Multidisciplinary Nature of Environmental Studies

- *Definition, Scope, Importance and Need for Public Awareness*

1

CHAPTER

Definition, Scope, Importance and Need for Public Awareness

Learning Objectives
Introduction • The Environment • Physical or Abiotic • The Biosphere • Environment And Man Relationship • Impact of Man on The Environment • Environmental Studies • Global Concern • Indian Initiatives • Environmental Education.

Introduction

'Environment' today is a global issue. Long before the advent of the man, the earth nurtured lush vegetation which bred and sustained life. Ruthless exploitation of nature by man to meet his ever increasing needs and greed, has brought about the ecological balance of our planet on the brink of collapse. It is now man's turn to pay the debt by keeping the earth lush and forests green so that we all live together and life sustains in a balanced environment. But, this concern and respect for nature is until now limited to a very small segment of the society. For tackling global environmental crisis, there is no other option except that the humanity must live within carrying capacity of the earth and must adopt life styles and developmental paths that respect and work within nature's limits. What is urgently needed is that this awareness of creating a more liveable environment becomes as universal as possible, and as far as environmental awareness is concerned, the whole population should become responsive.

Today, our lives have become so unbearable, if not quite tragic, largely because of our greed which far exceeds our needs, and our unwise view of progress and prosperity have ensured our blindness of approaching collective suicide. We have failed to live peacefully with the nature. No wonder, the beautiful face of nature has began to wear a frown and look ugly. We seem to be waiting to burn our fingers before waking up to the impending disaster.

And, unless we become seriously aware of the damage, how can we feel the urgency or even the necessity to take any remedial step? We are facing the painful facts squarely: vast areas in the world have lost their forest cover, our drinking water has become impotable, our food is contaminated with toxic chemicals, and even the air we breath is slowly poisoning and disabling us.

Let us, therefore, educate our new generation as well as grown ups, the sheer necessity of environmental harmony and realise that we will tragically perish if we foolishly continue to damage the environment. Let us not wait for any divine intervention to rescue us. Of what use will all our advancement, prosperity and power be if we gain all wealth but lose our lives or are reduced to the "living death"?

THE ENVIRONMENT

The term 'environment' is derived from a French word *environ* means to encircle or surround. Einstein has rightly defined environment as *everything that is not me*. In *Oxford Advanced Learners Dictionary*, the environment is defined as *natural world in which people, animals and plants live*. *Environ* means surrounding and *ment* means actioning, *i.e.*, environment is the interaction between man and the nature. Thus, the term 'environment' refers to our surrounding and variety of issues related with human activity and its impact on nature. Our environment is what we see around us. Thus, we infer that "final analysis of everything present outside an individual is the environment". Maharshi Patanjali, the founder of "Yoga Darshan", in his yoga aphorism, aptly termed it as "drishyam" and defined it as

'Prakaasha-Kriya-Sthitishilam
Bhutendriyatmakam
Bhogapavaragartham drishyam;

i.e., the seen (what is seen around) is composed of elements and organisms, is of the nature of illumination, action and inertia, and is for the purpose of satisfaction of needs and attainment of the ultimate. In our limited human existence, we are encircled by nature right from the time of our birth till death.

There are two ways in which the term 'environment' is used. In **one**, it refers to what surrounds an entity, say a person, a living organism, a company etc. There are other environments also. These include home environment, business environment, political environment and so on. In **other** way, the term environment recalls us of the natural environment:

the air, water, land, plants, animals and other organisms. An entity interacts with its environment and is influenced by it. We, as men, are surrounded by people, animals, plants and physical objects which are parts of our environment. Stockholm Conference, 1972, has declared *man is both a creator and moulder of his environment*. Fate of our life depends upon what do we do with and how we interact with the environment.

For practical reasons, the study of environment is confined to the system of interacting living and non-living components. The environment is the sum total of air, water and land interrelationship among themselves and also with the human beings, plants, animals and other organisms. Thus, environment includes all physical and biological surroundings and their interactions.

The three subdivisions of the earth: the atmosphere, hydrosphere and lithosphere, constitute physical (non-living or abiotic) environment of living (biotic) world and this biosphere provides not only habitats but also materials for their maintenance (fig. 1.1). The significant feature of the effect of the environment on life of an organism is the interaction of environmental elements. The physical and biological elements are dynamic in nature and interact with each other and show temporal as well as spatial variations.

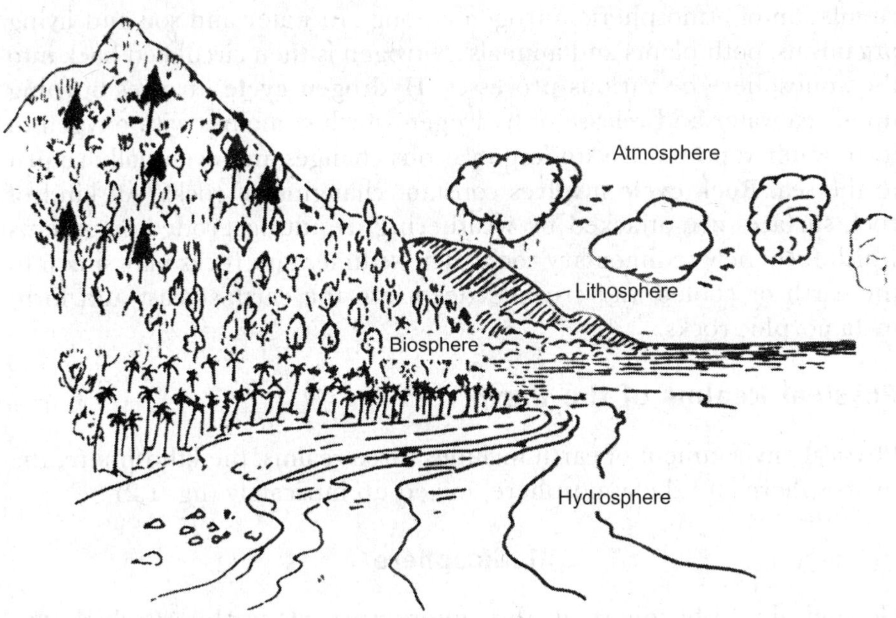

Fig. 1.1. Four components (atmosphere, hydrosphere, lithosphere and biosphere) of the natural environment.

Physical or **Abiotic Environment**

The physical or abiotic environment is made up of non-living components and includes air, water and land. *Atmosphere* is the realm of air comprising various gases that blanket the earth. Water bodies, both fresh and saline, found on the earth, make up the *hydrosphere*. *Lithosphere* includes continents and island which are made of solid rock and soil. Climate, tidal processes, chemical and geological processes, fossil fuel etc. are other integral parts of the environment. The sun is the most important and the only source of energy on the earth. It is also responsible for essential bio-geochemical cycles that involve and influence various components of the environment.

Carbon cycle involves circulation of carbon atoms in the nature. Plants take carbon dioxide from the air to carry out photosynthesis and make food which is used by themselves and by animals including man to make their tissues. During the process, plants release oxygen which is consumed by animals and plants during respiration. And when plants and animals die, bacteria and fungi use the dead tissues as food and, in the process, the carbon dioxide is released back into the atmosphere. Carbon dioxide is also given out during the respiration of all living organisms. Thus, a balance of percentage of oxygen as well as carbon dioxide remains constant in the atmosphere. **Nitrogen cycle** involves a continuous circulation of atmospheric nitrogen among air, water and soil and living organisms, both plants and animals. Nitrogen is then circulated back into the atmosphere by various processes. **Hydrogen cycle** involves burning up of sea water and release of hydrogen which combines with oxygen to form water vapours that undergo various changes and eventually return to the sea. **Rock cycle** involves constant changing of rocks. All kind of rock surfaces are attacked by weathering, resulting eroded fragments which form new sedimentary rocks. The sedimentary rocks may sink into the earth or cooled and crushed deep into the earth's crust and form metamorphic rocks.

Physical Realms of the Earth

Physical environment of earth includes three realms, the lithosphere, the hydrosphere and the atmosphere, mixed up intricately (fig. 1.2)

Lithosphere

Geologically, lithosphere is the outer crust of earth on which the continents and basins rest. It is heterogeneous and is made up of crust and upper mantle. Lithosphere includes different varieties of land masses and land forms with different types of ecosystems ranging from arid desert to the temperate forests and tropical rain forests. It consists of about 15 slowly drifting plates which move at a rate of 1.3-20 cm a year, on a layer of partially molten rock in the mantle called alhenosphere and these plates carry the continents and ocean floor with them. Mountains such as Himalayas are formed where two plates collide. Lithosphere covers about 30 per cent of the earth surface and is made up of solid rocks and soil. It is the thickest in the continental region where it has an average thickness of 40 km and is the thinnest in the oceans where it has maximum thickness of 10-12 km. The continental crust, which is made up of crystalline rocks, has undergone many changes such as continental drift, faulting and folding and other earth movements over a past millions of years of geological time. Technically the lithosphere includes both, the land mass and the ocean floor, but it is often used to indicate only the land surface. It is only upper few feet's of the soil which is the most important part of the lithosphere as it not only contains organic matter but also supports activities of the living organisms. More than a dozen elements constitute bulk of earth's crust (table 1.1).

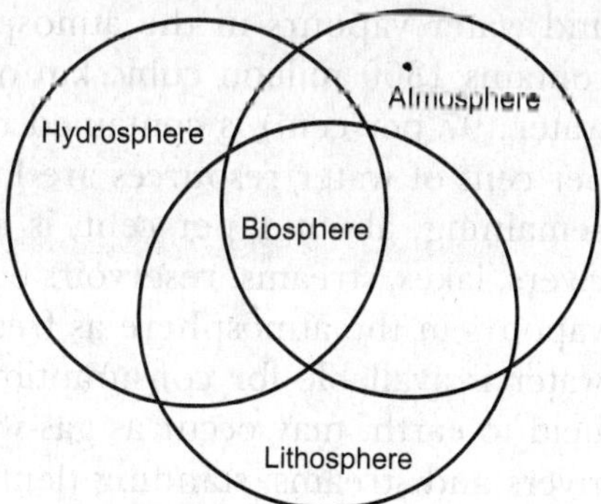

Fig. 1.2. Physical realms of the earth.

TABLE 1.1. Composition of Earth's Crust by Weight

Element	*Percentage*	*Element*	*Percentage*
1. Oxygen	49.85	8. Magnesium	2.11
2. Silicon	26.03	9. Hydrogen	0.97
3. Aluminum	7.28	10. Titanium	0.41
4. Iron	4.12	11. Chlorine	0.2
5. Calcium	3.18	12. Carbon	0.19
6. Sodium	2.33	13. Others	1.00
7. Potassium	2.33		

As regards percentage by volume, oxygen contributes 93.8 per cent of composition of the earth's crust. Among many minerals taking part in its composition, 58 per cent mass of the terrestrial rocks is made up of feldspars ($K_2 Al_2 Si_6 O_{16}$, $Na Al_2 Si_3 O_6$, $Ca Al_2 Si_3 O_8$).

Hydrosphere

Hydrosphere is the water domain as more than 2/3rd (70.8 per cent) of earth's surface is covered with water in the form of oceans, seas, estuaries, glaciers, polar ice, rivers, lakes, streams, reservoirs, shallow water bodies and water vapours in the atmosphere. It is assumed that hydrosphere contains 1360 million cubic km of water. By far the greatest volume of water (97 per cent) is contained in the oceans and inland seas. About 2 per cent of water resources are locked in the glaciers and ice caps. The remaining, about 1 per cent, is available as surface water resources in rivers, lakes, streams, reservoirs etc., ground water resource and as water vapours in the atmosphere as fresh water. Only a small fraction of fresh water is available for consumption for human beings. The water, being held to earth, may occur as gas-water vapour; as liquid-flowing (lotic) in rivers and streams, standing (lentic) in lakes, swamps and absorbed into rocks and soil; as ground water and as solid—ice in polar regions, mountain peaks and glaciers.

Interchange of water between earth's surface and atmosphere is governed by water cycle (or hydrological cycle). The water cycle encompasses the movement of water from the oceans to the atmosphere and back to the oceans by way of evaporation and precipitation. Evaporation of water also occurs in inland water bodies such as rivers, streams, lakes etc. and ground water flow. Oceans contribute water about 4,52,600 cubic km/year to the atmosphere through evaporation but in turn receive only, 4,11,600 cubic km/year as precipitation. From land surface only 72,500 cubic km/year water is evaporated but in turn it receives 1,13,500 cubic km/year as precipitation. Of this water received

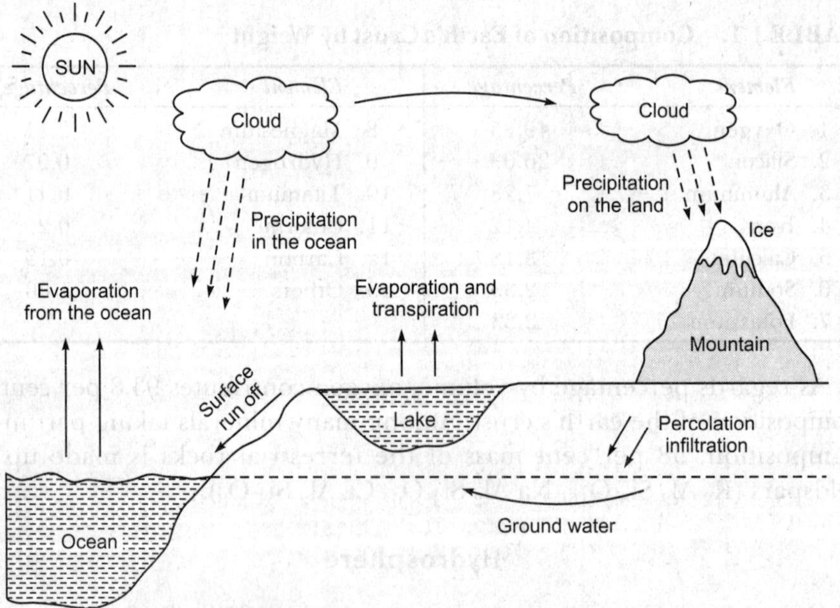

Fig. 1.3. A typical water cycle.

from precipitation about 41,000 cubic km/year goes back to the oceans as surface run off through rivers, streams, lakes and other tributaries to equalise the deficit (fig. 1.4). A small amount of water is retained on the land surface for use to the biosphere, replenishing ground water stock or locked up in ice deposits for considerable period of time.

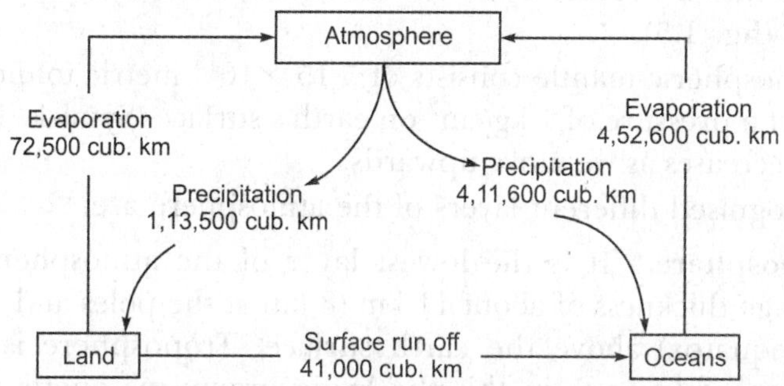

Fig. 1.4. Quantitative circulation of water in water cycle (hydrological cycle).

Continuous evaporation of water, both from sea and inland water bodies, causes it to rise as vapour and reach in the upper layers of the atmosphere where it cools and condenses and falls on the land in the form of rain (precipitation). The energy required for evaporation comes from sun. Both evaporation and precipitation occur nearly in a state of equilibrium. The precipitation that falls on the land runs off or stays on the ground surface or sinks into the soil and rocks to form ground water stock, the water table. Again the water gets evaporated and reaches the atmosphere, thereby operating water cycle (fig. 1.3).

Atmosphere

Atmosphere is the blanket of air including various gases, water vapours and variety of fine particulate matter that surround the earth and extend over 560 km to a zone characterised more by magnetic fields and ionised particles than by familiar air near the surface. The lower atmosphere is a mixture of molecules of three important gases: nitrogen, oxygen and carbon dioxide and trace amount of several other gases. The amount of gases in the atmosphere is normally stable but under some circumstances they react chemically to form new compounds. The atmosphere forms the major source of carbon dioxide to plants and oxygen to both plants and animals. The air is colourless, odourless and tasteless, mobile, compressible and expansible. Although, it can not be seen yet air is vital for life. It can, however, be felt when it is in motion. Layers of gases high in the atmosphere save us from harmful radiations, provide insulation of heat and stabilise climatic conditions. Destruction of layers of gases such as ozone layer, caused by pollution, is altering the earth's climate and causing global warming. It is difficult to ascertain vertical extent of the atmosphere as there is no sharp boundary with extra-terrestrial space. However, on the basis of temperature profile and other related phenomena, the atmosphere can be recognised/or categorised into a number of ill-defined layers, each layer having its own properties (fig. 1.5).

The atmospheric mantle consists of 5.15 $\times 10^{15}$ metric tonnes of gases which exert a pressure of 1 kg/cm^2 on earth's surface (Sytnick, 1985). The pressure decreases as we move upwards.

The recognised different layers of the atmosphere are:

(1) **Troposphere.** It is the lowest layer of the atmosphere with an average thickness of about 11 km (8 km at the poles and 14.5 km at the equator) above the earth surface. Troposphere is the most important layer as in this the living organisms operate. It is the densest layer because most of the gases are compressed. The layer contains about ¾th of the atmospheric mass. In being closest to the earth surface, it greatly influences biosphere. Strong air movements and storms and clouds are witnessed in this region. All weather activities occur in troposphere and hence it plays a significant role in the climates of various regions on the earth. It is the house of life sustaining gases: carbon dioxide, oxygen, nitrogen etc. and also contains water vapours and dust.

Troposphere is characterised by fairly uniform decrease in the temperature with increase in altitude which is about 6°C per km to a minimum of –50°C to –60°C or so. The zone marking the end of this temperature decrease is the top of troposphere called **tropopause**. This layer is of greatest interest for pollution control. Some of the recent pollution impacts are global warming and acid rains.

(2) **Stratosphere.** Above the troposphere, lies about 39 km thick stratosphere extending upto an altitude of 50 km above the earth surface. Ozone present in this region forms a layer called **ozonosphere** which functions as an umbrella and filters the harmful ultra-violet rays of sun reaching the surface of the earth. It also serves as a blanket in reducing the cooling rate of the earth. Thus, an equilibrium between ozone and rest of the gases in the air, is a significant feature of the environment. Most military and long distance air crafts operate in this region of the atmosphere.

In the region, the temperature is constant upwards to about 20 km and then increases as we move upwards from –80°C up to a maximum of 0°C near its outer limits, the **stratopause**.

(3) **Mesosphere.** Above the stratosphere, beyond the stratopause, is about 35 km thick mesosphere. It extends up to 85 km above the

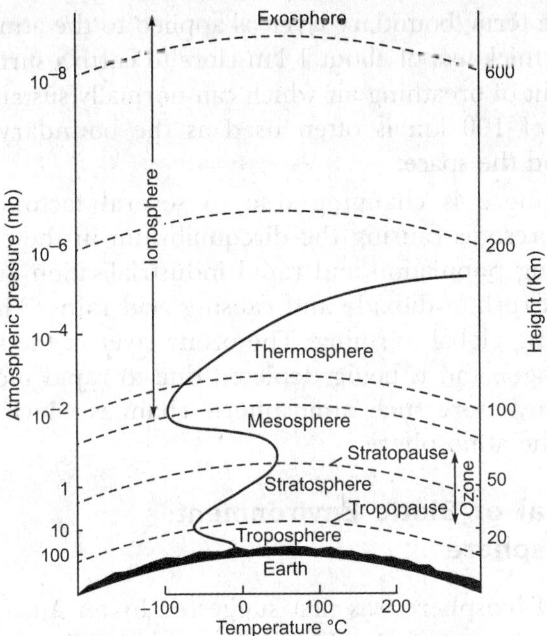

Fig. 1.5. Stratification of the atmosphere.

earth surface. This region is thick enough to burn and disintegrate most meteorites due to friction increasing as they pass through it.

In mesosphere, temperature decreases slowly from 0°C in its lower layer to about –95°C at its outer limits, the **mesopause**.

(4) **Thermosphere.** Thermosphere starts above mesopause. It is about 515 km thick which extends upto about 600 km above earth's surface. The temperature again rises to a great extent and at times, depending upon solar activity, reaching 2000°C or even more at about 500 km height. Although the gas molecules exhibit high kinetic energy resulting drastic increase in temperature but gas molecules are too sparse to transfer the significant amount of energy to an ordinary thermometer and thus the exposed hand of an astronaut would not feel hot in this region.

Coinciding with the lower portion of atmosphere, the thermosphere, is the **ionosphere** which is delimited on the basis of ionisation of molecules of oxygen and their effect on the propagation of radio waves which are reflected at great heights. This layer is important for operation of satellites.

(5) **Exosphere.** Atmosphere above the ionosphere which takes atmosphere except that of hydrogen and helium, extending up to 32190 km from the surface of the earth, is called exosphere or outer space. It has a very high temperature due to solar radiation.

At times, the term 'boundary layer' is applied to the atmospheric layer with a normal thickness of about 1 km close to earth's surface. It defines the normal limit of breathing air which can normally sustain the life. The **Karman line** of 100 km is often used as the boundary between the atmosphere and the space.

The atmosphere is changing due to several factors. Large scale clearance of forests is causing the disequilibrium in the composition of gases. Increasing population and rapid industrialisation are resulting in the increase of carbon dioxide and causing acid rains. The greenhouse gases are causing global warming. The ozone layer in the stratosphere is also facing danger and is being depleted due to rapid industrialisation. There are many more such atmospheric changes which are resulting disbalance in the atmosphere.

Biological or Biotic Environment: The Biosphere

The concept of biosphere was first suggested by an Austrian Geologist, Edward Suers long before. The idea, which failed to gain much significance at his time, later became not only necessary to understand but also to recognise its validity. In fact, biosphere is a thin shell that not only encapsulates the earth and encompasses lithosphere, hydrosphere and atmosphere but also supports life. This portion of earth, harbours all ecosystems and sustains all forms of life including human beings. Till now, more than 3,50,000 species of plants and 11 million species of animals have been identified and described. All life sustaining resources such as air, water, soil, food, energy etc. are procured and cycled in this region and gaseous, liquid and solid waste products are discharged into it. There occurs a continuous interaction among living organisms and in between them and their physical environment. Thus, this region affords sustenance to various life forms.

Biosphere, the fourth realm of the environment, supports micro-organisms, plants and animals including man. The living organisms are inseparable and interrelated and none lives alone. They interact with one another and with their physical environment and affect the structure and function of a living community. The interaction of an organism with others, arises on account of the organism's primary necessities such as food, reproduction, protection etc. The interacting organisms in a community cause changes in the physical environment and make the condition favourable for a new set of organisms. Such modification of community into a new one is termed as **succession**. Any unit in which there exists interaction between various organisms and their physico-chemical environment is referred to as **ecosystem**. The major ecosystems of the world are called **biomes**.

The upper limit of the biosphere is determined by availability of oxygen, moisture, suitable temperature and air pressure. Decrease in these factors, with increase in height, delimits upper limit of the biosphere. Bacteria have crossed this limit of about few hundred meters for most of the living organisms and are found upto 15 km or so. The lower limit of biosphere is determined by availability of required amount of oxygen and light which can sustain life. Over the land, it is upto the depth of deepest roots of trees or depth upto which the burrowing organisms can live whereas in ocean's, limit extends upto great depth of 9 km.

Appearance of free oxygen on our planet is traced back to 2.9-3.0 billion years ago with the onset of cyanobacteria which could split water molecules to obtain hydrogen atoms and evolve oxygen. It was the most significant step in the organic evolution. Some 450 million years ago enough oxygen could accumulate and paved a comfortable way for the development of aerobic forms of life which breath oxygen. The organic enrichment of earth's surface resulted in the formation of soils and dense strands of vegetation to harbour animal life.

Environment and Man Relationship

Man appeared on the planet earth barely a million year ago. During first 40 to 50 thousand years of his existence, he lived as wanderer, hunter and gatherer of food resources. The relationship between man and the environment is as old as history of the man itself. Man has enjoyed a dominant position over all other living organisms around him because of his erect posture, free hands and high degree of intelligence. Initially he gained mastery over his predators as his free hands enabled him to use self produced tools. Man's interaction with the environment became pronounced during **Pleolithic period** (50 thousand to 1 million year ago) when he gathered the knowledge of fire and tools to bring about change in the environment. Initially he was well associated with biotic community around him and was eager to have definite knowledge of the forces of the nature and of plants and animals. Despite the fact, he was not strongest, swiftest or the hardest among congners, he had effective social organisation, intelligence and manipulative power that made him well adaptive to his environment. During **Neolithic period** (10,000 to 6000 BC) only, man started domesticating plants as agriculture and animals as animal husbandry, used metals and harnessed energy. It is this period during which permanent villages, inter group cooperation and trade routes were developed and economics began at the expense of the environment. This proved the vital key to survival success as man's luster to master the nature for his vested interests. **Bronze age** (shortly after 3,000 BC) and **Iron age** (starting 1000 BC) are known for grievous environmental exploitation by large-scale deforestation, clearance of fields, mass scale grazing of pastures and ploughing, thus, carving the entire landscape within his reach to cope with his increasing economic demands. About 50,000 years ago, the total world population has been estimated to be not more than 10 million. The population has shown a tremendous rise from 1800 when it was around one billion to 1987 when it became about 4 billion and is expected to be 7 billion by 2010. This indiscriminate increase in human population is resulting over-exploitation of natural resources and pollution, causing serious damage to the environment. The main theme of the environmental studies focuses around man because all environmental problems are studied in relation to man and man himself is responsible for most of the environmental problems he faces.

Modern man's environmental perception has altogether a different one. A rapid socio-economic development of humanity has most vigourously stimulated changes in the form of interaction between the man and the environment with far reaching consequences for the nature. Man's rational behaviour towards the environment and the society changes with change in modern thought, attitude and culture. For humans involved in socio-economic revolution, the natural environment is a source of whatever he can take. In order to meet rising demands of the people to raise their standard of living and finally to enhance the power, mankind is engaged in a programme of self-defeat. It is estimated that today $2/3^{rd}$ of human population lives in underdeveloped areas of the world. Most of them drink unsafe water, eat inadequate and unsafe food, live in unfit dwellings, breath impure air and dispose wastes recklessly. The developed world is prudent to protect their own environment and conserve their own resources but they divert their problems on developing countries and encourage them to produce such requisites as they need or supply new materials as per their requirements for which they readily furnish loans and provide technical know-how.

Impact of Man on the Environment

In natural state, life forms keep maintaining an equilibrium with their environment. Abundance and activities of each species are governed by resources available to them. Interaction among species is a common feature. The product of one species may form the food of the other in the surrounding. Man alone has potential to gather resources from beyond his immediate surroundings and process them into different forms. This has enabled him to thrive and flourish beyond natural constraints and resulted overloading of pollutants in the system disturbing natural equilibrium of the environment. To satisfy his basic needs of food, shelter, clothing etc., primitive man used unprocessed natural resources readily available to him in his surroundings. The residues produced by use of these resources could easily be assimilated by the environment and left behind very few things which were readily cycled and left no significant impact on the environment. The development of skill of creating fire affected the environment by huge loss of forests and grasslands and consequently modified the environment. Agricultural development has also played a key role in environmental degradation. Green revolution, no doubt, increased the agricultural productivity, did a lot of damage to the environment. Man's ever increasing needs and greed have most vigorously stimulated changes in the manner of interaction between him and the nature with far reaching consequences for the environment. The balance between the world's population carrying capacity and natural resources is being disturbed tremendously. Man's increased demand of food, water, clothing, shelter and articles for luxury has resulted in the rapid industrialisation. Automobiles, household appliances, processed foods, beverages etc. have now become so popular as to seem necessities. These acquired needs, which are usually met by the items that must be processed, refined, distilled or manufactured, have become a major thrust of modern society. The production, distribution and consumption of these items usually result in more complex residuals or wastes or sometimes both. Many of these residues and wastes are not compatible with or readily assimilated by the environment. A large number of these by-products, which are not likely to be cycled, cause complexity in the production chain and affect various ecosystems. Industrial and automobile pollution is making the air abusive. The impact of man on the environment is a major global issue and man's fate depends largely on whether he succeeds in achieving new equilibrium between the environment he has conquered and the nature in which he was created. And if this equilibrium is not maintained, man's modern socio-economic programme may lead to the programme of his own defeat. Although in many ways man has made his environment more hospitable through rapid advancement in science and technology, yet in all his efforts to improve human life he has crossed all environmental barriers and disrupted the fragile intricately woven web of life and life supporting systems.

ENVIRONMENTAL STUDIES

Environmental studies is a branch of study of inherent or induced changes in the environment. However, the two terms **environmental studies** and **environmental science** are often used interchangeably. *Environmental science* is the scientific and systematic study of environmental system, damages incurred as a result of human interaction with the environment and our role in improving the environment. **Environmental education** deals with the environmental concerns and values and educating them to the entire human population by formal and non-formal education.

Environmental studies has a broader canvas. It includes not only the study of physical and biological characters of the environment but also social, economic, cultural and even political aspects of the environment and various issues such as safe and clean drinking water, hygienic living conditions, clean and fresh air, fertile land, healthy food and development that is sustainable. Environmental studies does deal with the science where necessary, but at level understandable to those not related to the science and hence made compulsory for all undergraduate studies in India. In 1991, Supreme Court of the country issued a directive to prepare all curricula environment oriented. The directive was, in fact, in response to a Public Interest Litigation (PIL) filed by M.C. Mehta, who is popularly known as '**One Man Environ-Legal Brigade**', and is a Supreme Court lawyer and hero of environmental case v/s Union Government of India (1988) that prompted the apex court to give a mandate for creating environmental awareness among all citizens of India.

Multidisciplinary Nature of Environmental Studies

The subject environmental studies is inherently multidisciplinary. The interconnections are many. Existence and behaviour of living and non-living components of the environment require inputs from physical, chemical, biological, social and geological science (fig. 1.6). Human behaviour and management of human societies are covered by social science, economics, psychology and political science. Rules framed for environmental management and protection fall under jurisdiction of national legislation. Agreements reached regarding common problems faced by two or more countries come under purview of international law and global governance. All these aspects of the environment and their respective disciplines of study are dealt with holistic approach. Study of environment, thus, requires attention of experts from different fields. Physical sciences including chemistry and physics along with geology, atmospheric science, oceanography, geography etc. increase our knowledge of physico-chemical structure of abiotic (non-living) components of the environment. Life sciences including botany, zoology, biochemistry, biotechnology, microbiology, genetics, ecology etc. help in understanding the biotic (living) components of the environment. Mathematics, statistics and computer science serve as effective tools in ecomodelling and resource management. Education helps in making people aware of various environmental problems and their solutions. Economics, sociology and mass communication furnish inputs for dealing with socio-economic aspects related to various developmental activities. Various branches of engineering form the basis of technologies dealing with the control of environmental pollution, waste management, industrial effluent treatment, understanding of natural calamities and erection and development of various projects for environmental protection. Meteorology adds our knowledge of weather forecasts. Environmental laws provide the tools for judicious management and protection of the environment. Political science helps in solving various environmental issues at global level. Thus, the knowledge of various disciplines helps us to understand the entire spectrum of the environment (fig. 1.6).

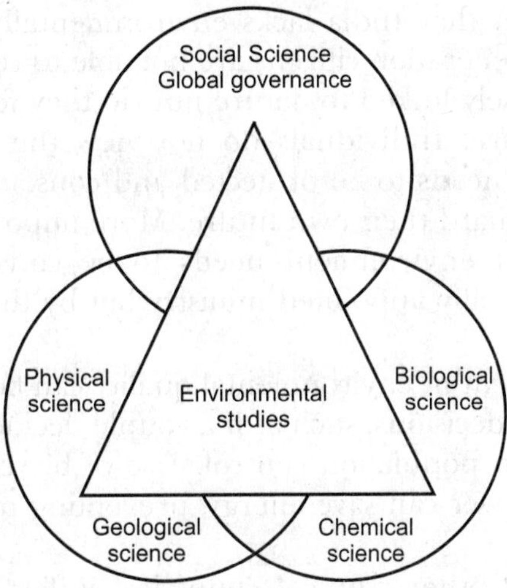

Fig. 1.6. Environmental studies as a multi-disciplinary subject.

The Tbilisi Conference, 1977, recommended that environmental study should consider the environment in its totality, be a continuous life-long process, be inter-disciplinary in its approach, examine major environmental issues from local, regional, national and international point of view, focus on current and potential environmental solutions, stress the values and necessity for local, national and international cooperation in approaching environmental issues, relate the studies to plans for development and growth, and help learners discover the symptoms and real causes of environmental problems.

Importance and Scope of Environmental Studies

Environment is global and belongs to all of us, irrespective of one's age, occupation, caste, religion or nationality. It has no boundaries and limitations. The pollutant produced at one place can be dispersed and transported to another place. Deforestation in Himalayas can cause floods in the plain. River water polluted due to industrial effluents, municipal discharge or agricultural wash outs at Kanpur can seriously affect the down stream aquatic life at Allahabad and Varanasi. Impact of mining and hydroelectric projects may spread far away. Construction of dam at Tehri can affect not only hills but also the foot hills and the plains by bringing about weather changes, temperature fluctuations and decreased precipitation. Some environmental problems, such as solid wastes, vehicular pollution etc., are localised and require to be solved locally whereas others such as global warming, ozone layer depletion, decrease in forest cover, depletion of energy sources, loss of global diversity and many others, pose serious threat to mankind and need to be tackled globally.

The problem lies that India lacks environmentally literate citizenry. Generation upon generation citizens are not able to realise how and why their deeds are closely linked to nature nor do they feel connected to or are a part of nature. Individuals do not view the environment as a personal asset that needs to be protected and conserved as a long term investment to safeguard their own future. More important is that people do not realise that environment needs to be conserved not by the government or a legally appointed ministry but by themselves and their life styles.

Much improvement in environmental quality can be brought about by individual life style decisions, such as if a couple decides to have only two children, it helps in population control. Use of bicycle can save petrol, economic use of power can save energy, preventing misuse of water can conserve water etc.

India, like most other tropical countries, is fast losing its natural resources due to the mounting demand of ever-increasing population and economic growth. Large stretches of forests are promptly clear-felled for development schemes. Rapid increase in human population is causing vertical and horizontal expansion of towns and cities. Horizontal growth requires clearing of forests or other land change patterns for building houses, residential complexes and industrial estates. The increased demand of food, is bringing more forest areas under cultivation of crops. Domesticated animals, reared for milk, food, wool etc., also require more grazing land. Deserts are being created and spread. Weeds and pests have multiplied to enormous number. Natural biotic communities are shrinking due to denudation of vast forest area. Many plant and animal species are becoming extinct. Floods inundating populated areas and large fertile areas have become an annual feature, resulting several million hectares of land water logged and enabling unfit for cultivation.

In order to respect, understand and collaborate with all these problems, it becomes necessary to view them in context of human influence looking at the economic, cultural, political structure and social equality. There is no just one right answer or solution for every environmental issue but there are many perspectives and much uncertainty. In long run, the socio-economic health of a nation depends on and is directly proportional to the health of its environment. A nation not able to effectively conserve and manage its natural resources eventually will have to pay a price for short-sighted policies. Now, world is slowly realising that sustainability is the most essential aspect of the education. Considering these facts, UNO (United Nations Organisation) has declared the decade (2005–2015) as a **Decade of Education for Sustainable Development**.

Scope of environmental studies has become broad based and enlarged as a result of widening of the dimensions of the environmental awareness. The subject is receiving a considerable attention world over. Increased pollution is creating tremendous demand of pollution control equipments, sewage and industrial effluent treatment plants, bioremedial treatment plants, fly ash management equipments, solar panels and cookers, bio-gas plants etc. and thus creating need for production units for them. Although, there has been arousing interest in the subject since 1960 but after first World Earth Day celebration in 1972, more and more people, including ecologists are linking themselves to the field of environmental studies and feel proud to be called environmentalists. Environmental studies provide many job opportunities such as environmental scientists, teachers, guides, natural resource conservators, lawyers, pollution control officers, forest officers, zoo incharges etc. An environmentalist along with coworkers can register a society and can form a NGO (non-governmental organisation) to deal with various environmental problems and play significant role in various environmental movements. Dimensions of environmental studies are increasing each year as government is becoming aware of the priorities of environment (fig. 1.7).

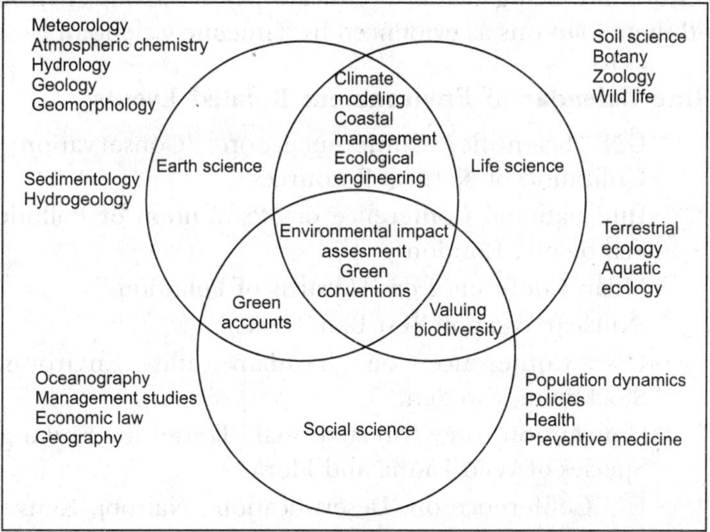

Fig. 1.7. Scope of environmental studies.

Need for Public Awareness

Progress of man is a story of his increasing materials needs that has resulted to such a tremendous change in the environment today. The magnitude of man's demand in pursuit of satisfying both his basic needs and greed is increasing proportionately to the continuing increase in world's population. Environmental degradation is also aggravated by growing inequalities in human societies and resulting threat not only to mankind but also to the living world. Man's economic development, coupled with technological and industrial development, has made him blind about the resulting environmental degradation and harms he is causing for his own existence. The goal of sustainable development can be achieved only when public has a participatory role in it. And, public participation is possible only when the people are aware about the environment and environmental issues. The people have to be educated about the fact that if they are degrading their environment they are actually harming themselves and their future generations. This is because the man is a part of the complex network of the environment where every component is linked to each other. People should also be made aware of the fact that most of the adverse impacts of environment are not noticed or experienced until a limit is crossed. Some of the adverse effects, however, out break in the form of disasters.

There are many factors that limit people's awareness of environment and environmental issues. People's need of existence are more important than the problems of environmental degradation. Most of the people do not pay the real cost of extent they have exploited the nature. A very few can look at the larger picture of what is happening and will happen to the nature. Efforts are being made to educate people about environmental issues and their solutions as evidenced by Timeline Calendar.

Timeline Calendar of Environment Related Events

1949 UN Scientific Conference on 'Conservation and Utilization of Natural Resources'.

1954 International Conference on 'Prevention of Pollution of Sea by oil', London.

1957 Milan Conference on 'Hazards of Pollution.'

1963 'Nuclear Weapon Test Ban Treaty.'

1972 UN Conference on 'Human and Environment', Stockholm, Sweden.

1973 Convention on 'International Trade in Endangered Species of Wild Fauna and Flora.'

1977 UN Conference on 'Desertification', Nairobi, Kenya.

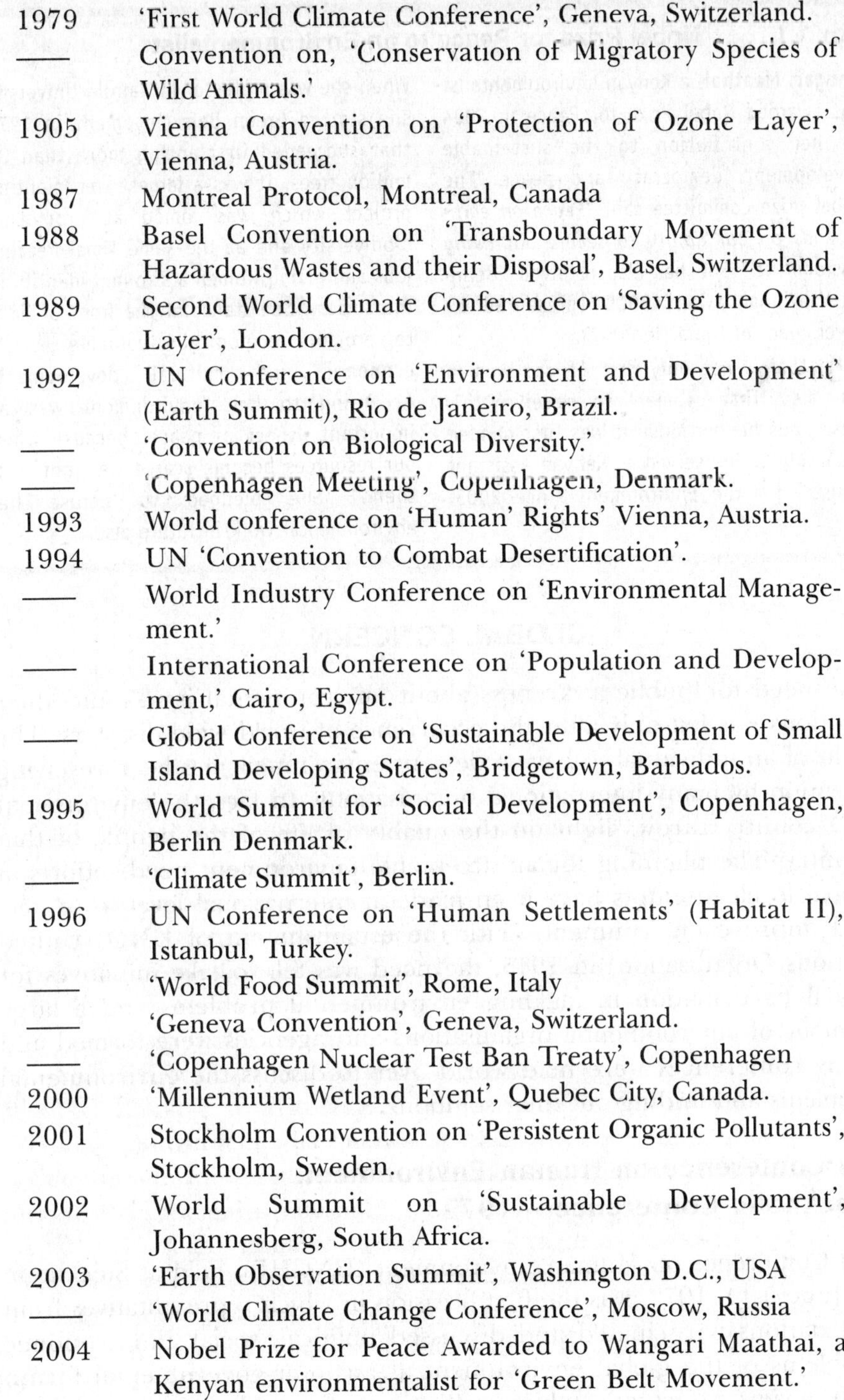

1979	'First World Climate Conference', Geneva, Switzerland.
——	Convention on, 'Conservation of Migratory Species of Wild Animals.'
1905	Vienna Convention on 'Protection of Ozone Layer', Vienna, Austria.
1987	Montreal Protocol, Montreal, Canada
1988	Basel Convention on 'Transboundary Movement of Hazardous Wastes and their Disposal', Basel, Switzerland.
1989	Second World Climate Conference on 'Saving the Ozone Layer', London.
1992	UN Conference on 'Environment and Development' (Earth Summit), Rio de Janeiro, Brazil.
——	'Convention on Biological Diversity.'
——	'Copenhagen Meeting', Copenhagen, Denmark.
1993	World conference on 'Human' Rights' Vienna, Austria.
1994	UN 'Convention to Combat Desertification'.
——	World Industry Conference on 'Environmental Management.'
——	International Conference on 'Population and Development,' Cairo, Egypt.
——	Global Conference on 'Sustainable Development of Small Island Developing States', Bridgetown, Barbados.
1995	World Summit for 'Social Development', Copenhagen, Berlin Denmark.
——	'Climate Summit', Berlin.
1996	UN Conference on 'Human Settlements' (Habitat II), Istanbul, Turkey.
——	'World Food Summit', Rome, Italy
——	'Geneva Convention', Geneva, Switzerland.
——	'Copenhagen Nuclear Test Ban Treaty', Copenhagen
2000	'Millennium Wetland Event', Quebec City, Canada.
2001	Stockholm Convention on 'Persistent Organic Pollutants', Stockholm, Sweden.
2002	World Summit on 'Sustainable Development', Johannesberg, South Africa.
2003	'Earth Observation Summit', Washington D.C., USA
——	'World Climate Change Conference', Moscow, Russia
2004	Nobel Prize for Peace Awarded to Wangari Maathai, a Kenyan environmentalist for 'Green Belt Movement.'

Box 1.1 Nobel Prize for Peace to an Environmentalist

Wangari Maathai, a Kenyan environmentalist was awarded Nobel Prize for Peace in 2004 for her contribution to the sustainable development, democracy and peace. The Nobel prize committee said *"Peace on earth depends on our ability to secure our living environment."* She has also received "Nehru International Award 2005" presented by Government of India (in 2007).

Maathai is not only first African woman but also first woman to receive this prestigious honour since it was first handed out in 1901. She served as Kenyan Assistant Minister for the Environment since 2003. When she was professor in Nairobi University she started Green Belt Movement in 1977 that succeeded in planting more than 30 million trees. It was a largest tree planting project which was aimed at promoting biodiversity and at the same time creating jobs and giving women a stronger identity in the society. She stands at the front of fight to promote ecologically valuable social, economic and cultural development. According to her "environment was as important aspect of peace, because when our resources become scarce we fight over them." She pledged to peruse her environmental work in future also.

GLOBAL CONCERN

The need for public awareness about environmental issues and their solutions was felt only after the occurrence of world wide disasters. The right of an individual to have a descent environment has been receiving attention by many international organisations. In fact, the environment of a country throws light on the quality of life of the people of that country. The planning for an acceptable environment needs efforts at global level. Attempts have been made at international level to protect and improve environment. With the establishment of UNO (United Nations Organisation) in 1945, the need was felt to take initiatives for global participation in tackling environmental problems and a large number of environmental organisations and agencies were formed and many conferences were held world over to discuss the environmental problems and finding out their solutions.

UN Conference on Human Environment: Stockholm Conference – 1972

UN Conference on Human Environment (UNCHE), held at Stockholm on June 5-14, 1972, was the first occasion in which representatives from 114 nations participated and discussed political, social and economic problems of the global environment at an inter-governmental forum with a view to actually taking collective action. The Conference was organised by United Nations Environmental Agency and adopted the motto "*Only One Earth*" for the entire humanity. The Conference declared *June 5* as the *World Environment Day* (Box 1.2). It proved a pivotal event in the growth of environmental movement and focused the attention of international community on environmental issues more sharply than ever before. The Conference was aimed to create a basis for comprehensive consideration within United Nations of the problems of the environment and to focus the attention of governments and public opinion to different countries on importance of defending the environment for present and future generation and emphasize the role of education in combating the threats to the human environment. The conference emphasized the need for a common outlook for common principles to inspire and guide the people of the world for preservation and enhancement of the environment. Sweden's Olof Palme and India's Indira Gandhi were the only heads of the states present. The conference resulted in the establishment of United Nations Environment Programme (UNEP) and caused a fundamental shift in the direction of environmentalism from emotional and occasional to more rational, political and global perspectives by posing a debate between poor and rich countries, with their different perceptions of environmental properties, into the open.

The Stockholm Declaration focused the following points:

(1) National resources should be conserved, the capacity of the earth to produce renewable resources should be maintained and non-renewable resources should be shared.

Box 1.2 June 5: World Environment Day

UN Conference on Human Environment held at Stockholm in 1972 declared June 5 as World Environment Day to undertake and make people more aware about the need for environmental protection all over the world. The Conference resolution urged the governments and organisations under the United Nations to undertake activities for the prevention of the environment on this day. Environment Day has become a major occasion every year to remind people of their responsibility to conserve the environment. The day emerged in the first row to boost the conservation spirit.

The period from November 19 to December 18 is known as *Environment Awareness Month* in India. This nationwide awareness programme was started in 1986 by Ministry of Environment and Forests with the theme 'Save Water'. Each year, there is a major environmental theme. Subsequent themes included 'Save Environment – Save Yourselves' (1990-91), 'People's Participation in Global Environmental Concern' (1991-92), 'Biodiversity (1992-93) 'Animal Welfare and Waste Management' (1993-94) 'Green Cities (Plan for the Planet)' (1994-95) and so on.

(2) Development and environmental concern should go together. Less developed countries should be given every assistance and incentive to promote rational environment management.

(3) Each state should establish its own standards of environmental management as it wishes but should not endanger other states. An international cooperation should be maintained at improving state of the environment.

(4) Pollution should not exceed the capacity of the environment to clear itself. Marine pollution should be prevented.

(5) Science, technology, education and research should all be used for promoting environmental protection.

The main principle of Stockholm Conference called **Magna Carta** of environment declared *every man has the fundamental right to freedom, equality and adequate conditions of life in an environment of a quality that permits a life of dignity*.

Montreal Protocol–1987

World's major producers of pollutants blamed of destroying ozone layer, which protects earth from cancer-causing radiation in sunlight, met for the first time at Montreal, Canada on September 16, 1987. Twenty four countries signed the Protocol that came into force on January 1, 1989. It sets out a schedule of freezing on consumption of five most severe chloroflourocarbons (CFCs) and three halons responsible for ozone depletion. Later, parties of Montreal Protocol met in Helsinki in May 1989 during which policy statement called *Helsinki Declaration* was agreed upon to phase out the production and consumption of controlled CFCs as soon as possible. Attending nations also agreed to phase out halons and to control and reduce other ozone depleting substances. India acceded Montreal Protocol and has prepared a national programme to implement the Protocol.

'Saving the Ozone Layer' Conference, London-1989

The 'Saving the Ozone Layer Conference', held at London during March 5-7, 1989, was attended by 123 countries and the European community. The conference expressed its concern for urgent need for effective action to safeguard the ozone layer and arrived on a general acceptance that the ultimate goal has to be the total elimination of production and consumption of chloroflourocarbons (CFCs) and halons.

Copenhagen Meeting-1992

Parties of Montreal Protocol had their fourth meeting in the last week of November in 1992 in Copenhagen. The developed countries agreed to phase out chloroflourocarbons (CFCs) by 1995. However, no compulsion of phase out time schedule for CFCs/ halons was posed on developing countries. India, which became party to Montreal Protocol on September 17, 1992, prepared an ambitious Rupees 6,000 crores programme to phase out CFCs and oth`r ozone depleting substances.

UN Conference on Environment and Development (UNCED): The Earth Summit –1992

The Earth Summit, held during June 3 to 14, 1992 at Rio de Janeiro, Brazil, was attended by 115 heads of states, 10,000 government missions and about 20,000 NGOs. It was a historic and the largest assembly ever held to ensure relationship between environment and development on a global partnership basis. PV Narsimha Rao, the then Prime Minister of India, had the honour of opening three day Earth Summit (for Heads of States and Government) with a lucid speech on June 12, 1992 and commented that India has been recognized as an activist on environment. The conference fostered '**Our Common Future**' and raised 6 basic conspicuous issues: *(i)* Greenhouse gas emission, *(ii)* Forests, *(iii)* Population, *(iv)* Technology transfer, *(v)* Finance (Global Environmental Facility), and *(vi)* degradation. The Earth Summit ended with the adoption of **Rio Declaration** and **Agenda 21:** a blue print for sustainable development. The Rio Summit culminated with a series of conferences that resulted a greatly expanded agenda of such global level threats as climate change and biodiversity loss, as well as long list of topics such as international cooperation in technology development and diffusion. Earth Summit came up with several documents including the Rio Declaration and *Environment and Development* listing 27 principles of sustainable development (**Earth Charter**). Agenda 21 is for sustainable development in twenty first century. Conventions on Climate Change, Biodiversity and on Forestry are the other outcomes.

The conference established important linkages between environment and development. It fleshed out the concept of **Sustainable Development** which has since dictated the international development agenda, and facilitated cooperation between government and non-governmental organizations on environmental issues. The Summit also brought home the reality of **Global Commons** and the urgent need to arrive at a global understanding to protect these shared resources. This informed the conception and the endorsement of two international conventions on biodiversity and climate change. The Rio meeting was an important milestone in defining the concept of sustainable development. This was spelled out in considerable detail in Agenda 21, which in effect was a plan of action for 21st century. The Rio declaration set out general principles for Sustainable Development and Climate Change Treaty. The convention on Biodiversity and the Framework of Principles on conservation and Use of Forests established important steps that needed to be taken to guarantee an environmentally stable and sustainable planet.

Over the last decade, many things have changed, much remains the same. The attitude of United States, for instance, has not changed significantly. Earth Summit recommended that industrialised nations had to change consumption patterns and life styles and eliminate the accumulation of greenhouse gases in the atmosphere. But United States continues to reject this perspective. The European countries, on the other hand, have been far more amenable and several countries have already met targets set for reduction of greenhouse gases. Unfortunately, obduracy of United States undercuts the effectiveness of these steps because of its contribution to global warming is far in excess than that of any other country in the world. The Agenda of Rio Summit was dictated by the developed countries and was, therefore, one-sided.

Rio Summit ended with the hope that the 21st century would be decisive for human species and the risks we face in common form mounting dangers to the environment would be minimised. India signed the Convention on Biological Diversity but the Convention came into force on December 29, 1993. It was further ratified in February 1994. Framework Convention on Climate Change was also ratified in November 1993 and came into force in 1994.

Berlin Mandate–1995

Berlin Climate Summit was held during March 28 to April 7, 1995 in Berlin. Participants attending the Summit had reviewed the obligations of the countries with heavy industrialisation. The United Nations Framework Convention on Climate Change (UNFCCC) commits these countries to reduce emission to 1990 levels by the year 2000. The meet had concluded on a modest note by recognising that the existing nature of the commitments is not adequate. Later the policies and measures were elaborated and timeframe was extended such as 2005, 2010, 2020. Besides, Berlin Mandate incorporated certain pointers to what should be the elements in a future protocol to be adopted at the Third Conference of Parties (CoP-3) in 1997 at Kyoto, Japan.

Geneva Convention-1996

Conference of Parties-2 (CoP-2) to United Nations Framework Convention on Climate Change (UNFCCC) slated for July 8-19, 1996 in Geneva, Switzerland. It may be recalled that UNFCCC was signed at the Earth Summit in Rio in 1992 and had called upon the industrialized nations to reduce their carbon dioxide emissions to 1990 levels by the year 2000. This commitment towards substantial reduction in emissions of greenhouse gases, was recognised as the first modest step in this direction.

Kyoto Protocol-1997

An International Conference of G-77 (group of 140 developing countries) was organised during December 1-10, 1997 at Kyoto, Japan to discuss global warming and sign an agreement to reduce it. Eighty four countries participated in this august event. The Kyoto Protocol, as the treaty is now called, is really an addendum to the UN Framework Convention on Climate Change (UNFCCC)—a legally binding Convention that was signed in 1992 following the Earth Summit in Rio in 1992. It came into effect on February 16, 2005, after more than 55 countries, including industrialised countries responsible for emitting 55 per cent of the global atmospheric levels of greenhouse gases, ratified. However, United States which alone accounts for 30 per cent of world's greenhouse gases, did not ratify the Kyoto protocol. Global warming is not imaginary but it is to be the most serious threat to the environment. The agreement to reduce global warming for atleast 10 years is called Kyoto Protocol or *Thermal Treaty*. It resolved reduction in emissions of greenhouse gases including carbon dioxide and establishment of Clean Development Fund funded by imposing fines realized from the countries which flout the Protocol. It was the first step to curtail emissions of the industrialised countries. Since Protocol is being used to set up a trading system to buy and sell carbon emissions, it is not being understood as environmental agreement but a trading agreement.

World Summit on Sustainable Development (WSSD)-2002

World Summit on Sustainable Development, popularly known as Rio + 10, was held during August 26 to September 4, 2002 at Johannesberg, South Africa. It was attended by over 4000 delegates from about 100 countries and members of NGOs. It reviewed the progress on sustainable development made since Earth Summit, 1992 held in Rio and recognized that implementation of Rio agreements had been poor. The preparative committees of this Summit emphasized largely on new development models. The chairman of Preparative Committee of WSSD Emir Salim of Indonesia stated *new ways and means have to be found to move towards sustainable development*. The Summit marked a shift from agreements in principle to more modest but concrete plans to action. The key issues of WSSD were poverty, globalisation, financing for development, fresh water and governance. According to UN Secretary General Kofi Annan *WSSD will provoke global opportunity for world leading in the direction of global governance*.

Box 1.3 2007-09: The International Year of Planet Earth (IYPE) Earth Science for Society

The International Year of Planet Earth (2007-2009) aims to contribute to the improvement of everyday life especially in the less developed countries, by promoting the societal potential of the world's Earth Scientists as expressed in the Year's subtitle Earth Sciences for Society. Ambitious outreach and science programmes constitute the backbone of the International Year, politically endorsed by all 191 member states of United Nations Organisation when it proclaimed 2008, the central year of the triennium, as the UN Year of the Planet Earth. This initiative was piloted by International Union of Geological Science (IUGS) and UNESCO. It has a science outreach programme with ten themes such as Groundwater: reservoir for a thirsy planet; Hazards: minimising risks, maximising awareness; Earth and Health: building a safer environment; Climate Change: the 'stone' tape; Resources: towards sustainable use; Megacities: our global urban future; Deep Earth: from crust to core; ocean: abyss of time; Soil: Earth's living skin; Earth and Life: origin of diversity.

INDIAN INITIATIVES

Beginning with the mid-sixtees, it was felt that the public awareness is the key issue for saving the environment. During last few years, there has been a tremendous development in both public and government initiatives in environmental issues in India. There are large number of Governmental Organisations (GOs), industries and Non-Governmental Organisations (NGOs) engaged in arousing public awareness in various aspects of the environment. But, probably, the most striking aspect of this mass awareness programme of environmental concern is that it is being shared even by those voluntary organisations who are, in fact, less concerned about the environment and are more concerned about the economic development of the people.

Government of India is playing its major role by enacting environmental legislation and make sure people and organisations concerned with environmental issues to abide by them. However, public participation is the key to success to bring about any environmental movement/programme and ensure a better life for present and future generations.

Indian Environment Policy

Environment policy of Indian government focuses on the following areas:

1. Conservation of natural resources by direct action such as declaration of resource forests, biosphere reserves, wetlands, mangroves and protection of endangered species.
2. Check further degradation of land and water through wasteland management and restoration of river water quality programmes.
3. Monitoring development through environmental risk assesment studies of major project proposals.
4. Penalty measures for industries which violate Pollution Control Act.

Role of Ministry of Environment and Forests, Government of India

Soon after attending Stockholm Conference, 1972, Indira Gandhi, the then Prime Minister of India, constituted a National Committee on Environmental Planning and Coordination (NCEPC), the apex body in the Department of Science and Technology (DST). Keeping in view the need for comprehensive recognition of environmental issues, Department of Environment succeeded NCEPC and with the addition of Forests (including Wildlife) was raised as Department of Environment and Forests and later Ministry of Environment and Forests.

The Ministry of Environment and Forests is perhaps most instrumental in public participation in various programmes. Ministry looks after envi-

Box 1.4 NEERI: The Environment Watchguard of India

National Environmental Engineering Research Institute (NEERI), Nagpur, India has been instrumental in keeping check on environmental degradation by managing toxic wastes, reclaming barren land and designing environmental technologies in the country. This CSIR's National Institute established in 1958 works as the Environmental Watch Guard. NEERI's mandate includes environmental monitoring, environmental biotechnology, toxic wastes management, environmental systems modeling, environmental impact and risk assessment, environmental audit and environmental policy analysis. Besides, it also plays an important role in National and Societal Missions from time to time. Revegetation efforts by scientists of NEERI have transformed some of the barren landscapes into lush green forests.

ronmental monitoring and control; survey and conservation of natural resources; enforcing environmental laws and policies; environmental education, awareness and information; management of forests and wildlife; international relation on environmental issues etc. Conservation of flora and fauna including forests and wildlife, prevention and control of pollution, afforestation and regeneration of degraded areas and protection of environment are the mandates of Ministry. And these tasks are fulfilled by the Ministry through Environmental Impact Assessment (EIA), eco-regeneration, assistance to organisations implementing environmental and forestry programmes, promotion of environmental and forestry research, extension education and training to augment the requisite manpower, dissemination of environmental information and creation of environmental awareness among all sectors of country's population. These objectives are well supported by legislative and regulatory measures, which are aimed at the preservation and protection of the environment.

Ministry has also been designated as the nodal agency of United Nations Environment Programme (UNEP), South Asia Cooperative Environmental Programme (SACEP), International Union for Conservation of Nature and Natural Resources (IUCN), International Centre for Integrated Mountain Development (ICIMD) and European Union (EU). Ministry also functions as nodal agency and participates actively in the international agreements relating to environment such as the Convention on International Trade in Endangered Species of Wild Fauna and Flora (CITES), Convention on Wetlands of International Importance especially as waterfowl habitat, Convention on the Conservation of Migratory Species of Wild Animals, Vienna Convention for the Protection of the Ozone Layer, Montreal Protocol on substances that deplete the ozone layer, Convention on Biological Diversity (CBD) and Convention on Climate Change, the Basel Convention on Transboundary Movement of Hazardous Substances, Convention to Combat Desertification etc. Annual financial contributions when needed are given to these organisations. Ministry also handles bilateral issues and matters pertaining to multilateral bodies such as Commission on Sustainable Development, UN Environment Programme, UN Development Programmes, Environmental Support Programme and several regional bodies. In addition, Ministry also deals with the work of National Environmental Council and the India Canada Environment Facility. India has been pursuing its commitments under various conventions vigorously by initiating various measures nationally and by taking several important initiatives in the region.

Various awareness programmes initiated by Ministry of Environment and Forest are as follows:

(1) *National Environment Awareness Campaign* (NEAC)

National Environment Awareness Campaign was started in 1986 with an objective to create environmental awareness among people all over the country. The Ministry works through appointed Regional Resource Agencies (RRAs) in various parts of the country that assist it in coordinating, implementing, monitoring and evaluating NEAC activities at the regional levels. Many of these RRAs are Non-Governmental Organisations (NGOs). Thus the campaign reaches the remotest part of the country and caters to local specific needs of environmental awareness and protection. NEAC is a multimedia campaign using all types of conventional and non-conventional media of communication, thereby spreading various messages relating to environmental issues among different categories of people.

A large number of NGOs, women and youth organisations, school, colleges, universities, army units etc., all over the country, are allocated financial assistance for organising awareness oriented activities such as 'padyatras' (street walks), rallies, public meetings, exhibitions, folk dances, 'nukkar nataks' (street plays), essay/debate/ poster competitions for school children, seminars, symposia, workshops, conferences, training course etc. as well as for the preparation and distribution of environment education resource material. Diverse target groups from school children to teachers, rural population, tribals, journalists and other professionals are covered by the campaign. To start with only 127 organisations participated in this awareness campaign but till 2005 there were about 9566 organisations from across the country that took part in this campaign.

(2) *Paryavaran Vahini* (*Environment Brigade*)

Paryavaran Vahini scheme was launched by the Ministry in 1992 with the objective to create environmental awareness and encourage involvement of the people through active participation as well as to report illegal acts pertaining to forests, wildlife, cruelty to animals, pollution and environmental degradation. *Paryavaran Vahinis* also encouraged to give feedback regarding reforestation, survival of plants and animals, monitoring ambient air and water quality and vehicular pollution. Selected *Paryavaran Vahinis* have been provided with water testing kits. One *Paryavaran Vahini* is constituted for each district especially identified for this purpose. The selection of the district is made on the basis of high incidence of pollution, density of tribal population and forest cover. Till 2000, 194 districts were selected for this purpose and so far, *Paryavaran Vahinis* have been constituted in 144 districts. Each *Paryavaran Vahini* has 20 active members coming from all categories of professions including NGOs.

(3) Eco-Clubs

Keeping in view that the school children are the future decision-makers of the country, a special scheme Eco-Club, to encourage and mobilise students' participation in various environmental conservation and preservation activities in their localities, has been launched by the Ministry. Ministry provides financial assistance to NGOs and professional bodies for setting up Eco-Clubs (also to Nature Clubs or Environment Clubs). Each Eco-Club has 20-25 members taken among students of the school. An Eco-Club in one or more geographical areas is serviced by a coordinating agency which may be an educational institution, a NGO or a professional body. Upto the year 2006, 72,000 Eco-Clubs were set up in different parts of the Country. Apart from imparting environmental education, the Eco-Clubs are aimed at involving students in mobilising and participating actively in environmental protection efforts at the local community level.

(4) *Production of Films and Audio-Visuals*

The ministry also collaborates with print and electronic media to ensure coverage of environmental issues. To solve these purposes Ministry uses newspapers, magazines and television channels. Widespread importance and tremendous reach of electronic media has made Ministry to commission the production of good quality of films, serials, video spots and other audio visuals on various environmental themes and make them available to various TV channels for telecasting.

(5) *Natural History Museum*

The ministry has also set up Natural History Museums in New Delhi and Mysore. Till 2000, two more museums were underway in two different cities of the country. The Natural History Museums have been set up to spread awareness about nature and natural resources of the country. The National Museum of Natural History is located in New Delhi. It provides a unique facility of promotion of non-formal environmental awareness among people in general and school children and students in particular. This Museum and the other one at Mysore looks after a wide range of activities for diverse target groups round the year contributing significantly to the Ministry's efforts towards creation of environmental awareness.

(6) 'ECOMARK' *for* Environment Friendly Products

The ministry has introduced a scheme called Environmental Friendly Products in 1991. Under this scheme, some products, considered unsafe from pollution point of view, are tested before they enter the market.

These products will have to be labelled 'ECOMARK'. The logo of this label is 'Earthen Pot'. In initial stage 16 consumable products are required to get ECOMARK. A notification to this regard was issued on February 21, 1991 covering four articles namely soap, detergents, paper and paints.

(7) Centres of Excellence

To cope with the needs of environmental education and conduct variety of programmes to spread awareness and interest about environmental issues among the people, especially NGOs, women, youth and children, Ministry has established following centres of excellence till 2006.

(1) Centre for Environment Education (CEE), Ahmedabad (Linked with Nehru Foundation for Development, Ahmedabad—established in 1984).

(2) CPR Environment Education Centre (CPR-EEC), Chennai (Linked with sir C.P. Ramaswamy Aiyer Foundation Chennai—established in 1988)

(3) Salim Ali Centre for Ornithology and Natural History (SACON) Kalayampalayam, Coimbatore (Linked with Bombay Natural History Museum, Mumbai)

(4) Centre for Ecology Research and Training (CERT), Banglore (Linked with the Indian Institute of Science, Banglore).

(5) Centre for Mining Environment (CME), Dhanbad (Linked with Indian School of Mines, Dhanbad).

(6) Centre for Environmental Management of Dedraded Ecosystem (CEMDE), New Delhi (Linked with Department of Environmental Biology, Delhi University, South Delhi Campus)

(7) The Tropic Botanic Garden and Research Institute (TBGRI), Thiruvanthapuram, Kerala (Linked with State Government of Kerala)

(8) Madras School of Economics (MSE), Chennai (Linked with Madras University, Chennai)

(9) Foundation for Revitalisation of Local Health Traditions (FRLHT), Banglore.

(8) Awards and Fellowships

The Ministry has made every effort to recognise an individual or group's contribution in different areas of environment by way of instituting a number of national level awards and fellowships to encourage people and to initiate them to add themselves in the area of environment and environmental issues. Some of the awards are major and some are minor but all of them are for afforestation, wasteland development, promotion of healthy trees, pollution prevention, clean technology etc. Some awards are as follows:

(i) **Indira Gandhi 'Paryavaran Puraskar':** The 'Puraskar', instituted by the Ministry in 1987, is awarded every year since 1991 for the outstanding contribution in the field of environmental protection. Puraskar carries a cash component of Rs 1.0 lakh, a silver trophy, a scroll and a citation.

(ii) **Indira Priyadarshni 'Vriksh Mitra' Award:** The awards, instituted by the Ministry in 1986, are given in recognition to pioneering an exceptional contribution of individuals or organisations annually in the field of afforestation and wastelands development. Awards are given under twelve different categories. Each award carries a cash component of Rs 1.25 lakh for first prize and Rs 1 lakh for second prize (from the year 2004), medallion, scroll and citation.

(iii) **'Mahavriksha Puraskar':** 'Puraskar', instituted by the Ministry in 1993-94, is given in recognition of individuals or organisations annually for preserving or protecting trees of notified species. A roaster of notified tree species is proposed for five years. Each award carries a cash component of twenty-five thousand, a plaque and a citation.

(iv) **Pitambar Pant National Fellowship Award:** The award is given annually in recognition of significantly important research and development contributions in the field of environmental sciences to encourage talented individuals to devote themselves to research and development in the area of environment. The fellowship is for two years. Twenty seven fellowships have been awarded so far.

(v) **B.P. Pal National Environment Fellowship Award for Biodiversity:** The award is given annually to talented individuals to devote themselves to research and development pursuits in the field of Biodiversity. The duration of fellowship is for two years. So far nine fellowships have been awarded.

(vi) **'Jile Ki Sabse Hari Panchayat':** The award, instituted by the Ministry, is to motivate 'Gram Panchayats' to undertake Rural Employment Generation and other plantation schemes. The efforts in afforestation activities of 'Gram Panchayats' are evaluated after every four years in every district and the best 'Gram Panchayat' is awarded a cash component of Rs 1.0 lakh.

(vii) **Salim Ali and Dr. Kailash Sonkhala National Wildlife Fellowship Award:** The fellowship award, instituted by the Ministry, is to give recognition to the eminent officers and field workers for exemplary work in the field of Wildlife Conservation and Research.

(viii) **Amrita Devi Wildlife Protection Award:** The national award, instituted by the Ministry in the name of Amrita Devi Vishnoi, is given annually to the village communities showing valour and courage for protection of wildlife. It carries a cash amount of Rs 1.0 lakh apart from citation and medallion.

(ix) **National Award for Prevention of Pollution:** The award, instituted by the Ministry, is given every year to the industries and operations which make a significant contribution towards development or use of clean technologies and products that prevent pollution and finding innovative solutions to the environmental problems. The significance of the contribution both in qualitative and quantitative terms is the major criteria for selection. Only industries and operations adhering to the standards stipulated by authorities is considered for awards. Eighteen awards, one each for identified category out of highly polluting industries and five to small scale industries such as tanneries, pulp and paper, dye and dye intermediates, pesticides and pharmaceuticals are given in every financial year for adopting cleaner technologies and products that prevent pollution.

(x) **Rajiv Gandhi Wildlife Conservation Award:** The award is given each year to a recognised institution, individual or government and forest staff of India for significant contribution in the field of wildlife which is recognised as having made or has the potential measurable and major impact and conservation of wildlife in the country. The award carries Rs 1.0 lakh in cash along with medallions, trophy and citations.

(xi) **'Vishist Vaigyanik Puraskar':** The 'Puraskar' under the scheme 'Paryavaran aur Van Mantralaya Vishist Vaigyanik Puraskar' was introduced in 1992 by the Ministry and its associated officers to encourage original and applied research.

(xii) **Medini 'Puraskar Yozana':** The 'Puraskar' was introduced to encourage the Indian writers to write more and more books originally in Hindi on environment and forests and its related subjects. Four cash prizes of Rs 31, 25, 20 and 15 thousand along with citations are given every year judged as the best books by the Ministry.

(9) *Ministry's Library*

Library of Ministry acts as a document repository for dissemination of the information in the field of environment and its associated areas. Till 2000 it had a collection of 21,000 books including proceedings of various conferences/workshops/seminars covering different areas of environment and sustainable development. Besides, it also has over 3,000 scientific and technical reports, more than 100 national and international journals covering diverse areas of environment so that it could disseminate its information to its users. The library has also procured a wide range of general books and magazines both in Hindi and English for the use. It also provides references/referral services to those who need.

(10) *Environmental Information System* (ENVIS)

The Ministry initiated its Environmental Information System (ENVIS) in 1982 which performs the activities of information collection, collation, storage, retrieval and dissemination to its users in the field of environment and its associated areas. Besides the focal point located in the Ministry, the ENVIS network consists of twenty-five nodes, known as ENVIS Centres located throughout the country on various environment related subject-specific areas and the activities of these centres run in coordination with the focal point in the Ministry.

(11) *Logo* or *Slogan Competition*

The Ministry organises a national contest for creating a logo or slogan for various aspects of environment. The selected ones are awarded prizes and certificates.

(12) *Assistance for Organizing Seminars/Workshops*

The Ministry extends financial assistance to NGOs/Academic institutions/Colleges/Universities/Organisations to organise symposia, conferences, seminars or workshops to provide professionals a common forum for sharing uptodate knowledge on various issues related to environment and to create awareness about various environmental issues.

(13) *Celebration of Environment Dates*

The Ministry issues instructions for celebrating certain days such as World Earth Day, World Environment Day, World Water Day, World Wetland Day, World Forestry Day, Van Mahotsav Week etc. all over the country to create mass awareness about various environmental issues. The people's interest is aroused by various activities such as public meetings, video and audio cassetes, seminars, workshops, plays, exhibitions, poster shows etc.

ENVIRONMENTAL EDUCATION

Environmental Education (EE) is an approach of learning various aspects of the environment. It endeavours to create a way of thinking and it is through this process that individuals and community gain awareness of their environment, acquire knowledge, skill, experience and determination to act individually and collectively to solve various

Box 1.5 **April 22: Earth Day**

First Earth Day was observed in the United States on April 22, 1970. The day was funded by Gayland Nelson and organised by Denis Heyes. Twenty million Americans participated in this event. It marked the beginning of the environmental movement. Its message was *Give Earth a Chance* and it focused especially on greenhouse effect and depletion of ozone layer. Since then, April 22 is observed as Earth Day every year in many parts of the world. The Earth Day Network promotes environmental awareness and year round progressive action throughout the world. It has reached over 12,000 organisations in 174 countries. Earth day is the only event celebrated round the globe simultaneously by people of all nationalities, backgrounds and faiths. Every year more than half-a-billion people participate in the campaigns. The day has been an annual event for people all over the world to celebrate the earth and our global responsibility towards it. The Earth Day date is April 22, but in most of the countries the events are often scheduled for weekends before or after April 22. Many groups and communities choose to organise the environmental activities over the long period, the earth week, the week before Earth Day or even Earth Month: the whole month of April.

environmental problems and tackle various environmental issues. Soon after the deliberations in UN Conference on Human Environment, held at Stockholm in 1972, the issue of environmental education has thoroughly been discussed at several national and international platforms by organising seminars, workshops and conferences.

The first inter-governmental Conference on Environmental Education held at Tbilisi, Russia in 1977 discussed the broad environmental issues and recommended Environmental Education (EE) *to cater to all ages and socio-professional groups in the population*. The World Summit on Environment and Development held at Rio in 1992 laid emphasis on 'sustainable development for the future survival'. The objectives and guiding principles for developing environmental education at all levels in both formal and non-formal levels to help individuals and social groups formulated at Tbilisi Conference, are as follows:

Objectives

(1) **Awareness**, *i.e.*, acquire an awareness of and sensitivity to the total environment and its allied problems.

(2) **Knowledge**, *i.e.*, gain a variety of experiences and acquire a basic understanding of the environment and its associated problems.

(3) **Attitude**, *i.e.*, acquire a set of values and feelings of concern for the environment and the motivation for the active participation in environmental improvement and protection.

(4) **Skill**, *i.e.*, acquire skills for identifying and solving environmental problems.

(5) **Evaluation Ability**, *i.e.*, evaluate environmental measures and education programmes in terms of ecological, economic, social, aesthetic and educational factors.

(6) **Participation**, *i.e.*, provide an opportunity to be actively involved at all levels in working towards the resolution of environmental problems.

Guiding Principles

Following are the guiding principles:-

(1) To consider the environment in its totality (natural, artificial, technological, social, economic, political, moral, cultural, historical and aesthetic).

(2) To consider a continuous life process (from pre-school to all higher levels: formal as well as non-formal).

(3) To be interdisciplinary in nature.

(4) To emphasise active participation in prevention and solution to environmental problems.

(5) To combine major environmental issues from local, national, regional and international point of view.

(6) To focus on current potential environmental situations.

(7) To consider environmental aspects in plans for growth and development.

(8) To emphasise the complexity of environmental problems and need to develop critical thinking and problem solving skills.

(9) To promote the value and necessity of local, national and international cooperation in the prevention and solution of environmental problems.

(10) To utilise diverse learning about environment and different approaches of teaching and learning about environment.

(11) To help learners to discover the symptoms and the real causes of environmental problems.

(12) To relate environmental sensitivity, knowledge, problem solving and values clarification at every grade level.

(13) To enable learners to have a role in planning their learning experiences and provide an opportunity for making decisions accepting their consequences.

Environmental education, under environmental studies, is concerned with environmental damages and minimisation of their impacts by bringing changes through societies. Depending on varied environmental problems in different parts of the world, there appears no possibility of a common environmental education programme. In India, Government's main aim is that environmental education should play a key role in creating awareness and better understanding of environmental problems.

Environmental Education in India

Environmental scenario of India is quite different from rest of the world. The country is highly diverse geographically, climatically, geologically, edaphically, floristically, faunastically, ethnically, culturally, socially, economically and politically. Therefore, environmental education has to be location and level based. In India, environmental education is aimed at involving people's relationship with their cultural, social and man-made environment. In nature, it is multidisciplinary and requires involvement of all disciplines and of the entire human population inhabiting the country. It is one's duty to protect air, water, soil, flora, fauna and social environment so that he may live in harmony. The purpose of environmental education is to include skill, attitude and value necessary to understand, appreciate and improve the environment. There are two main ways of imparting environmental education.

(I) *Formal Environmental Education*

Formal education given in schools, colleges, universities and elsewhere is limited to a specific period. The best approach to create awareness about the environment and associated problems is to impart environmental education to children right from the beginning at primary school level as they take the new idea quickly. And as such, the curriculum should be constructed taking into account the class and the age of the students. The best way is to make the environmental education compulsory for students in all classes. The course content at the primary school level must be easily accessible whereas at secondary and higher secondary levels it should be multidisciplinary because at this stage the child becomes conscious about the physical, social and aesthetic aspects of the environment and requires increased knowledge about environmental problems, conservation and sustainable development. The students should not be taught only through books but should also be given a chance for field activities. National Council of Educational Research and Training (NCERT) has developed a curriculum framework that proved very useful for students. University Grants Commission (UGC) has made the subject environmental studies compulsory to undergraduate levels. For undergoing higher education in the subject there are more than 200 departments of environmental studies in colleges and universities all over India.

The Ministry of Environment and Forests interacts actively with National Council of Environmental Research and Training (NCERT),

University Grants Commission (UGC) and Ministry of Human Resource and Development (MHRD) for introducing and expanding environmental concept, themes, issues etc. in the curricula of schools, colleges and universities. The two centres of excellence, the Centre for Environment Education, Ahmedabad and the CPR Environment Education Centre, Chennai of the Ministry are also fully involved in the activities of NCERT, UGC, and MHRD related to formal environmental education. Based on State Ministers Conference, **Revitalisation of Environmental Education** in schools and strategies formed by CEE, Ahmedabad, environmental education has been included in the syllabi of school systems. Under the directive of Supreme Court environmental studies has been made compulsory to all undergraduates irrespective of faculty of arts, science, commerce, agriculture, medicine, engineering or any other branch of study.

(II) Non-Formal Environmental Education

Non-formal environmental education is for a majority of population that still does not have adequate access to formal education. Such people can be made aware of the environment and its management by non-formal environmental education. Environmental education needs to be life long affair. It should be designed for all age group and include activities outside the framework of the established formal education system. The process is experience based, involving exercises of solving environmental problems. Such education gives students an out of school exposure which involves them in the natural process of enquiring, exploring, comparing, concluding, evaluating and decision making about the environmental problems in their locality. The fundamental characteristics of non-formal environmental education is the flexible approach. Its objective should be organisation of extra-curricular participatory programmes such as eco-development camps, eco-clubs, poster and essay writing competitions, exhibitions, seminars, audio visual slides, mobile exhibition etc. Eco-development camps aim at creating awareness about basic environmental principles, identifying the causes of the environmental problems and ways to solve them. The camp activities include tree plantation, trenching, fencing, clearing water bodies, collecting polythene bags, hygiene and promoting non-conventional energy sources. Arts and crafts, folk dances, puppet shows, street plays, arranging public meetings, organising workshops and seminars can also be used to promote non-formal education. Ministry of Environment and Forests has initiated activities like celebrating important environmental dates, organising 'padyatras', rallies, public meetings, conferences etc. and for many of these activities relating to mass awareness Ministry provides funds. Under the *Samvardhan* project launched with the support from Field Studies Council, UK, efforts are being made to add dimensions of conservation and sustainability to rural development studies in Rural Higher Education Institutions (RHEI) for capacity building of teachers and youth and facilitation of adoption of community based projects through workshops on project planning communication etc.

Role of NGOs

Non-Governmental Organisations (NGOs) have a far better understanding of the surroundings and life style of people, than the government agencies. In being close to grass roots or even more of the people and the problems, NGOs action programmes fulfil the needs of healthy environment and reach the target even to poor and powerless to which government agencies cannot. Keeping in view diverse social and environmental problems, role of NGOs in India proves to be logical as they hold a great potential to be efficient and effective alternatives to the government agencies in delivery of programmes and projects. In India, within last few years, a lot of work, with regards to environmental conservation and development, has been made by a number of voluntary agencies, community groups, academic bodies, NCC cadets, corporate entities and even individuals. More than 5000 such voluntary organisations are being funded by the government for the purpose. The number of NGOs in India is larger than any Third World Country. Many NGOs are networking eco-clubs and various other environmental programmes. They are displaying a far greater level of professional competance and proving alternatives to official development programmes.

Despite the fact that everybody talks of environment, only a few have a clear idea as to what is needed to be done and still a fewer have the actual experience or expertise in the field. The reason being, the environmental awareness campaigns have very often been exploited for political propaganda rather than being an integral part of our educational programmes in theory and practice.

Chapter Summary

Long before the advent of the man, environment was in a balanced state. Man has exploited the environment and brought the earth on the brink of collapse. The environment is everything outside an individual. Two components of the environment are the physical or *abiotic* including air, water, land etc. and the biological or *biotic* including plants, animals and other organisms. The physical realms of the earth are *lithosphere, i.e.,* rock and soil, the *hydrosphere, i.e.,* water domain and *atmosphere*. Atmosphere is the blanket of air that surrounds the earth. From earth's surface to its upper limits atmosphere is spread over *troposphere,* the layer in which living organisms operate because of oxygen, carbon dioxide and other gases; the *stratosphere*, having ozone layer; *mesosphere; thermosphere* and *exosphere* on the basis of temperature profile. *Biosphere* encompasses, lithosphere, hydrosphere and the atmosphere and is the most important for sustenance of life. Biosphere harbours all living organisms, contains all resources and operates various bio-geo-chemical cycles. Release of free oxygen in the atmosphere by cyanobacteria, proved a significant step and gave the way to the origin of breathing organisms. Man has caused a substantial damage to the environment since his origin. Still he is acting upon the environment and is causing a serious damage. Socio-economic and industrial development has greatly damaged the environment. Environmental studies deals with environmental issues and is multidisciplinary in nature. Importance and scope of environmental studies are manifold. Its importance lies in understanding environmental problems and finding out their solutions. Its scope is broad based and provides many opportunities to work for various environmental aspects. There is an urgent need for public awareness about environmental degradation and its conservation. World over held environmental conferences have tried to find out solutions of many environmental issues. Many international organisations are working to solve various environmental problems.

In India, there has been a tremendous development in both public and government's initiative in environmental issues. Indian environment policy is aimed at conserving natural resources, control pollution and sustainable development. Role of Ministry of Environment and Forests is worth appreciating. It performs understanding of the environment and helps for awareness camps, 'Paryavaran Vahinis', use of electronic media, Centres of Excellence, Natural History Museums, instituting awards and fellowships and many other ways to make people aware of various environmental issues. Environmental education both formal and non-formal, has played a vital role to bring about environmental consciousness and awareness among children, youth and general masses. NGOs have played a significant role in understanding various environmental issues and solutions to the problems. Environmental activists are also struggling for various public interest issues of the environment.

Study Questions

1. Define environment. What are the various components of environment? Why healthy environment is needed?
2. What is the scope and importance of environmental studies?
3. What are various steps taken by the Ministry of Environment and Forests, Govt. of India to make people aware of the environmental degradation and its management?
4. Write short notes on:

   (1) Atmosphere (2) Multidisciplinary nature of environmental studies (3) Water cycle (4) Environmental education (5) Role of NGOs (6) Conferences related with environment (7) Biosphere (8) Environmental challenges faced by India (9) Global environmental problems

Objective Questions. *Select the correct answers*

1. Which of the following elements is present in the largest percentage in the biosphere?

(1) Carbon (2) Hydrogen
(3) Nitrogen (4) Oxygen

2. Which of them is fittest for survival of humans?
   (1) Lithosphere (2) Hydrosphere
   (3) Atmosphere (4) Biosphere
3. The first international conference on environment was held at–
   (1) Rio de Jeneiro (2) Stockholm
   (3) Johannesberg (4) New Delhi
4. Average thickness of troposphere is :
   (1) 5 km (2) 7 km
   (3) 15 km (4) 11 km
5. Atmospheric layer important for operation of satellites is :
   (1) Stratosphere (2) Thermosphere
   (3) Mesosphere (4) Troposphere
6. Man's interaction with the environment become pronounced during :
   (1) Iron age (2) Bronze age
   (3) Pleolithic period (4) Neolithic period
7. 'One Man Environ–Legal Brigade' is a popularly known designation for :
   (1) Anna Hazare (2) M.C. Mehta
   (3) Sundardarlal Bahuguna (4) Medha Patkar
8. Tbilisi Conference on Environmental Education was held in the year :
   (1) 1967 (2) 1977
   (3) 1987 (4) 1997
9. A decade declared as 'Decade of Education for sustainable Development is :
   (1) 2005–2015 (2) 2000–2010
   (3) 1995–2005 (4) 1990–2000
10. U.N. Conference on Environment and Development was held at :
   (1) Rio de Janeiro (2) Johannesberg
   (3) Stockholm (4) Montreal
11. World Environment Day is celeberated on :
   (1) September 5 (2) July 5
   (3) June 5 (4) May 5

Answers

1. (3) *2.* (4) *3.* (2) *4.* (4) *5.* (2) *6.* (3)
7. (2) *8.* (2) *9.* (1) *10.* (1) *11.* (3)

Natural Resources: Renewable and Non-Renewable Resources

- *Natural Resources and Associated Problems*
- *Forest Resources*
- *Water Resources*
- *Mineral Resources*
- *Food Resources*
- *Energy Resources*
- *Land Resources*
- *Role of Individual in Conservation of Natural Resources*
- *Equitable use of Resources for Sustainable Life Styles*

2

CHAPTER

Natural Resources and Associated Problems

LEARNING OBJECTIVES
Introduction • Types of Natural Resources • Depletion of Natural Resources • Conservation of Natural Resources • Problems Associated with Natural Resources

Introduction

The natural resource is the stock that can be drawn from nature. The earth is a storehouse of various resources such as land, water, soil, minerals, vegetation, forests, fuels, solar energy etc. that are of immense importance to mankind. Thus, natural resources are the components of the environment that can be drawn upon for supporting life. Any such material which can be used directly or after transformation to sustain life is resource. Without these natural resources, life on earth would not have been possible. Man satisfies his primary needs of food, shelter and clothing and fulfils his other requirements with the resources made available to him from nature.

The term '**resource**' is dynamic but its meaning got changed with the advancement in science and technology. Previously, natural resources were defined as those materials which were of value to a particular human culture (Dasmann, 1968). But, now everything on this earth is useful to man. Some of the resources such as land, water, air, light, plants and animals are readily available whereas others such as minerals, coal, oil etc. are hidden under earth's surface and are to be taken out.

TYPES OF NATURAL RESOURCES

The earth's natural resources can be categorised on the basis of their chemical nature, availability or abundance, area of occurrence, origin and utility etc.

(I) Natural Resources Based on Chemical Nature

Based on chemical nature, natural resources may be of the following types:

1. Inorganic resources. These natural resources are inorganic in nature. Inorganic resources include air, water, rocks, mineral ores, minerals etc.

2. Organic resources. These natural resources are organic in nature. Organic resources include organisms such as plants (forests), animals, microbes and their products like fossil fuel (oil, coal, natural gas etc.).

3. Mixed resources. These natural resources are a combination of both inorganic and organic components. Soil is the best example of mixed resource as it is formed of both inorganic components, *i.e.*, inorganic matter of the soil which results from weathering of rocks and organic components, *i.e.*, humus from the plants and animals formed by the activity of microorganisms.

(II) Natural Resources Based on Availability or Abundance

Based on availability or abundance natural resources may be of the following types:

1. Inexhaustible resources. These natural resources are found in such abundance so as to unlikely deplete or get exhausted. They are not likely to be exhausted by man's consumption. Inexhaustible resources include solar energy, air, water, sand, clay, tidal energy, hydropower, wind, temperature, precipitation etc.

2. Exhaustible resources. These natural resources are likely to be depleted and then exhausted upon their continuous exploitation as they have a limited stock on the earth. Exhaustible resources may be non-renewable or renewable.

(a) Non-renewable resources. Non-renewable natural resources lack ability of recycling and replacement or are replaced after a very long time. They are unevenly distributed in the world. Some are found abundant in one country and poorly present or even absent in another country. Similarly a country may be rich in one kind of resource and poor in other kinds of resources. Non-renewable resources may be natural, *e.g.* fossil fuels or man made, *e.g.* polythene. Naturally occurring non-renewable resources include mineral ores, minerals, metals and fossil fuels. These occur in earth's crust and are difficult to be taken out. Fossil fuels, such as coal, oil, natural gas, have a limited stock and require millions of years to reform. Their consumption rate is very high. At this rate of consumption it is likely that these may be exhausted before they are reformed. They cannot come back to their entity after being used and cause severe pollution. Minerals are also being used at a very fast rate. The reformation period of minerals and fossil fuels is very long. Similarly, soil is often considered as non-renewable resource as it can be degenerated only in few years but need several hundred years for its formation. Genetic biodiversity is also a non-renewable natural resource. Care should be taken in use of these resources so that they may not be exhausted and that nothing is left for future.

***(b)* Renewable Resources.** Renewable natural resources can be replenished, reproduced or recycled. They have the inherent capacity of reappearing or replacement within a reasonable time and maintain themselves. They can last indefinitely. A judicious balance is required to be maintained between their exploitation and replenishment. Over-use or improper management can diminish or exhaust renewable resources. The chief renewable resources are soil fertility, water and living organisms: plants (forests), animals and microorganisms. If consumption of these resources exceeds their rate of renewal, not only their quality becomes affected but they may become non-renewable. For example, excessive felling of trees results in deforestation and overgrazing causes denudation of the land and desertification. Over consumption of ground water may lead to lowering of water table. Thus, these resources need judicious management and conservation so that future generations may not be deprived of them.

(III) Natural Resources Based on their Occurrence

Based on occurrence, natural resources may be of the following types:

1. International resources. These natural resources are available to all countries and have no boundaries. Atmosphere with sunlight and air including many gases is common to all countries.

2. Multinational resources. These natural resources are shared by more than one country. Certain rivers and lakes may be shared by more than one country, *e.g.* Brahmaputra river-common to Tibbet, India and Bangladesh. The migratory birds also form multinational resource.

3. National resources. These natural resources are restricted to a particular country. Physical features, geological structure, river systems, flora, fauna and human resource constitute the national resource of a country.

(IV) Natural Resources Based on their Origin

Based on origin, natural resources may be of the following types:

1. Biotic resources. Biotic resources are basically organic in nature. These include forests and forest produce, vegetation with variety of plants, wide variety of animals such as birds, fish and other marine life forms and microorganisms. Fossil fuels also belong to this category since they originat from organic matter. Some biotic resources such as plants (forests) and livestock are renewable whereas fossil fuels are non-renewable natural resources.

2. Abiotic resources. Abiotic resources are basically inorganic in nature. They are composed of non-living matter. Land, water and minerals, such as iron, copper, lead, gold etc. are common examples of abiotic resources. Some abiotic resources occur as nodules such as nodules of copper and manganese.

(V) Natural Resources Based on their Utility

Based on utility, natural resources may be forest resource, water resource, food resource, energy resource, land resource or any other resource which may be useful for man.

Certain resources such as energy resources belong to several categories and can be inexhaustible, *e.g.* solar energy, exhaustible renewable, *e.g.* forests and exhaustible non-renewable, *e.g.* fossil fuels-coal, oil, natural gas etc.

DEPLETION OF NATURAL RESOURCES

Man as the superconsumer of resources has overexploited natural resources to serve his primary needs as well as his basic amenities that make his life comfortable. Scientific and technological discoveries have also increased man's demand for natural resources such as fossil fuels, minerals and other sources of energy. The continued and careless use of natural resources will result in their degradation, *i.e.*, loss of quantity and quality leading to its end.

India, after 500 years of colonial exploitation, 50 years of growth model development and 15 years of corporate rule is depleting its natural capital : land, water, forests, air, fossil fuels etc. at an alarming rate and those communities that depend on them for their livelihood are being displaced or marginalised. Undoubtdely the people of the country are extracting natural resources at a rate far greater than their capacity to regenerate.

Life Time and Depletion Time of Resources

Life time of a resource is a period along time scale upto which its availability for human use is assured on a global basis. Depletion time is the period along the time scale when the availability of the resource will decrease to such an extent that it is not available to meet human requirements. Depletion patterns of resources and their managements are of three types (fig. 2.1).

(1) Rapid depletion time (Boom and Bust). It is due to policy of extract-use-and-throw away. It was prevalent during early years. There was unrestricted economic extraction, involving wastage at every step. The after use materials were thrown away.

(2) Extended depletion time. It involves proper management at the point of extraction, concentration and manufacture so that the wastage can be reduced. Partial recycling extends the depletion period of resource.

(3) Indefinite depletion time. It involves various conservation strategies, as such depletion period of the resource can be extended indefinitely. Conservation techniques involve prevention of wastage, recycling and substitution.

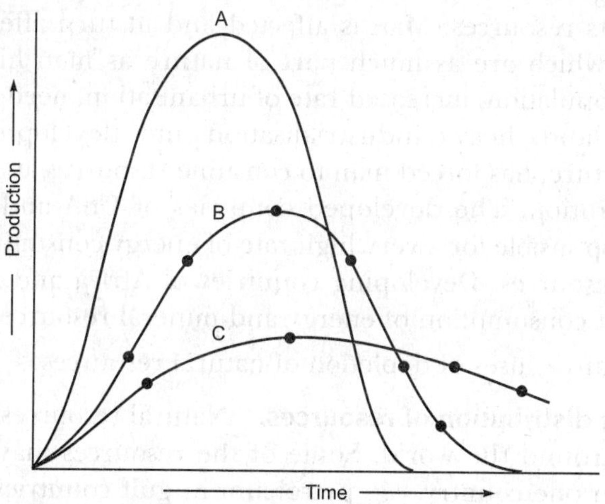

Fig. 2.1. Depletion patterns of resources. A-rapid depletion time, B-extended depletion time, C-indefinite depletion time.

CONSERVATION OF NATURAL RESOURCES

It becomes necessary to use the natural resources in such a way that they can also be saved for future use and are not lost. In other words, allocation of resources have to be done in such a manner so that they are well managed and conserved. Management of resources is controlling its use in such a way as to maintain its sustainable use and equitable availability. Conservation of resource is its prevention from loss, waste or degradation so that it lasts indefinitely, *e.g.* conservation of wildlife, minerals and fossil fuels. Both management and conservation are important as some of the natural resources are undergoing rapid depletion. Thus, it becomes necessary to protect, manage and conserve natural resources so that they are not exhausted. It does not mean that their use should be stopped but they should be used in such a way that they are saved enough for the future generation. Rapid increase in human population is causing an increasing demand for natural resources. If this pace continues there will be a time when the natural resources will be depleted and future generations will be deprived of them. According to Brundtland report, entitled *Our Common Future* published in 1987 *natural resources are not inexhaustible and development process should be aimed to meet the needs of the present generation without compromising the ability of future generations to meet their own needs.*

Problems Associated with Natural Resources

Man is closely associated with nature as he is a part of it. A striking feature of man's progress through centuries lies in his outright dependence on nature and its resources. Man is affected and in turn affects the other constituents which are as much part of nature as man himself. Rapid increase in population, increased rate of urbanisation, need for increased agricultural land, heavy industrialisation and development without caring the nature, has forced man to consume resources at a rate beyond their regeneration. The developed countries of USA and Europe and Japan are responsible for a very high rate of energy consumption and use of mineral resources. Developing countries of Africa and Asia have far lower rates of consumption of energy and mineral resources.

There are many causes of depletion of natural resources.

(I) Uneven distribution of resources. Natural resources are unevenly distributed around the world. Some of the resources may be found in abundance in one country, *e.g.* petroleum in gulf countries and may be poorly represented or even lacking in another country. Similarly a country may be rich in one kind of resource and poor in other kinds of resources. For example, South Africa contains most of the world's gold and platinum but has little of silver. North America is rich in molybdenum, Malaysia and Indonesia are rich in tin, tungston and manganese.

Large countries such as USA, Russia, China and Australia have a wide diversity of natural resources and these had been efficiently used for the development. Africa and Asia, although rich in natural resources, were exploited by foreign rulers for many years and much of their mineral and forest wealth has been depleted. Also, the countries in Asia and Africa lack the money and technologies to develop and use them optimally to bring about the progress in economy. Underdeveloped countries fall into debt traps of developed countries and try to furnish their debts either by large scale deforestation by exporting timber or overexploitation of their mineral resources, both harming the less developed countries.

(2) Population growth. Rapid increase in population growth has resulted into expanding needs of man. Continuous increase in the population has caused an increasing demand for resources. A large population requires to be fed. In order to grow more crops, forested areas are being converted into agricultural lands. The need for huge quantities of food crops, resulted in the intensive farming methods that soon deplete the soil of its nutrients. Addition of fertilizers to boost the crop production and use of synthetic pesticides to control pests, destroy the soil quality in the long run. Vast quantities of fresh water is diverted to the agricultural fields for irrigation and fulfilling the needs of man for drinking, cooking and other purposes. A large amount of wood is used as fuel and foliage is used as fodder for animals. Forested areas often give way to human habitats. Transport and communication network further depletes natural habitats and natural resources. Rapid and uncontrolled urbanisation, especially in developing countries, results in environmental pollution and health problems due to lack of infrastructure and proper awareness.

(3) Industrial development. Rapid industrial development, without regard for environmental standards, consumes large quantities of minerals, burns huge amount of fossil fuel, uses large quantities of water and consumes plenty of energy-electricity. Setting up of industrial areas and estates results in clearing of forest areas and loss of habitats. Production of cheap electricity for industries requires setting up of hydroelectric projects across the rivers and streams in the upper course. Construction of dams for water results in the incision of forests and uprooting of wildlife and tribal communities.

(4) Over exploitation for economic development. Over exploitation of natural resources for fulfilling human demands is causing a great shortage and non-availability of natural resources. Thus, prices of resources are increasing tremendously. The price rise adversely affects economic conditions of many countries. Prices of resources, especially of petroleum, diesel and mineral oil, are undergoing abrupt hike. Intensive agricultural practices are causing decrease in ground water, thereby lowering the water table. Mass scale deforestation, poaching and animal trade is causing threat to biodiversity. Loss of forests and improperly managed agriculture is causing desertification.

Chapter Summary

Natural resource is the stock that can be drawn from the nature and can be used directly or after transformation to sustain life. Some of the resources such as air, water, land, plants, animals etc. are readily available whereas others such as minerals, fossil fuels-coal, oil and natural gas, are hidden under earth surface. Natural resources have been classified variously on the basis of their chemical nature, availability or abundance, occurrence, origin, utility etc. Air, water, land, solar and tidal energy, rain etc. are inexhaustible resources and are found in such abundance that they are not likely to be depleted. Exhaustible resources are likely to be depleted and are either non-renewable—lack ability of recycling and replacement such as minerals, coal, oil, natural gas etc. or renewable resources—can replenish, reproduce and are recycled within a limited time and maintain themselves such as animals, plants (forests), water, soil fertility etc. If consumption of renewable resources exceed their rate of renewal they may become non-renewable or may be lost, Man's ever increasing needs and greed are resulting their depletion. Thus resources require to be used in a sustainable manner. Depletion of natural resources has posed a serious threat to mankind. Natural resources need to be consumed in a sustainable pattern and conserved for future generation. Conservation of resource is its prevention from loss, waste or degradation so as to last indefinitely. Problems, associated with natural resources, are many such as their uneven distribution; population growth causing pressure on their consumption; industrial development causing increased demand of sources of energy, water, minerals, wood etc.; and over exploitation for economic growth etc.

Study Questions

1. What are natural resources? Discuss their various types.
2. Write an account of depletion of natural resources.
3. What are various methods of conservation of natural resources. Describe.
4. Discuss various problems associated with natural resources.
5. Explain Exhaustible and Inexhaustible Resources.
6. Describe Renewable and Non-renewable Resources.
7. Write a note on Life Time and Depletion Time of Resources.

Objective Questions. *Select the correct answers.*

1. Fossil fuels are:

(1) Renewable Resources (2) Non-renewable Resources
(3) Inexhaustible Resources (4) Non-Renewable and exhaustible

2. Most of the world's gold resources occurs in:

(1) North America (2) South America
(3) Africa (4) India

Answers:

1. (4) *2.* (3)

3

CHAPTER

Forest Resources

LEARNING OBJECTIVES
Introduction • Forest Distribution • Importance of Forests • Deforestation • Conservation of Forests • Indian Forest Scenario

Introduction

Forests are one of the most abundant natural resources on this earth. They not only produce innumerable material goods, but also provide various environmental services that are essential for life. The epic 'Vedas' and 'Upanishads' were composed of sages singing *may the god, the water, the plants and the forests, the trees accept our prayers.... May the blessings of our 'simul' (tree) protect us for ever*. The 'Puranas' have said *one tree is equal to ten sons because tree means water, water means bread and bread is life*. According to 'Chandogya' Upanishad, a book thousands of years old *water is the essence of earth and plants are the essence of water*. Gunnar Poulson quotes *These are useful to man in two different ways: as producers of a wide variety of goods, commonly called 'forest produce', and as custodians of favourable environmental conditions. It would not make sense to try to qualify one of these functions as more important than other. Both are indisputable to all well beings, indeed to the survival of man*.

Perhaps, the word 'forest' has been derived from the Latin word *Floris*, which means outside, *i.e.*, uncultivated area covered with tall and dense tree growth outside the settlement. In general, the term **forest** is referred to an area, set aside for production of timber and other produce. Ecologically speaking, the forest is a biotic community predominantly of trees and other vegetation. United Nations Food and Agriculture Organisation (FAO) has defined forest land as *all lands bearing vegetation dominated by trees of any size, exploited or unexploited, capable of producing wood or other forest products*.

FOREST DISTRIBUTION

Forests contain about 90 per cent of biomass and cover about one-third of the land area of the earth. Nearly 50 per cent of total forest land is tropical forest. The FAO estimates world's land area as on 1994 to be 144.8 million sq km (or about 29 per cent of the surface of the globe), of which forests and woodlands account for 30 per cent (fig. 3.1). The FAO defines forest and woodland as land under natural or planted stands of trees, whether productive or not, including land from which forests have been cleared but will be restored in near future. Thousands of years ago, before large-scale human disturbances of the world began, forests and woodlands probably covered about 6.0 billion hectares of land. Since then, about 16 per cent of forest land area has been converted into cropland, pasture, settlements or unproductive wastelands. The FAO estimates the world's area under forests and woodlands as on 1994 to be about 4.7 billion hectares. Closed forests occupy about four-fifths of the forest and the rest are open forests. Closed forests are those in which tree crowns spread over 20 per cent or more of the ground and have potential for commercial timber harvesting. Open forests or woodlands are those where tree crown covers less than 20 per cent of the ground.

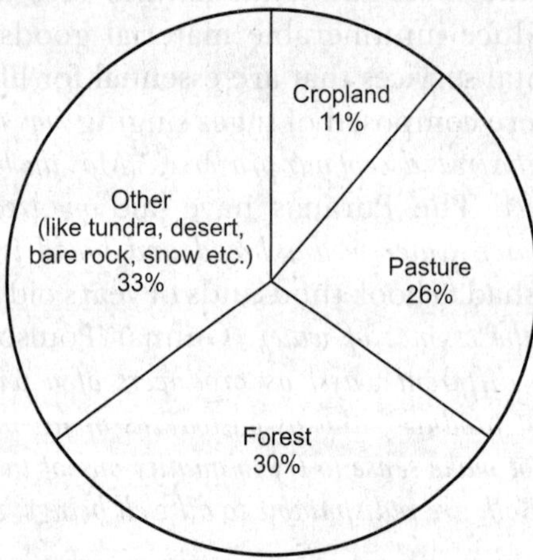

Fig. 3.1. World land use distribution.

World distribution of forests by regions depicts (fig. 3.2) that Latin America and Russia share between themselves about 39 per cent of world's forests. Africa possesses about 33 per cent forest land. North America and Europe share 11 per cent and 3 per cent, respectively. Asia with almost half of the world's population shares only 14 per cent and India with about 16 per cent of world's population possesses only 1.5 per cent of forest land. Over the years, there has been a decline of forests world over. The greatest loss has been in tropical Asia where one-third of the forests have been destroyed. India lost some 26,000 sq km of dense forests in between 2001-2003.

Fig. 3.2. World forests by region.

Distribution of Forests in India

India had abundant forests in ancient periods of Chandra Gupta Maurya who took keen interest in the management of the forests. Ashoka's stone edicts prescribed tree planting. At the beginning of twentieth century, about 30 per cent of land in India was occupied by the forests. But by the end of twentieth century the forest cover declined to 19.4 per cent (table 3.1). This is much less than the goal of having 33 per cent forest cover (20 per cent for the plains and 60 per cent for the hills) as inundated in the National Forest Policy (1988). Dense forest cover, with tree canopy covering 40 per cent or more of land area, decreased tremendously and by 1997 accounted for only 11.5 per cent. It has gone down further and is posing an alarming situation. Of the existing forests, about 65 per cent are dense forests and rest are open degraded forests. Presently, per capita forest area available in India is 0.06 hectares, which is much below the world's average of 0.64 hectares per person. Among sixteen different types of forests in India, composition wise, tropical dry deciduous forests occupy 38.7 per cent, tropical moist deciduous 30.9 per cent and rest are of other types of forests. Open forests and mangrove forests occupy 7.8 per cent and 0.1 per cent, respectively. Nearly 96 per cent the forests are owned by the government, 2.6 per cent by corporate bodies and the rest are in private sector. The forest area in different States and Union Territories of India is given in the (table 3.2). According to the Annual Report 2005-2006 of Ministry of Environment and Forests all the forest area occupied by Madhya Pradesh is, 76,429 km^2, highest among all states, and by Andaman and Nicobar Island is 6,964 km^2 highest among all

TABLE 3.1. Forest Cover in India (1999 Estimate)

Class	*Area (sq km)*	*%Geographic area*
1. Dense forest[1]	3,77,358	11.5
2. Open forest [2]	2,55,064	7.8
3. Mangrove [3]	4,871	0.1
4. Sub total	6,37,293	19.4
5. Scrub	5,896	1.6
6. Non-forest (other land use)	25,98074	79.0
Total	32,87,263	100.0

[1] Canopy cover >40% of land [2] Canopy cover 20-40% of land [3] Canopy cover 0.1% of land

Union Territories. Union Territory, Delhi has the forest cover of only 170 km^2. An overall increase in forest cover from 2001 to 2003, is 2,795 km^2. The total forest cover of the country as on 2003 was 678,333 km^2 (20.64 per cent). According to State Forest Report 2003, mangrove cover in the country occupied an area of 4,461 sq km (0.14 per cent) of total geographic area.

TABLE 3.2. Forest Cover in Different States and Union Territories (as on 2003)

Sl. No.	*State/ U.T.*	*Forest cover as on 2003 (km^2)*	*Sl. No.*	*State/ U.T.*	*Forest cover as on 2003 (km^2)*
1	Andhra	44,419	19	Mizoram	18,430
2	Arunachal Pradesh	68,019	20	Nagaland	13,609
3	Assam	27,826	21	Orissa	48,366
4	Bihar	5,558	22	Punjab	1,580
5	Chhattisgarh	55,998	23	Rajasthan	15,826
6	Delhi	170	24	Sikkim	3,262
7	Goa	2,156	25	Tamil Nadu	22,643
8	Gujarat	14,946	26	Tripura	8,093
9	Haryana	1,517	27	Uttar Pradesh	14,118
10	Himachal Pradesh	14,353	28	Uttarakhand	34,651
11	Jammu & Kashmir	21,267	29	West Bengal	12,343
12	Jharkhand	22,176	30	Andaman and Nicobar Islands	6,964
13	Karnatka	36,449			
14	Kerala	15,577	31	Chandigarh	15
15	Madhya Pradesh	76,429	32	Dadar & Nagar Haveli	225
16	Maharashtra	46,865	33	Daman and Diu	8.34
17	Manipur	17,219	34	Lakhsdweep	23
18	Meghalaya	16,839	35	Pondicherry	40

Components of Forest

A forest is essentially a community-an ecosystem with social grouping of plant species in which trees are dominant part of vegetation but in which other organisms and the soil, water and atmosphere play an important role. Thus flora (vegetation) and fauna (animals)-the biotic components and soil, water and atmosphere-the abiotic components are integral parts of a forest. The biotic components, *i.e.,* flora including variety of plant species and fauna including animals of all kinds such as mammals, birds, reptiles, insects etc., interact with each other and in turn with abiotic components, *i.e.,* air, water, soil, temperature, humus, rain, humidity, wind etc. and maintain a balance.

IMPORTANCE OF FORESTS

Forests play a significant role in maintaining the balance in nature. Gautam Buddha quotes *A tree is a peculiar organism of unlimited benevolence. It makes no demand for its sustenance and extends generously its products of life activity. It affords protection to all beings, offering shade even to the axeman who destroys it.* Forests provide important environmental services to mankind and are storehouse of biodiversity. They are the home of innumerable species of plants and animals. Forests have also housed some of the oldest indigenous tribes and played a vital role in the life of forest dwellers. Their main role, in the nature, is to maintain gaseous balance. The services rendered by forests fall under three categories.

I. Productive Functions

Productive functions of forests include production of wood and variety of produce which are important for sustenance of life. Forests are of immense importance to the life and prosperity to human beings and the nation.

1. Production of fuel wood. Fuel wood is the major forest produce useful to affluent many societies and the rural poor. In developing countries forests are looked upon mainly for the fuel wood. In fact, fuel wood accounts for almost half of all wood harvested worldwide. About 1.5 billion people largely depend on fuel wood for their primary energy requirement. About 58 per cent of total energy used in Africa and about 42 per cent in South-East Asia comes from fuel wood. The world consumption of fuel wood is estimated to be 2,600 million cubic meters till 2005. In India, fuel wood requirement in the year 2000 was 320 Mm^2 and estimated to be 350 Mm^2 by 2010 approximately.

2. Production of industrial wood. The unprocessed logs, the round wood and timber obtained from forests is used for commercial purposes such as construction of houses, furniture, ship building, boat making and in making of carts, carriages, railways poles, agricultural implements, packing cases, sports goods, toys etc. The wood also finds its use in making paper, plywood, lumber, veneer and other articles of industrial importance. Together, all these purposes account for the figure approaching one-half of the worldwide wood consumption. The, developed countries produce less than half of industrial wood but account for about 80 per cent of its consumption. On the other hand, less developed and underdeveloped countries, mainly tropical countries, produce more than half of industrial wood but use only about 20 per cent of it. Important forest wood based industries are plywood, furniture, paper, match box, sports goods etc. The FAO estimates that total removal of industrial wood from global forests in 1991 was 3,429 million cubic meters.

3. Production of non-wood Produce. Forest also contribute in production of many non-wood products which play a significant role in man's life and economy of the country. Vast varieties of food article, fibre, honey, medicinal plants, spices, resins, tannins, gums, lac and many articles of minor importance such as *kutch, ratti, ritha, sikakai, tendu patta, rudraksh* are produced by the forests. Bamboo, commonly called as poor man's wood, is also obtained from the forests. Some forests are also source of minerals. Non-wood produce fetch a big business in the country.

4. Other contributions. Forests provide a wide variety of employment to local people by way of enganging themselves in many forest oriented work plans and industries. Forests add to our knowledge by way of education, research and scientific study. Many of them protect our cultural heritage. Besides, the forests give us aesthetic pleasure and spiritual solace.

II. Protective Functions

Protective functions of the forests include conservation of soil and water; prevention from flood, drought and pollution; and protection against wind, cold, radiation etc.

1. Control of floods, drought and soil erosion. Forests control some of the recurrent natural disasters such as floods, landslides and avalanches. The roots of these plants keep the soil intact. The aerial parts of the plants intercept rains, decreasing their erosive power. Ground flora and thick layers of litter and humus in the forests act as sponge and retain the water received in the form of rains. By decreasing the velocity of water coming down from upper streams, forests help in absorption of water by soil, preventing floods. Extensive roots of the forest trees not only keep the soil particles intact but also induces structural development in the soil by adding humus. After being intercepted by aerial parts of the plants, the rain water reaches the ground level where it is easily absorbed by the soil. Thus trees along rivers, streams and lakes prevent bank erosion and silt deposition. Importance of forests also lies in preventing draughts leading to desertificaiton.

2. Prevention from pollution. Forests absorb many toxic gases and help in keeping the air pure and fresh. They operate carbon cycle and convert carbon dioxide into oxygen which is needed for all living beings in the forests and elsewhere. Many trees act as duet filters. Forests are the best tool for abatement of intolerable and high level of noise pollution which has become synonymous with the present day civilised societies with increasing number of automobiles on roads and industries all round. Properly placed screens of trees and shrubs decrease traffic noise along busy streets and highways. Many trees prevent and reduce water pollution in rivers, streams, lakes, reservoirs and estuaries.

III. Regulatory Functions

Regulatory functions of forests include absorption, consumption and release of gases like oxygen and carbon dioxide and minerals, water, radiant energy etc. Forests bring about gaseous exchange by releasing oxygen and consuming carbon dioxide. They also bring about moderation of climate and improve temperature conditions and enhance the economic value to landscape. Forests also provide habitat to tribals and wildlife.

1. Maintenance of gaseous balance. The main role of forests in nature, is to maintain gaseous balance. All breathing organisms, including man, take oxygen from the atmosphere and release carbon dioxide. Increasing population coupled with increasing number of automobiles, trains, aeroplanes and industries and many other activities release carbon dioxide and other pollutants in the atmosphere. Forests act as 'lungs' as they convert carbon dioxide into oxygen through photosynthesis. A mature *Ficus religiosa* (Peepal) tree, is capable of absorbing 2252 kg of carbon dioxide and releasing 1772 kg of oxygen per hour. Therefore, to save us from carbon dioxide and for maintenance of healthy gaseous balance in the atmosphere still more green plants (forests) are needed.

2. Habitats of flora and fauna. Forests are important habitats for a large diversity of flora and fauna. They have also housed and are still housing some of the tribals. They play a vital role in the life and the economy of forest dwellers. Forest plants provide fruits, roots, tubers, leaves and medicinal plants for forest based sustenance patterns. Forests are the homes to innumerable species of plant and animals and wildlife.

They bring about balance in the forest ecosystem and maintain equilibrium between various forms of life. They maintain wide variety of biodiversity.

3. Regulation of hydrological cycle. Forested watersheds act like giant sponges that they absorb rain water and slow down water run off. The absorbed water is slowly released for recharge of springs. About 50 to 80 per cent of the moisture in the air, in tropical forests, comes from the transpiration of plants and trees and this moisture brings about rains. Forest plants also help in attracting clouds and causing rains.

4. Moderation of climate. Forests are thermostat of nature and great moderators of climate. By releasing water vapours from their surfaces, plants increase humidity in the atmosphere. Increased rate of release of water as vapours, brings the atmospheric temperature down. By promoting seepage, forests increase the rate of recharge of ground water. They play a significant role in weather generation and reducing wind velocity. Forests also cause rainfall. Trees along rivers, streams, lakes and other water bodies lower water temperature and temperature of the atmosphere. Large trees provide shade to trespassers.

Among tangible benefits, the productive value of Indian forests is not very encouraging but their regulatory and protective values can not be ignored.

Aesthetic Value of Forests

Forests have a lot of aesthetic value. People all over the world appreciate beauty and tranquillity of forests. Forests are the centres of recreation, joy and spiritual renewal. They provide pleasant weather, pure and fresh air and beautiful scenery and thus refresh both body and mind. Every year lakhs of people visit national parks, sanctuaries and innumerable picnic spots in the forest areas all over the world for recreation. The joyful thrill of spotting a wild animal in the forest can be rightly appreciated only by those who have experienced it. Hill stations which attract tourists not only from India but also from abroad owe most of their charm to the existence of dense forests. India being very rich in its forests and wildlife has a great tourist potential and has a vast scope of eco-tourism.

DEFORESTATION

Deforestation refers to removal of trees or other such plants from a land area without intention of reforesting it. It started long back when man invented fire. He also exploited forests for food, shelter, agriculture and to meet his other needs. Considerable forests were felled during First and the Second World War to meet the needs of war. In India, tribals in

North-East Hill States and some part of Central India started *jhum cultivation* or *shifting cultivation, i.e.,* slash and burn the forests for agricultural land, discard it after two or three years and then move to the new area. Each year upto 15 million hectares of forests are cleared in the tropics and the land is used for other purposes. World's forest cover is shrinking rapidly. In developing tropical countries, about 40 per cent forest land has been degraded (fig. 3.3). However, temperate forests have lost only one per cent. The current deforestation rate in tropics is estimated to be more than 10 million hectares per year. If this rate of deforestation continues, it is estimated that the remaining tropical rain forests may disappear within a century.

Fig. 3.3. Destruction of rainforests.

Forests, apart from maintaining ecological balance, play a key role in terrestrial carbon cycle. Rapid rate of deforestation is posing dangers for avalanches, land slides, floods, drought and soil erosion. The causes of deforestation are well known. These include population pressure for agricultural land, economic incentives, inappropriate government policies, forest settlements, construction of dams, river valley projects and other population issues. Increased demand of forest produce is such that a large number of tropical countries exporting wood products have decreased their forest cover from 33 to 10 per cent by the end of the 20th century. In tropical South-East Asia, rapid population growth appears to be the major factor causing deforestation. India is loosing more than 1.5 million hectare of forest cover each year.

Causes of Deforestation

Causes of deforestation are many. Some outstanding causes of deforestation are:

1. **Population pressure for agricultural land.** Unprecedented growth in the world population is demanding increased agricultural production to sustain high birth ratio. In developing tropical countries there is a trend of reversion of forests to agricultural land and land for other uses. Shifting cultivation (*jhum*), a 9000-year old practice, is a more specific cause of deforestation. This land use changes pattern of slash and burn for agriculture, practiced by landless indigenous people in the tropics, is causing a great damage to the forests. The percentage of this practice is about 50 in South-East Asia including India where it is common in northern hill states and some parts of Madhya Pradesh. Local tribal people practice this type of agriculture in which they clear some forest area, cultivate it for two or three years, abandon it and move to new areas of the forest. Abandoned forest patches (*forest fallows*) may revert back to forests, if left undistributed. But, due to population pressure, little of forest fallow is allowed to revert back. It is estimated that about 50 million people that constitute about 10 per cent of world's population and about 240 million hectares of closed forests are involved in this practice and the rate is increasing at an annual average of 1.25 per cent each year.

2. **Fuelwood requirement.** Fuelwood demand is tremendously increasing in the poor people of developing tropical countries. As such several million hectares of forest land has to be stripped to meet this demand. About 1.5 billion people world over largely depend on fuelwood for their energy requirement. In India, demand for fuelwood has increased sharply. Rapid population growth has caused the demand of fuelwood to shoot up 300 to 500 million tons in 2001.

3. **Industrial demand for wood.** Forests provide wood for construction of houses, making furniture, plywood, packing cases, pulp for paper industry, match box, pencils etc. At present plywood is in great demand for making doors, windows, packing cases for tea and apple, furniture etc. In India, industrial requirement of wood is estimated to be about 40 million m^3, out of which about 50 per cent is met by bamboo plantation. It has been observed that poorly performed commercial logging causes degradation of the forests.

4. **Development projects.** Forests have undergone massive destruction for various development projects such as dams, river valley projects, reservoirs, hydroelectric projects, construction of highways, road, rail roads, irrigation canals etc. Dams, hydroelectric projects and reservoirs submerge large forest tracts causing harm to the plants and animals of the area. They uproot thousands of forest dweller not only from their area of residence but also from their livelihood. Much before construction of these development projects, a lot of forest land is cleared for residence of their workers for which wood and other produce are used up.

5. Overgrazing. The poor in the tropics mainly rely on clearing the forest land to convert it into grazing land for their cattles. Overgrazing leads to further degradation of these lands. India has a large livestock population but grazing land is limited to only 13 million hectares. This much of grazing land is insufficient to meet the demand and as such the livestock naturally grazes in the forests and causes a lot of grazing pressure and loss of the forests. The animals trample the seedlings and cause compaction of soil so that its water storing capacity is decreased and run off capacity is increased. The combined effect of these activities leads to soil erosion and deforestation.

6. Quarrying and mining operations. In forested areas, especially in hilly regions, quarrying and mining operations are extremely harmful as they cause great damage to vegetation over large areas due to mine dusts, transportation of ore and mine wastes. In India, large forest land areas are clear felled and laid barren as a result of open cast mining of mica, coal, manganese, limestone etc.

7. Forest fires. Sometimes forest suffers from forest fires which may be natural or man-made. A forest fire is any unexpected or unregulated combustion of vegetation spreading over a forest area. The forest fires are of three types:

1. **Ground Fire:** The fire spreads when dry grasses, lichens or top dry humus etc., burn up. Usually ground fire does not damage full grown trees, but poses the danger of transforming it into a crown fire. It normally destroys herbs, shrubs and bushes. Ground fire initiates usually after a prolonged drought of the forest floor during second half of the summer.
2. **Tree Crown Fire:** The fire in the upper portion of the tree is normally derived from ground fire as there appears no reason of combustion of tree canopies without the support of ground fire.
3. **Underground Fire:** The fire results due to ground fire. Peat soils, which have undergone prolonged drought in the forests are prone for ground fire. After the upper peat soil cover is burnt, smouldering continues in patches covering dry earth mounds and the fire penetrates into the peat horizon of the soil. Underground fire results funnel-shaped pits. Since this fire spreads under the ground in the horizontal manner, tree roots are damaged.

Most of the forests fires are deliberate burning of trees by people in timber trade with the conspiracy of forest guards so that the burnt trees are auctioned at cheaper rate and traders may buy burnt trees.

Forest fires usually start by lightening and release the minerals locked in the organic matter. The mineral rich ashes, necessary for the growth of plants and vegetation, usually flourishes after fire. Fire removes the plant cover and expose the soil. Exposed soil stimulates germination of certain type of seeds. Occasional fires burn away some of the dry organic matter and prevent from more destructive forthcoming fires. Thus for all these reasons, prevention of fires is not necessarily good for the forests. Human induced fires, however, cause serious damage and such fires have become major problems in many countries including India. In developing countries, fires are caused for land clearing with many adverse effects. Climate change and global warming also cause forest fires. Some other causes of forest loss include suffering from natural enemies like termites and phyto-pathological diseases.

Methods of controlling forest fires are aimed at arresting the fire at initial stage. Once the fire has taken place, the flame edge of the fire can be beaten with green branches or covered with soil. Quenching the fire with water or fire extinguishers is very effective but their transportation and cost becomes limiting factor. Construction of fire barriers or control belts and trenches, either manually or with mechanical tools or by blasting, will go a long way in controlling forest fires.

Eco-social and Environmental Impacts of Deforestation

Deforestation affects human life and environment in many ways (fig. 3.4). The impacts may be direct or indirect, regular or occasional and mild or severe. Loss of forests causes both flood and drought and results in soil erosion, *i.e.*, loss of the top fertile soil causing loss of fertile land. Denuded land mass gradually gets converted into sand due to action of winds laden by fragmented rock dust in rain scarced areas and causes expansion of deserts. Deforestation in temperate forest areas convert these areas into hot houses and affects their climate. Deforestation threatens many wildlife species and their natural habitats causing the loss of biodiversity and ultimately the genetic diversity. Altered hydrological cycle affects rain fall. In India the entire Himalayan ecosystem is threatened and under severe imbalance as snow lines have thinned and perennial springs have dried up. In hilly areas landslides and avalanches have become the regular features. Deforestation also lowers down the ground water table and loss of commercial and industrial produce of the forests. It also brings about the change in the environmental quality by increasing carbon dioxide and other pollutants in the atmosphere. Loss of mountain forests increases amount of run off water into rivers causing frequent floods. Deforestation also causes loss of soil fertility, threatens tribals, brings about regional and global climate changes, change in rainfall pattern and contributes to global warming.

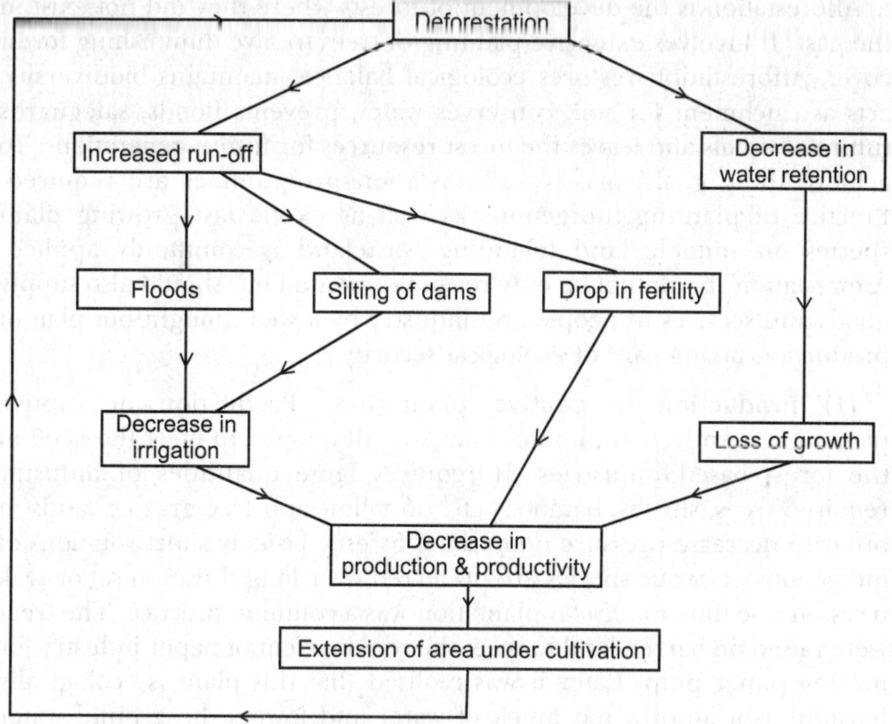

Fig. 3.4. Impact of deforestation.

CONSERVATION OF FORESTS

Conservation of forest includes the development and management of forests in such a way as to improve climate and water flow, prevent soil erosion and provide optimum sustainable yield both for present and future generations. In their hymn to earth, the sages of "Atharwa Vedas" chanted: *what of thee, I dig out let that quickly grow over. Let me not hit thy vitals or thy heart*. In India, increasing destruction and degradation of forests, especially in mountaneous and hilly areas, lead to heavy deforestation. Thus, forests are to be developed and managed judiciously. These programmes are aimed at sustainable supply of tree products and services to people and industry and maintenance of long term ecological balance through production and protection. National Forest Policy (1988) has also set a goal of having 33 per cent of forest on the total land area of the country, out of which 20 per cent forest cover should be in plains and 60 per cent in the hilly areas. To achieve these goals following methods are commonly used:

I. Production or Commercial Forestry: Afforestation

Production or commercial forestry aims to fulfil commercial demand through intensive plantation and production or captive plantations in available land without causing denudation of natural forests.

Afforestation is the development of forests where they did not exist in the past. It involves extensive planting of trees to save diminishing forest cover. Afforestation restores ecological balance, maintains biodiversity, acts as catchment for soil, conserves water, prevents floods, safeguards future of tribals and leaves the forest resources for future generations. To achieve these goals, massive afforestation programmes are required. Practice of planting indigenous as well as exotic fast growing plant species on suitable land including wasteland is commonly applied. Afforestation should not be only revenue-oriented but should also supply goods and services to people and industry by a well thought out-plan of production taking care of ecological security.

(1) Production or **captive plantation.** Production or captive plantation is entirely commercial forestry, developed to meet the need of the forest based industries. It requires large quantities of industry required trees, shrubs, bamboos etc. on fallow and free grazing lands in order to decrease pressure on existing forests. Usually short rotations of indigenous or exotic species are preferred over long duration sal or teak trees. At one time *Eucalyptus* plantation was a common practice. The trees were raised on barren land to meet the requirement of paper industry for making paper pulp. Later it was realised that this plant is ecologically harmful as it absorbs too much of water and lowers the ground water table. It is important that neither forests should be transformed into agricultural lands nor are agricultural lands into forests as both are critical to our future. Any dynamic and successful production plantation should be such that it should reduce stress on the forests. It should be aimed at increased productivity of wood per unit area. If needed, planting stock should be achieved by tissue culture.

(2) Intensive plantation. Intensive plantation refers to raising the plants on all available land from villager's field to community land, to roads, rail road sides etc. in every available space. For the purpose, indigenous, exotic or a combination of both can be used for plantation. Such programmes will meet the needs of timber, fuel, fodder, food, fibre, medicines, and thus prevent denudation of natural forests.

(3) Social Forestry. Social forestry is more a programme relating to institutional development and involves integration of forest resource, labour and capital in such a coordinated way so that it improves the quality of natural life. It has a potential to bring about economic revolution in the rural areas. Social forestry is a new concept which has been taken seriously only about 20 years back and is being implemented on a massive scale. It is a multidisciplinary activity which involves diverse land uses and different groups of people warranting different approaches under different socio-economic environments. The objectives of social forestry are to be attained through farm forestry, rural forestry and urban forestry. Farm forestry includes combining agriculture with forestry and is aimed at fulfillment of need of fuelwood, small timber, fodder and extra income to the farmer. The main objective of farm forestry is to create fuelwood resource so as to release cow dung for its legitimate use as manure. Farm forestry involves raising of wind breaks, shelter belts, farm wood lots, raising trees on village's common land etc. The farmers grow their own timber along field bunds, wasteland and marginal land unfit for cultivation. Rural forestry, also called extension forestry, is entirely a new activity of raising tree crops on community lands, panchayat lands, degraded forests along roads, rail road sides, canal banks, afforestation of fore-shore areas of tanks, ponds and reservoirs and plantations to reclaim wastelands unfit for agriculture. Rural forestry requires mass scale involvement to rural people. It can encourage cottage industries such as silk, honey, household furniture, agricultural implements, toys etc.

Too much success has been achieved by social forestry in India, bringing it into limelight in the forestry world. The credit for this goes to the people of India and the devoted and dedicated functionaries. There is yet a long way to go and new approaches have to be adopted. New institutions will have to be developed in future in light of the experience gained to grow and sustain itself. The social forestry requires political and bureaucratic support. At the same time, barriers between both land uses – agriculture and tree culture should be removed as both of them are the pillars of strength of our economy.

In real sense, social forestry is a programme of land transformation to reestablish a green umbrella over the country. There can be no improvement in the environment without the tree restoration activity. The programme being labour oriented, would help in generating employment opportunities to the very door-step of unemployed. 20 years is too short to reach a goal of self-reliant and self-sufficient people's movement, supported by well developed institutions that are responsive and responsible to the people. Everyone should have an opportunity to

Box 3.1 **Van Mahotsava**

In 1950, Government of India initiated *Van Mahotsava* programme to make people aware of plantation and afforestation. The programme was started by KM Munsi. People all over the country were motivated to raise plants and trees atleast in two occasions, *i.e.*, the first week of February and July. The programme soon became popular and by now a large number of fallow areas have been converted into green cover. But, the people still need to be educated so that the care is taken of newly planted trees or such other plants.

contribute to the activity and should add to this novel programme. Because if Himalayas are affected, even the south India cannot escape of the consequences. Without worrying about who does and who does not, our country needs trees, more trees and still more trees as there can not be excess of trees in India.

(4) Urban forestry. Concept of urban forestry is the aesthetic development of urban areas. Flowering and fruit trees of ornamental values which flower and fruit at different seasons, are planted along roads, on round abouts, in city parks and on camp sites and picnic spots. It also includes beautification of private compounds, farm houses, reserve lands etc. Green belts around cities, industrial and commercial complexes are also part of urban forestry. Trees along roads have aesthetic and ornamental values. City forests and parks provide source of recreation to dwellers of city and elsewhere. Urban parks expose urban society of our rich heritage of wildlife. Many trees act as dust filters and keep us away from dust and pollutants.

(5) Agro forestry. Agroforestry involves variety of land uses where woody species are grown in combination with herbaceous crops either at the same time or in time sequences. Actually it is a modified version of social forestry. In agroforestry, woody perennials are deliberately used on the same land management units as annual agricultural crops. Agroforestry is indeed a new scientific version of ancient land use where land was used for agriculture, forestry and animal husbandry. It has many advantages than traditional forestry. Agroforestry responds to population pressure and needs no surveillance, no unfamiliar technology and consequences to the environment. It results in the produion of fuel wood, timber, fodder etc. Depending upon the situations and needs, agroforestry may be agri-silvicultural, agri-pastoral or agri-silvi-pastoral, raising multipurpose species. It would generate employment also.

II. Protection or Conservation Forestry

Protection or conservation forestry involves protection of degraded forests to bring back its biodiversity. Well stocked forests are managed scientifically for producing timber and other forest produce without causing any negative environmental impact on the forest. Forest areas of national parks and sanctuaries are protected from human interference. Protection or conservation of forest is achieved by the following methods.

1. Reforestation. Reforestation involves development of the forests where they have been destroyed. It is achieved through protection or conservation of degraded forests. The areas with water regimes such as Himalayas, Western Ghats, Eastern Ghats along with their catchment areas and national parks, sanctuaries, sacred grooves, biosphere reserves and all ecologically fragile areas are not allowed for commercial exploitation. For this, public support is generated to make the people aware of the consequences of degradation and loss of the forests.

2. Production forests. Production forests are the forests on flat land that are managed for high degree of production. The working of these forests is based on scientific lines and involves proper logging techniques involving no environmental problems. Efforts are made at generating forests and not the plantations. The forest thus maintained exhibits its three stories – the tall trees, small trees and shrubs coupled with soil micro-flora that forms a dynamic living system.

Limited production forests are less fertile areas at more than 1000 meters altitude with hilly topography. A part of annual increment may be harvested from them in a very careful and controlled manner so as to avoild soil and tree damage. Thus, in these forests, the basic forestry stock and the forest health remain unaffected.

3. Sacred forests. Sacred forests are forest patches of varying dimensions protected by tribal communities due to religious sanctity accorded to them. They represent islands of pristine forests, *i.e.*, the forests most undisturbed and unaffected by human activity. These forests have remained free from all disturbances despite being surrounded by highly degraded landscapes. In India, sacred forests are located in various areas such as in Maharashtra, Kerala, Madhya Pradesh etc. They are serving as refugia for a number of rare, endangered and endemic plants and animals.

4. Sustained yield block cutting. In sustained yield block cutting, a forest is divided into blocks depending upon the period required by trees to mature. For example, 100 blocks of trees requiring 100 years to mature, in one year, trees of only one block are felled and the block is reforested soon. Thus annual deforestation is compensated by annual reforestation so that the forest is conserved indefinitely and gives sustained yield.

5. Reserve forests. Reserve forests are normally raised over water sheds, slopes and other ecologicaly fragile areas. The forests are not allowed to be disturbed by any kind of human activity. There is total ban on grazing and felling of trees, even litter collection is not allowed from these forests.

General Conservation Strategies

The following measures, in general, can be adopted to conserve the forests:

1. A tree removed from the forest for any purpose, should be soon replaced by a new tree, thus compensating tree felling with tree planting.

2. Plantation should be done in areas unfit for agriculture along highways, roads, rivers, canals, playgrounds etc.
3. *Van Mahotsava*, a programme of tree plantation in first week of July and February, every year, should be popularised and made effective among common people.
4. Maximum economy should be observed in the use of timber and fuel wood and their wastage should be minimised.
5. As far as possible, use of fuel wood should be discouraged and alternative energy sources such as solar cooker, natural gas or bio-gas should be made available to the users.
6. Forests should be protected from fire and modern fire fighting equipments should be used to extinguish forest fires.
7. There should be an effective programme of control of pests and diseases of forest plants preferably through biological methods of control.
8. Grazing in the forests should be discouraged.
9. Modern methods of forest management should be adopted.
10. Mycrorrhizal inoculation, breeding of elite trees and tissue culture techniques should be promoted.
11. Rural and tribal communities should be involved in forestry programmes.
12. Commercial felling of trees, in ecologically fragile areas, should be stopped for certain specified period.
13. Forest conservation policies should be made more efficient and effective in India. National Afforestation and Ecodevelopment Board (NAEB), established in August 1992, looks after afforestation and restoration and development of forests in Ecologically fragile areas.

INDIAN FOREST SCENARIO

Forests are considered as green gold of a country. They are the renewable resources and contribute substantially to the economic development and enhancement of quality of environment. The forest cover–6,78,333 sq km of land in the country constitutes 20.64 per cent of its geographical area. Of this total forest cover, very dense forests constitute 51,285 sq km covering 1.96 per cent, moderately dense 3,39,279 sq km covering 12.32 per cent and open forests 2,87,669 sq km covering 8.7 per cent of the geographic area of the country. Despite rich forest wealth, the annual increment of forests is one of the lowests in the world. It is only 0.5 cubic metre per hectare (as against world average of 2.1). A comparison of forest cover assessment of 2003 with that of 2001 reveals that there is an over all increase of 2,795 sq km. The total tree cover for the country (forest area of 70 per cent canopy density) has been estimated as 99,896 sq km (about 3.04 per cent).

Forests extend from Himalayas in the north to Kanyakumari in the south and from North-East Hill States in the east to the deserts of Rajasthan in the west. Composition wise 93 per cent forests are tropical (both dry and moist) and the remaining 7 per cent are temperate. Among tropical forests, 80 per cent are deciduous, 2 per cent evergreen and 8 per cent others. Among 7 per cent of temperate forests 3 per cent are coniferous and 4 per cent are with broad leaved species. Broadly speaking, India has 6 types of forests:

1. **Evergreen** (Tropical Forests) in area with 200 cm to 300 cm annual rainfall, average annual temperature 20°C to 27°C, and average humidity more than 80 per cent.
2. **Deciduous** (Monsoon Forests) in areas with lesser rainfall between 150 cm to 200 cm per annum, mean annual temperature between 24°C to 28°C and average humidity 75 per cent.
3. **Dry Forests** in areas with scanty annual rainfall between 75 cm to 100 cm, mean annual temperature between 23°C to 29°C and average humidity 50-60 per cent.
4. **Hill forests** occurring commonly in Himalayas and south India.
5. **Tidal Forests** (Mangroves) occurring in coastal submerged plains of Ganges (Sundarbans), Mahanadi, Godavari and Krishna.
6. **Grasslands** in hilly high Himalayan and Deccan Hills, in lowlands in Punjab, Haryana, Uttar Pradesh, Bihar, North-West Assam.

The types of Indian forests range from evergreen tropical rain forests in Andman and Nicobar Islands, the Western Ghats, which fringe the coastline of Indian peninsula, and in greater Assam region in the north east to small remnants of rain forests found in Orissa. Semi-evergreen forests are more extensive and occur between the two-north and south extremes of the country in which also occur deciduous monsoon forests, thorn forests, sub-tropical pine forests in the lower montane region and temperate montane forests at higher altitudes in Himalayas. The human interference has degraded many evergreen forests either by replacing them with semi-evergreen forests or bringing them to other land uses.

Productive Functions of Indian Forests

Apart from protective functions of controlling gaseous balance in the atmosphere; forests bring about moderation of climate, prevention of floods, droughts and soil erosions, provide habitats for wildlife, and play a significant role in the production of a wide variety of articles. Annual contribution of forests to our national income is about 2.7 per cent. They furnish numerous major and minor forest produce and provide habitat for the tribals. Forests also provide employment to wood cutters, sawyers, carters, carriers, craftsman and in other forest based industries. About 1.34 lakh professional, non-professional and non-technical staff is employed in various forest departments. About 3.8 crore tribals find their homes in the forests all over the country and derive their necessities from the forests. The total standing volume of timber produced is 85.694 m^3 out of which 93 per cent is non-coniferous and 7 per cent coniferous. About 450 species of trees provide wide varieties of commercially important products including valuable drugs. About 63 per cent of wood produced is used as fodder. Forests also provide fodder to livestock including cattles. Some forest based industries such as broom making, silkworm rearing, lac, toy making, leaf plate making, paper and pulp, pencil making, saw mills, match, plywood, tea, chest, fibre board, chip board, resin and turpentine obtain raw materials from the forests. Besides, forests produce also includes bamboo, canes, drugs, spices, edible fruits and vegetables, fibres, grasses, gums, resins, rubber, latex, incense and perfume woods, dying and tanning materials "tendu" leaves for "bidi" and oil seeds, "kutch" and "kuthha", essential oil aromatic grasses like lemon grass, khas etc. Important varieties of timber produced in forests are teak (*Tectona grandis*), sal (*Shorea robusta*), sissoo (*Dalbergia sissoo*), mulberry (*Morus alba*), Chir (*Pinus roxbughii*), deodar (*Cedrus deodara*), mango (*Mangifera indica*) etc.

Total increment in wood production from all forests is about 60 to 70 million m^3 out of which only 44 million m^3 is the annual cut and rest is being wasted due to faulty modes of extraction, conversion and damage caused by accidental fire, phytopathogens and other causes.

Forest Policy of India

India is one of the few countries which has a forest policy since 1894. It was revised in 1952 and again in 1988. The goal of National Forestry Policy (1952) of optimum forest cover could not be achieved due to continuous deforestation for various reasons. With a view to conserve the forests, Government of India could enact the Forest (Conservation) Act, 1980. Under the provision of this Act, prior approval of Central Government is required for diversion of forest lands for non-forest purposes. The Government has notified Forest (Conservation) Rules, 2003. National Forest Policy, 1988 envisages people's involvement in the development and protection of degraded forests as a permanent resource base to fulfil forest based requirements of local communities as well as to develop the forests for improving the environment. The main plank of the forest policy is protection, conservation and development of the forests.

Aims of Indian Forest Policy

Aims of the Indian forest policy are:

1. Maintenance of environmental stability through preservation and restoration of ecological balance.
2. Checking of soil erosion and denudation in catchment area of rivers, lakes and reservoirs.
3. Conservation of natural heritage.
4. Checking of extension of sand dunes in desert areas of Rajasthan and along coastal tracts.
5. Substantial increase in forest tree cover through massive afforestation and social forestry programmes.
6. Steps to meet requirements of fuel wood, fodder, minor forest produce and timber for rural and tribal populations.
7. Increase in productivity of forests to meet the national need.
8. Encouragement to efficient utilisation of forest produce and optimum substitution of wood.
9. Steps to create massive people's movement with involvement of women to achieve the objectives and minimize pressure on existing forests.

In year 1990, Ministry of Environment and Forests issued guidelines to involve village communities in the development and protection of degraded forests on the basis of their taking share of usufruct from such areas. The concept of Joint Forest Management (JFM) was accordingly initiated in 1998. Joint Forest Management was pursued vigorously and as a result JFM revolution has now been adopted in all states. An additional component entitled 'Strengthening of JFM' in all 1,73,000 forest fringe villages have been included in National Afforestation Programme scheme w.e.f. August 2004. By now 84,632 JFM committees have been formed. About 58.28 lakh families are involved in this programme all over the country.

In the year 1992, Ministry set up National Afforestation and Eco-developments Board (NAEB) to promote the task of afforestation, tree planting, ecological restoration and Eco-development activities in the country, with special attention to degraded forest areas and land adjoining forest areas as well as ecological fragile areas like Western Himalayas, Aravalies, Western Ghats etc. NAEB, in its meeting held on 19th January 2006, has prepared a Draft Action Plan to achieve the target of 33 per cent forest cover in the country by 2012. In the year 1999, Ministry formulated National Forestry Action Plan (NFAP) for sustainable development of forests in the country in the next 20 years. For Tenth Five Year Plan, an Integrated Forest Protection Scheme has been prepared.

During 2005-06, a restricted scheme 'Grants-in-Aid for Greening India' has been launched to promote high quality plantation.

Forestry Research and Development Scheme

Ministry of Environment and Forests in its Central Plan Scheme has included promotion of research in multidisciplinary aspects of protection, conservation and development of forests in India. The objective of the Scheme is to generate information required to develop technologies and methodologies for better forest management. It is also aimed at achieving solutions to practical problems of forest resource management, conservation and regeneration. The Scheme also seeks to strengthen facilities to promote research and scientific manpower development. In order to achieve such objectives, Ministry has set up Indian Council of Forestry Research and Education (ICFRE), at Dehradun, which is coordinating with various forest research institutions all over the country. Some of these institutions are:

1. Forest Research Institute, Dehradun
2. Institute of forest Genetics and Tree Breeding, Coimbatore.
3. Institute of Wood Science and Technology, Banglore
4. Tropical Forest Research Institute, Jabalpur
5. Institute of Rain and Moist Deciduous Forests Research, Jorhat
6. Arid Forest Research Institute, Jodhpur
7. Himalayan Forest Research Institute, Shimla
8. Institute of Forest Productivity, Ranchi
9. Centre for Forestry Research and Human Resource Development, Chindawara.
10. Wildlife Institute of India, Dehradun.

ICFRE also coordinates with the following institutions:

1. Indira Gandhi National Forest Academy, Dehradun
2. Indian Plywood Industries Research Training Institute, Banglore
3. Indian Institute of Forest Management, Bhopal.

People's Initiative and Social Movements

A forestry programme can be successful only when the local community feels the responsibility for the protection of the forests. In 1787, Maharaja Jodhpur ordered wood cutter to fetch firewood from Vishnoi village. Vishnoi's as a community, abhorred the killing of any form of life including trees. Wood cutter went to the village to fetch the wood when the men folks were away. The women, led by Amrita Devi protested and stood hugging (clinging tightly) the trees to save them. Their protest

Box 3.2 **V D Saklani**

V D Saklani was awarded prestigious Indira Priyadarshini Vrikshmitra Award in 1986 for planting and foundly nurturing more than 2 lakh trees on once barren land in and around Pujargaon, a remote Garhwal village about 50 km from Dehradun.

He has planted these trees in last 48 years and covered more than 100 hectares of land with trees like oak, rhododendron, cedar, walnut etc. He has brought life back in its myriad form to this area. He has made the hill sides green which were once indiscriminately used for quarrying and tree felling. The soil of the terraced fields has stabilized. The villager's traditional sources of fodder and fuel have been restored. Birds have returned to the area. According to him *if you don't cover land with trees, the soil will get washed away and there will be no more land left for you, for me, for any one* (from Reader's Digest, 1997).

resulted in that 363 of them were hacked to death but not a single tree could be cut (Box 3.3).

'Chipko' Movement (Chipko Aandolan)

In India, Alaknanda valley in Himalayas, undergone a massive degradation of forests and faced unprecedented flood in 1970. In 1972, the "Chipko" movement (Chipko means to hug or stick tightly) led by Gaura Devi was launched by women activists. The women of the village Advani of district Tehri Garhwal (Uttarakhand) used to tie the sacred thread ('Kalawa') around the trees and hugged the trees, protested and courted arrest against tree felling (fig 3.5). In 1973, Chandi Prasad Bhatt and Sundar Lal Bahuguna took keen interest in the movement against large scale tree felling. A unique feature of the movement was that local hill women from the villages were organised and made aware of the ecological threat to the hills.

Sundar Lal Bahuguna became active leader of Chipko movement and in 1979 he established Navjeevan Ashram in Tehri Garhwal to speed up the movement against indiscriminate felling of trees in hills of the region and elsewhere. This movement of saving hill forests and greenery soon spread all along the hill regions. Bahuguna marched more than 3000 km from Kashmir to Siliguri (known as 'Kashmir to Kohima Chipko Foot March') to bring about awareness of the aims of this movement and the need to start similar movements in all hill parts of Himalayas. During his march, environmentalists from all over the world namely France, Germany, Switzerland, Sweden etc. met him to offer moral support. Bahuguna presented the plan of this movement of protection of soil and water through ban on tree felling in the Himalayas at the United Nations Environmental Programme (UNEP) meeting held in London in 1982.

Box 3.3 They Saved Trees by Gifting their Heads

Controversies arising about poaching of deer by film stars around village near Jodhpur, Rajasthan have consequently raised the debate about the conservation efforts and enforcement of laws for punishment for those found guilty. While the case is pending in courts, it is time to look at the efforts at conservation by 363 Visnois lad by Amrita Devi, the pioneer of *Chipko* movement, of village Khejadi, who sacrificed their lives on September 12, 1787 to save Khejadi (*Prosopis cynararia*) *Kalpataru of Thar* trees from the soldiers of the then Maharaja of Jodhpur Abhay Sigh Rathore. It is because of these Vishnoi people that the country has come to know of such heinous crimes. The movement of Vishnois for wildlife conservation is replete with a wonderful mixture of love and compassion. The selfless service of the Vishnois and the determination with which they have been fighting for innocent wild animals, gives us lesson that if we can not go to that extent of supreme sacrifies, we should at the very minimum try to support their cause as actively as we can.

The Chipko plan is infact a slogan of planting five Fs-food, fodder, fuel, fibers and fertilizer plants to make communities self sufficient in all their basic needs to generate a decentralised, self-renewing and long term prosperity. It will protect the environment and bring peace, prosperity and happiness to mankind. The movement soon became popular and gained appreciation from all over the world. Today the movement has grown to more than 4000 groups working together to save Indian forests. In the course of time Indian Chipko movement crossed the

Fig 3.5. Chipko movement.

geographical barriers and April 29, 1983 was observed as "Chipko Day" in New York, USA during which school children, supported by some adults, assembled and hugged a big tree in Union Square Park. On a 5th June, while celebrating World Environment Day in an poster exhibition at Stockholm (Sweden) following account was written and displayed about the Chipko movement *A powerful environmental movement has grown up on slopes of mountains of Himalayas. Villagers have created an effective non-violent way to stop the devastation by forest industries. When the axeman comes, the people form a ring (circle) around trees and embrace the trees. This has given the movement its name Chipko Aandolan - the tree hugging movement.*

'Apikko' Movement

Chipko movement flew from Himalayas to Sirsi region of Karnataka in the south in 1983. Prior to 1947, eighty two per cent of this area was richly forested with teak trees. Later, due to construction of certain dams and development of wood based industries, forest cover reduced to 20 per cent only by 1983. Farmers non-violently protested clear felling of trees. Initially 160 men and women Marched 8 km to the Kelase forest where the contractor's axemen, under the order of forest department, have turned the area into slaughter house. They hugged the remanant trees of the area. The movement named as "Apikko" movement (Apikko means stick tightly to), became popular state wide and later people joined this movement willingly in many forest areas of Western Ghats from Koorg to Goa. To speed up the movement, "padyatras", slide shows, street plays and other ways were adopted to make people aware of the consequences of tree felling. The movement has three objectives: "uliso" (conservation), "belesu" (growing or plantation) and "balasu" (rational use of forest vegetation).

Silent Valley Movement

Silent Valley, occupying an area of 8950 hectares at an altitude of 3000 feet in Palghat district of Kerala, is perhaps, the only remaining undisturbed tropical rainforest in Indian peninsula. This tropical rain forest in Western Ghats is a precious reservoir of biodiversity where many plant and animal species have survived for centuries.

History of the Silent Valley movement is related with the proposed construction of a river valley project to fetch the need of electricity and irrigation to the people of the area. From the begining, the project was started, it was opposed by many NGOs and environmentalists as it could cause serious damage to this "cradle of evolution" – the home to many rare and unique species of plants and animals. With the intervention of Friends of Trees Society, the controversy of the proposed hydal project was over in 1979 and it was declared as a National Park by the then prime minister Rajiv Gandhi in 1985. The proposal was seriously criticized by Kerala Shastra Sahithya Parishad – coming the forefront of the campaign to save Silent Valley. Some other societies such as Kerala Nature History Society and NGOs are taking keen interest to save this valley from destruction due to invasions, construction and other such activities. This ecologically fragile area storing rich stock of biodiversity has been declared as biosphere reserve and now is under the category of Hot Spot.

Chapter Summary

Forests are one of the most abundant resources on this earth. They not only produce wood and other produce but also furnish various environmental services which are essential for life. Forests cover about $1/3^{rd}$ of total land area on this earth. Nearly 50 per cent of total forest land is the tropical forest. Among total forest land in the world's $4/5^{th}$ is covered by forests with tree cover more than 20 per cent and the rest are open. India possesses 1.5 per cent of total forest land on the earth. Contry's 19.4 per cent of land is covered by different types of forests which is much less than the goal of having 33 per cent (60 per cent for hills and 20 per cent for plains). In India, tropical forests dominate and are spread over Andaman and Nicobar Islands, Western Ghat, North-East Hill regions of Assam etc.

Forests are important because of their productive, protective and regulatory functions. They produce fuel wood, timber and variety of produce such as resin, gum, *lac, tendu patta*, bamboo, honey, drugs etc. Forests protect us from flood, drought, soil erosion, pollution, desertification etc. Regulatory functions of forests overweigh their productive and protective functions. Their most important regulatory function is to maintain gaseous balance in the atmosphere. Forests are called *lungs* of earth as they give us oxygen and take carbon dioxide for making their food on which all of us depend. Besides, they form the habitat of flora and fauna including wildlife, regulate water cycle, bring about moderation of the climate and render aesthetic values.

Deforestation is the removal of trees or other such plants from a land without intension of reforesting it. Man has been engaged in felling a considerable amount of forests to meet his requirements since historic times. Causes of deforestation include population pressure for need of agricultural land, fuel wood requirement, industrial demand for wood, various development projects such as dams, river valley projects, over grazing, quarrying and mining operations, forest fires and many other reasons. Deforestation affects human life by way of flood, drought, soil erosion, loss of biodiversity and destruction of habitats of tribals and wildlife. In hilly areas, massive deforestation results land sides and in the plains lowering of water table. Large scale deforestation is causing global warming and many other problems.

Conservation of forests includes development and management of forests. There is an urgent need for conservation of forests in India which is loosing its forest cover very rapidly. Forest conservation can be achieved by production or commercial forestry involving afforestation through production or captive plantation, intensive plantations, social forestry, urban forestry, agro-forestry and protection and conservation forestry involving reforestation, production forests, sacred forests, sustainable yield block cutting, reserve forests and several other methods. A forestry programme can be successful only when the local community feels its responsibility for the protection of the forests. Many people's movements such as *Chipko*

Aandolan Apikko movement, *Silent Valley* Movement, have brought tremendous change in man's intention of forest felling. Forests are the *green gold* of our country covering about 20.64 per cent of its geographical area. Composition wise 93 per cent forests are tropical and 7 per cent temperate. Forests include more than 5000 species of trees which produce wood for fuel and commerce and wide variety of non-wood produce. About 3.8 crore tribals find their home in forests.

India has its own Forest Policy (1952, 88) and Forests (Conservation) Act, 1980 that are aimed at protection, conservation and management of the forests. Ministry of Environment and Forests undertakes evaluation, protection, conservation and management of forests by its various plans and projects including establishment of Indian Council of Forestry Research and Education (ICFRE) at Dehradun for promotion of research, education and training in different aspects and coordinating with forest institutions all over the country.

Study Questions

1. Write an account of importance of forests.
2. What are various methods of conservation of forests? Describe.
3. Discuss briefly state of forests in India.
4. Describe various social movements for afforestation in India.
5. What are various efforts made by Government of India to save country's forest cover? Explain.
6. Briefly describe the distribution of Forests in India.
7. Explain the role of forests in control of flood and drought.
8. Write short notes on:

   (i) Deforestation
   (ii) Afforestation
   (iii) Forest Fire
   (iv) Agroforestry
   (v) Forest Policy of India
   (vi) Chipko Movement
   (vii) Silent Valley Movement

Objective Questions. *Select the correct answers:*

1. Indian's 11.5 per cent forest cover is of the type:
   (1) Open forests (2) Dense forests
   (3) Mangroves (4) Temperate forests

2. The State of India with maximum percentage of its area covered by forests is:
   (1) Rajasthan (2) Karnataka
   (3) Bihar (4) Madhya Pradesh.

3. Precipitation, the function of forests is a:
   (1) Productive Function (2) Aesthetic function
   (3) Regulatory function (4) Destructive function

Answers

1. (1) *2.* (4) *3.* (3)

4

CHAPTER

Water Resources

Learning Objectives

Introduction • Types of Water Sources and their Uses • Available Water for Human Use-The Fresh Water • Fresh Water Requirements in India • India's Fresh Water Budget • Indian Riverine System • Interlinking of Indian Rivers: Projected Gains in v/s Hydrological Suicide • Use and Over-utilisation of Surface and Ground Water • Floods • Drought,Conflicts Over Water • Dams: Benefits and Problems.

Introduction

Water is a wondrous gift of nature. Literally, water is elixir of life. It is the only inorganic liquid that occurs naturally on this earth. Water is also the only chemical compound on this planet that exists in all three physical states-solid (ice), liquid and gaseous (vapour). Sanskrit equivalent of water is "jeevan" or life. Hydrosphere is the water domain. A, once abundant resource that existed in oceans, seas, rivers, streams, lakes, reservoirs, glaciers, polar ice-caps, shallow ground water bodies and as ground water, is increasingly being wasted, misused or polluted.

People say that world is moving towards water wars. World is facing an impending crisis of water most visible in many countries. Fresh water resources of the world are finite in nature and at the same time only a small fraction of it is accessible and readily usable to human and ecological systems. The hydrological cycle controls the temporal and spatial distribution of the renewable fresh water in the form of evapotranspirations, precipitation and run off. Climatic factors and climatic changes further complicate the predictability of this distribution, particularly in heavily populated areas of the world. The total stock of ocean water and fresh water on this earth has been fairly constant throughout geological history but the ratio between ocean water and fresh water always changed according to climatic changes, mostly brought about by human activities. Evidently, global climate is getting warmer and sea level, during the last decades, is rising slowly.

Despite the fact that most part of India is blessed with abundant monsoon rains, its water supplies are stranded by unchecked population growth and rapid developments most visible in many of its cities. Water crisis can also be seen in its rural areas but the reasons are different. India and China are considered among the water hotspots, primarily because of their large populations that have to be provided with drinking water. By 1994, there were more than 55,000 villages in India which did not have a source of drinking water. Apart from increasing population, steady deterioration, disuse, and disappearance of traditional tanks, ponds and wells, are the other causes of water crisis in India. In recent years the problem got aggravated due to extraction of ground water using bore well operarted by electric, diesel or petrol pumps. It is estimated that the water extraction rate is already twice the recharge rate.

Wisely used, water means harvests, health and prosperity for people and the nations on the earth. When badly managed or out of control, water contributes to economic underdevelopment, poverty, disease, flooding, drought, soil erosion, salinisation, water logging, silting, environmental degradation and human conflict. At the end of century the world faced a number of challenges affecting the availability, accessibility, use and sustainability of its fresh water resources. Consequently this could adversely impact upon the needs of present and future generation. Equally critical are the needs of ecosystems that depend on fresh water for their existence. The effective management of world's water resources will contribute to the strenghtening of peace, security, cooperation and friendly relations among nations. Because water is most vital of all natural resources, it has the power to promote economic and social advancement of all peoples.

Types of Water Sources and their Uses

In nature, water occurs in oceans, on the land, underground, in the atmosphere and in the biomass. Of the total volume of water available on this earth, 97 per cent is in the vast oceans, 2 per cent is locked in the form of ice-sheets and only less than 1 per cent is available as fresh water for which all people, animals and plants compete.

Water in the Oceans. Oceans contain 97 per cent of water but their water contains high concentration (3.5 per cent) of salts in solution. The main source of salinity of the oceans is due to falling of rivers that carry with them billions of tons of dissolved minerals every year.

Most of the ocean water is contained in Pacific Ocean, covering about one-third of earth's surface. It is deepest of all oceans and its area exceeds the total land area of the earth. Atlantic Ocean is roughly half of the size of the Pacific Ocean. The Indian Ocean, with an average depth of 4000 metres, is smaller than Atlantic Ocean. The Arctic Ocean, surrounding North Pole, is completely frozen and unnavigable.

Salinity and temperature are two important features of ocean water which determine the movement of large masses of water and their characteristics and also the type of marine flora and fauna. The huge amount of the salinity in ocean water makes it uneconomical to make it potable, *i.e.*, fit for drinking and cooking. But, at places where sea water is the only source available, the potable water is obtained by desalinising or demineralising the sea water, as is done in ships on the high seas.

Water on the Land Surface. Water falling in the form of precipitation runs down the slope of land in the form of lotic water bodies (in which water flows) like rivers, streams etc. or lentic water bodies (in which water remains stagnant) such as lakes, reservoirs, ponds etc. or may get stored on the land in the form of ice-sheets. Water is also used by plants, animals and man in the biosphere. Some water percolates and is trapped underground below the land surface. Surface water available on the land is most accessible source of water for human needs. The amount of surface water in a region depends on total precipitation and its seasonal distribution and also on the nature of rocks and soils of which land is made.

In India, according to an estimate, to 1,150 cubic km of fresh water, which appears as surface water, may be added 200 cubic km of surface flow which comes from outside the country. The surface flow is further enhanced by addition of about 450 cubic km of fresh water from ground water flow while about 50 cubic km are added as run off from irrigated areas. The surface also loses about 50 cubic km of its water which percolates down to the ground water resources. The total surface flow per year is about 1,800 cubic km which is distributed among a number of river basins. Various surface water sources are:

1. **Rivers and Streams:** The precipitation, that does not evaporates or infiltrates, runs off over the ground surface in the form of streams and rivers which flow towards the sea. Rivers and streams are the major source of water supply. They have long been used for discharging waters. Most of the civilisations have grown and flourished on the banks of the rivers. In our country most important cities and religious places are located near rivers. Their water is generally more variable in quality. It is less satisfactory than water from lakes and impounded reservoirs. The water quality of rivers depends on various factors such as the characteristics and the area of watershed, geology, topology, seasonal variations, weather conditions, disposal of sewage, pollutants, agricultural washouts, industrial effluents etc. Most of Indian rivers especially Ganga and Yamuna, have become polluted to such an extent that their water has become unfit for human use.

2. **Natural Lakes and Ponds:** Natural lakes are the inland depressions that hold fresh water for long periods. Ponds are generally small and permanent or temporary fresh water bodies. The water of lakes and ponds is much more accessible than the ground water. They are the important sources of water supply in limited quantities. Their water is more uniform in quality than water from rivers and streams. Unfortunately most of Indian lakes are shrinking due to man caused pollution and ponds are disappearing due to increased land demand for agriculture or housing to accommodate huge population.

3. **Artificial Impounding Reservoirs:** Artificial impounding reservoirs are formed by constructing hydraulic structures such as dams across river valleys. Deeper and narrower valleys are most suitable for construction of dams. Their water is much more accessible than ground water as their water quality is similar to that of natural lakes and ponds. They have played a crucial role in improving the economy of our county by providing enough water for irrigation and domestic use and generating hydro-power but they have left a serious problem of resettlement and rehabilitation of oustees from the areas where dams are constructed.

Most surface water originates directly from the precipitation or snowfall but it is also contributed by ground water from springs and seeps. A large amount of surface water is also contributed by melting of glaciers due to global warming.

GROUND WATER

The rainwater and water from melted ice that neither runs off along the surface nor evaporates but reaches under the surface of the land through pores and cracks, is called ground or underground water. Some ground water may originate from the streams rising from the molten rock materials deep within the earth. Springs owe their existence due to flow of ground water. Although ground water occurs in small amounts everywhere in the soil but great amount of it reaches the sea after moving through underground channels, becoming a part of hydrological cycle.

About 9.86 per cent of the total freshwater resources is in the form of ground water. Till sometimes back, ground water was considered to be pure but now many ground water aquifers are being contaminated by leaches from sanitary landfills and dumping of many undesirable substances.

A layer of sediment or rock that is highly permeable and contains water is, called aquifer. Sand and gravel layers are good aquifers while clay and crystalline rocks, because of low permeability, are not so. Aquifers may be unconfined or confined. Unconfined aquifers are overlaid by permeable earth materials and are recharged by water seeping down from above in the form of precipitation and snow melt. Confined aquifers are sandwitched between two impermeable layers of rocks or sediments and are recharged only in those areas where the aquifer interacts the land surface. Sometimes their recharge is hundreds of kilometers away from the location of the well. The two aquifers are shown in the ground water system (fig. 4.1). The Ground water is not static but it moves. The rate of its movement is very slow, *i.e.*, about one meter or so in a year.

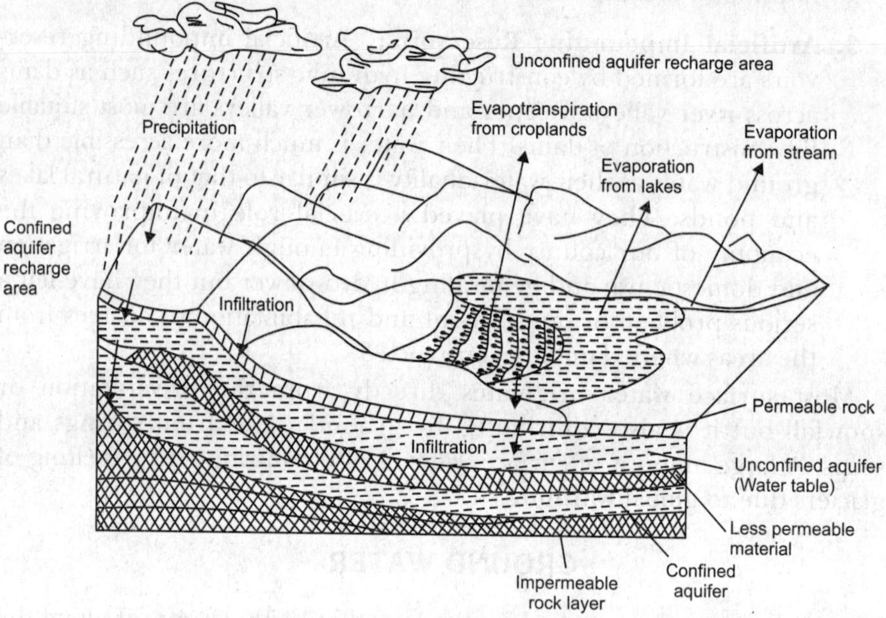

Fig. 4.1. Ground water system.

Ground water is one of the largest fresh water resources, next to surface water, for agricultural and domestic use in many areas of the world, particularly in areas having insufficient surface water resources.

Advantages of Ground Water: Ground water offers many advantages as compared to surface water reservoirs. Ground water reservoirs do not suffer seepage losses and losses due to evapotranspiration. Ground water sources are relatively simple to develop quickly near the place of use. The water from ground sources is colourless, clean and pollution free. Its quality is generally uniform which makes it as the major source of agricultural and domestic use in many areas of the world. Groundwater works as a remedy for water logging, especially where water table is high. Groundwater is harder than the surface water of the region in which it occurs. It is less likely to be contaminated.

Effects of Excessive Ground Water Uses: Excessive extraction of ground water, especially in arid and semi-arid regions, causes a sharp decline in the level of water table. Excessive irrigation with brackish water gradually raises water table and causes water logging and salinity problems. When the ground water withdrawl rate exceeds its recharge rate, the sediments in the aquifer get compacted, a phenomenon known as "ground subsidence". It may cause huge economic losses because it results in the sinking of overlying land surface associated with structural damage to buildings and sewage and pipelines. It may also increase tidal flooding. Excessive pumping of ground water results porous formations and makes the shallow well to become dry. Over use of ground water and fresh water reservoirs along coast lines often allows salt water to intrude into aquifers that are used for agricultural and domestic purposes. There are many aquifers with low recharge rates which once emptied require thousands of years to refill them. The "fossil water", when pumped out from such aquifers can not be refilled in our lifetime.

Available Water for Human Use: The Fresh Water

Available water for human use, the fresh water, is only 0.3 per cent of the total volume of water in the hydrosphere. Man has to share this usable water with the plant and animal world. According to World Water Vision, the "blue water" or renewable portion of rainfall that enters into streams and recharges ground water is the traditional focus of water management whereas "green water" or soil water is the portion of rainfall that is stored in the soil and then evaporates or is incorporated in plants and organisms. It is estimated that the total amount of available fresh water on the earth is only about 84.4 million cubic km. Much of the water on earth's surface and ground water represents deposits, which have accumulated over a long span. About 60,000,000 cubic km of fresh water is found as deposits under the land surface, about 24,000,000 cubic km locked as ice in snow caps over mountains, ice-sheets and glaciers and about 35,000 cubic km is present as soil moistures on which plants and subsoil organisms survive. Nearly 1,200 cubic km of fresh water occurs in our rivers and streams (fig 4.2).

These deposits are interrelated and interconnected. Ice sheets and glaciers regularly feed rivers, streams and underground deposits at lower altitudes when ice melts and water trickles down. Water lying in high altitude lakes and underground strata also seeps down and trickles out under the influence of force of gravity and further adds to rivers and streams. The rivers and the streams flow over long distances catering to the need of people living all along their course. Also they transport large quantity of sediments rich in plant nutrients which are deposited in low

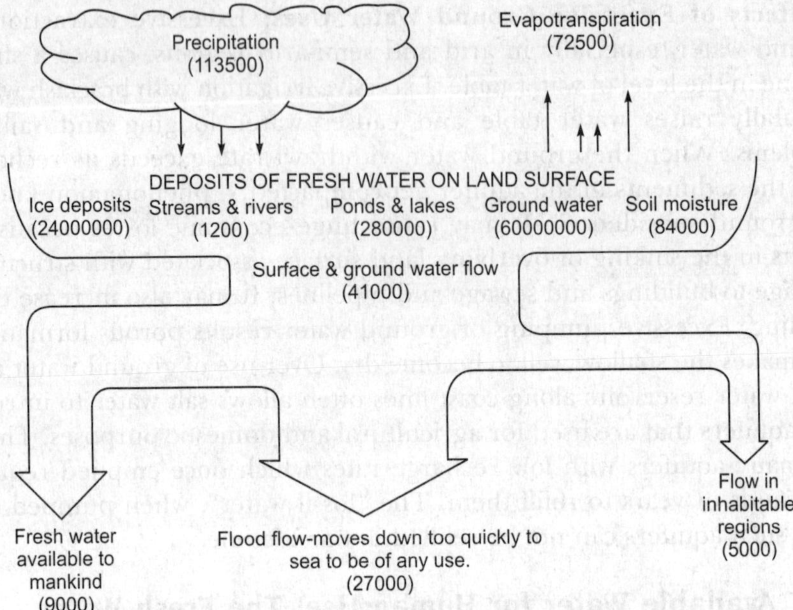

Fig. 4.2. Fresh water system (values in cubic kms.)

lying areas in a broad zone all along their course. Water seeping down the beds of rivers replenishes and recharges ground water table on either side to a reasonable distance even in dry months making ground water available at convenient depths. It is these deposits which fulfil most of human needs of fresh water. Based on data available, it is estimated that in the year 2000, about 9605 cubic km of water was drawn for human use from these deposits. And this quantity shall be about 14,102 cubic km by the year 2025.

Global fresh water resources have their own limitations. Their withdrawl beyond the limits shall diminish this natural resource base and bring about the adverse changes in the environment: the greenery shall disappear, flora and fauna shall undergo changes and desertifications will succeed. Data regarding global availability of fresh water indicate that South America, South Asia and North America occupy first, second and third ranks, respectively. Among regions of the world facing scarcity of fresh water are Kuwait, Quatar, Malta and Lybia. Among global withdrawal of fresh water, Middle East is at the top. Future estimates of water consumption depict a grim picture. By 2000 we have drawn more water than the total amount of renewable water available to us. By 2025 fresh water supply and demand situation would appear pretty grim and will compel us towards water economy, to take steps to recharge and replenish the ground water deposits and to raise our surface storage or rainwater harvesting capacity.

The rate of increase in global water use accelerated in the twentieth century (particularly after 1940) as compared to earlier periods (fig. 4.3).

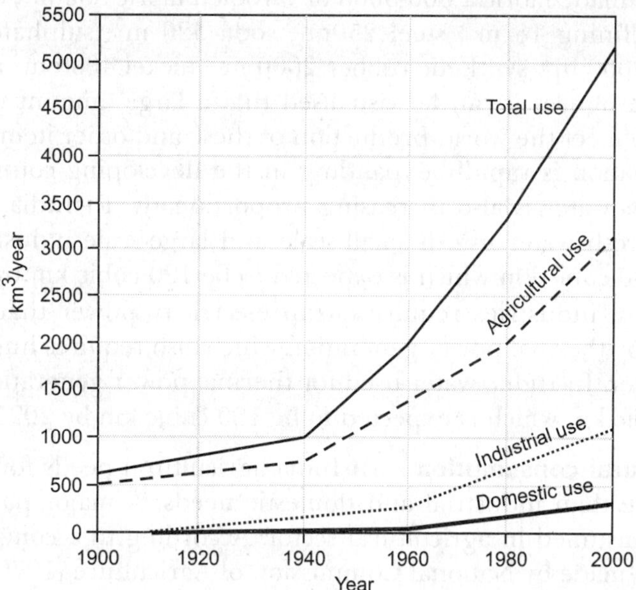

Fig. 4.3. Trend of total water use during twentieth century.

Fresh Water Requirements in India

Adequate fresh water supply of standard quality is essential to meet the individual, domestic, industrial, agricultural, power generation and livestock management needs of Indian people. India being agricultural country, needs its much of fresh water for agriculture sector followed by livestock management and domestic use.

Individual's needs. On an average, a human being needs drinking water of about two litres to compensate its loss from the body through urine, breath and sweat. India being a tropical country, needs more than this amount of water for an individual for his sustenance.

Domestic needs. To live standard healthy life, a family of six persons needs daily a minimum of 600 liters of water for drinking, bathing, cooking, washing etc. In addition, it needs water for watering plants, lawns or garden, to meet the needs of domestic animals and to wash vehicles for personal use. The potable water supply, at present, is very limited. In India estimated fresh water requirement for fulfilling domestic needs as on 2000 was 26.6 cubic km. It is expected that in 2025 it would be about 39.0 cubic km Water needs for livestock management is expected to increase from 4.7 cubic km to 11.0 cubic km from 2000 to 2025.

Industrial needs. The industrial and commercial needs for water are tremendous, *e.g.* production of one ton of cotton fabric needs 250 m^3 water, man-made fabric 2,500-5000 m^3, woolen fabric 580 m^3, cane sugar 4m^3, oil refining 18 m^3, steel 250m^3, soda 320 m^3, sulphate 240 m^3, ammonia 1000 m^3, synthetic rubber 2000 m^3, nickel 4000 m^3 and so on. From these figures it can be visualised that a large amount of water is required to meet the world production of these and other items. Further, industrialization is rapidly expanding in the developing countries. The demand for water is also increasing proportionatly. In India, estimated values of fresh water use in small-scale and large-scale industries as on 2000 was 30 cubic km which is expected to be 120 cubic km by 2025.

Moreover, industries require cheap electricity power that is chiefly produced by thermal power generation which also requires huge amount of freshwater. In India, water used for thermal power generation in 2000 was 60 cubic km which is expected to be 160 cubic km by 2025.

Agricultural consumption. In India agricultural needs for water are much more than industrial and domestic needs. A major part of fresh water is consumed in agricultural sector. According to a comprehensive assessment made by National Commission of Agriculture (1976), the total annual basic water resources of the country are 185 million hectare metres (M ham) comprising 135 M ham of surface water and 50 M ham of ground water resources. On full development, total annual basic ground water resources would increase to 85 M ham and surface water to 185 M ham including 45 M ham generated by ground water. Besides, some credit can be taken from underground fossil water, surface flows from glaciers and permanent snows and expensive desalinated water from the sea. The utilizable flows aggregate to about 104 M ham :70 M ham from surface water and 34 M ham from ground water. But, not all this water is available for irrigation as there are other demands like industrial, municipal etc. It is expected that by 2025 about 770 M ham of water would be required for irrigation and rest for other uses (table 4.1).

TABLE 4.1. Expected Fresh Water Requirements in India

Water requirement for	*1974*	*2000*	*2025*
1. Irrigation	350.0	630.0	770.0
2. Thermal power generation	11.0	60.0	160.0
3. Industries	5.5	30.0	120.0
4. Domestic use	8.8	26.6	39.0
5. Livestock management	4.7	7.4	11.0
	380.0	754.0	1100.0

Of the three major fresh water uses, *i.e.*, for agriculture, industry and for fulfilling domestic needs, agriculture is the greatest consumer of water in India. Agriculture has been to a great extent responsible for over-exploitation, leading to depletion of water in most part of the country. In response to green revolution, which began in India in 1960s, great stress was laid on expansion of agriculture. Agriculture accounts for about 95 per cent of total water consumption. Only 3 to 4 per cent is needed by industries and 1 per cent to fulfil domestic needs. Construction of dams have affected our rivers by way of destroying their natural flow and alter ecosystem of river basin. The growing demand of water has led to over-exploitation of rivers to fulfil the needs of the people. As a result, most of the rivers have reduced in volume drastically. The growing population, heavy industrialisation and rapid development have accelerated the demand and use of water. This has caused a great stress on the resources particularly water of our country even in humid regions. The strain is more acute because of deterioration of water quality which has been brought about by agricultural and industrial washouts and domestic pollution.

India's Water Budget

India receives about 3 trillion m^3 of water from rainfall which amounts to about 105 to 1117 cm annually-a huge water resource, perhaps the largest in the world. Almost 90 per cent of this precipitation falls between mid-June and October. Further major river systems, *viz.*, Indus, Ganga, Brahamputra, Godawari, Mahanadi, Krishna, Brahmani, Vaitarini, Cauveri, Subernrekha, east flowing rivers; Narmada, Tapti, Mahi and other west flowing rivers; 44 medium and 55 smaller river systems that are fast flowing monsoon fed originating from coastal mountains, account for 85 per cent of surface flow and share 83 per cent of drainage basin. They serve 80 per cent of the total population of India. The water storage capacity is 3.65 million m^3. Of the total precipitation, country utilises only 10 per cent, which may increase to about 26 per cent by 2025. Of the total water used in India, 95 per cent is for agriculture and 5 per cent for industrial and domestic use. Although there is a lot of precipitation but sufficient care has not been taken in India for unused river flows and about 1,677 billion cubic meters of river water, of which Ganga and Brahmputra together account for 1,089 billion cubic metres, are left to the seas annually. Thus surface water resources need to be exploited from available potential sources. Despite, rivers and streams are the major source of water supply, the water from these sources is generally more variable in quality as well as less satisfactory than water from lakes and impounded reservoirs.

In India, on account of broad regional differences, little is known regarding the water budget, annual recharge and ground water levels. The Central Ground Water Board has estimated that the available ground water in India is 210 billion m^3 and the annual utilization potential is about 42.3 million hectare. The ground water availability is not adequate in states like Tamil Nadu and Andhra Pradesh (fig 4.4). Large scale deforestation and monsoon failure is causing regular occurrence of drought in Kalahandi and some other districts of Orissa, Karnataka, Rajasthan and Maharashtra.

In India, major portion of fresh water, which goes to earth's crust, is retained by its upper layers in the form of soil moisture amounting about 1,650 cubic km Only 500 cubic km of water percolates down to the ground water deposits. Large amount of fresh water (about 120 cubic km) applied to agricultural fields moves down to ground water table while small amount (about 50 cubic km) of surface flow also surrenders itself to ground water. Thus a total of about 670 cubic km of fresh water enters the

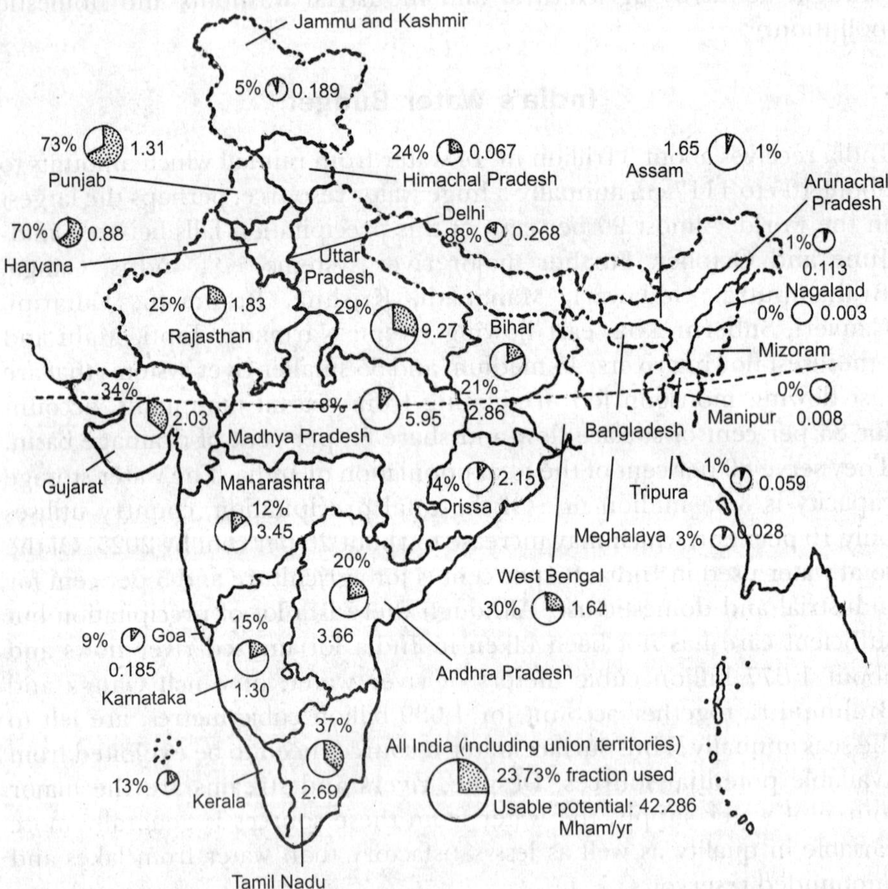

Fig. 4.4. Groundwater potential and use in India (before Uttarakhand, Chhatisgarh and Jharkhand states were formed).

Box 4.1 Cherrapunji: Once Wettest Place on the Earth-Now Has No Water to Drink

At this place in Meghalaya in North East India, monsoon did not rain but used to pour water continuously even for two months. With an annual rainfall of 11.5 m. it is listed in every geography book as the place with the highest rainfall in the world. In 1974, Cherrapunji had a record rainfall of 24.5m and on one day it was 1563 mm, yet it faces severe drought conditions for the rest of the year. The reason why this place remains to quench the thirst of people is the destruction of the forests. There was a time when hills around Cherrapunji were covered with dense forests which used to soak up the heavy rainfall and released it slowly during the rest of the year. Over the years forests were cut down. The heavy rains washed away the topsoil, turning the slopes into deserts. Now mining operations in this area for coal and chalk are making it dusty and it is difficult to breath over there. There is no reservoir to store water and for over 20 years residents have depended on a piped water supply that comes from a far. This supply is always unreliable, erratic and nondependable. Although rains do come to this area for four months yet people do not have enough water for the next eight months.

ground as ground water annually. We can withdraw fresh water up to this amount from our sub-surface deposits and any withdrawl above this limit shall be detrimental to the resource base.

The Indian sub-continent receives most of its fresh water (about 75 per cent) during monsoon months. About 4000 cubic km of fresh water is received as precipitation annually. Rest of the months are usually drier which necessiate the use of ground water or stored water during the dry spells. The uneven distribution of precipitation in different months of the year is matched by its equally uneven distribution over different regions of the country.

Indian River System

A river is a large stream of water that flows from high land to low land. The water in rivers comes from rains, snow melt, lakes, springs and water falls. The river water eventually flows into oceans.

Indian river system can be classified into four groups, *viz.*, (i) Himalayan rivers (ii) Deccan rivers (iii) Coastal rivers and (iv) Rivers of the inland drainage systems. The Himalayan rivers are formed by melting snow and glaciers and, therefore, flow continuously throughout the year. During monsoon, Himalaya receives heavy rainfall and rivers swell, causing frequent floods. The Deccan rivers on the other hand are rainfed and, therefore, fluctuate in volume. Many of these are non-perennial. The coastal streams, especially on the west coasts, are short in length and have limited catchment areas. Most of them are non-perennial. The streams of inland drainage basin of western Rajasthan are few and far between. Most of them are of an ephemeral character.

The main Himalayan river systems are those of Indus-Brahamputra-Meghna system. Indus, one of great rivers of the world, rises near the Mansarovar in Tibet and flows through India and thereafter through Pakistan and finally falls in the Arabian sea near Karachi. Its important tributaries flowing in Indian territory are the Sutluj (originating in Tibet), the Beas and Ravi, the Chenab and the Jhelum. The Ganga-Brahmputra-Meghana is another important system of which the principal sub-basin are those of Bhagirathi and the Alaknanda, which join at Devprayag to form Ganga. It traverses through Uttrakhand, Uttar Pradesh, Bihar and West Bengal states. Below Rajmahal hills, the Bhagirathi, which used to be the main course in the past, takes off, while the Padma continues eastwards and enters Bangladesh. The Yamuna, the Ramganga, the Ghagra, the Gandak, the Kosi, the Mahananda and the Sone are the important tributaries of the Ganga. Rivers Chambal and Betwa are the important sub-tributaries, which join Yamuna before it meets the Ganga. The Padma and Brahmputra join inside Bangladesh and continue to flow as Padma or Ganga. The Barahmputra river in Tibet, where it is known as Tsangpo, runs a long distance till it crosses over into India in Arunachal Pradesh under the name of Dihang. Near Passighat, the Debang and Lohit join the river Brahmputra and the combined river runs all along the Assam in a narrow valley. It crosses into Bangladesh downstream to Dhubri. The principal tributaries of Brahmputra in India are Subansiri, Jia Bharli, Dhansiri, Puthimari, Pagladiya nad Manas. The Brahmputra in Bangladesh receives the flow of Tista etc., and finally falls into Ganga. The Barak river, the head stream of Meghna, rises in the hills of Manipur. The important tributaries of river are Makku, Trang, Tuivai, Jiri, Sonai, Rukni, Katakhal, Dhaleswari, Langachini, Maduva and Jatinga. Barak continues in Bangladesh till the combined Ganga-Brahmputra join it near Bhairab Bazar.

In Deccan region, most of the major river systems, flowing generally in east direction, fall into Bay of Bengal. The major east flowing rivers are Godavari, Krishna, Cauvery, Mahanadi etc. Narmada and Tapti are major west flowing rivers. The Godavari in the southern peninsula has the second largest river basin covering 10 per cent of the area of India. Next to it is Krishna basin in the region, while the Mahanadi has the third largest basin. The basin of the Narmada in the uplands of the Deccan flowing towards the Arabian sea, and of the Kaveri in the south falling into the Bay of Bengal, are about of the same size though with different characters and shape.

There are numerous coastal rivers, which are comparatively small. While only handful of such rivers drain into the sea near the delta of east coast, there are as many as 600 such rives on the west coast.

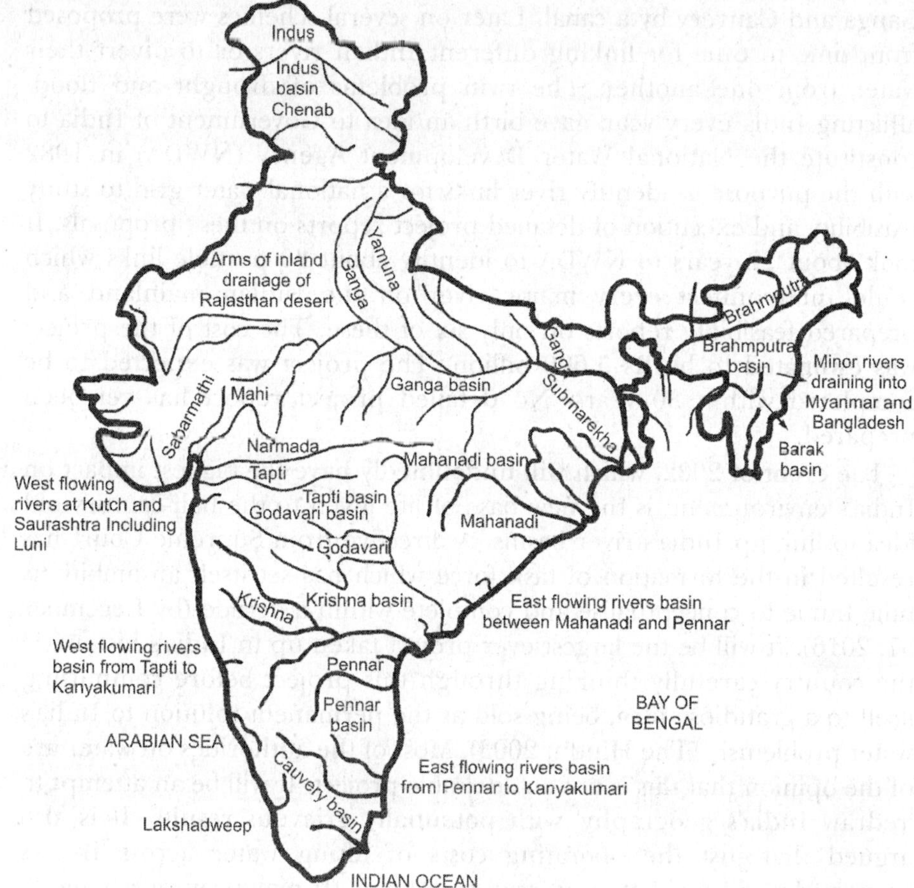

Fig. 4.5. Indian river system and river basins (Andaman and Nicobar not included).

A few rivers in Rajasthan do not drain into the sea. They drain into salt lake and get lost in sand with no outlet to sea. Besides these, there are the desert rivers which flow for some distance and are lost in the deserts. These are Luni and others namely Machhu, Rupen, Banas and Ghaggar. Ganga has been declared as National River of India (2008).

Interlinking of Indian Rivers: Projected Gains v/s Hydrological Suicide

India has witnessed various attempts leading to hydraulic manipulations in past decades. It was in the mid-nineteenth century that the first scheme of interlinking of Indian river systems was put forward. This first river interlinking scheme was proposed by Sir Arthur Cotton, a British ruler and a reputed hydrological engineer, with the hope that the scheme would solve the transportation problem of people and goods in India. But the scheme was dropped and was given a new shape in 1960 to link river

Ganga and Cauvery by a canal. Later on several schemes were proposed from time to time for linking different Indian rivers or to divert their water from one another. The twin problems of drought and flood, afflicting India every year, gave birth an idea to Government of India to constitute the National Water Development Agency (NWDA) in 1982 with the purpose to identify river links for a national water grid to study feasibility and execution of detailed project reports on these proposals. It took about 22 years to NWDA to identify about 30 possible links which could interconnect every major river on the Indian mainland and prepared feasibility reports on only six of these. The cost of the project was estimated to be Rs 5,600 billion. The project was expected to be completed within 30 years. No detailed project report has yet been prepared.

The event of 2002, which will undoubtedly have the biggest impact on India's environment, is the new base of life given to the half-century old idea to link up India's river basins. A directive from Supreme Court has resulted in the formation of task force which has set itself an ambitious time frame to conceptualise and complete within a decade (by December 31, 2016). It will be the largest ever project taken up in India's history. Is the country carefully thinking through this project before committing itself to a grandiose idea, being sold as the permanent solution to India's water problems? "(The Hindu, 2003). Most of the authorities on water are of the opinion that this is not an advisable project. It will be an attempt to 'redraw India's geography' with potentially grievous results. It is also argued that just the operating costs of lifting water across India's topography will make it an unrealistic scheme. It may prove a scheme as India's final step towards hydrological suicide. There can be other and less expensive alternatives for dealing with India's water problems. The international experience with similar such projects does not suggest any room of optimism with the Indian experiment. Ramaswamy R Iyer (2003), former Secretary Water Resources in the Government of India, quotes *The proposed interlinking of rivers is touted as the answer to India's water problems, however, that it will be a terrible intervention in nature and that better solutions are available*. He also quoted that *Rivers are not human artefacts; they are not pipelines to be cut, turned around, welded and re-joined*. Rohan D'Souza (2003), Visiting Fellow, Centre for World Environmental History, University of Sussex, quotes *At a time when countries are beginning to explore ecologically sound avenues of water management, India is bent upon pursuing a project that will mark the final plunge into destroying country's rivers*.

Promises of the Project. The benefits which shall accrue from the proposed project are:

1. The project will prove as an answer to recurrent problems of floods and drought in India.

2. The project will transfer 173 billion cubic metres of water from river basin in the north uphill to water stressed regions in central and south India.
3. As much of peninsular India consists of huge higher plateau, the water shall be lifted by pumps to elevations several hundred metres above the level of Ganges basin.
4. A network of 11,000 km of canal shall be built.
5. The project will irrigate about 34 million hectares of dry parched land of the country.
6. The project will provide drinking water to nearly 101 districts and five metro cities.
7. The project would generate about 34,000 M W of cheap hydro-electric power.
8. Above all the project would boost GDP growth by 4 per cent.

The Pitfalls of the Project. Among various dimensions of river linking projects, most important is the cost involved in its execution which will be much more than the total annual expenditure in India's budget and so also will be the cost of its operation and maintenance. Some of the projected pitfalls of the project are:

1. The canal system and reservoirs built to collect water shall submerge a large forest area as well as usable productive land.
2. There shall be substantial loss of water through evaporation, leaks, seepage etc.
3. Change in flow regime and temperature shall destroy many of the sensitive ecosystems.
4. Water logging and salinisation shall adversely affect large productive land area.
5. Higher sedimentation rates could shorten the life span of so constructed costly canals and reservoirs to nearly half of their projected life span.
6. Millions of people will have to rehabilitated as the canal and reservoir network comes into being.
7. The project shall give birth to many inter-state water disputes and conflicts with neighboring countries such as Bangladesh, Nepal and Pakistan.
8. A substantial power, involving money, would be needed to pump water uphill.
9. The most serious challenge would be of raising funds for such a megaproject.

It appears that High Court of India has taken an unprecedented step in favour of right of the people for drinking water without considering the concept of eco-relation of nature. Many of us may not be there to see the completion of the project. This is quite likely that future generations may have to pay in their future.

USE AND OVERUTILISATION OF SURFACE AND GROUND WATER

The total global water use has steadily increased throughout recorded history. People throughout the world are least hesitant about use and misuse of water. A clear and detailed analysis of the total global water use in recent decades indicated that the rate increase accelerated markedly after 1940 in the twentieth century as compared to earlier periods (fig. 4.3). The trends observed in the twentieth century are alaming. Currently there are no visible indications that this rate will decrease in the near future. Since fresh water is a limited resource, such high rate can not be sustained for a long period in the future. There will be a stage, more likely within next 10 to 40 years, when the global fresh water use may likely level off as a result of physical, environmental, economic and political constraints, firstly in certain individual country and then globally. Nearly 5000 cubic kms annual total global fresh water use is extraordinarily high as compared to past consumption. There is no limit of total demand of usable water, which is increasing rapidly.

It also indicates that total global fresh water use has increased about 10 times from 1900 to 1999. The situation is even worse for major water consuming developing countries such as countries of South East Asia. According to World Water Vision, France, "In South and East Asia, irrigated area under the business as usual scenario grows only slightly between 1995 to 2025, while irrigation efficiency improves. The effect is in decrease in water used for irrigation from 1,359 to 1,266 cubic km a year. At the same time economic growth leads to more material possessions and greater water use by households, increasing water withdrawals for domestic use from 144 to 471 cubic km a year. This economic growth also requires larger quantities of water increasing from 153 to 263 cubic km for Asian industry. The sum of these trends is an overall increase in water withdrawls between 1995 to 2025. Thus the pressure on water resources will become even greater than was experienced in 1995, when about 6.5 million square kilometers of river basin were under high water stress. As the area increases to 7.9 million square kilometres in 2025, the number of people living in these areas also grow tremendously from 1.1 billion to 2.4 billion."

Despite the differences in estimated values of global fresh water consumption by various agencies, the net outcome of all these calculations, indicates that we are moving ahead towards an era of serious water crisis.

FLOODS

Floods have been an integral part of human experience ever since the beginning of agricultural revolution when people built the first permanent settlement on great river banks of Asia and Africa. Seasonal and casual floods deliver valuable top soil and nutrients to farm land and bring life to otherwise infertile regions of the world such as the Nile River Valley. However, flash floods and large floods, which are mainly because of mishandling of the environment, are responsible for mass-scale casualities and damage to property, much more than the tornadoes or hurricanes. Floods are the most common and widespread of all natural disasters. A flood is defined as *overflow of inland and tidal waters causing rapid and usual accumulation or run off of surface water from any source or a mudflow in a considerable area of land*. In the last hundred years alone, flood has claimed millions of lives on our planet.

The flood is a natural phenomenon associated with hydrological cycle and not an accumulation of large quantities of water flowing through nerve channels spilling over the banks. Since the land surface receives more water than it loses, this surplus amount of water has to flow back to oceans. Excessive rain water which falls down on the land surface, flows under the influence of gravity and comes to meet the streams and rivers along with silt and sediments which deposit in the low lying areas and nearby streams and rivers. Floods are caused under the influence of both-natural and anthropogenic factors which vary from place to place. The cause most responsible for floods is the removal of plant cover resulting soil erosion. Soil erosion causes the flood flow to carry more silt and sediments, thereby making streams and rivers shallow. Thus, the carrying capacity of streams and rivers is reduced and when large volume of water flows down in them, it spills over the banks, flooding the plains on either sides. The natural factors include high amount of precipitation under the influence of altered wind current, the monsoon wind system or unusual tidal activity extending reach of the ocean farther inland than normal. Among the anthropogenic causes, the chief is breaking of dams. Various other reasons responsible for floods are high intensity rains in the catchment area of river; shallow channels and extensive flood plains; sudden change in channel-gradient or blockage of river; curves, bends and meandering course of the river channel; extensive deforestation; impact of urbanisation and construction activities; overgrazing; mining; rapid industrialisation; global warming etc. Million of hectares of land area has been affected by floods in India since last fourty years (fig. 4.6).

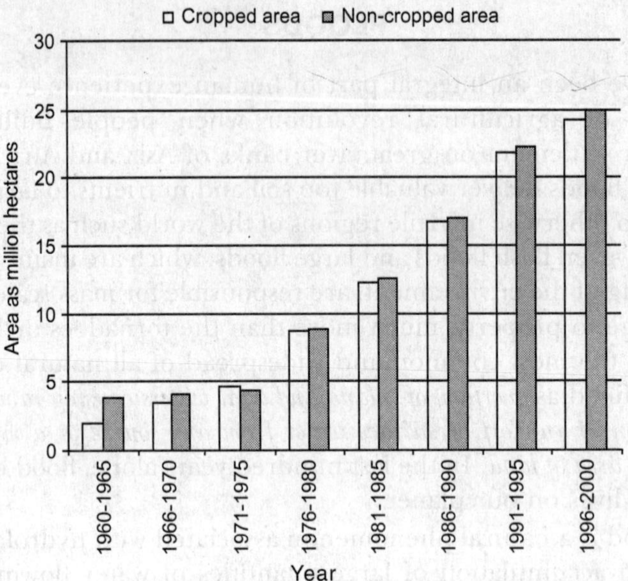

Fig. 4.6. Area affected by floods in India.

People of India are accustomed to moderate flooding during monsoon period. They utilise the flood water for growing paddy in their fields and are benefited by the increased fertility of the soil. But recurrent floods and severe floods such as during 1988 and 1991, resulting from excessive Himalayan run off and storms, have caused disastrous impacts including heavy loss of human life and damage to properties. The flood caused in 1970, brought about drowning of about one million people. In India, river floods, primarily caused due to peculiarities in rain fall, are the most frequent and often most devastating distaster. Nearly 75 per cent of the total rainfall is concentrated over a short monsoon season of hardly four months (June to September) resulting the rivers to witness a heavy discharge during this period leading to widespread floods. Moreover, the problem of flood is compounded by sediment deposition and drainage congestion and in the coastal plains synchronisation of river floods with sea tides. The rivers originating in the Himalayas carry with them a large amount of sediments causing erosion of banks in the upper reaches and overtopping in the lower segments. In India, the most flood-prone areas are the Brahmaputra-Ganga-Meghana basins which carry 60 per cent of the country's total river flow into the Indo-Gangetic-Brahmaputra plains in the north and the northeast part of the country. This basin is one of the largest in the world and is spread over 15 states. It covers a geographical area of 1.75 million square km of which 75.8 per cent lies in India and the remaining in Bangladesh and elsewhere. About 47 per cent of population of the country resides in this basin. The basin is spread over into four regions, *viz.*, Himalayan zone, the plains, the hilly tracts of northeast and southwest hilly tracts along fringe states in India. The other flood prone areas are the northwest regions of west flowing rivers, *viz.*, Narmada and Tapti, the Central India, and the Deccan plateau with major east flowing rivers, *viz.*, Krishna, Cauveri and Mahanadi.

In lack of proper flood policy and flood control schemes in the country, flood damage is increasing and larger populations are subjected to distress in flood-prone areas. Presently the locus has shifted away from the Gangetic belt with worst-hit being in Andhra Pradesh, Karnataka, Kerala and Tamil Nadu in the south; Maharashtra, Gujarat and Rajasthan in the west; Uttar Pradesh in the North; and Bihar and the West Bengal in the east. The problem has become recurrent in 10 out of 19 states exposed to the flood. Of these four Assam, Bihar, Punjab and West Bengal alone lie in the Gangetic flood plains and the other six lie in the peninsular India or in the Himalayan ranges. Since 1965, Kalahandi, Koraput, Bolangir, Malkangiri, Phulbani and Rayagada districts of Orissa have witnessed recurrent floods.

Box 4.2 Kalahandi

Kalahandi in Western Orissa, is a typical rare and classical case of environmental degradation leading to poverty and deprivation. The monsoon has always been liberal in this area. The area receives an average annual rainfall of 1250 mm. Water table, in some places, is very high. Yet Kalahandi is known to be for its extreme poverty and deprivation as it is often under drought and sometimes has floods. The inhabitants are compelled to migrate to other parts of India looking for work and survival. The place was not always centre of hunger and deprivation as it had mass of hills and was forest till most of the nineteenth century and just a few decades ago it was all green here. The forests used to provide livelihood for six months. And the next six months used to give way for agriculture for raising a large diversity of crops as it was one of richest area in the eastern India. In the past, there was a network of many traditional water harvesting structures including ponds, lakes, check-dams and even tanks with paddy fields and the whole system was designed to suit the topography of the land so that no part of rainwater went waste. Moreover, the system was under community control which ensured proper maintenance and water sharing through 'jal sabhas' (water councils).

The present state of Kalahandi is mainly because of failure of rainfall. After 1947, government took over many of the structures which remained ill maintained. Water bodies such as ponds were converted into crop-lands by the people and there was lack of proper irrigation water supply. Forests were being cut down for timber and river beds were being silted up leading floods to downstreams. This all led Kalahandi to sid into a vicious circle, from which it has never recovered.

DROUGHTS

Drought is one of the most serious environmental setbacks as it directly affects our water and food supply. Although drought is a normal recurrent feature of nature occurring almost everywhere but its extent varies from region to region. It is difficult to define drought because it depends on difference in regions, needs and disciplinary perspectives. The definition of drought may be different in Jaisalmer of Rajasthan where it hardly rains than in Cherrapunji of Meghalaya, once considered the wettest place on the earth. Commonly drought involves shortage of water originating from deficiency of precipitation over an extended period of time. The drought can not be viewed solely as a physical phenomenon but in terms of its impacts on society particularly needs of water for the people of the area affected (fig. 4.7).

Fig. 4.7. A drought scene: Trees, fields and wells become dry.

The drought may be meteorological, *i.e.,* associated with the degree of dryness and duration of dry period, hydrological, *i.e.,* associated with the degree or period of precipitation and short rainfall supplying water to surface and subsurface, or agricultural associated with various characteristics of meteorological or hydrological droughts to agricultural impacts focussing on decrease in the amount of precipitation, differences between actual and potential evapotranspiration, soil water deficits, reduced ground water and so on (fig. 4.8).

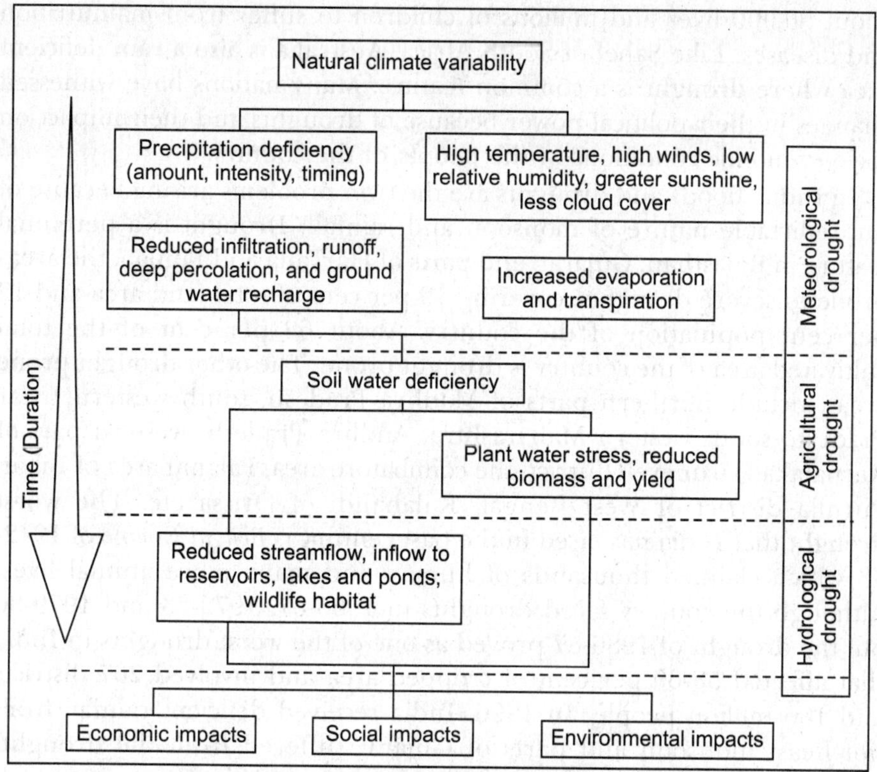

Fig. 4.8. Types of drought and their inter-relationship.

Drought may also be due to a limited number of people in a locality using unlimited water leading to its acute shortage.

Drought affects all life forms because all plants, animals, man and even microorganisms need water for their survival. Prolonged droughts cause impacts on ecological, demographic, economic as well as political conditions of the area affected. The biotic components are affected by elimination of many species of organisms in absence of conditions suited to their survival; migration of some species, *e.g.* water fowls to other places due to water scarcity resulting decline in the number of species or number of individuals of a species in the area; death of many organisms including man due to starvation; and tough competition for resources among individuals of the same species or between different species. There are about 80 countries in arid, semi-arid or sub-humid regions of the world that experience frequent spells of droughts a year long duration or more than that. The worst effect of drought results in economic losses due to reduced agricultural production resulting starved people to migrate temporarily to other locations. A large number of inhabitants of Sahel region of South Africa have left the area because of persistent drought. In Ethiopia, the drought caused in 1968 persisted till 2001 and claimed about 50,000 lives and millions of children to suffer from malnutrition and diseases. Like Sahel of South Africa, Australia is also a rain deficient area where drought is a common feature. Many nations have witnessed changes in their political power because of droughts and their impact on socio-economic conditions of the people of the country.

In India, floods and droughts are the twin problems arising because of unpredictable nature of monsoon and rainfall. Drought is a perennial feature in Rajasthan, Gujarat, and parts of Haryana and Punjab, the areas prone to severe droughts, covering 19 per cent of total land area and 12 per cent population of the country. About 68 per cent of the total cultivated area of the country is drought prone. The other drought prone areas include northern parts of Madhya Pradesh, south western Uttar Pradesh, south western Maharashtra, Andhra Pradesh, western part of Karnataka, Tiruneveli district, the coimbatore area, Palamu area of Bihar, Purulia district of West Bengal, Kalahandi of Orissa etc. The worst drought that India has faced in the past century is *Bengal Famine* of 1942-43 which claimed thousands of human and many more animal lives. Although the country faced droughts in 1966-67, 1972-73 and 1979-80 but the drought of 1986-87 proved as one of the worst droughts in India that affected 58-60 per cent of cropped area and involved 267 districts and 166 million people. In 1996, India received deficient rainfall from northeast monsoon and parts of Gujarat suffered from the drought. Droughts create unemployment and under-employment and as such government of India has launched Jawahar Rojgar Yojna, the largest programme in the country, aimed at generating gainfull employment for unemployed and under-employed men and women in the rural areas. The Employment Assurance Scheme launched in drought prone areas of the country also helps generate employment opportunities in the country.

CONFLICTS OVER WATER

Increased demand of water for agricultural, industrial and domestic use appears to be the main cause of conflict between the various beneficiaries in most parts of the world. The conflict may arise from drive to possess or control another country's water resource and thus making the water systems and resources as political goal. Conflicts may also stem when water systems are used as instruments of war, either as targets or tools. The national water scarcity may also escalate existing tensions between nations and lead to flare ups of long-term international water conflicts. If today, countries are poised to go to war over oil, the catalyst for future armed conflict would be water. For example King Hussein in Jordan cited a dispute over water as one issue that could provoke his country to start fighting with Israel again. Egypt and Ethiopia also have serious disputes over water. Strife over water is also erupting throughout Middle East from water sheds of Nile to Tigris and Euphrates rivers due to Turkey's power to curtail the water flow into Iraq through these rivers.

India also suffers from conflicts over water within some of its states and with some neighboring countries. A tussle is swimmering in South Asia's Ganges-Brahamputra basin where countries, *viz.*, India, Bangladesh and Nepal are interested in exploiting basin's huge hydro power generating potential, whereas Bangladesh wishes that water be managed in such a way as to minimise flooding during monsoon months and cope with shortages during dry months.

Some Major River Water Disputes in India

Some of the river water disputes of India are discussed below:

1. Cauvery Water Dispute: The river Cauvery originates in the hills of Western Ghats, runs through the rain shallow regions of Karnataka in South India and flows eastwards to the Thanjavur delta region, the rice bowl of Chennai. The rapids and falls of the river are used to generate hydro-electric power in Karnataka. The down stream of the river is enjoyed by Tamil Nadu. Tamil Nadu wishes water use regulated in the upstream in Karnataka which refuses to do so and claims its primacy over the river as upstream user. The river water is almost fully utilised and both the states pose increasing demands for agriculture and industry. However, consumption is more in Tamil Nadu than Karnataka where catchment area is more rocky. Stanley reservoir and Tamil Nadu's Mettur-dam control the flood flow of the river. An attempt was made to solve the dispute by Government of India and on June 2, 1990, the Cauvery Water Dispute Tribunal was set up which, through an interim award, directed Karnataka government to ensure that specified amount of water was made available in Tamil Nadu's Mettur dam every year, till a settlement was reached. Recently the Tribunal (05-02-2007) decided that Tamil Nadu will get 419 TMC waterly. In 1991-92, there was a good monsoon and there was no dispute, but in 1995, the situation turned into a crisis due to delayed monsoon rains and an expert committee was set up to look into the matter which found that there was a complex cropping pattern in Cauvery basin, *i.e.*, Sambra paddy in the winter, Kurvai paddy in the summer and some cash crops round the year which demanded intensive supply of water, thus aggravating the water crisis. Some measures, to solve the problem suggested, were selection of proper crop varieties, optimum use of water, rational sharing patterns and pricing of water.

2. Indus Water Treaty. The Indus, one of the mightiest and world's largest rivers, is dying a shallow death due to construction of dams and barrages over it. The river has resulted in severe shrinkage of its delta due to Sukkur barrage (1932), Ghulam Mohammad Barrage at Kotri (1958) and Tarbela and Chasma dams on Jhelum, a tributary of Indus. In 1960, Indus Water Treaty was set up to allocate Indus, Jhelum and Chenab to Pakistan and Satluj, Ravi and Beas to India. Being the riparian state, India has the primitive right to construct barrages a cross all these rivers in its territory. The treaty, however, requires that three rivers allocated to Pakistan may be used for non-consumptive purposes, *i.e.,* without changing its flow and quality by India. With gradual improvement of political relations between the two countries it is hoped that desirable techno-economic details would be worked out by both the countries and they would go for an integrated development of the river-basin in a sustainable manner.

3. Sutluj-Yamuna Link (SYL) Canal Dispute. The dispute of sharing the Ravi-Beas water and Sutluj-Yamuna Link issue lies between Punjab and Haryana. The issue has been discussed from time to time but the case is again and again in the Supreme Court. The Eradi Tribunal (1985), based on previous records, recommended a specified amount of water to Punjab which argued that there has been consistent decline. On January 15, 2002 the Supreme Court directed Punjab to complete and commission SYL within a year, failing which Government of India was directed to complete the link. Till now, neither SYL has been completed nor the conflict over sharing of Ravi and Beas water is resolved.

4. Godavari Water Dispute. Godavari, the largest among the Indian peninsular rivers, is held in the reverence as “Vridha Ganga”. It rises in the Nasik district of Maharastra, joins the Bay of Bengal after traversing a length of about 1965 km. Of its total catchment area, 49 per cent lies in Maharashtra, 21 per cent in Madhya Pradesh, 1 per cent in Karnataka, 5 per cent in Orissa and 24 per cent in Andhra Pradesh. The dispute between beneficiary states is very old and has been tried to be solved by Godavari Tribunal (1969), which in 1980 formulated trans-basin-diversion, in which each of the state concerned will be at liberty to divert any part of the share of the Godavari waters allocated to it from the Godavari basin to any other basin.

Over these disputes and others, violence, riots, property destruction and even arrests have been made several times. Police were reported to have gun down crowd at Falla village near Jamnagar against the diversion of water from Kanakvati dam to Jamnagar town. During summers, city pipes are often empty or are discharging waters for a very limited period, for the people living in these area.

DAMS: BENEFITS AND PROBLEMS

The dams are the means to store the water in the river valleys and to release it for hydro-electric projects, irrigation or other purposes as and when required. River channels usually run in valleys with the hills on the either sides. If the river channel of deep and narrow valleys is blocked by raising artificial barrier, the river water is easy to store as huge reservoir. The water storing capacity of a dam depends on the height of the hills, the distance between two lateral hills and the slope of the river bed. Besides river water, reservoirs can also store rain water as surplus. Dams are considered as the most efficient method to harness and utilize surface water. According to World Commission on Dam Report (2000), there are 45,000 large dams in 140 countries. Of these India alone has 4391 dams which is 9 per cent of the world's total. River valley projects with dams, *viz.*, Bhakra-Nangal, Heerakund, Nagarjuna Sagar, Damoder etc. have played a key role in shaping socio-economic conditions of India during last fifty years. Geographic conditions of India have made it distinct in having largest number of river valley projects. The first Indian Prime Minister Jawahar Lal Nehru, called these dams as the temples of modern India which store precious rain water to irrigate lands, generate electricity, supply drinking water and save human beings from clutches of floods and to certain extent from droughts. Various benefits of the dams are:

1. The potential energy of the stored water can be used to generate hydro-electric power-the cheap and cleanest source of energy.
2. The water stored in dams can be slowly discharged and made available for irrigation during dry periods for needy areas.
3. Rain water deposited in reservoirs can be discharged in a regulated way and downstream floods can be checked.
4. The water can be transferred to dry areas by using canals to partially manage the droughts.
5. Year round water supply can be ensured.
6. Reservoirs can be used to breed fishes.
7. Multi-river-valley projects can be used for inland water navigation.

Despite various benefits, dam also create a number of environmental problems, such as-

1. The enormous weight of water behind the dam could trigger seismic activity that might crack the dam and cause floods.
2. The dam expansions themselves submerge large fertile land area, forests and human settlements.
3. Submergence of forests may result habitat destruction of tribals and wildlife.
4. Rehabilitation and resettlement problem for local inhabitants may be created for displaced persons.
5. A number of water related diseases, such as malaria, may occur, particularly where water impoundments provide breeding sites for the vectors such as mosquitoes.

6. Catchment areas may suffer from water logging and siltation.
7. Loss of free flowing river currents, decrease in their breadths and volume of water.
8. Incidence and spread of onchocerciasis disease in populations living near the dams.
9. Growth of snail population in shallow permanent canals that distribute water to fields and human settlements.
10. Salts left behind evaporation of water of reservoirs, increase the salinity of rivers and make their water unusable when it reaches downstream cities.

Environmentalists are of the opinion that several small dams, instead of one large dam will be more useful. These cause less damage to environment. But, it is not convincing as to how these small dams would fulfill the needs of water for irrigation, domestic uses and generation of electricity etc.

Chapter Summary

Water is a wonderful gift of nature. Literally, water is elexir of life because survival of all organisms depends on water. It is the only inorganic liquid occurring naturally in all three physical states, *i.e.,* solid, liquid and gas. A once abundant resource, water is being wasted, misused and polluted in such a way that world is moving towards water wars. There is a great water crisis most visible in many countries and many parts of our country. Of the total water present on this earth, only less than 1 per cent is available as fresh water for which all people, animals and plants compete. Oceans contain 97 per cent of water but their water is saline and is of no use to mankind. The usable water comes from precipitation and is stored in the rivers, lakes, streams etc. or percolates underground. For our water supply, we depend on rivers, lakes, streams etc. or ground water which has its limited stock. Unfortunately, our rivers are shrinking or are being polluted and our reserve water source, the groundwater is being depleted. Water is largely needed for agriculture, industry and for fulfilling domestic needs. Large proportion of water is consumed in agriculture, particularly after green revolution. Indian water budget is at an alarming stage. Despite the fact that India is endowed with a rich river system, most of the water of these rivers goes to the oceans unused. An attempt is being made to interlink Indian rivers, but the project is too expensive, time consuming and is being doubted that will it be that beneficial as is thought about it. The global water consumption rate is much higher than the recharge rate of ground water. There will be a stage, more likely within 10 to 40 years, when the accessible water will be labeled off. Flood, a natural phenomenon associated with hydrological cycle, is caused under the influence of natural and anthropogenic factors. Floods cause serious losses to mankind both economic and physical and sometimes result heavy casualities. Floods are of common occurrence in India during monsoon period. Drought is one of the most serious environmental setbacks because it directly affects our food and water supply without which man can not live. Drought affects all life forms in various ways. People are forced to leave the area and require resettlement and rehabilitation. Conflicts over water arise due to increased demand of water between beneficiaries. Conflicts may arise between different states of a country or between different nations. Cauvery Water Dispute, Indus Water Treaty, Satluj-Yamuna-Link Canal Dispute are some common disputes before India which need to be tackled carefully. Among various development projects, dams play an important role in livelihood of people and economy of a country. Many river valley dam projects in our country have provided adequate water for irrigation and hydropower for the benefit of the people. They have been able to provide year round water to many areas. Despite various benefits, dams also create a number of environmental problems and problems of resettlement and rehabilitation of oustees.

Study Questions

1. What are the various types of water sources? What are their uses? Explain.
2. Write an account of ground water resources and its future.
3. Discuss the available water to man and man's requirements for the water.
4. What are various methods to conserve our water resource?
5. Write an account of use and overutilisation of surface and ground water
6. Write short notes on-
   (i) Dams: Benefits and problems
   (ii) Droughts
   (iii) Floods
   (iv) Conflict over water
   (v) World Water Vision
   (vi) Interlinking of Indian Rivers
   (vii) Cauvery Water Dispute

Objective Questions. *Select the correct answers*

1. An example of lotic water body is:
   (1) Pond (2) River
   (3) Lake (4) Reservoir

2. Which of them is most responsible for world water crisis?
   (1) Dams (2) Floods
   (3) Drought (3) Population Growth

3. Indus Water Treaty is in between:
   (1) India and Bangladesh (2) India and China
   (3) India and Pakistan (4) India and Afghanistan

4. In India, fresh water requirement is higher on:
   (1) Industries (2) Irrigation
   (3) Domestic use (4) Thermal power generation

5. Percentage of the volume of fresh water available for Ruman use in the hydrosphere is:
   (1) 0.3 (2) 1.3
   (3) 1.7 (4) 2.1

Answers

1. (2) *2.* (4) *3.* (3) *4.* (2) *5.* (1)

5

CHAPTER

Mineral Resources

LEARNING OBJECTIVES

Introduction • Formation of Mineral Deposits • Types of Mineral Resources • Mineral Wealth of India • Mineral Exploration • Methods of Mineral Extraction • Use and Impacts of Over Exploitation of Minerals • Environmental Effects of Extraction and Use of Mineral Resources • Conservation of Mineral Resources • Indian Legislation on Mineral Rights.

Introduction

Mineral resources are the non-living naturally occurring chemical compounds of both organic and inorganic origin, that are concentrated in the earth's crust. Geographically localised mineral deposits of economic importance are called ores. The earth's crust is composed of about 88 elements. Ninety nine per cent of the earth's crust is composed of Oxygen, Nickel, Sodium, Potassium and Uranium. All other elements together, form one per cent. Besides the earth's crust, oceans and atmosphere are also potential sources of some elements. There occurs a great variation in the distribution patterns of minerals in the earth's crust and in the degree in which these elements are concentrated in the ores. Wide range of use of minerals has caused enormous depredations in the stock of minerals in the accessible part of the earth. The onset of the industrial revolution opened a new area in the utilisation of minerals. India has rich reserves of iron, titanium and thorium ores and mica and a surplus of bauxite and manganese ores and silica. Mineral extraction and processing cause a wide range of environmental impacts on land, water, atmosphere and socio-economic environment of people and need remedial measures. Mineral resources are valuable natural resources of exhaustible and non-renewable nature. They constitute the vital raw materials for many industries and form a major source of development.

Formation of Mineral Deposits

Formation of mineral deposits is a very slow process, taking millions of years. They are formed by any of the processes such as concentration of minerals during cooling of molten rock material; by evaporation of lake or sea water; by action of intense heat and pressure; by concentration of minerals during weathering, transport or sedimentation; or by microbial activity.

Types of Mineral Resources

Minerals can be metallic, *e.g.* iron, copper, gold etc. or non-metallic, *e.g.* sand, stone, salt, phosphates etc. Metallic minerals are found basically in the form of ores whereas non-metallic ones occur as minerals. Coal has been quoted as mineral because it, along with crude oil (petroleum) and natural gas, comes under the category of fuel minerals. Fuel minerals constitute nearly 88 per cent (coal 42 per cent, petroleum 33 per cent, natural gas 10.5 per cent and lignite 2.5 per cent) of the total value of mineral production whereas metallic and non-metallic minerals constitute 6-7 per cent only.

Mineral Wealth of India

India is richly endowed with minerals such as iron and coal whereas it is poor in many other minerals. Except for iron ore and bauxite, our share of world reserves of every other mineral is 1 per cent or less but there has been a phenomenal growth in extraction since independence. And if the present rate of extraction continues, we will exhaust our resources of all important minerals and fuels except iron ore, coal, limestone and bauxite in coming 25-30 years. In India, under the constitution, mineral rights and administration of mining laws are vested in the respective state governments. The Centre, however, regulates the Department of Minerals under the Mines and Minerals (Development and Regulation) Act, 1957 and the rules and regulations framed there-under. The number of minerals mined in India is more than 80 belonging to metallic, non-metallic and fuel minerals (table 5.1)

India has an unevenly distributed huge resource of iron ore, extensive deposits of coal and mineral oil, rich deposits of bauxite and has a virtually monopoly in mica but is deficient in petroleum, tin, lead, zinc and nickel. Great plains of northern parts of the country are almost devoid of minerals but south Bihar and Orissa, on the north eastern part of the peninsular India, hold large concentration of mineral deposits accounting for nearly 75 per cent of country's coal deposits as well as rich deposits of iron ore, manganese, mica, copper and bauxite that are also scattered elsewhere in the country.

Iron Ore. India has a huge *in situ* reserves of iron ore distributed unevenly. Of the total iron ore reserves, about 12317 million tonnes is haematite and 539.5 million tonnes is magnetite. Resources of very high grade ore are limited and are restricted mainly in Bailadila sector of Chattisgarh and to lesser extent in Bellary-Hospet area of Karnataka and Barajamada sector of Jharkhand and Orissa. Haematite resources are located in Orissa, Jharkhand, Chattisgarh, Karnataka, Goa, Maharashtra, Andhra Pradesh and Rajasthan whereas magnetite resources are located in Karnataka, Andhra Pradesh, Goa, Kerala, Jharkhand, Rajasthan and Tamil Nadu.

Bauxite. The total *in situ* reserves of bauxite in the country are placed 3075 million tonnes. The conditional resources of bauxite are about 0.59 million tonnes. Orissa, Andhra Pradesh, Chattisgarh, Gujarat, Maharashtra and Jharkhand are the principal states where bauxite deposits are located. Major reserves are concentrated in east coast of Orissa and Andhra Pradesh.

Coal and Lignite. Resource wise, coal occupies the pride place in the inventory of mineral resources in India. In India, coal occurs in rock sequences mainly of two geological ages, namely Gondwana, little over 200 million years in age and in Tertiary deposits which were found at a much later geological epoch (about 55 million years ago). The major resources of Gondwana coal are located in coalfields occupying the Indian heartland in the states of West Bengal, Jharkhand, Orissa, Chattisgarh, Madhya Pradesh, Maharashtra, Uttar Pradesh and Andhra Pradesh. Tertiary coals occur in Assam, Arunanchal Pradesh, Meghalaya and Nagaland. Besides, brown coal and lignite occur in coastal areas of Tamil Nadu, Gujarat and inland basin of Rajasthan. The total known geological reserves of all types of coal in Gondwana and Tertiary coal fields stand estimated at 247.64 billion tonnes as on 1 January 2004. Resource wise coalfields of Orissa contain the largest coal resources, though mostly of inferior quality. Metallurgical coal is restricted to parts of Damodar Valley Coalfields in West Bengal and Jharkhand and in the Central Indian Coalfields, namely Sohanpur, Sonhat, Jhilmill and Satpura. The reserves of lignite have been estimated as litter over 37.15 billion tonnes as on 1 April 2005, out of which the major contributor is the lignite basin of Tamil Nadu. Geographic distribution of some important minerals in India is shown in table 5.1 and fig. 5.1.

TABLE 5.1. Important Minerals, their Distribution and their Total *in situ* Reserves in India on April 1, 2000

Mineral resource	*Distribution*	In situ *Reserve*
1. Iron	Madhya Pradesh, Goa, Chhattisgarh, Jharkhand, Karnataka	12,317 MT
2. Copper	Singhbhum district of Jharkhand, Balaghat district of Madhya Pradesh, Jhunjhunu and Alwar districts of Rajasthan	721.5 MT (9.44 MT Metal)
3. Chromite	Orissa, Jharkhand, Madhya Pradesh, Karnataka, Maharashtra, Manipur and Chhattisgarh	114.37 MT
4. Gold Ore	Kolar Gold Fields in Kolar district, Hutti Gold Field in Raichur district, both in Karnataka and Ramgiri Gold Field in Anantpur district of Andhra Pradesh	222.4 MT (116 T Metal)
5. Lead -Zinc Ore	Rajasthan, West Bengal, Andhra Pradesh, Gujarat, Uttar Pradesh, Madhya Pradesh, Orissa, Maharashtra, Meghalaya, Tamil Nadu and Sikkim.	231.22 MT (5.1 MT Lead and 17.01 MT Zinc metals)
6. Manganese Ore	Karnataka, Orissa, Madhya Pradesh Maharashtra and Goa	406.193 MT
7. Tungsten Ore	Maharashtra, Haryana, West Bengal and Andhra Pradesh	43.15 MT
8. Nickel	Orissa, Jharkhand and Nagaland	188.71 MT
9. Diamond	Panna belt in Madhya Pradesh, Munnimadugu-Benganapalle in Kurnool district, Wajrakarur in Anantpur district and Krishana river basin in Andhra Pradesh	2.6 M Carate
10. Dolomite	Madhya Pradesh, Chhattisgarh, Orissa, Gujarat, Karnataka, West Bengal, Uttar Pradesh, Maharashtra.	7.348 MT
11. Fireclay	Jharia and Raniganj Coalfields in Jharkhand and West Bengal, Corba Coalfields in Chhattisgarh, Neyveli Lignite Field in Tamil Nadu	706 MT
12. Limestone	Madhya Pradesh, Chhattisgarh, Andhra Pradesh, Gujarat Rajasthan, Karnataka, Tamil Nadu, Maharashtra, Himachal Pradesh, Orissa, Bihar, Uttar Pradesh, Uttarakhand.	169,941 MT
13. Phosphate mineral	Madhya Pradesh, Rajasthan, Uttarakhand, Uttar Pradesh and Gujarat	139.23 MT

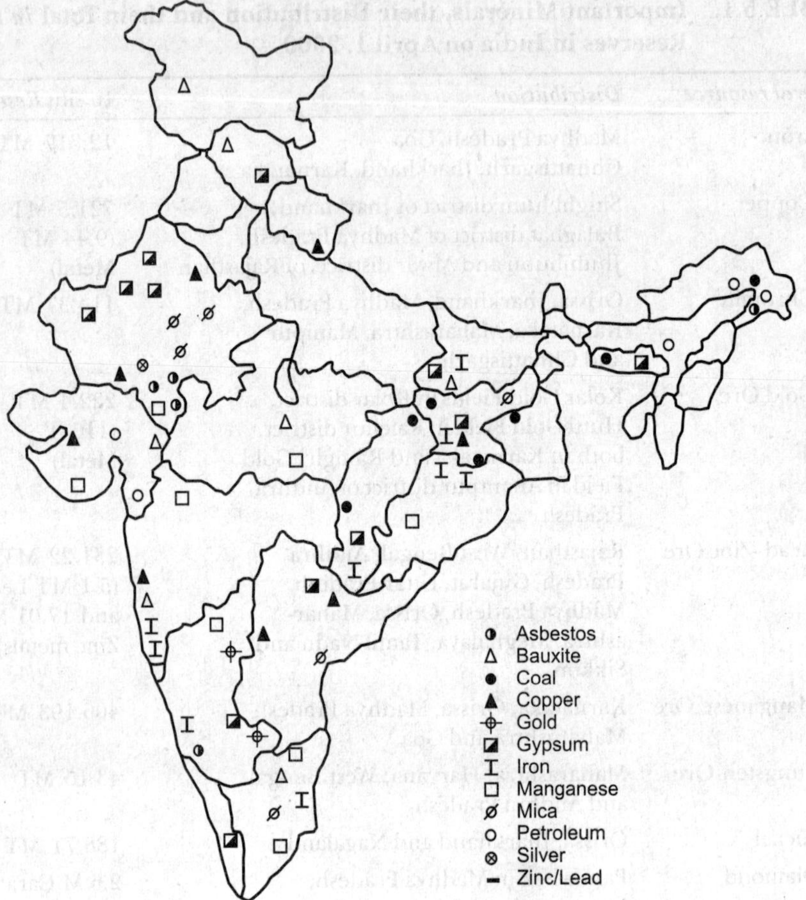

Fig. 5.1. Geographical distribution of some important minerals in India.

The Geological Survey of India (www.gsi.gov.in), set up in 1851, is the premier national scientific survey and research organisation that functions for the development of minerals in India. Indian Bureau of Mines, established on March 1, 1948, as a multidisciplinary scientific and technical organisation under Ministry of Mines, looks after conservation and systematic exploration of mineral resources other than coal, oil and natural gas, atomic minerals and minor minerals. Among public sector undertakings are Mineral Exploration Corporation Limited, National Aluminium Company Limited, Hindustan Copper Limited, Bharat Gold Mines Limited etc.

Mineral Exploration

In the past, miners used to discover minerals simply looking at the surface of the rocks. However, the modern exploration is much more advanced and a complex process. Field survey and remote sensing are the two techniques most commonly used today. Field survey involves simple observation, analyses of the rock and geophysical study whereas remote sensing is the study of an object using instruments placed at a distance from the object, more commonly using satellites. Many indicator plants also help in exploration of minerals.

Methods of Mineral Extraction

Mineral extraction is done through dredging, surface mining or mining by other methods. In surface mining, sand, gravel and other surface deposits, which are covered with water, are collected by special buckets. Strip or open cast mining involves removal of overlying layers of the soil and digging out the minerals situated close to the surface. Although the process is economically cheapest but environmentally it is very destructive. Quarrying involves digging out the minerals or blasting with the explosives to recover the minerals. This process is commonly used for obtaining stones for construction of buildings or roads. Adit mining involves tunneling a horizontal shaft of the mineral present in a horizontal seam. Shaft mining involves shucking a vertical shaft deep into the earth from which a series of gallaries are made at different depths. Once removed from the ground, the mineral must be processed to remove uneconomic waste minerals. The production processes range from simple washing and screening to electrolysis.

Use and Impact of Over-Exploitation of Minerals

Use of minerals and metals by man started as early as 4000 BC the Age of Metals. Copper was the first to come into a widespread use and was followed by gold and silver, the rare elements found in their metallic state. The extraction of metals from ores, however, could only be possible around 1100 BC when furnaces, capable of attaining high temperatures to reduce ores into metals, were developed. This gave way to extraction of iron from its various ores. To produce harder steel, a little amount of charcoal (carbon), obtained from burning trees, had to be mixed to the iron. Depending on their use, mineral resources can be divided into several broad categories such as elements for metal production, technology, building materials or minerals for chemical industry and for agriculture. In general, the term 'mineral resource' is applied to metal but with the exception of iron, the predominant mineral resources are not metallic. Of all the metallic minerals, iron makes up 95 per cent of all the metals consumed. In terms of annual world consumption, iron is mainly used for industry. Iron along with sodium, man's basic salt requirement, is used at the rate of 0.1 to 1.0 billion metric tonnes per year. Nitrogen,

TABLE 5.2. World's Important Minerals and their Uses

No.	*Mineral*	*Uses*
	METALLIC	
1.	Aluminium	Building materials, electrical wiring, utensils, aircraft, rockets, packaging, food items
2.	Beryllium	Refractories, copper alloys
3.	Chromium	Refractory, metallurgy, chemicals, textile, tanning.
4.	Cobalt	Alloys, radiography, catalysts, therapeutics
5.	Columbium	Stainless steel, nuclear reactors
6.	Copper	Alloys, electrical products, electronic goods.
7.	Gold	Monetary purposes, jewellery, dentistry
8.	Iron	Steel, building materials, numerous industrial uses, transport vehicles
9.	Lead	Batteries, paints, alloys, public health fittings, gasoline, ammunition
10.	Magnesium	Structural refractoriness, alloy steels
11.	Manganese	Alloy steels, disinfectants
12.	Molybdenum	Alloy steels
13.	Nickel	Used in over 3,000 alloys, chemical industry
14.	Thorium	Nuclear bombs, electricity generation
15.	Tin	Soldering, chemicals, tin plates
16.	Tungsten	Alloys, chemicals
17.	Titanium	Alloys, pigments, aircraft
18.	Uranium	Nuclear bombs, electricity generation, tinting glass
19.	Vanadium	Alloys
20.	Zinc	Galvanizing, chemicals, soldering, die-casting
21.	Diamond	Jewellery
22.	Platinum	Catalytic converter, electronics, medicinal use, automobile, Jewellery
23.	Silver	Photography, jewellery, electronics
	NON-METALLIC	
24.	Asbestos	Roofing, insulation, ceramics, textiles, gasoline, solid propellants.
25.	Corundum	Abrasives
26.	Felspar	Ceramic flux, artificial teeth
27.	Fluorspar	Flux, refrigerants, propellants, acid
28.	Nitrates	Fertilizers, chemicals
29.	Phosphates	Fertilizers, chemicals
30.	Potassium	Fertilizers, chemicals
30.	Salt	Chemicals, glass, metallurgy
31.	Sulphur	Fertilizers, acid, iron and steel industry
32.	Limestone	Concrete building stone, neutralizing acidic soils, cement.

sulphur, potassium and calcium are used as fertilizers at a rate of 10 to 100 metric tonnes per year. Gold and silver, the rare metals commonly used for jewelery and as a sign of richness, are used at rate of 10 thousands metric tonnes per year. The non-metallic minerals are consumed at much greater rates than the elements which are used for their metallic properties. Some important minerals and their common uses are shown in the table 5.2.

Ever increasing demand of minerals in the country for the industry, agriculture, transport, defence preparations and various development projects are cause of concern. With such a huge demand in the last 20 years, it is quite likely that the stock of the mineral deposits may undergo depletion. Since most of the mineral resources are non-renewable, taking millions of years to reform, a situation may arise when they are exhausted. Country's economy largely depends on the extraction, processing and suitable use of large amount of minerals to make the products. Thus it becomes necessary to adopt a low waste sustainable use approach to deal with them and make use of recycle and reuse benefits so that our environment is not affected.

Consequences of over-exploitation of our mineral wealth are likely to cause a serious damage to the entire biosphere. Rapid rate of consumption of high grade minerals shall deplete our good quality resources, increase their costs and affect the economy of the country. Mining being a dirty industry, is causing enormous pollution, loss of usable land and resulting some of the worlds greatest environmental disasters.

Environmental Effects of Extraction and Use of Mineral Resources

The quality of ore, mining procedures, local hydrological conditions, climate, rock type, dimension of operation, topography and so on, cause direct or indirect effects on the environment. Exploration and extraction of minerals cause significant impact on the land, water, air and the biosphere. Various operations involved also cause social impacts such as demand for resettlement and rehabilitation for the oustees and settlement for workers involved. Some major environmental effects of mining, processing operations and use of minerals are:

1. Pollution of surface and ground water sources due to release of harmful trace elements such as cadmium, cobalt, copper, lead etc.
2. Pollution of air due to emission of mine dust, harmful gases and transport vehicles emitting various pollutants.

3. Physical change in the land causing it to be degraded. The land that has been destroyed is known as *derelict* or *mine spell*. This land becomes unusable for agricultural purposes.
4. Adverse effect on biological environment including checking of growth of vegetation and loss of the forests.
5. Loss of biodiversity and chances of extinction of wildlife species due to deforestation.
6. Harmful effects on the historical monuments and places of religious importance in the surrounding areas.
7. Rehabilitation problems for those who have lost their habitats, especially for tribals.
8. Land use change from open range, forest and agriculture to urban pattern.
9. Stress on local services including water supplies, sewage, solid waste management and living conditions etc.
10. Operations in ecological fragile areas causing increased stress due to additional people and their activities.
11. Accidental hazards during the operations, due to shafts that are not filled and old quarries and open cast pits.
12. Causing a wide variety of diseases due to dust or dirty water.
13. Giving ugly appearance to the locality.

Many of the these causes resulting from ruthless exploitation of mineral resources have compelled the closing of many of the mines and

Box 5.1 **Mining Ban in Aravalli**

In order to preserve the ecology of sensitive Aravalli Hills, the Supreme Court rejected pleas for lifting the ban on mining activities in the range. The Supreme Court had for the first time banned the mining activities in Aravalli Hills on May 6, 2002 on noticing that several illegal mining activities in the range were causing a great damage to the environment. The court has now constituted a high level Monitoring Committee to suggest ways and means for the overall restoration of this range. That ecology of Aravalli Hills has to be preserved at any cost, the court directed the Committee to go into all issues related to mining in the area and suggest remedial measures for eco-restoration and also look into previous matters. It would also inspect and assess the mines that were operational till the ban orders came into force in May 2002 and recommend whether these could resume operation on the basis of sustainable development principle. The court also made it clear that each mine owner, before applying for renewal of mining lease, will have to obtain Environmental Impact Assessment clearance from the authority concerned. At last the committee suggested that already present legally permitted mining in the area was adversely affecting the ecology of Aravalli range which is spread over about 692 km in the North West India covering Rajasthan, Haryana and Delhi, harbouring rich biodiversity as well as treasure of mineral resources.

mining operations or abandoning them Many of the mining operators are not interested in rehabilitation of local inhabitants, especially the tribals because of heavy expenditure involved. The result is that many of the mines now look deserted and are left to fall in ruins.

Conservation of Mineral Resources

As stock of mineral resource is limited and they are being depleted very fast, their conservation needs an urgent attention. Various steps to be taken in the economy of their use include their recycling, reuse, substitution, use of the waste etc. New technologies to recycle them need be developed. Used and discarded items be collected, remelted and reprocessed into new products. Use of scarce metals such as gold, silver, platinum, mercury etc. should be substituted by more abundant minerals and man made products such as plastics, ceramics, high strength glass fibers and alloys. In order to reduce demand for pure metals, alloys should be developed, *e.g.* alloys of magnesium are replacing steel and reducing the demand of copper, lead and tin. Alternate non-conventional energy sources such as wind, nuclear energy, hydro-power etc. should be developed to replace the coal in mineral extraction process. Mining areas need be reclaimed. Use of the waste products of one manufacturing process should be made the raw material for another industry. It will prove as an effective measure in conservation of mineral resources. Exploitation of untapped deposits such as deep-sea mining, would also solve the problem to some extent. To make extended supply of minerals for longer period, their consumption must be decreased. A data bank, on the availability and use of mineral resources to regulate their consumption, should be maintained. Now-a-days, there is heavy demand of minerals due to armament race. Reduction in this race as well as reduction in defense production will greatly reduce the rate of depletion of minerals.

Indian Legislation on Mineral Rights

Under the constitution, mineral rights and administration of mining laws are vested in the respective state governments. The Centre, however, regulates the development of minerals under the Mines and Minerals (Development and Regulation) Act, 1957, and the rules and regulations framed there-under. This statute empowers the Central Government to formulate rules for: 1. grant renewal etc. of reconnaissance permits, prospecting licenses and mining leases for major minerals (*viz.*, Mineral Concession Rules, 1960), 2. the conservation and development of minerals (*viz.*, Mineral Conservation and Development Rules, 1968 for major minerals except atomic fuel and minor minerals) and 3. the modification of old leases. The Acts came into force on 1 June 1972 and amendments in 1986, 1994 and 1999.

Chapter Summary

Mineral resources are the reserves of organic or inorganic nature that are concentrated in the earth's crust which is composed of about 88 elements. Geographically localised mineral deposits of economic value are called ores. There occurs a great variation in distribution pattern, degree of their concentration in earth's crust and use of mineral resources. Wide use of minerals has caused an enormous depredations in their stock. which is limited and non-renewable. Mineral extraction and processing cause a wide range of environmental impacts on land, water, atmosphere and socio- economic environment of the people. Formation of mineral deposits takes millions of years. Minerals may be metallic (copper, iron etc.), non-metallic (sand, salt etc.) or fuel (coal). India is richly endowed with huge deposits of several minerals including iron, coal, bauxite and mica. Mineral exploration is done by field survey, remote sensing or other such methods. Mineral extraction is made through dredging, surface mining or mining. Minerals find many uses in industries, construction of infrastructures, making of steel and many household articles. Environmental impacts of mineral extraction include degradation of land, water quality, atmospheric pollution, loss of the forests and habitats of man and wildlife, diseases, accidental hazards, ressettlement and rehabilitation of oustees etc. Since minerals are non-renewable resources with their limited stock, they require their sustainable use and need conservation. To make their extended supply for longer period, consumption of minerals require their wise use.

Study Questions

1. Write a brief account on types of minerals found in India and their uses.
2. Discuss the impacts of over-exploitation of minerals and methods of their conservation.
3. What are various environmental impacts of mineral extraction? Explain.
4. Discuss briefly:
   (i) Formation of Mineral Deposits
   (ii) Mineral Wealth of India
   (iii) Indian Legislation on Mineral Rights
   (iv) Methods of Mineral Extraction

Objective Questions: *Select the correct answers:*

1. The mineral that virtually has monopoly in India is:
   (1) Uranium (2) Thorium
   (3) Mica (4) Tin

2. A metal largely used in the electricity generation is:
   (1) Gold (2) Thorium
   (3) Tin (4) Mica

Answers

1 (3) 2 (2)

6

CHAPTER

Food Resources

LEARNING OBJECTIVES
Introduction • Sources of Food • World Food Problem • Undernourishment • Malnutrition • Changes Caused by Overgrazing • Changes Caused by Agriculture • Pesticide Problem • Water Logging • Soil Salinity

Introduction

Among three basic necessities – food, shelter and clothing, food has been the utmost priority of human beings. It has received attention since the time man became growers from being mere gatherer of food. This transition took place nearly 11,000 years ago, when man adopted agriculture by cultivating plants of his choice but he faced problems relating to soil, crop health and climatic conditions. Everyday now world has to find food for more than six billion people and this number is steadily growing. Ever since Thomas Malthus, 1778 in his, *Essay on the Principle of Population* proposed that *human fertility outstrips the ability to produce enough food*. Plants and the animals are the main source of human food. About 76 per cent of the food comes from the crops mostly from cereals and the remaining from grazing livestock and fishes. Each of us need sufficient food to derive energy to carry out our functions and the materials needed to built a healthy body. Food and its associates are indispensable for health and satisfactory growth at all stages of life, *i.e.*, infancy, childhood, adolescence, adulthood and old age. Human diet is not restricted to any special category but includes a variety of foods of plant and animal origin as no single food provides all the nutrients and sufficient calorie – the unit of food value.

There is enough food in the world but the problem is that many people are too poor to buy readily available food. At least 700 million people do not have food to eat. Every year hunger kills 12 million children worldwide. Acute food scarcity witnessed in the 1950s in China, Bangladesh, Ethiopia and the Sahelian countries in Africa. India faced the great *Bengal Famine* in 1942-43 when 1.5 million people died. There are two major steps to eliminate hunger problem. The first relates to producing adequate quantities of food and other agricultural commodities within the country and the second is launching a concerted attack on endemic hunger. During last 50 years, India, by and large, has controlled famines through a multi-pronged strategy comprising of increased food production, building grain reserves, maintaining a public distribution system, promoting employment generation and extending help to unemployed and senior citizens. But endemic hunger, resulting from under nutrition and malnutrition, persists on a large scale. Future food security can be achieved only when the rapidly growing population is brought under control very soon otherwise all other measures to increase the food production would come naught. Equally important would be emphasis on sustainable utilisation of our food resources.

SOURCES OF FOOD

Primitive societies used to obtain food by gathering and hunting. Domestication of plants and agriculture started at later stage. But now great majority of people obtain the food from cultivated plants, domestic animals and from fresh water or marine animals including fishes. The croplands provide about 76 per cent of the total food, mostly food grains. Rangelands produce meat mostly from grazing livestock, accounting for about 17 per cent of the total food. The fisheries supply the remaining about 7 per cent. The majority of food for human population comes from traditional land based agriculture. The meat is mainly consumed by more developed nations of North America, Europe and Japan that consume about 70 metric tonnes of high quality proteins of the world's diet. Harvesting of fishes for food from many oceans of the world and fresh water bodies is also very common. The milk, fruits, vegetables, eggs, butter etc. are consumed all over the world.

Crops

As many as about 3,000 species of plants have been tried as agricultural crops but only about 100 constitute the food for humans. Most of the world's food comes from about 35 crop species including wheat, rice, barley, oats, sorghum, millet, maize, potato, cassava, sweet potato, sugarcane, pulses and about 20 common fruit and vegetable species. Amongst these, wheat, rice and maize are the major grains, producing about half of all agricultural crops on which human population depends for its nutrition and calories. In developing countries, about 4 billion people thrive on wheat and rice which form the staple food in India. Rice is the principal crop and is followed by wheat. Pulses, vegetables and fruits also make a large contribution to human diets. Vegetable oils and fats and condiments are used as cooking medium and food adjuncts, respectively.

Livestock

The domesticated animals are also an important source of food since they provide meat or other products such as milk, butter, *ghee*, fat etc. Major domesticated animals, used as food by humans, include ruminants that convert cellulose part of the plants – earth's most abundant organic compound indigestible to people, in human food. Cattles including cow, buffalow, goat, sheep, camel, rendeer etc. are commonly used for obtaining meat for human food. During 2001-06, Government of India has taken several measures for *white revolution* to increase milk production. Now India has become the largest producer of milk in the world. During 2003-04, per cent availability of milk was estimated to be 231 gm per person per day. Currently India ranks fifth in egg production in the world.

Aquaculture

Aquaculture includes production and harvesting of food from aquatic habitats, both marine and fresh water. Fishes, some other fresh water animals, and seafood contribute about 70 million metric tonnes of high quality protein rich food which is about one-half as much as to the many parts of the world including Asia and Europe. In Japan, fish and sea food contribute up to one-half of the animal protein and one-fourth of the total dietary protein. In fisheries sector India is major contributor of foreign exchange earning through export. Export of fish and fishery products during 2003-04, was 4.12 lakh tonnes valued at Rs 6091.95 crore.

WORLD FOOD PROBLEMS

The concern about the relationship between people and food supply is not new. Agriculture is the most important production practice of the world. Despite the fact that during last five decades world food production has increased almost threefold, population has increased at such a rate in less developed countries that it overtripped the food production. The Food and Agriculture Organisation (FAO) estimates that about 840 million people remain chronically hungry, nearly 800 million of them living in developing countries. Every year hunger kills 12 million children world wide. Although the number of people suffering from hunger has been decreasing, 2.5 million per year over the last decade, but the World Food Summit's, 1996 target of cutting half the number of world's chronically hungry and undernourished people by 2015 will be met 100 years late if the present trend of increase in population coupled with poverty continues. There is enough food in the world to provide at least 2 kg per person including 1 kg of grains or its substitute per day and other eatables to make people fit but the problem is that many people are too poor to buy readily available food. India now produces 180-210 million tonnes of food grains, 5 million tonnes of meat products and 6 million tonnes of fish annually. The increase in grain production since 1960 owes to the *Green Revolution*. In India alone 300 million people are food insecure and poverty driven as they do not possess adequate purchasing power to buy food which could fulfil the minimum calorie requirement of a human body. The reason for this insecurity is inequitable distribution of income among people. Thus it is evident that the problem is not of production but of access and distribution. The average calories available, *i.e.*, 2807 per person per day exceed average requirements, *i.e.*, 2511 per person per day world wide. In India, per capita per day protein consumption is less than the amount usually considered adequate. Apart from rapid population growth, unequal distribution of food resources and lack of power to buy the food requirements, other causes of food shortage are floods, droughts, earthquakes, water scarcity etc.

Box 6.1 October 16: World Food Day

World Food Day is observed on October 16 every year since founding of Food and Agriculture Organisation of United Nations in, 1945. FAO is the lead agency of the UN system for technical assistance, research and policy making in world agriculture, fishing, forestry and rural development. The day is observed worldwide to increase awareness, understanding and information year around on the long term action on the complex issues of food security for all. World Food Day is a 'tripartite' effort by the private voluntary organisations, governments and the international system that acts in many different ways to build the public will to struggle for hunger alleviation and world food security. Each year World Food Day highlights a specific theme on which to focus the activities.

India had a severe food problem in 1964 when the then Prime Minister Lal Bahadur Shastri had to give slogan, *Jai Jawan Jai Kisan*. Thereafter, there was a boom in agriculture yields. Stocks of the food were built up and India produced more food than it need to be consumed. In all, it was due to green revolution. M S Swaminathan made significant contribution in this are and is popularly known as the *Father of Green Revolution in India*.

The global problem of food insufficiency is causing undernourishment and malnutrition.

Undernourishment

Undernourishment refers to the lack of sufficient calories in available diet, so that one has little or no ability to move or work. The reason being, the body begins to breakdown its own stored fats and proteins. Undernourishment is more common in poor countries. People receiving less than 90 per cent of their minimum dietary intake, on a long-term basis, are considered as undernourished. While not starving to death, such people do not have enough energy for an active and productive life. These people are more suceptible to diseases and deficiency diseases like beriberi and anaemia due to lack of nutrients. Such people are weak and can not work because of poor diet. On account of inadequate income they can not afford good food. Since they cannot work, they can not buy food for their dependant children. Thus thcir children fail to grow properly and are likely to face impoverishment in their adulthood. Those receiving less than 75 per cent of their minimum daily caloric intake requirements are considered seriously undernourished. The undernourished children are likely to suffer from permanently stunted growth, mental retardation and other development disorders.

Undernourishment is a global problem but it is more prevalent in the developing countries. In 1992, in USA, there were an estimated 12 million children getting inadequate food. In 1996 alone, an estimated 195 million children, under the age of five, were undernourished in the world. According to United Nations figures, in 1996, there were about 200 million Africans suffering from undernourishment. Scale of undernourishment in the world, in 1999-2001, is depicted in fig. 6.1. Millions of the people, including children under the age of five, die due to hunger each year. In developing countries, one child in four (around 13 million children per year), dies of diseases that could be prevented with a better diet and other requisites.

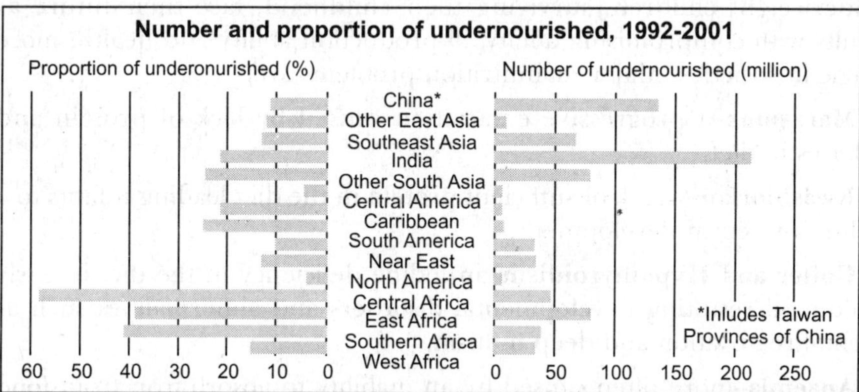

Fig. 6.1. Scale of undernourishment in the world.

The 2006 was year during which progress made during 2000 – 05, in achieving Millennium Development Goals, among which elimination of hunger and poverty occupies the first place, was reviewed. In India's missions 2007–*Hunger Free India,* is at the top priority. A hunger free India can only be achieved by (I) redesigning the delivery of nutrition support programme on a life cycle basis, starting with pregnant women and infants and ending with old and infirm persons, (ii) enhancement of small farm productivity through integrated action in improving soil and plant health, water conservation and use, and post-harvest technology and (iii) promotion of diversified livelihood opportunities through marked driven micro-enterprises supported by micro-credit. This work would be assisted by larger companies through self help groups. Bringing India at *Zero Hunger* level, like Brazil, Kenya and a few other countries, can be achieved by developing National Food Guarantee Act, to provide food to all.

Malnourishment

Malnourishment refers to lack of specific components of food, such as proteins, vitamins or certain essential elements, required for the sound health and development of human body. Malnourishment can be caused both by over nourishment to have excess food without a proper nutritional imbalance caused by lack of a specific dietary constituents or an inability to absorb or utilise essential nutrients from the food. Poor diet may result faulty nutrition causing lack of appetite and abnormal absorption of nutrients by the gastrointestinal tract. Poor nutrition and ill health, in the long run, result in overall quality of life and in the levels of development of human potential. Since people get afflicted by nutritional or related disorders, there is lowering of working capacity and productivity potential resulting in economic losses. The children become weak or sick and their educational carrier is affected. Malnutrient and underweight children, surviving their childhood, face their future as adults with compromising ability, ill-production ability and health, more tragic in societies. Major malnutrition problems are:

Marasmus–a progressive emaciation caused by lack of protein and calories in diet.

Kwashiorkor–a lack of sufficient protein in the diet leading infants to a failure of neural development.

Goiter and Hypothyroidism–an iodine deficiency in the diet in early childhood resulting developmental disorders and abnormalities such as mental retardation and deep mutism.

Anaemia–more often caused by an inability to absorb iron from food than lack of iron in the diet.

Chronic Hunger–occurs when people have enough food to stay alive but not enough to lead proper productive lives.

Pellagra–occurs due to deficiency of amino acids such as tryptophan and lysine.

These overlapping types of malnutritions are more common in developing countries. But in developed and richer countries, people eat too much of meat, fat and processed food and too little of fibres, and vitamin and mineral rich vegetables and fruits. Thus they take to much of calories as result of over nutrition. Over nutrition results overweight, high blood pressure, diabetes, causing heart attack and other cardiovascular diseases, the leading causes of death in most developed countries. The daily caloric intake in North America and Europe is above 3,500 calories per day, nearly 30 per cent more than is needed for adequate nutrition.

Balanced Diet

Malnutrition and its ill effects can be avoided by consuming balanced and varied diet that includes all essential groups of foods. In general, the balanced diet includes plenty of whole grains, pulses, vegetables, fat, sugar, milk, fruit etc. for vegetarians and meat and egg in addition or as a replacement, for non-vegetarians. Cereals like wheat and rice, the staple food of mankind, provide a fraction of nutrient supply and need be supplemented with other foods that provide proteins, fats and traces of minerals, vitamins and amino acids. A balanced diet simply means a diet that will supply all the nutrients necessary for the growth and development of the body. The healthiest combination for a balanced diet, in general, is that about 50 per cent of calories should come from complex carbohydrates, about 30 per cent from all fats and about 20 per cent from proteins. Conclusively, expanded diet sheet in a suitable proportion and variation in menu in our daily meals will prove better for the health.

The amounts of various foods that will make up a balanced diet for the average Indian is given in table 6.1. The quantity of the food varies according to the age and the work of diet consumer.

TABLE. 6.1. Balanced Diet for an Average Indian

Food Group	*Food Stuff*	*Amount Per Day (g)*
1.	Wheat, Rice and Millets	350
	Oil, ghee, Butter etc.	35
	Sugar and Juggary	40
2.	Milk, Curd, etc.	255
	Pulses, Dried Beans, Nuts etc.	45
	Meat, Fish, Egg	60
3.	Fruits	30
	Green Leafy Vegetables	150
	Other Vegetables	125

CHANGES CAUSED BY OVERGRAZING

Livestock wealth plays a significant role in the rural life and food supply to urban areas of our country. India is among one of the leading countries in the livestock population in the world. And this large population of livestock needs to be fed. Unfortunately the grazing lands and pasture areas in our country are not adequate. Very often it is observed that livestock grazes upon a particular grassland or pasture and surpasses the carrying capacity, *i.e.*, the maximum population that can be supported by the said land area on a sustainable basis. Sometimes the grazing pressure is such that the grazing land's carrying capacity is crossed, sustainability fails and overgrazing results.

The changes caused by overgrazing are:

(1) Land Degradation: Overgrazing removes the vegetation cover of the soil making it exposed and to get compact. Thus the operative soil depth declines and hinders the roots going deep into the soil and make them unable to get enough soil moisture. The organic cycling in the ecosystem is also hindered because not enough detritus or litter remains on the soil to be decomposed. Thus the soil becomes organically poor, dry and compacted. Moreover, trampling by cattle makes the soil to lose its infiltrate capacity which reduces water percolation into the soil. Thus, ecosystem loses its water which runs off the surface. Conclusively, overgrazing leads to multiple losses including loss of soil texture, hydraulic conductivity and soil fertility.

(2) Soil Erosion: Overgrazing results removal of vegetation cover of the land, exposes the soil, causes multiple soil losses including erosion by the action of strong winds, rain fall etc. The removal of grasses makes the soil loose because the grass roots are very good binder of the soil. Thus the soil gets prone to erosion.

(3) Loss of Useful Plant Species: Overgrazing adversely affects the diversity of plant species. It brings about negative effects on the composition of plant population and their regeneration capacity. When the livestock grazes upon the original grassland with good quality grasses and foliage with high nutritive value, vegetation gets destroyed and some new species appear in their place. These secondary species, though hardier, are less nutritive in value. Thus, nutritious juicy fodder giving species are replaced by unpalatable and sometimes undesirable thorny plant species lacking good soil binding capacity and thus making the soil more prone to erosion. Vast areas in Arunachal Pradesh and Meghalaya are getting invaded by low fodder value thorny bushes, weeds etc. due to overgrazing. Thus, overgrazing makes the grazing land lose its regenerating capacity and good quality pasture ecosystem is converted into poor quality thorny vegetation.

CHANGES CAUSED BY AGRICULTURE

Agriculture is world's oldest and largest trade. More than half of world's population still lives and depends on agriculture. In India, 60 per cent of the population is engaged in agriculture. Production, processing and distribution of food all bring about the changes in the environment which may be both primary and secondary. The primary effect is onsite impact on the area where agriculture takes place whereas secondary effect is off-site, *i.e.*, an effect on the environment away from the agricultural site. Agricultural impacts on environment may be local, regional or global. Local changes occur near the site of farming and include soil erosion and increase in sedimentation down streams in local rivers. The fertilizer loaded sediments can cause eutrophication of local water bodies and polluted sediments can transport toxins and destroy local fishery. Regional changes may result from combined effects of farming practices in some large region and include deforestation, desertification, large-scale pollution, sedimentation in major rivers and estuaries at the mouth of rivers. Regional Changes also affect chemical fertility of soil over large areas. In tropical waters, sediments entering the oceans, may destroy coral reefs that are near the shore. Global changes include climate changes and potentially extensive changes in chemical cycles.

Traditional agriculture involves small plots, simple tools, naturally available water, organic fertilizer and a mix of crops. The practice, still in operation by about half of the world, is more near to natural conditions and usually results in poor yield. Impacts of traditional agriculture are deforestation, soil erosion, depletion of soil nutrients etc. Slash and burn cultivation or *jhum* cultivation (shifting cultivation), still prevalent in many tribal areas of North East Hill States of India, results loss of the forests and low production. Clearing of the forest cover exposes the soil to the action of wind, water etc. and results in the loss of the top fertile soil, *i.e.*, soil erosion. During slash and burn, the organic matter in the soil gets destroyed and most of the nutrients are used up by the crops within a short period. Thus the soil becomes nutrient poor and the cultivators are compelled to shift to the other area.

Effects of Modern Agriculture

The modern agricultural practices have both positive and negative impacts on the environment. Though they have created a revolution in agriculture in short term but their long term effects have proven extremely undesirable. Modern agricultural practices involve the use of high yielding varieties, *i.e.*, the use of hybrid seeds of selected and single crop variety, high-tech equipments and lot of energy subsidies in the form of fertilizers, irrigation water and pesticides.

Green Revolution and its Impacts

Green Revolution, a term coined in 1960, refers to the rapid increase in the world food production, especially in developing countries, in the second half of the twentieth century, primarily through the use of lab-engineered High Yielding Varieties (HYVs) of seeds. Though hailed as a success story of agricultural science, it has also brought in its wake many problems. The period during which this programme was launched, witnessed severe food shortage in many developing countries. With growing populations, countries like India had to import food. The answer to this problem came from Mexico where in 1940s wheat yield was low and the country was importing 50 per cent of its food. The Rockfeller Foundation in the US set up a research programme to increase the production of food grains. An American agricultural scientists Norman E. Borlaug joined the team and developed high yielding variety of wheat through mutation and plant breeding. The variety was capable of responding to high impacts of fertilizers and irrigation and increased the Mexican wheat yields 400 per cent between 1950 and 1965, giving birth to green revolution. The new HYVs were Sharbati Sonara, Lerma Rojo, Sonalika etc.

By 1965, the green revolution was fully adopted in India. Efforts made by a team of Indian agricultural scientist M S Swaminathan made him to be known as *Father of Green Revolution in India*. Punjab was the first state to achieve this goal. The new high yielding varieties showed a tremendous increase in production and made multiple cropping possible. As a result, India achieved self-sufficiency in food to the extent that warehouses were so full that in some of the stores grains got rotted. The green revolution proved to be a clear shift from traditional agriculture and gave way to modern agriculture. It came as a package of high yielding varieties along with high inputs of inorganic fertilizers, pesticides, water and agricultural machinery such as tractors and water lifting pumps. Overall, it was an energy intensive programme and required a lot of energy to run the machinery and to pump the water. Green Revolution agriculture consumed about 8 per cent of the world's oil. Government of India had to subsidise many of the inputs to keep the green revolution on track.

The green revolution, the gateway of modern agriculture, benefited large land owners and not the subsistence farmers. It proved as an energy oriented programme and required heavy expenditure on use of fertilizers, pesticides and irrigation. The farmers had to buy seed every year. And year after year, the inputs had to be increased to maintain productivity levels. The number of varieties being used by the farmers went down drastically and a large number of traditional varieties have disappeared leaving just a handful of high yielding varieties. The soil has been degraded and drained off its nutrients as a result of excessive use of chemicals (fertilizers, pesticides etc). The green revolution widened the gap between the rich and the poor farmers.

Impacts Related to High Yielding Varieties (HYVs)

The use of High Yielding Varieties (HYVs) has encouraged monoculture, *i.e.*, the same genotype is grown over vast areas and in case of attack by some pathogen, there is a total devastation of the crop by the disease in the whole lot. Moreover, the disease spreads rapidly as the conditions are exactly uniform all over the growing area. There has been a great reduction in the genetic diversity both in crop and livestock.

FERTILIZER PROBLEMS

Fertilizers are the materials having definite chemical composition with a high analytical value that supply plant nutrients in available form. Most of the chemical fertilizers are inorganic (excpt urea – the solid nitrogenous organic fertilizer) in nature and are manufactured by industries and sold by trade name. They are added to the soil to restore and enhance the soil fertility and to improve the plant growth to increase crop yield. The most important elements required by plants for growth are nitrogen, phosphorus and potassium (NPK) along with calcium, magnesium, sulphur and some others. The chemical fertilizers are advantageous over organic fertilizers in that they can be stored, handled, transported and applied easily and have lower risk of pathogenic contamination.

In early days farmers used to supplement the soil with organic manure derived from livestock refuse which could fulfil the inorganic requirement of the soil to increase production. In hope of increased yield now farmers are applying more fertilizers than the crop requires, as a result a great amount of them is leached out of the soil and contaminate the sub-soil water resources. Most of these chemical fertilizers are non-biodegradable, *i.e.*, as such they persist for long periods and accumulate to reach objectionable levels as they pass through various trophic levels of the food chain and can be serious source of pollution and cause of a number of problems.

The Problems Caused by the Use of Fertilizers

Some of the problems caused by use of fertilizers are:

1. Micronutrient Imbalance. Indiscriminate use of excessive amount of macronutrients (nitrogen, phosphorus and potassium etc.) causes micronutrient (elements required in traces) imbalance. For example, excessive use of fertilizers in Punjab and Haryana has caused deficiency of micronutrient zinc, which is affecting soil fertility.

2. Nitrate Pollution. Nitrogenous fertilizers, such as urea, applied in the fields often leach deep into the soil and invariably contaminate ground water resource. The nitrates get concentrated in water and when their concentration exceeds they become the cause of serious health hazards. Nitrate is toxic to infants. At concentration of fourty-five parts per million, nitrate consumed through potable water can cause fatal *methaemoglobinema* sickness or *blue baby syndrome*. Nitrate consumed through drinking water and leafy vegetables are reduced to nitrites by bacterial action. In intestine, on reaching the blood stream, nitrogen oxides become attached to haemoglobin forming methaemoglobin which retards the oxygen transporting capacity of the blood and skin develops bluish colour and in young babies it results even to death. This problem is more common in England, France, Germany, Netherlands and Denmark and in many parts in India. In adult humans, gastric ulcer is caused if the nitrate is further converted into amines and nitrosoamines.

3. Impact of Phosphates: Eutrophication. Excessive use of fertilizers, particularly rich in phosphorus in agricultural fields affects water bodies like lakes, reservoirs, ponds etc. Nutrient such as phosphorus, and nitrogen get washed off from the fields and along with run off water reach the water bodies causing them to become overnourished and give way to the growth of undesirable aquatic plants, the eutrophication. As a result, water bodies get invaded with trouble causing harmful algae and water weeds such as water hyacinth (*jalkumbhi*). Also the water bodies get shrinked. These troublesome weeds check the growth of other plants and adversely affect the life of aquatic animals. Decrease in the amount of oxygen in the water, increases *Biological Oxygen Demand* (BOD) of water bodies and cause them to become polluted. The algal blooms and pathogenic anaerobic bacteria enter the water cycle and food and cause serious harm to man and other animals.

If not applied with caution, many artificial fertilizers cause contamination and fail to give their full effect towards increased yield. There are several uncertainties about long-term impact of chemical fertilizers on farmland. Continued monocroping and use of artificial fertilizer and not practising crop rotation and use of livestock manure, result fertility problems of loss of organic matter of the soil.

Fertilizer problems can be reduced by using organic manure, live stock refuse, biofertilizers, integrated nutrient supply etc.

PESTICIDE PROBLEMS

Agriculture suffers a considerable loss from pests–the undesirable competitors, predators or parasites. The major agricultural pests are insects which feed mainly on leaves, young stems or sometimes the whole plant. Nematodes and small worms feed mainly on roots or other plant parts. Bacteria, viruses and fungi cause various plant diseases. Weeds–commonly flowering plants compete with the crops. Vertebrates—mainly birds and rodents feed on fruits or grains.

Pesticides are the compounds used to kill, deter or disable pests inorder to improve appearance of crop, control of losses made by diseases, check consumption of biomass by pests and to maximize crop yield. Benefits claimed for pesticide use are many in terms of improved harvest, reduced storage losses, human and livestock disease control but the risks involved are more. In early days farmers used to control pests by application of salt, smoke and insect repelling plants. Use of oil sprays, ash and sulphur ointments for the purpose was also a common practice. Spices were used to deter spoilage and pest infestations; predatory ants to control caterpillars in orchards; ducks and geese to catch insects and control weeds; and burning fields and rotation of crops to reduce crop diseases.

The first generation pesticides include chemicals like sulphur, arsenic, lead or mercury which are used to kill the pests. The modern era of pest control began with the second generation organic pesticide discovery of DDT (Dichloro-Diphenyl Trichloroethane), initially synthesised in 1874. Its insecticidal properties were discovered by Paul Mueller in 1939. It was adopted as a pesticide and in public health systems from 1950s. It is a typical chlorinated hydrocarbon which is usually long-lived, of low toxicity and bio-concentrating, *i.e.*, persisting in biological tissues and getting more concentrated at each trophic level of the food chain. Slowly, the effects of long-term exposure to these chemicals revealed to be mutagenesis and carcinogenesis. Residues of these toxins are still present all over the planet in all populations. Though chlorinated hydrocarbons have been largely replaced by safer organo-phosphates, which although are not prone to bio-concentration but they are much more acutely toxic, so that every minute quantity is harmful. Investigations are suggestive that these short-lived unstable compounds, once below the biological active soil or even when trapped in topsoil micropores, can become extremely tenacious.

The HCB (Hexachlorobenzene), a suspected carcinogen used as a fungicide and as a soil fumigant on stored grains and milling equipment had been used in America since 1948. The pesticide started apearing up in ground water in some parts of America following the Pillsburry cake mix-scare, as a result the use of these chemicals was banned. The DDT and other chlorinated hydrocarbons and parathione are most likely to leave persistent residue and creep into the food chain. It is shocking to realise that pesticide-caused diseases have overtaken endemic diseases in the Third World Countries including India. This becomes clear that the profit of the richer nations of the world received from the export of pesticides and the accompanying agricultural technology comes at the expense of the health of the people living in the poor nations.

The most ethical problems related with use of pesticide in agriculture are those that affect the fields themselves. Some of the impacts associated with agricultural pesticide use are:

1. Creating Resistance in Pests and Producing New Pests. Some individuals of the pest species usually survive even after pesticide spray and the survivors give rise to highly resistant new pest generations. They set up agriculture's dangerous addiction to pesticides, creating an increasing need and use for these chemicals, failing which chances of intensified pest outbreak increase. About 20 pest species are now known which have become immune to all types of pesticides and are called *super pests*.

2. Death of Non-Target Organisms. Many insecticides, in being broad spectrum in their action, not only kill target species but also kill several non-target species that are useful to us. Due to killing of beneficial predators, that previously checked on number of pests, many new pest species are created, thus predator-prey population relation is being disturbed.

3. Biological Magnification. Many of the pesticides, in being non-biodegradable in nature, keep on accumulating in the food chain-biological magnification and ultimately reach the body of human being that occupies a position higher in the trophic level and cause serious harms to mankind.

4. Impact on Human Health. Impacts of pesticides on human health are both short-term and long-term. Short-term impacts include acute poisoning and illness caused by relatively high doses and accidental exposures. Long-term impacts, caused even by very low doses, include growth of cancer, birth defects, immunological problems, Parkinson's disease and other chronic degerative diseases. High suspect of DDT in human beings may result in cerebral hemorrhage, hypertension, cancer, liver damage etc. DDT has now been banned in many countries.

Pesticide problems can be reduced by **(1)** banning dangerous compounds; **(2)** developing alternatives-biological control or integrated pest management; **(3)** restricting trade of pesticide-contaminated produce; **(4)** controlling pesticide usage by monitoring, inspecting and licensing to ensure safer and sensible producer **(5)** developing less dangerous and safer pesticides; **(6)** controlling prices of pesticides to discourage excessive use; **(7)** education to discourage unsound strategies; **(8)** rotation of crops to upset pest breeding and access to food; **(9)** weeding or non-chemical weeding; **(10)** encouraging funding agencies to cut financial support for pesticides treated drinking water to remove pesticides etc.

WATER LOGGING

Water logging occurs due to surface flooding because of high water table or more commonly due to over-irrigation of croplands by the farmers in hope for good growth and yield of their crops. Inadequate drainage causes excess of water to accumulate underground and to form a continuous column with water table. In water logged soils, the pore spaces get fully drenched with water thereby replacing the soil air and changing soil texture. Plant sensitivity to wet conditions depends on the temperature. Under warm weather, crops suffer more from water logging due to increased oxygen consumption by the roots of the plant than under cold weather. Water logging in humid tropical regions may result due to intense rainfall for prolonged period and due to accumulation of snowmelt water in many cold and moderate climatic regions. Since water table rises, roots of the plants do not get adequate air for their respiration. The soil loses its mechanical strength resulting crop plants to get logged and thus lowering the crop yield. The productivity of the water logged soils is very poor due to less oxygen and excess of water. Sometimes seepage or undesirable rise in water level of canal systems also result water logging in the area (fig. 6.2).

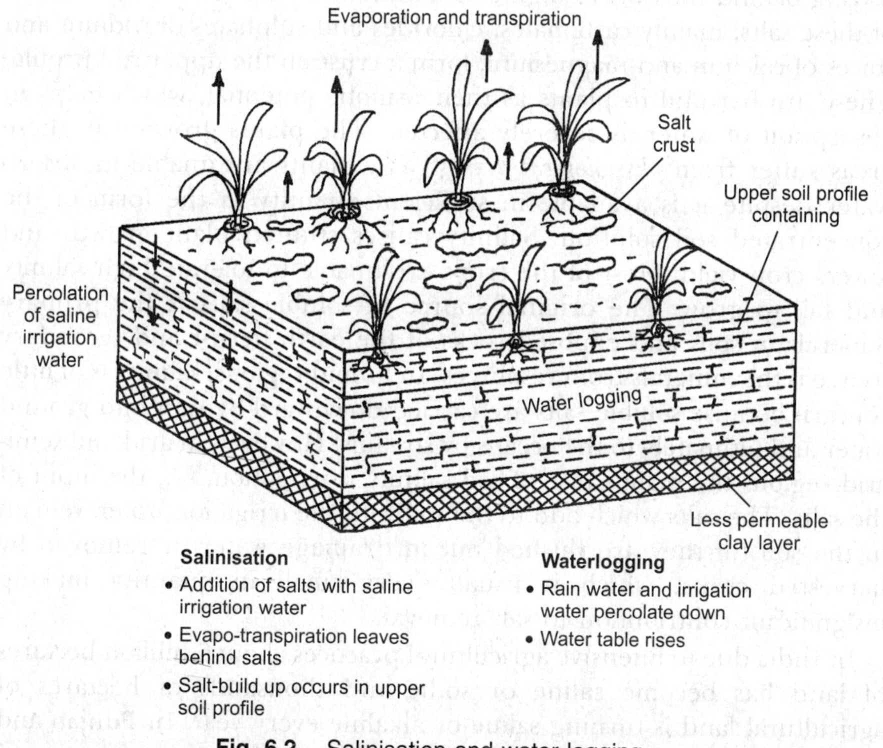

Fig. 6.2. Salinisation and water logging.

In India, estuarine delta of river Ganges, coastal areas of Kerala and Andaman and Nicobar Islands are some areas which are frequently water logged. Wet and water logged areas are also common in *Terai* regions of Uttar Pradesh and foot hill regions of Uttarakhand. In Punjab and Haryana, with extensive network of canal system and innumerable bore wells, extensive areas have become water logged because farmers use excess of water from these tube-wells and canal systems under the impression that more irrigation would result higher crop yield.

Water logging problem can be solved or at least reduced by preventing excessive irrigation, developing suitable drainage systems, using sub-surface drainage technology and bio-drainage with trees such as Eucalyptus.

SOIL SALINITY

Soil salinity refers to increased concentration of soluble salt content in the soil. It results due to intensive agricultural practices involving excessive irrigation with canal systems. Due to poor drainage under dry weather, water evaporates leaving behind salts in the upper layers of the soil. In arid regions with low rainfall, poor drainage of irrigation, flood water and high temperature, the water evaporates quickly from the soil leaving behind the salts in high concentration on the soil surface. Excess of these salts, mainly carbonates, chlorides and sulphates of sodium and traces of calcium and magnesium, form a crust on the upper soil profile. These are harmful to plants as their osmotic potential, which helps in absorption of water, is severely affected. The plants growing in these areas suffer from *physiological dryness, i.e.,* plants are unable to absorb water despite it is available in sufficient quantity in the form of the concentrated soil solution. Salinity causes stunted plant growth and lowers crop yield. Most of the crops are unable to tolerate high salinity and fail to grow. The original source of soluble salts is the primary minerals in soils and exposed rocks of the earth's crust and secondary source is the minerals from resources. In humid regions, salinity is of little concern because soluble salts are carried downward by rain into ground water and ultimately to the oceans by streams. However, in arid and semi-arid regions the main cause of soil salinity is irrigation, *i.e.,* the input of the salts. The salts which add to the soil with the irrigation water, remain in the soil till they are flushed out in drainage water or removed by harvested crops, which is usually very small in quantity making insignificant contribution to salt removal.

In India due to intensive agricultural practices about 6 million hectares of land has become saline or sodic and thousands of hectares of agricultural land is turning saline or alkaline every year. In Punjab and

Haryana alonc, thousands of hcctarcs of land area is affected by soil salinity and alkalinity. Tsunami in Andhra Pradesh and Tamil Nadu has increased soil salinity in their coastal regions.

The most common method of getting rid from salinity is to improve drainage to flush salts out by heavy irrigation by good quality water to such soils. The other method is laying underground network of perforated drainage pipes for flushing out the salts slowly-the sub surface drainage system. This system has been tried on experimental basis in some places in Haryana.

Chapter Summary

Among the three basic necessities of human beings, food is of utmost priority because without the food no one derive the energy and survive. Plants and animals are the main source of food for man. About 76 per cent of food comes from crops, mostly from cereals. Food is required at all stages of life right from childhood till old age for satisfactory growth and health. Most of the food comes from about 100 crop plants. The food is also derived from livestock and fishes. Today the world is facing a severe food problem. According to Food and Agriculture Organisation (FAO) about 840 million people in the world remain chronically hungry and every year hunger kills about 12 million children worldwide. In India alone, 300 million people are food insecure. The main reason of food insecurity is inequitable distribution of money among people. India had a serious food problem during 1960s. Undernourishment, *i.e.*, lack of sufficient calories in available diet is another cause of food insecurity. Undernourishment makes the people weak and diseased, leading to poverty. Undernourishment and malnutrition are more common in developing countries. Malnutrition results many diseases like marasmus, goiter, anaemia, pellagra etc. Balanced diet can solve the ill effects of malnutrition. A balanced diet includes the healthy combination of all food groups.

Over grazing brings about many ill effects such as land degradation, soil erosion, desertification, loss of useful plants species etc. Use of new technology coupled with use of chemical fertilizers and pesticides, especially during green revolution has resulted many problems to the environment and human health. It has resulted various kinds of pollutions and eutrophication in water bodies. Pesticide pollution has decreased the quality of the soil and is causing various diseases in man. Water logging, surface flooding because of high water table, reduces the productivity. Soil salinity, *i.e.*, the increased concentration of soluble salts in soil water, lowers the crop yield.

Study Questions

1. Write an essay on world food problem giving emphasis on undernourishment and malnutrition.
2. What are the various sources of food and their values to man.
3. Write an account of changes caused by overgrazing.
4. Discuss ill effects of the use of chemical fertilizers and pesticides in agriculture.
5. Briefly describe:
   (i) Balanced Diet
   (ii) Water logging

(iii) Soil salinity
(iv) After effects of Green Revolution
(v) Nitrate Pollution

Objective Questions. *Select the correct answers*

1. The great Bengal Famine in India occurred during the year:
(1) 1939-40 (2) 1940-41
(3) 1941-42 (4) 1942-43

2. India faced worse food problem during:
(1) 1940s (2) 1950s
(3) 1960s (4) 1970s

3. Most of the world's food comes from:
(1) Cereals (2) Crops
(3) Grazing livestock (4) Fish

4. World Food Day is celebrated on:
(1) July 16 (2) August 16
(3) September 16 (4) October 16

5. The solgan **Jai Jawan Jai Kisaan** was given by:
(1) M.S. Swaminathan (2) Jawahar Lal Nehru
(3) Lal Bahadur Shastri (4) B.P. Pal

6. Pellagra disease is caused due to lack of:
(1) Amino acids (2) Vitamins
(3) Carbohydrates (4) Minerals

7. Scientist known as 'Father of Green Revolution in India' is:
(1) B.P. Pal (2) T.N. Khosoo
(3) M.S. Swaminathan (4) Norman E. Borlaug

8. 'Blue baby syndrome' disease is caused due to:
(1) Phosphate pollution (2) Sulphate pollution
(3) Chloride pollution (4) Nitrate pollution

Answers

1. (4) *2.* (3) *3.* (2) *4.* (4) *5.* (3) *6.* (1)
7. (3) *8.* (4)

7

CHAPTER

Energy Resources

Learning Objectives

Introduction • Growing Energy Needs • Renewable and Non-Renewable Energy Sources • Renewable (Non-Conventional) Energy Sources • Renewable Energy Sources: Solar Energy, Hydro Energy, Wind Energy, Ocean Energy • Indian Scenario • Non-Renewable (Conventional) Energy Sources: Coal, Crude Oil, Natural Gas, Nuclear Energy • Alternate Emerging Energy Sources • Hydrogen Energy • Fuel Cells • Alcohol as A Source of Energy • Energy Plantations • Energy From Agricultural Waste Biomass • Biogas • Energy From Urban Waste • Dendrothermal Energy • Petroplants • Bio-diesal • Power Development In India.

Introduction

The term 'energy', is the capacity to do work, has been coined by Thomas Young (1737-1829) and is applied to what it is now called kinetic energy. The two laws of thermodynamics describe the behaviour of energy. The first law states that energy can be transformed from one form to the other, but cannot be created or destroyed. The second law states that some energy is always dispersed into unavailable heat energy. No spontaneous transformation of energy from one form to another (in context of protoplasm) is 100 per cent efficient.

Energy is an essential input for the economic development and improvement of quality of life. Every work in this world is accomplished with the help of energy. Infact, all living beings operate by means of energy. Energy moves the universe. About 99.8 per cent of our energy comes from solar radiation. It is this solar energy that plants use to make food which gets stored in plants as biomass. Other energy inputs to the earth are wind energy, ocean thermal and tidal energy, geothermal energy, nuclear energy and energy from fossil fuels such as oil, coal and natural gas. Many other alternate sources of energy are being explored and used.

About 24 per cent of energy consumed globally, is used for transportation, 40 per cent for industries, 30 per cent for domestic and commercial purposes and the rest 6 per cent for other uses including agriculture. About 2 billion people, one third of global population living in developing countries, lack access to adequate energy supplies. Three billion people depend on fuel wood, coal, charcoal, dung and kerosene etc. for cooking and heating. On the other hand, industrialized nations, with only 25 per cent of global population, account for 70 per cent of commercial energy consumption. Most of it comes from oil that is used for transportation. The United States, with just 4.6 per cent of world's population, consumes 24 per cent of the total commercial energy produced and is the largest consumer of energy in the world. India, with 16 per cent of world's population, accounts for just 3 per cent of the total energy consumption of which about one-third comes from biomass. There is a wide disparity in energy consumption in developed and industrialised countries and developing countries. The energy consumption of a nation now-a-days is usually considered as an index of its development.

Today, energy is the primary input for almost all economic activities and has become vital for improvement in the quality of life. Infact, the whole infrastructure rests upon energy. Use of energy in various sectors such as industry, commerce, transport, telecommunication, agriculture and household services has compelled us to focus our attention to ensure its continuous supply to meet our ever increasing demands. The modern civilisation is much dependant on energy availability but the energy resources, mainly oil, coal, natural gas and hydroelectric power, are becoming scarce and costlier. The prices of the oil, the most common source of energy, are going very high. To cope with ever increasing demand, efforts are being made to develop new approaches in conventional and non-conventional energy sources and new measures of energy conservation.

In India, power development commenced at the end of 19th century with the commissioning of electricity supply in Darjeeling in 1897 followed by commissioning of a hydropower station at Sivasamundram in Karnataka during 1902. In pre-independence period the supply was in private sector that too was restricted to urban areas. After independence, with the formation of State Electricity Boards during Five Years Plans, a significant step was taken to bring about systematic growth of power supply in industries all over the country. National Electricity Policy, 2005 aims at accelerated development of power sector, providing electricity supply to all areas and protecting interests of consumers and other stakeholders with the view of availability of energy resources, technology available to exploit these resources, economics of generation using different resources and energy security issues.

Growing Energy Needs: Gap Between Demand and Supply

History of human existence is related with the ever increasing energy needs of man. For the primitive man, the only source of energy was muscular energy which was regularly replenished through his intake of food. This conversion of energy, bestowed in the food into muscular energy, was confined to the natural process of energy flow in the biosphere. Through his progress, from primitive to modern technological man, entered into newer pathways of energy flow in nature to cope with his ever increasing demands of energy for cooking and warming. This he could procure by converting wood into fire. The agricultural man, in addition, turned grass into the muscular energy of the livestock to plough his fields and to transport his materials. As he progressed further, man utilised wind and water to move wind mills, sail and operate water mills. The advent of steam engine brought about the first industrial revolution. Passing through various stages, man finally succeeded in fabricating high speed electronic computers. The enormous amount of energy needed by man for his domestic purposes, transport, telecommunication, agriculture, commerce and industry was later obtained by the exploration of fossil fuels-petroleum, coal and natural gas. Not only oil, coal and natural gas but also other energy sources, especially hydro and nuclear energy were used to generate electricity. The solar energy by direct conversion, tidal energy from ocean and geothermal energy from volcanoes and hot springs are also being utilised lately, though in small amounts.

Presently, development in various sectors relies largely upon energy. With increasing per capita demand of growing population, the world is facing acute energy deficit. Increase in per capita consumption of energy is related to the extent of development. It shows wide variation throughout the world, being maximum in developed countries and minimum in developing countries. For a 100 units consumed by an American citizen for transportation, a Japanese citizen consumes 30 units and an Indian consumes only 2 units. The case is more or less similar with other developmental activities. The countries like US and Canada, hardly constitute about 5 per cent of global population, but consume one fourth of world's energy as in them an average man consumers 300 Giga Joules (GJ)- equal to 60 barrels of oil per year. In contrast, an average man in a poor country, namely Bhutan, Nepal or Ethiopia, consumes less than one GJ in a year. There is so great variation in the life styles and standard of living in rich and poor countries that a person in a rich country consumes almost as much energy in a single day as one person in whole year in a poor country. Strong correlation, occurring between per capita energy use and Gross National Production (GNP) is shown in fig. 7.1. USA, Norway, Switzerland etc. with high GNP show high energy use while India, China etc. have low GNP and low energy use. Exceptionally UAE

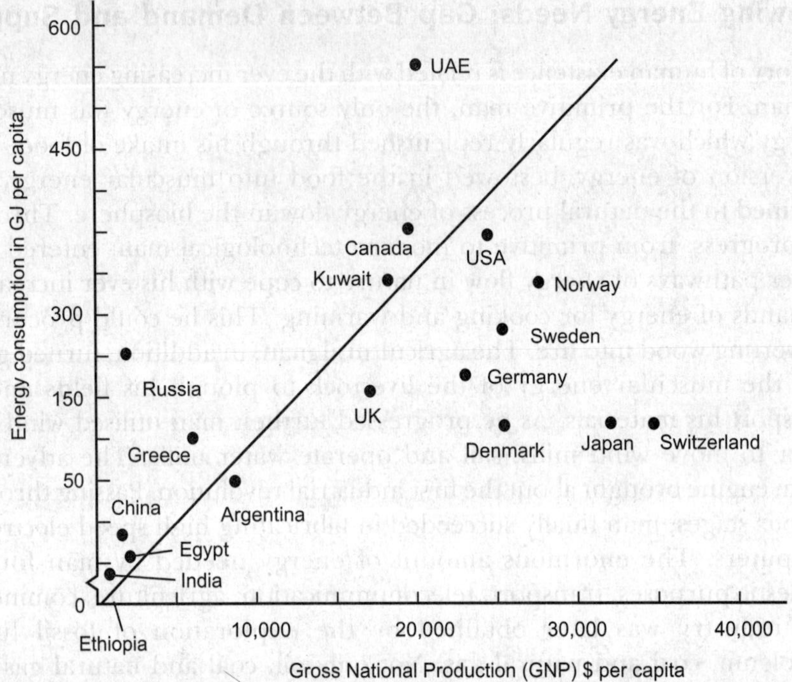

Fig. 7.1. Per capita energy use and GNP (Data from World Resources Institute, 1997).

States like Bahrain, Quatar etc. are oil rich and hence their GNP and energy consumption are more but their development is not at par.

India is an agricultural country. Majority of its population lives in villages and uses wood, agricultural wastes, livestock, dung etc. as energy source. In urban areas, in industries, transportation, telecommunication, domestic establishments etc., the energy consumed is derived from oil, coal, natural gas, hydroelectricity or nuclear power. The industries claim a large share (about 38.5 per cent) of total energy followed by transportation (about 31.2 per cent), domestic establishment (about 13.7 per cent) and the rest in agriculture. Share of various energy sources in the commercial consumption of energy in India mostly comes from the coal (about 56 per cent) and petroleum (about 32 per cent). The other sources are hydroelectric, nuclear power, natural gas etc. The commercial energy consumption has gone from 130.7 Million Tonnes of Oil Equivalent (MTOE) in years 1991-92 to 176.08 (MTOE) in 1997-98. The accompanying structural change in economic growth, industrialisation and a rise in population coupled with rapid urbanisation are the main causes of this increase. Petrol, high speed diesel, gasoline and CNG are largely used in transport, accounting for about 50 per cent of the total consumption. In India, annual growth of commercial energy in the form of electricity has been increasing at the rate of 10 per cent. The installed generation capacity is not matching the demand. India's living standards, raising national programmes for the people, are causing further increase in demand, especially in rural sector. To cope with the increasing demand, alternate energy programmes have been developed on priority basis. In farm sector, the share of electrical energy consumption has increased from a mere 3.9 per cent in 1950-51 to about 32.5 per cent in 1996-97. In domestic sector, the consumption of natural fuel, mostly wood, is very high. Around 70 per cent of rural and about 30 per cent of urban households depend on fuel wood. In years between 1970-71 and 1994-95, annual consumption of electricity per household went up from 7 kwh to 53 kwh, of kerosene from 6.6 litre to 9.9 litre, and cooking gas from 0.33 kg to 3.8 kg. It is estimated that the coal may take about 200 years to be exhausted but the oil shortage is acute and by 2010 oil from current oil wells will be insufficient to meet the demand and there will be a massive disruption in transportation and the world economy. One more reason for the gap between the demand and supply, is widespread mismanagement in power distribution. Moreover, there will not be enough time to switch over from oil to other available sources of energy and there will be global food insecurity due to lack of fertilizers and chemicals and many more goods.

RENEWABLE AND NON-RENEWABLE ENERGY SOURCES

Broadly, energy sources can be recognised as renewable and non-renewable. The renewable energy sources are regenerated in natural process so that they can be used indefinitely. The renewable energy sources generally causes little or no environmental impact as compared to non-renewable sources, such as fossil fuels and nuclear energy. Keeping in view, the current state of technology, the generation of renewable energy is often more expensive than energy produced by fossil fuels or nuclear energy. However, with the technology advances, the cost of renewable energy is expected to decrease. Renewable energy sources are inexhaustible, *i.e.*, they can be replaced after we use them and can produce energy again and agian. Among the renewable energy sources, the most commonly and easily accessible is solar energy. The other renewable energy sources are hydropower, wind, geothermal, ocean wave and tidal energy. Non-renewable energy sources include various fossil fuels including petroleum products, coal and natural gas and nuclear energy. Nuclear energy is mainly obtained from the nuclear fission of uranium or thorium. The global resources cf fossil fuel and uranium and thorium are limited and will eventually be depleted. Moreover, use of fossil fuels for energy has negative environmental consequences, such as air pollution, global warming, acid rains and oil spills. Thus, it has become essential to minimize the use of fossil fuels and to replace them with renewable resources. Keeping in view, the growing energy needs, use of non-coventional energy sources, over conventional, should be promoted.

RENEWABLE (NON-CONVENTIONAL) ENERGY SOURCES

Renewable energy sources are regenerated in natural processes and can be used indefinitely. These are available in great amount in nature and develop in a relatively short period of time. These include solar energy; wind energy; ocean thermal power; wave and tidal energy; geothermal energy; hydropower etc. They are capable of solving the twin problems of energy supply in decentralised manner and helping in sustaining cleaner environment.

Solar Energy

Among renewable energy sources, the easily accessible and most important is the solar energy. It can be used directly or indirectly for human welfare. The direct solar energy is the radiant energy whereas the indirect solar energy is the energy obtained from the materials such as biomass, in which sun's radiant energy has been incorporated by the plants.

(i) Direct Solar Energy. On global scale, 15 days of solar energy is roughly equivalent to the energy stored in all known reserves of fossil fuels on the earth. The continuous input of the energy from the sun is 1,67,000 times greater than the current consumption. As such many countries now are in effort to harness the solar energy for domestic, commercial or industrial purposes. Solar energy can be used for direct heating. Alternatively, the heat can be converted into the electricity-the thermal electric generation. Photovoltaic cells-the solar cells or solar batteries covert direct solar energy into electricity. Other types of solar energy devices are solar power

Box 7.1 Magic Lanterns at Besantnagar Beach, Chennai

As an escape from the humid and polluted air of the city, hundreds of people come at Basantnagar beach, Chennai, in the evening, for taking in the fresh air. They are faced by the pollution caused by kerosene run petromaxes used by hawkers. But since 2001, there is an option for the hawkers to take a rented solar lantern for evening. Hawkers have to pay Rs. 100 as a refundable security and Rs. 15 as the rent for four hours for each lantern to the enterprise which has a small office near the beach where the lanterns are charged during day using solar pannels. Hawkers prefer using solar lanterns because they spend the same amount as before without being harmed and harming the visitors from pollution. This is an unique example that promotes renewable energy, generates employment and saves people from pollution.

pumps (fig. 7.2) and solar pond electric power plants (fig 7.3). Direct utilisation of solar energy is also being tried by using part of natural oceanic area as a gigantic solar collector and constructing large scale ocean thermal plants. The only need is to develop a massive scale back-up system to store and generate electricity when solar power is not operative at night or during the cloudy days.

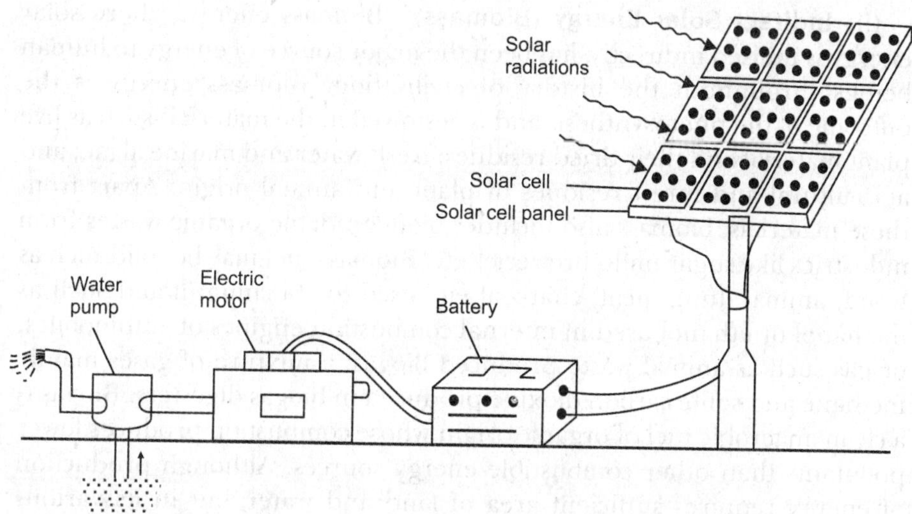

Fig. 7.2. A solar pump run by electricity produced by solar cells.

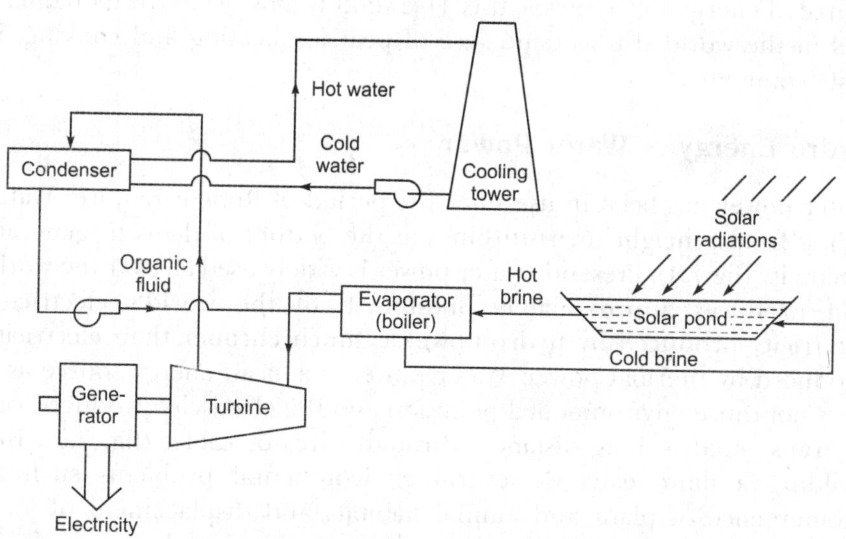

Fig. 7.3. Solar pond electric power plant.

In tropical countries like India, the use of solar energy could prove most suitable in the long run as sunlight is available in plenty, almost throughout the year. India has been one of the major producers of photovoltaic systems, currently used in Street Lighting Systems (SLS), Domestic Lighting Systems (DLS), Community Lighting Systems (CLS), water pumping systems, small power plants, solar cockers etc. India installed largest solar steam cooking system for cooking food for 15,000 people per day at Tirupati Temple in Andhra Pradesh.

(ii) Indirect Solar Energy (Biomass) Biomass energy, where solar energy is utilised indirectly, has been the major source of energy to human beings throughout the history of civilisation. Biomass energy is the outcome of the photosynthesis and is bestowed in the materials such as live plant material and their dried residues; fresh water and marine algae; and agricultural and forest residues of plant and animal origin. Apart from these materials, biomass also includes biodegradable organic wastes from industries like sugar mills, breweries etc. Biomass fuel may be solid such as wood, animal dung, peat, charcoal etc. used for burning; liquid such as methanol or ethanol used in internal combustion engines of automobiles; or gas such as animal waste produced biogas: a mixture of gases mainly methane and some carbon dioxide produced in biogas digesters. Biogas is a clean anaerobic fuel of organic origin whose combustion produces fewer pollutants than other combustible energy sources. Although production of energy requires sufficient area of land and water, but in its various forms, biomass energy appears to have a bright future as a source of energy. At least half of the global population relies upon biomass as main source of energy for domestic use. Firewood is most widely used biomass fuel in the world. In Scotland, use of peat for heating and cooking, is quite common.

Hydro Energy or **Water Power**

Water power has been in use since the period of Roman Empire. Water falling from a height turns turbines at the bottom of dams to generate electricity (fig 7.4). Presently water power is widely used all over the world and produces approximately one-fourth of the world's electricity. Electricity produced by hydro-power is much cheaper than electricity produced by thermal power. Water power is a clean energy source as it does not cause environmental pollution and the electricity produced can be transmitted to long distances through wires or cables (fig. 7.5). But building a dam leads to several environmental problems such as submergence of plant and animal habitats and displacement of local inhabitants including tribals.

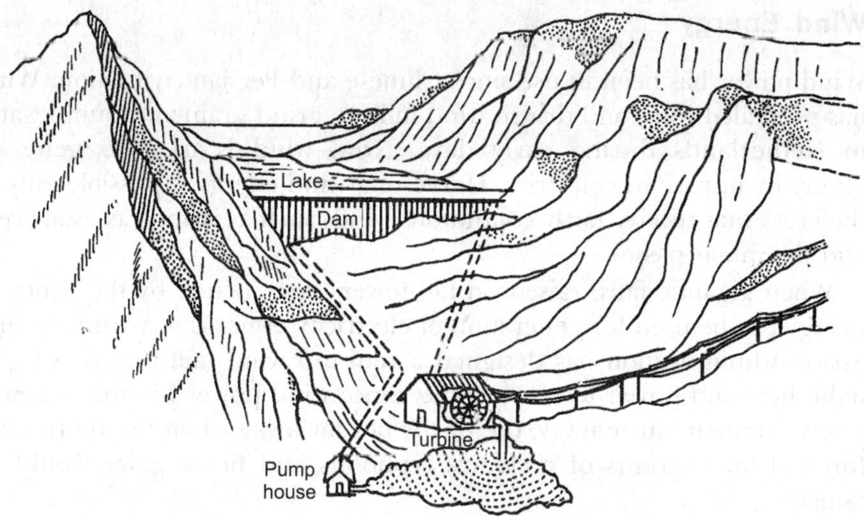

Fig. 7.4. Generation of hydroelectric power.

About 75 per cent of the total electricity consumed in South America is derived from hydro-power. Japan, US and Russia are the leading countries in the production and consumption of hydroelectric power. Canada, Norway, New Zealand, Switzerland and many other European countries have trapped most of their water sources and harnessed energy to generate electricity.

In India, generation of hydroelectric power was emphasised in the First Five Year Plan when a number of multipurpose projects such as Bhakra Nangal project on river Sutluj, Bokaro in Panchet and Talaiya in Damodar Valley and Hirakund, Rihand, Nagarjuna Sagar, Kosi, Koyana were launched to generate hydroelectric power, apart from their use for irrigation and other purposes.

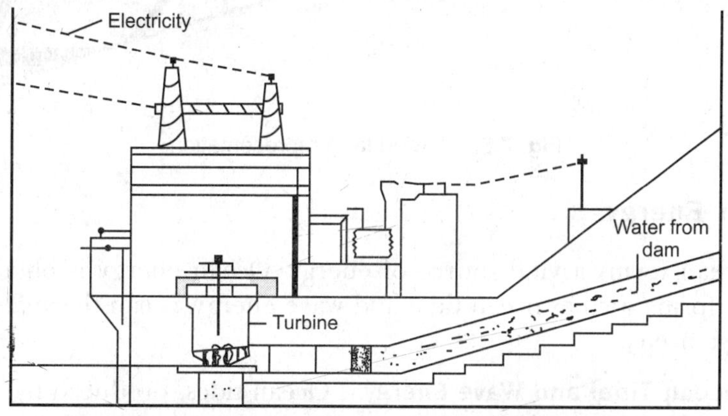

Fig. 7.5. Hydroelectric power plant.

Wind Energy

Wind power has been in use since Chinese and Persian civilisation. Wind has propelled ships and driven wind mills to grind grain and pump water in Netherlands during early days. Now, wind is used to generate electricity in various countries. Harvesting wind energy is possible only in the areas that receive fairly continuous winds such as islands, coastal areas and mountain passes.

When gigantic fans, raised on tall towers, are rotated by the wind, its energy can be used for generation of electricity. American Astronauts and Space Administration has designed a roof top wind mill that could give light, heat and power to every home. Since wind power has the potential to give tremendous energy, the day is not far away when the destructive force of air currents of cyclones, typhoons and fierce gales would be tamed.

India occupies the fifth position, after Germany, USA, Denmark and Spain, in wind power generation adding 2483 MW to the power sector. Wind mills are used in Rajasthan to draw subsoil water for irrigation. Recently two wind farms of 10 MW each have been established in Tamil Nadu and Gujarat with international cooperation from Denmark (fig. 7.6).

Fig. 7.6. A wind farm representation.

Ocean Energy

The ocean forms a vital source of energy. Ocean energy is obtained in various forms such as ocean tidal and wave energy, ocean thermal energy conversion etc.

(i) Ocean Tidal and Wave Energy. Ocean tides, produced by gravitational force of sun and moon, contain enormous amount of energy. The high tide and low tide refer to the rise and fall, respectively of the water in the oceans. Tidal energy is harnessed by construction of a tidal barrage. Dams built across the mouth of a river confluence with oceans, permits sea water to flow through small opening filled with propellers, connected to electric turbines. During high tide the sea water flows into the reservoir of barrage and turns the turbines which in turn produces electricity by rotating the generators (fig. 7.7 A). During low tide, when the sea level is low, the sea water stored in the barrage reservoir flows out into the sea and again turns the turbines (fig. 7.7. B).

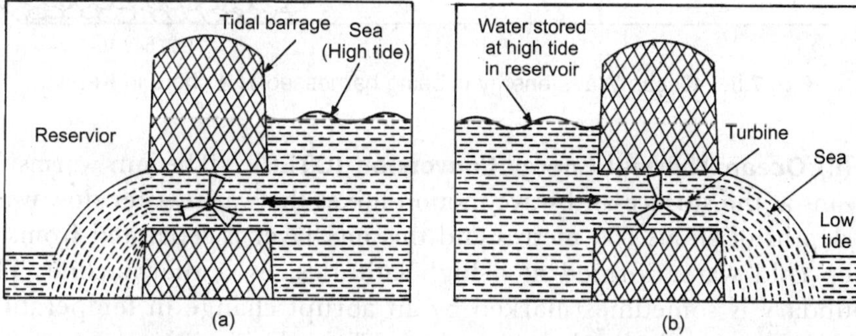

Fig. 7.7. **(a)** Water flows into the reservoir to turn the turbine at high tide, and **(b)** Water flows out from the reservoir to the sea, again turning the turbine at low tide.

In India efforts are being made to identify suitable sites along coastline to explore the possibility of utilisation of tidal energy to produce electricity. A feasibility report for setting up a tidal power project at Durgaduani Creek in Sundarbans area of West Bengal, has been examined. It is found that current technology levels can not produce electricity at economically acceptable rates. The tidal power potential of India is estimated to be about 15,000 MW. The potential sites identified are Gulf of Kutch (1,000 MW), Sundarbans (100 MW) and Andaman and Nicobar Islands, Lakshdweep Islands, the coasts of Orissa, Kerala, Tamil Nadu, Karnataka and Maharashtra.

Power of ocean waves, which operates on the principle of oscillating water column, has not been exploited to its full potential execpt as power supplies for navigational aids. India has initiated the wave energy project off the Vizhinijam Fishery Harbour near Trivendrum in Kerala as an indigenous effort (fig 7.8). It was expected that on the completion, the project would be able to derive an energy output of 4.45 lakh units per year. The project resulted in a strict reality in 1991 when it started generation of electricity to be fed to the grid of Kerala State E'ectricity Board.

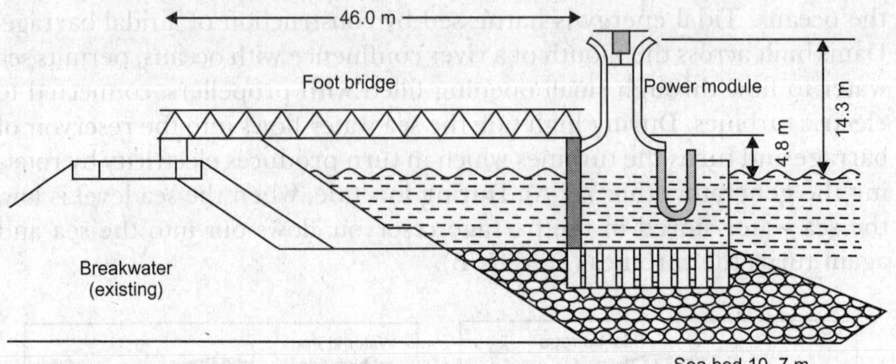

Fig. 7.8. Ocean Wave energy is being harnessed at a shore in Kerala.

(ii) Ocean Thermal Energy Conversion (OTEC). The sun warms the oceans at the surface and wave motion mixes the warm water downward to depths of about 100 metres and this mix layer is separated from the deep cold water formed at a high latitudes, by a thermocline. This boundary is sometimes marked by an abrupt change in temperature, more often the change being gradual. Thus, the resulting temperature distribution consists of two layers separated by an interface with temperature differences between them ranging from 10°C to 30°C. The higher values are found in equatorial waters. And these two enormous reservoirs in some oceanic regions, provide the heat source and heat sink required to operate heat engine (fig. 7.9). The engine using this energy is referred to as OTEC which makes use of the difference in temperature between the two layers of the sea to harness energy which in turn is used to drive turbines for generating electricity. OTEC can also be used to desalinate water, support deep water mariculture and provide refrigeration and air conditioning facilities and can prove as an aid to mineral extraction.

The concept of OTEC was first demonstrated in 1979 when a small plant mounted on a barge off Hawai, USA produced 50 KW gross power. India possesses a huge potential of OTEC which could be of the order of about 500,000 MW, about 150 per cent of the present total installed power generating capacity of the country. Some of the best global OTEC sites are situated off the Indian mainland and near the islands of Lakshadweep, Andaman and Nicobar. In India, an ocean energy cell has been developed at IIT, Chennai to keep pace with the international developments in this area. A US company, M/S Sea Solar Power Inc., is promoting the use of OTEC and the world's first plant in India is proposed off the coast of Tamil Nadu with capacity of 100 MW.

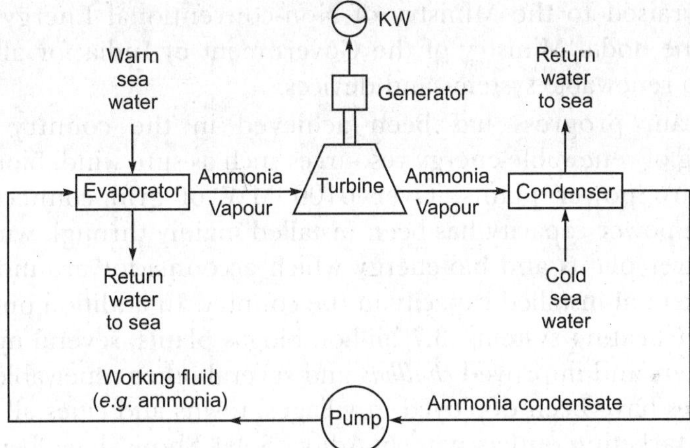

Fig. 7.9. Closed cycle (Rankine cycle) OTEC system.

Geothermal Energy

The natural heat from the interior of the earth can usefully be converted into energy. This natural heat comes from the fission of radioactive material present in the rocks in the interior of the earth. The idea of harnessing geothermal energy, utilising dry steam, was developed in Italy as early as in 1904. Natural internal heat of the earth was harnessed by geo-hydro-thermal conversion, hot igneous and geo-pressured systems. Presently, there are several geothermal plants working successfully in USA, New Zealand, Russia, Japan, Mexico and California. Heated groundwater flowing upwards as hot water stream and hot springs-the nature's geysers, can be used to turn turbines and generate electricity.

In India, natural geysers are common in Kullu and Manali, Manikarn, Sohana and some other places. Assessment of geothermal energy potential of selected sites in Jammu and Kashmir, Himanchal Pradesh, Uttarakhand, Jharkhand and Chhattisgarh is being undertaken. Some more field investigations, including deep drilling at potential geothermal sites, would be required before these site can be taken up for development for geopower generation.

INDIAN SCENARIO

Importance of renewable (non-conventional) energy sources was recognised in the country in the early 1970s. Today, India has large programme for renewable energy. The activities cover all major renewable energy sources, such as biomass(biogas), solar, wind, ocean, small hydro power plants and other emerging technologies. Several renewable energy systems and devices are now commercially available. The Department of Non-conventional Energy Sources, set up in 1982, has been raised to the Ministry of Non-conventional Energy Sources (MNES)-the nodal Ministry of the Government of India for all matters relating to renewable systems and devices.

Significant progress has been achieved in the country towards harnessing of renewable energy resources such as sun, wind, biomass and small hydro power plants. Over 6100 MW of grid connected with renewable power capacity has been installed mainly through wind, small hydro power plants and bio-energy which accounts for around 5.5 per cent of the total installed capacity in the country. In addition one million solar water heating systems, 3.7 million biogas plants, several millions of solar cookers and improved *chullhas* and several other renewable systems and devices have been deployed in villages, towns and cities all over the country. Marketing outlets namely Aditya Solar Shops, have been set up in major cities and towns with a view to promote sale, servicing and repair of these renewable energy systems and devices. During last two decades, several renewable energy technologies have been deployed in rural and urban areas. Some of the estimated potential and achievements, in this area, in India are given below (table 7.1).

TABLE 7.1. Renewable Energy Potential and Achievements in India

No.	*Source/System*	*Estimated Potential*	*Achievements (as on 31 March 2005)*
(A)	**Power from Renewable Systems**		
1.	Solar Photovoltaic Power	–	2.72 MW
2.	Wind Power	45,000 MW	2,483.00 MW
3.	Small Hydro Power (up to 25 MW)	15,000 MW	1,705.63 MW
4.	Biomass Co-generation Power	19,500MW*	613.43 MW
5.	Biomass Gasifier	-	60.20 MW
6.	Energy Recovery from Wastes	2,700 MW	41.43 MW
	Power from Renewables (Total)	82,200 MW	4,906.41 MW
(B)	**Decentralised Energy Systems**		
7.	Family-size Biogas Plants	12 million	3.77 million.
8.	Community/Institutional Biogas Plants	–	3,950 Nos.
9.	Improved Wood-stoves	120 million	35.2 million
10.	Solar Photovoltaic Systems	20 MW /sq km	
	(i) Solar Street Lighting Systems	–	52,129 Nos.
	(ii) Home Lighting Systems	–	3,18,522 Nos.
	(iii) Solar Lanterns	–	5,38,718 Nos.
	(iv) SPV Power Plants	–	930.0 KWp
11.	Solar Water Heating Systems	140 million sq m Collect or area	1.00 million sq m Collector area
12.	Solar Cookers	–	5,80,000 Nos.

13.	Solar PV Pumps	–	6,774 Nos.
14.	Wind Pumps	–	1015 Nos.
15.	Hybrid Systems	–	386 Nos
16.	Battery Operated Vehicles	–	300 Nos.

sq km = square kilometer sq m = square meter MW = Mega Watt
KW = Kilowatt
KWp = Kilo Watt peak

Government has established many Solar Energy Centres, Sardar Swaran Sing National Institute of Renewable Energy and Centre for Wind Energy Technology for research and development and related applications.

NON-RENEWABLE (CONVENTIONAL) ENERGY SOURCES

Non-renewable energy sources are those natural resources which are exhaustible and can not be replaced once they are used. These are available in limited amount and are replenished over a long period. Fossil fuels- coal, oil and natural gas and nuclear fuels are the non-renewable energy sources. Fossil fuels are formed by the incomplete biological decomposition of the remains of the plants and animals which got buried under the earth's crusts millions of years ago. Since their stock is limited and they take a long time to be formed, fossil fuels become precious and need to be used with great care otherwise we will lose them very soon.

Industrial growth in the world today is totally dependant upon fossil fuels. According to one estimate, from the initial reserve of fossil fuels which was about 64,000 million kilowatt hours, upto 1970, about 1,600 million kilowatt hours of energy has been consumed. Among fossil fuels include 95 per cent coal, 4 per cent oil and 1 per cent natural gas. Oil and gas account for 65 per cent of energy consumption in the world today. Further energy consumption is increasing at the rate of 60 per cent per decade. The awareness of this fuel crisis has resulted in the periodic price hikes by the oil producing countries. Likewise world reserves of uranium and thorium, used for producing nuclear energy, are also limited and will eventually be exhausted.

Coal

Coal was formed around 255-350 million years ago, during the Carboniferous age, in hot damp regions of the earth. The plants and animals that occurred during this period, along the banks of rivers and swamps, got buried alive or after their death into the soil and due to heat accompanied by pressures gradually got converted into peat and coal over a millions of years of period. Partially decomposed vegetation deeply buried in sedimentary environments slowly transformed into solid, brittle, carbonaceous rocks commonly known as coal. The coal is the most abundant fossil fuel with a total recoverable resource of about 6,000 billion tonnes in the world. With present rate of consumption, the coal reserves are likely to last during next 200 years and if the use rate increases by 2 per cent per year, then it will last within next 65 years. The coal is mainly of three types: anthracite (hard coal), bituminous (soft coal) and lignite (brown coal). The anthracite coal has maximum amount of carbon (90 per cent) and has highest energy contention, *i.e.*, a caloric value of 8700 k cal/ gm. The bituminous and lignite coal and the peat contain 60 per cent of coal and a caloric value of 8,070 k cal /gm.

China is the leading producer of coal. USA is the second largest producer of coal. In India, coal occurs in rock sequences mainly of two geological ages, namely, Gondwana, little over 200 million years in age and in Tertiary deposits which were found at a much later geological epoch (about 55 million years) ago. The major resources are located in central and eastern parts of the country. In India, the major coal fields are Raniganj, Jharia, Bokaro, Singrauli and Godavari valley in states of Jharkhand, Orissa, West Bengal, Madhya Pradesh, Andhra Pradesh and Maharashtra. The proven coal reserves of the country as on January 1, 2005 is 247,847 million tonnes, the highest proven amount being 35417 million tonnes in Jharkhand. The coal production in India from 1985 to 1999, is shown in the fig. 7.10. The total coal production in the country during the year 2004-05 was about 376.63 million tonnes by Coal India Limited, Singrauli Collieries Company Limited and captive collieries. About 65 per cent of coal produced in India is used to produce electricity and rest in industries and for other purposes.

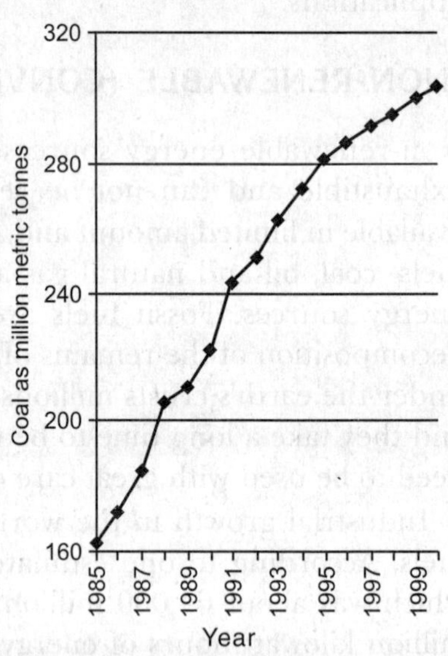

Fig. 7.10. Coal production in India from 1985-99.

On buring, coal produces heat which is converted into various forms of energy by different technologies. At the same time it produces several air pollutants including carbon dioxide which is a green house gas and causes global warming. Besides, there are many problems with coal use. It has to be mined from underground or from the surface. Underground mines, besides being dangerous, cause several damages to the environment and lung diseases in the miners. In addition, coal releases huge amount of radioactive particles into the atmosphere which are harmful to human beings.

Crude Oil

It is believed that petroleum has been formed over a period of millions of years through conversion of remains of microorganisms, living in sea, in hydrocarbon by heat, pressure and catalytic action. Extraction of crude oil from the bowels of the earth is not so simple. When a new oil field is tapped, the oil gushes out, but not for long and one has to wait for the oil to slowly seep into the well from the surrounding rocks which provide maximum not more than 10 per cent annual production. Crude oil or petroleum, the lifeline of global economy, is contained mainly in 11 countries forming OPEC (Organisation of the Petroleum Exporting Countries). About 70 per cent of the global petroleum reserves occur in these countries. Of these, Saudi Arabia alone has one fourth of the oil reserves. OPEC was formed in 1960 with the purpose of jointly regulating the production and the control of the price of the exported oil. Currently it has country members namely Algeria, Indonesia, Iran, Iraq, Kuwait, Libya, Nigeria, Quatar, Saudi Arabia, UAE and Venezuela. OPEC discussions have a major impact on world oil price which largely depends on political factors, global demand and the production policies of the organization. Per barrel oil price has increased from US$ 4 in 1973 to US$ 50 in 2004. Such a price rise directly affects transportation cost and leads to inflation in many countries including India. It is expected that world production of oil (except Middle East) will be in peak in 28-30 years and if consumption rate increases within 8-10 years only, the peak will occur between 2010 and 2020.

Petroleum is a cleaner fuel as compared to coal as it burns completely and leaves no residue. It is also easier to transport and use. Petroleum is largely used in transportation, operating water lifting engines, generators etc.

In India, crude oil was first recovered from Makum in North East Assam. Later, drilling for crude oil was done at Digboi, Dibrugarh, Narharkatiya and Surma valley in the north east. The oil field also lies around Bay of Cambay, Gujarat. The most important achievement was the exploration of oil in Bombay High on the continental shelf of Maharashtra, located at a distance of 167 km north-west of Mumbai. Recently oil has been located in the off-shore areas of deltaic coasts of Godavari, Krishna, Cauvery and Mahanadi. Oil prospects in India are not so high as coal. But the demands are very high and the country has to import oil from OPFC countries at higher rates.

Indian oil reserves are about 4.45 billion tonnes out of which about 1257 million tonnes are recoverable. Since 483 million tonnes have been extracted, the recoverable balance is now about 774 million tonnes which shall last within next 25 years with the condition that crude oil production rate of about 30 million tonnes per year is maintained. In India, the Ministry of Petroleum and Natural Gas looks after exploration and production of oil and natural gas (including Liquefied Natural Gas-LNG), refining, distribution, marketing, import and export and conservation of petroleum products. Keeping in view, the oil demand scenario vis-a-vis domestic production level, the Government is encouraging oil sector. The India is a member of International Energy Forum (IEF), with permanent secretariat in Riyadh, which provides a platform for biennial meeting of the ministers from the energy producing and consuming countries. Two national oil companies, Oil and Natural Gas Corporation Limited (ONGC) and Oil India Limited (OIL) and private and joint venture companies are engaged in the exploration and production of oil and natural gas in the country. Crude oil production by these organisations during the year 2003-04, was 30.96 MMT (Million Metric Tonnes) and for the year 2004-05, the target was set at 33.64 MMT. The natural gas production during 2003-04 was 30.96 Billion Cubic Metres (BCM) and for the year 2004-05 the target was set at 31.07 BCM. The total quantity of oil imported during 2003-04 was 90.43 MMT of crude oil and 7.897 MMT of petroleum products valued at 83,528 crore and 9677 crore, respectively. In the same period 14,620 MMT of petroleum products valued at 16,781 crore were exported.

Natural Gas

Natural gas, a fossil fuel gift from nature, is composed of methane (96 per cent) with small amounts of propane and ethane. Natural gas deposits often accompany oil deposits or may occur independently. It is the cleanest source of energy among fossil fuels. Natural gas can easily be transported through pipelines. It has a high calorific value and burns without any smoke. Natural gas can be used as a source of energy for domestic or industrial use. It can also be used for power generation and as a raw material for petrochemical industries and fertilizer plants. It results as a byproduct during crude oil refining and from fractional distillation plants. About 40 per cent of total natural gas is found in Kazakhistan, Russia.

India has a huge reserve of natural gas of which a large amount flares up due to lack of adequate storage, compression and transportation facilities as a result about 17 million cubic meters of gas a day is wasted or burnt. Now the gas is distributed from Bombay High to Rajasthan,

Gujarat, Madhya Pradesh and Uttar Pradesh by a 1730 km pipeline, the Hazira-Vijaipur-Jagdishpur pipeline. A similar pipeline is proposed for South India to feed the natural gas of Bombay High and the gas imported from West Asia to southern states. A gas grid is also proposed for Assam. Gas Authority of India Limited, a 'Navratna' enterprise established in 1984, is primarily an integral National Gas Company which focuses on all aspects of the gas supply and value chain including exploration, production, transmission, petrochemical, processing, distribution and marketing of natural gas and other related products and services. In 1994, India produced 2.8 million tonnes of natural gas against a demand of 3.4 million tonnes. This deficit in 2001-02, is now gone upto to 3.30 million tonnes.

LPG: Liquidified Petroleum Gas (LPG), widely used as a domestic fuel for cooking, has its main content as oduorless butane to which other gases like propane and ethyl mercaptan are added to give fowl smell to identify leakage. It is obtained by converting petroleum into liquid form under pressure. Indane and Bharat Petroleum are the chief distributing agencies of LPG.

CNG: Compressed Natural Gas (CNG) is used as an alternative to petrol and diesal for transport of vehicles. Delhi transport has totally switched over to CNG where buses and auto rikshaws run on this fuel. It has greatly reduced vehicular pollution in the city. CNG is a cleaner fuel than diesal, used currently in many cities and long distance transport across the country. It contains mostly methane, compressed to 80 atmosphere. LPG emits negligible levels of particles which diesal does and even those emitted are not as toxic as those emitted from diesal. It is an ideal cleanest burning alternative fuel. CNG also works out cheaper (one-third) than diesal in long run because of stable price. Moreover, it is readily available, its carcinogenic potential is lesser, it can not be adulterated and gives higher mileage, *i.e.,* 35-40 km per kg.

Shale Oil: It is a fine grained sedimentary rock which contains organic matter. Upon heating, shale oil gives significant amount of hydrocarbons that are otherwise insoluble. Total global shale oil resources are estimated to contain about 3 trillion barrels of oil of which US alone possesses 2 trillion barrels.

Tar Sands: They are sedimentary rocks or sands impregnated with tar oil, asphalt or bitumen from which tar oil can not be recovered by usual commercial methods such as by wells because the oil is too viscous to flow easily. The total global resource is 1000 billion barrels, of which seventy five per cent deposits are found in Canada.

Nuclear Energy

Known for its high destructive value, as evidenced from nuclear weapons, non-renewable nuclear power can also be harnessed to produce energy of commercial value. Nuclear energy can be generated either by nuclear fission in which nucleus of certain isotopes with large mass number is splitted into lighter nuclei on bombardment of neutrons in order to release a huge amount of energy through a chain reaction or by nuclear fusion in which two isotopes of a light element are forced to form a heavier nucleus releasing enormous energy in the process. The heat energy produced as a result of either of the processes is used to produce steam which runs electric turbine (fig. 7.11). The process of nuclear fission is difficult to initiate but releases more energy than nuclear fusion. Nuclear energy has a tremendous potential but any leakage from the nuclear reactor may cause devastating nuclear pollution including world's most hazardous Chernobyl disaster. Disposal of nuclear waste is also a troublesome process. Uranium, a rare element, is the primary source of nuclear energy although thorium can also be converted into uranium isotope and used as a fuel. Today, nuclear energy accounts for about 6 per cent of total commercial energy.

In India, the genesis of nuclear power can be traced back to 1945 when Tata Institute of Fundamental Research was set up near Mumbai under the leadership of Homi J Bhaba. The Atomic Energy Commission was set up in 1948. Later, in 1954, Department of Atomic Energy was set up to look after all activities related to nuclear energy. The first nuclear power station was raised at Tarapore, near Mumbai in 1969. The nuclear power is still not well developed as there are only five nuclear plants, namely, Narora (UP), Rawatbhata (Rajasthan), Kakrapar, Tarapur (Maharashtra) and Kalpakkam (Tamil Nadu). At present, country produces 2250 MW

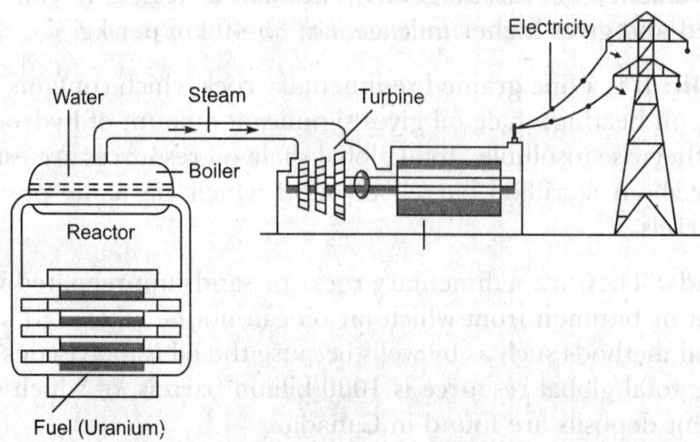

Fig. 7.11. Nuclear power reactor generating electricity.

(2.6 per cent of the country's total installed power generation capacity) from these nuclear power generation plants.

The primary source for nuclear energy generation, the uranium and thorium, are non-renewable rare elements. Their stock is very limited. United States and Canada are the leading producers of uranium but uranium also occurs in Russia, Germany, Spain, Sweden, South Africa, Australia etc. Thorium occurs in Brazil, Australia, Malaysia, Sri Lanka etc. In India, uranium deposits are found in Gaya and Singhbhum districts of Bihar; Udaipur and Jaipur districts of Rajasthan; Nellore district of Andhra Pradesh; and Palghat district of Kerala. Reserves of thorium are found in Rajasthan and Andhra Pradesh and coastal parts of India.

ALTERNATE EMERGING ENERGY SOURCES FOR THE FUTURE

The future scenario of energy supply appears to be grim as it is not yet clear as to how the transition from fossil fuels to other energy sources will be accomplished. Energy policy in developing countries today seems to follow *hard path, i.e.,* leading to further development of the so called hard technologies which involve in finding greater amounts of fossil fuels and building even larger power plants and are continuing as they have been for a number of years. This hard path seems to be more comfortable to them as it requires no new thinking or realignment of political, social and economic conditions. The other way, the *soft path* involves renewable energy alternatives that are flexible and do not harm the environment. The renewable energy alternate sources include sun, ocean (tidal), wind and biomass. These sources are diverse and individually tailored for maximum effectiveness under specific conditions and are flexible, *i.e.,* based on relatively simple technologies that are accessible and understood by many people. They are matched with geographical distribution and actual use of energy and there is good matching between quality and actual use of energy. With the approaching depletion of coal, oil and natural gas the scientists were forced to develop alternate sources of energy that should be renewable and free from pollution.

In India, since 1970, attempt has been made to concentrate on exploitation of vast potential of alternate energy sources. These sources can augment power in specific areas in decentralised manner, especially for local use. Indian government has always been effortful to develop various methods to generate energy from alternate sources. Alternate energy programmes have been developed on a priority basis to cope with the increased demand of commercial energy and pressure on foreign exchange reserves by the increasing use of commercial fuels. This programme seeks to develop new sources of renewable energy besides the existing ones. In order to improve energy situations in rural areas, the national programme on biogas development, waste recycling resource recovery systems, hydrogen power, fuel cells, alcohol etc., are being developed, which besides creating energy, will help improve sanitary conditions in villages and cities.

Indian Renewable Energy Development Agency (IREDA), established under Ministry of Non-conventional Energy Sources, Government of India, looks after the promotion of renewable sources of energy. It operates a revolving fund for promotion and development; assists in rapid commercialisation technologies, systems or devices; helps in upgradation of technologies; and extends financial support to the energy efficiency and conservation projects and schemes in the area of renewable resources of energy. Some of these schemes of IREDA in operation are biomass power generation, biomass gasifiers, biomethanation of industrial effluents, small scale biogas plants etc. The currently used emerging technologies, systems or devices in renewable energy generation in India are discussed here.

Hydrogen Energy

The hydrogen burns in the air, combines with oxygen to form water and releases a large amount of energy (150 KJ per gram). Taking into consideration of this large amount of the energy produced in the process, hydrogen is used as an ideal source of fuel. Moreover, hydrogen is non-polluting and can be readily produced by thermal dissociation and photolysis or electrolysis of water. Involving thermal dissociation hydrogen (H_2) can be obtained directly from water at 3000°C or above or thermo-chemically whereby hydrogen is produced by chemical reaction of water with some other chemicals in 2-3 cycles, not needing high temperature as in direct thermal dissociation. Photolysis of water involves splitting of water in presence of sunlight to release hydrogen, as is done by green plants in photosynthesis, and efforts are being made to trap this hydrogen molecule. The electrolytic method dissociates water in hydrogen (H_2) and oxygen by making a current flow through it.

In nature, hydrogen is inflammable and explosive and thus using it as fuel requires safe handling. Moreover it is difficult to transport and store in bulk as it is very light. Presently hydrogen is used as a fuel in spaceships in the form of liquid hydrogen. A large-scale introduction of hydrogen as a fuel would reduce consumption of fossil fuels and keep the air clean and free from pollution. The hydrogen as a source of energy is advantageous in that it is a clean fuel that can be produced from water and used for broad range of applications. India has undertaken a programme covering research and development pertaining to production of hydrogen, its storage, safety, applications etc. with a view to create an alternate source of energy. A National Hydrogen Energy Board has been set up to guide and oversee the preparation of a Hydrogen Energy Road Map and its implementation through a National Programme on Hydrogen Energy. Applications of hydrogen directly in internal combustion engines for use in transport, in the fuel cells, for stationary and mobile and for decentralized power generation have been demonstrated. Hydrogen powered two wheelers, three wheelers, catalytic combustors and power generating sets have been developed and demonstrated. A pilot project for field testing hydrogen fueled motor cycle is under progress. A pilot plant for production of hydrogen from distillary waste has also been set up.

Fuel Cells

Fuel cells are electrochemical devices that convert the chemical energy of a fuel directly and very efficiently into electricity and heat in combination. The fuel cells are used for power generation, vehicular applications, uninterrupted power supply etc. They have been used in space flight and for combined supply of heat and power. Fuel cell powered vehicles – the electric vehicles are the best option to check the urban pollution. Fuel cell powered buses are already in operation in Canada and parts of USA. The most suitable fuel for such cells is hydrogen or a mixture of compounds containing hydrogen.

In India, Research and Development Projects for development of different types of fuel cells such as Proton Exchange Membrane Fuel Cells (PEMFC), Phosphoric Acid Fuel Cells (PAFC), Solid Oxide Fuel Cells (SOFC), Direct Methanol Fuel Cells (DMFC), Direct Ethanol Fuel Cells (DEFC) and Molten Carbonate Fuel Cells (MCFC) and component and materials for fuel cells, including control and instrumentation systems, are being supported.

Alcohol as a Source of Energy

The use of alcohol as a fuel for internal combustion engines, either alone or in combination with other fuels, has been given much attention. This is because of its possible environmental and long-term economical advantages over fossil fuels. Both ethane and methane have been used up for this purpose. Despite the fact that both ethanol and methanol can be obtained from petroleum or natural gas, ethanol finds superior place because many believe it to be a renewable source. Proposal for use of alcohol as a fuel may suit to transportation purposes as a partial or total replacement of gasoline in combustion engines in cars, buses etc. But it could not gain much popularity because of other less conventional approaches, such as use of alcohol in fuel cells directly or as feedstock for hydrogen production have been advanced. Fuel alcohol can be produced from a variety of crops such as sugarcane, sugarbeet, maize, barley, potato, cassava, sunflower, eucalyptus etc.

Gasohol, a mixture of ethanol and gasoline is a common fuel used in Brazil and Zimbabawe for running cars and buses. In India, gasohol is proposed to be used in transport vehicles in some parts of the country.

Methanol is produced by bacterial anaerobic digestion of organic waste. Contribution of industrial effluents in the form of methane gas along with carbon dioxide and hydrogen sulphide is also very useful. Since it burns at a very low temperature than gasoline or diesel, methanol is quite useful as a fuel. It would facilitate replacement of bulky radiators by sleek design radiators in the cars. Moreover it is a clean and non-polluting fuel.

Ethanol, can be derived from corn, wheat, potato wastes, cheese, whay, rice straw, urban wastes, paper mill wastes, yard clippings, molasses, sugarcane, seaweed, surplus food crops and other cellulose wastes. Industrial ethanol is produced from petroleum. Ethanol, being the same chemical as the alcohol in alcoholic beverages, can reach 96 per cent purity by volume by distillation and is as clear as water. To be used as fuel, nearly 100 per cent pure ethanol is being produced using additional industrial process. In being flammable, pure ethanol burns more clearly than many other fuels. The combustion products of ethanol are only carbon dioxide and water which are usual by-products of regular cellulose waste decomposition. And only for this reason, ethanol is preferred as a fuel for environmentally conscious transport schemes such as operation of buses.

Energy Plantations

Making into use the photosynthetic activity of green plants in production of biomass, fast growing trees such as cottonwood; poplar; *Leucaena*; herbaceous grasses; crop plants like sugarcane, sweet sorghum and sugar beet; aquatic weeds like water hyacinth and sea weeds; and carbohydrate rich potato, cereals etc. are grown as energy plantations. Energy can be produced from them directly by burning or getting them converted into burnable gas or converting them into fuel by fermentation.

Energy from Agricultural Waste Biomass

Many agricultural wastes such as crop residues, bagasse-the sugarcane residues, coconut shells, peanut hulls, cotton stalks etc., are used to produce energy by burning. Animal dung, fishery and poultry waste and even human refuse can be used to produce energy. In Brazil, about 30 per cent of electricity is obtained from burning bagasse. In India, about 80 per cent of rural heat energy is met by burning animal dung cake, agricultural wastes and wood in *chullahas* which produce enormous amount of smoke. These are not efficient as their efficiency being less than 8 per cent. Development of *Improved Chullahs* wilh tall chimney are smokeless and have more fuel efficiency. The *chullahs*, apart from smoke, produce a lot of ash and waste residue. The burning of dung causes long run loss of soil nutrients like nitrogen and phosphorus. To increase efficiency and reduce pollution, biomass needs to be converted into biogas.

Biogas: In India, successful experiments have been done for the development of gobar gas plants. The history begins with Uttar Pradesh where the first gobar gas plant was set up at Ajitmal in district Auraiya during 1960s. Biogas is the mixture of methane (50-60 per cent), carbon dioxide, hydrogen, nitrogen, hydrogen sulphide, the major component being methane. Biogas is produced by anaerobic degradation of animal wastes and also sometimes plant wastes, in presence of water to undergo biological breakdown of organic matter by certain bacteria in the absence of oxygen. Plenty of animal waste and agricultural waste is produced every year. This can be used to produce biogas as cogenerative source of energy. Generation of biogas is an environmentally clean technology. Moreover, biogas is a low cost fuel which is very useful for rural areas where it proves advantageous in that it is clean, cheap and non-polluting and can be supplied to place of use directly from biogas plant and thus there is no storage problem. The left over sludge is rich in fertilizer containing nutrient rich bacterial biomass. To avoid health hazards occurring in direct use of dung due to direct exposure to faecal pathogens and parasites, the digestion of animal waste is done in air tight gasifires.

Different countries follow different types of biogas generation devices. The Chinese use an underground design. In India, biogas is produced commonly by floating gas holder type of or fixed dome type of biogas plant.

(i) Floating Gas Holder Type Biogas Plant. This type of biogas plant (fig. 7.12) has a well-shaped digester tank made up of bricks and placed

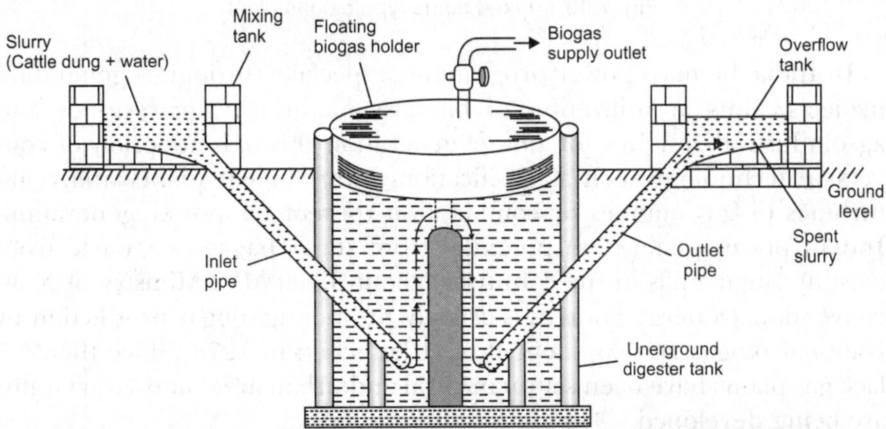

Fig. 7.12. Floating gas holder type biogas plant.

underground. An inverted steel drum is floated over the dung slurry to hold the biogas produced. In order to check movement of gas holder, a gas outlet pipe is placed at its mid top and the gas outlet is regulated by a valve. The digester tank is partioned so that one side of it can receive the dung water mixture through inlet pipe and the other side discharge the spent slurry through outlet pipe. After long use often steel gas holder is subjected to corrosion causing leakage of biogas which can be controlled by patch sealing. The tank requires time to time painting which increases its cost and hence another device is designed.

(ii) Fixed Dome Type Biogas Plant. Structural plan of this type of biogas plant (fig. 7.13) is basically the same as of the floating gas holder type, but instead a steel gas holder, there is a fixed dome-shaped roof of cement and bricks. There being no partioning, the plant is a single unit in the main digester tank but it has inlet and outlet chambers attached laterally. (fig. 7.13).

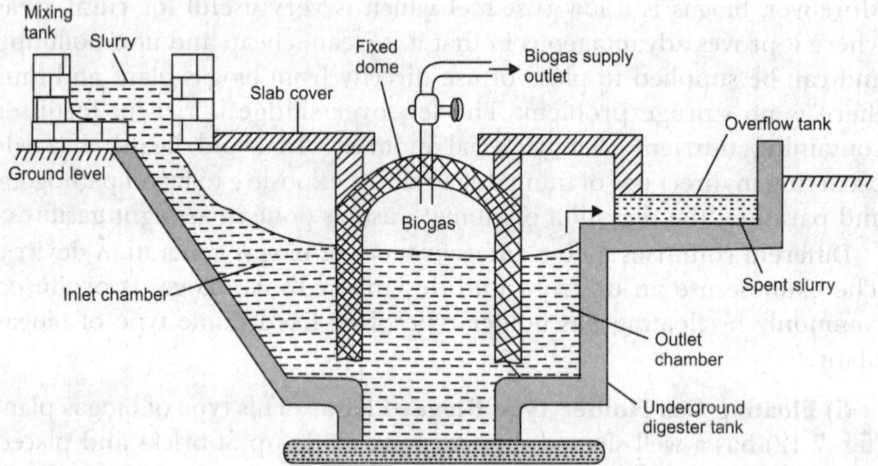

Fig. 7.13. Fixed dome type biogas plant.

In India, biomass power programme, especially through cogeneration projects, aims at utilization of biomass materials, agroresidues and agroindustrial residues for power generation through adoption of conversion technology such as gasification. Large biogas plants utilize gas turbines or gas engines or combination thereof for power generation. Indian power generation potential from sugar bagasse- a waste from existing sugar mills in the country is about 3,500 MW. Ministry of Non-conventional Energy Sources is actively promoting biogas production by National Biogas Development Project launched in 1978. Since then 6.5 lack gas plants have been set up in rural and urban areas and Urja Grams are being developed.

Energy from Urban Waste

Urban waste can also be utilised for production of energy, which with the help of turbines, can be converted into electricity. A pilot plant for demonstration, already been set up in Delhi to treat solid municipal waste, is producing around 4 MW energy every year. Sewage generated biogas is being produced at Okhla, Delhi.

Dendrothermal Energy

Plantation of fast growing shrubs and trees on denuded wasteland can also be a good source of energy. These energy rich plants can be grown for fuel wood, charcoal etc. The method has been experimented in Philippines and many other countries. In India, attempts are being made in Rajasthan and Madhya Pradesh for energy plantations because in these areas only such plants can be grown which can thrive even in dry conditions and cater the energy needs of local people.

Petroplants (Petrocrops)

Attempts are being made to identify potential plant species as source of liquid hydrocarbons as a substitute for liquid fuels. The hydrocarbons present in such plants can be converted into petroleum hydrocarbons. About 385 species of latex yielding plants belonging to Euphorbiaceae, Asclepiadaceae, Apocynaceae, Urticaceae, Convolvulaceae and Sapotaceae have been screened for hydrocarbon contents. Fifteen species have been identified suitable for the purpose. There is an urgent need to increase the biomass of these potential plant species by raising them as petrocrops and their hydrocarbons into petroleum fractions. Certain 'Euphorbias' and oil palms are rich in hydrocarbons and can yield an oil like substance under high temperature and pressure. The oily material of *Jatropha* can be burned in diesel engines directly or may be refined to form gasoline. Indian Institute of Petroleum, Dehradun, has made a significant contribution in hydrocracking of the crude products and derived a conclusion that the product obtained from the latex of petroplants possesses many compounds present in petrol hydrocarbons.

Biodiesel

Biodiesel production, based on the concept of using vegetable oil as fuel, was adopted as early as 1895 when Rudolf Diesel developed the first diesel engine to run on vegetable oil. In many European countries, to the fossil diesel is mixed about 25 per cent biodiesel. France is the leading country in production of biodiesel.

In India, first diesel engine was run in 1930 by British Institute of Standards in Kolkata using eleven non-edible oils. In 1986, IIT, Delhi started research work on biodiesel with different species of forest trees like 'Mahua' and 'Neem'. Thereafter, many plant species have been tried upon by different organizations and a list of more than 100 plants species has been finalised but planning commission of India has recommended two species jatropha (*Jatropha curcas*) and 'Karanj' (*Pongamia pinnata*) for biodiesel extraction. In 1994, Biodiesel Development Board came into existence in India. It is estimated that India will be able to produce 280 metric tonnes of biodiesel by 2012 which will supplement about 41.14 per cent of the total demand of diesel consumption in the country. The Planning Commission of India has launched a biodiesel project in 200 districts in 18 States of India where 'Jatropha' will be cultivated. Apart from biodiesel, more than 19 lakh rural people will get the opportunity of employment. Biodiesel is important not only in affecting savings in the economy but also for keeping our environment pollution free. Advantages of biodiesel are many.

1. It is the only alternative fuel that runs in may conventional diesel engines.
2. It can be used alone or mixed in any ratio with petroleum diesel fuel, the most common blend being 'B2O'-20 per cent biodiesel with 80 per cent petroleum diesel.
3. It produces about 80 per cent less carbon dioxide emission and almost 100 per cent less sulphur dioxide.
4. It is 11 per cent oxygen by weight and contains no sulphur.
5. It extends the life of diesel engines because it is more lubricating than petroleum diesel fuel.
6. It is biodegradable and is thus safe to handle or transport.
7. It has a high flash point of about 300°F compared to petroleum diesel fuel which has flash point of 125°F.

Power Development in India

Commercial power development in India started in 1897 with the commissioning of electricity supply in Darjeeling. Later on the power commissioning was made in other parts of the country. As on March 31, 2006 the power generation capacity of the country was 1,12,058.42 MW (77,968.5 MW thermal, 29,500.23 MW hydro, 1,869.66 MW wind and 2,720 MW nuclear). A capacity addition programme was fixed for the year 2004-05. Importance of increasing the use of renewable energy sources was recognized in the country as far back as 1970s. The country has developed a large programme for renewable energy from sources such as biogas, biomass, sun, wind, small hydro power and other emerging technologies. The Ministry of Non-conventional Energy Sources is now actively engaged in the development of newer renewable sources of energy such as biodiesel, hydrogen power etc.

Chapter Summary

The term 'energy' was coined by Thomas Young (1737–1829) and applied to what it is now called kinetic energy. Energy is an essential input for sustenance of life of all living organisms, improvement of quality of life of man and for economic development of a country. Infact energy is the backbone of prosperity and progress of a nation. Growing energy needs due to rise in population growth coupled with raised standard of living have widened the gap between the demand and supply. Presently development in various sectors relies largely upon energy. United States and Canada are the highest consumer of energy. Energy sources fall under two categories, *viz.,* renewable and non-renewable energy sources. Among renewable energy sources (non-conventional energy sources) are solar energy (direct or indirect), hydro energy, wind energy, ocean energy (tidal and wave and OTEC), geothermal energy etc. India has a great potential of renewable energy and has set up the Department of Non-conventional Energy Sources which has made a significant progress in this area. Among non-renewable energy sources (conventional energy sources) are included exhaustible sources such as coal, oil, natural gas on which the industrial growth of many countries depend. But the stock of coal, oil and natural gas is limited and is likely to be exhausted within few decades. Moreover their use causes pollution and brings about deterioration in the quality of the environment. Nuclear energy is a newer and cleaner source of energy but the stock of the fuel uranium is limited and disposal of nuclear wastes is a tedious problem. Moreover, nuclear hazards such as Chernobyl, are likely to occur. India has about half a dozen nuclear power stations, the first being established at Tarapore. India is now promoting many alternate renewable energy sources such as biogas energy, hydrogen energy, alcohol energy, agricultural and urban waste energy, biodiesel etc. In India commercial power generation started in 1897 with the commissioning of electricity supply in Darjeeling. At present the power generation capacity of the country is about 1,12,058.42 MW.

Study Questions

1. Write an account of the growing energy needs with special reference to India.
2. Discuss briefly the various renewable (non-conventional) energy sources and energy production processes.
3. Throw light on non-renewable energy sources and discuss how energy produced by fossil fuels causes environmental pollution.
4. Describe the uses of new alternate renewable sources of energy in India.
5. Briefly describe:
   - (i) Power Development in India
   - (ii) Nuclear Energy
   - (iii) Biodiesel as a Source of Energy
   - (iv) OPEC
   - (v) Biogas Plant
   - (vi) Solar Cooker
   - (vii) OTEC
   - (viii) Dendrothermal Energy.

Objective Questions: *Select the correct answers*

1. A cleaner source of energy is:
(1) Coal (2) Oil
(3) Natural Gas (4) Hydro Power

2. CNG is preferred over LPG because it is:
(1) Costlier than LPG (2) Lighter than LPG
(3) Easy to transport than LPG (4) Causes less pollution

3. The first nuclear power station established in India is:
(1) Narora (2) Tarapore
(3) Kalpakkam (4) Rawatbhata

4. About 99.8 per cent of our energy comes from:
(1) Electrical energy (2) Solar energy
(3) Thermal energy (4) Tidal energy

5. Largest percentage of energy in developed countries is used for:
(1) Agriculture (2) Domestic use
(3) Industries (4) Transportation

6. Non-renewable source of energy is:
(1) Solar energy (2) Wind energy
(3) Hydro energy (4) Coal energy

7. Major coal resources in India are located in:
(1) Himalayas (2) South India
(3) Central and Eastern India (4) Western India

8. Major component of biogas is:
(1) Butane (2) Propane
(3) Methane (4) Ethane

9. The energy which can reduce global warming is:
(1) Geothermal energy (2) Coal
(3) Petroleum (4) Natural gas

Answers

1. (4) *2.* (4) *3.* (2) *4.* (2) *5.* (3) *6.* (4)
7. (3) *8.* (3) *9.* (1)

8

CHAPTER

Land Resources

LEARNING OBJECTIVES

Introduction • Pattern of Land use • Land as a Resource • Land Degradation and its causes • Man Induced Landslides • Soil Erosion • Prevention of Soil Erosion • Desertification

Introduction

Land is a major constituent of lithosphere and is one of the main components of biosphere, besides air and water. The biosphere forms a natural environment for plants and animals including man. Land forms about 29.1 per cent of the earth surface covering an area of about 148,950,800 sq km. It is the most precious resource because it is put to diverse uses by man and is the source of many materials essential to man and other organisms. Performance of various natural activities need space for their location and development which is provided by land. Many life sustaining biogeochemical cycles take place in the land. Land supports large forest areas and many water bodies such as rivers, lakes, ponds etc. Various purposes, for which man uses land, include agriculture and horticulture for food production; energy production; extraction of minerals, oil, coal, natural gas; human dwellings; commercial and industrial purposes; waste disposal etc. India with a land area of about 2.4 per cent of the global land area, supports about 15 per cent of world's population.

Pattern of Land Use

The pattern of land use varies from country to country. The pattern of land use distribution in the world is 30 per cent forest land, 26 per cent pasture land, 11 per cent cropland and the rest 33 per cent as land like tundra, desert, bare rock, snow etc. In India, the so called agricultural country, more than two-fifth of the land is agricultural land. The pattern of land distribution in India is 43.6 per cent agricultural land and cultivated land, 14.6 per cent permanent pasture and meadows, 12 per cent culturable wasteland, 11.5 per cent forest, 8.0 per cent barren and unculturable land, 5.3 per cent urban land and 5 per cent for which no proper information is available (fig 8.1).

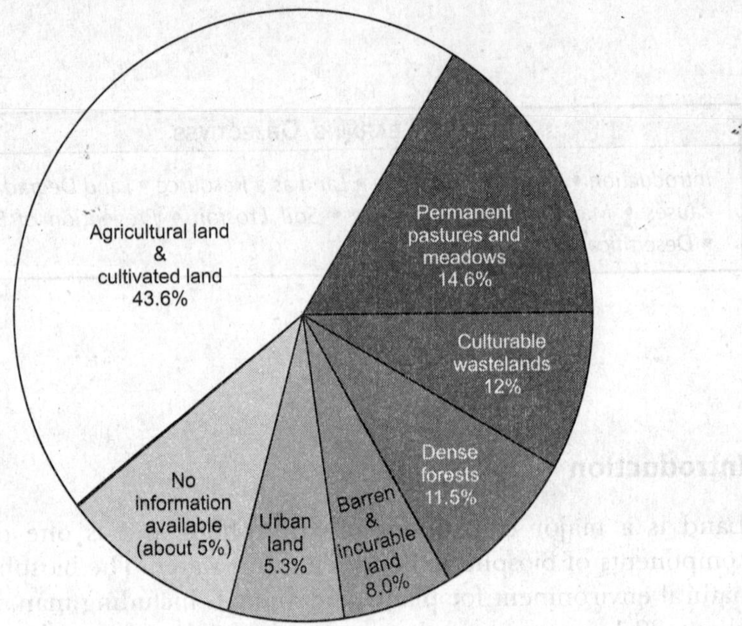

Fig. 8.1. Pattern of land distribution in India (approximates)

Land as a Resource

Land is a storehouse of valuable resources upon which man depends for food, shelter, clothing, movement, security and many other of his needs and greed. The surface layer of the land is called soil. About four-fifth of the land area is covered by the soil. The word soil has been derived from the Latin word *solum* – means upper crust of the earth. The soil in general is defined as the upper layer of the earth which is differentiated into various horizons and is capable of supporting plant life. The top horizon-the fertile region of the soil is classified as a renewable resource because it keeps on continuously regenerating, changing or developing by natural processes at a very slow rate. The soil is formed by wear and tear of the earth's crust, a process which has been going on through the ages. Many biogeochemical cycles operate in the soil due to presence of a variety of microorganisms including many forms of bacteria, fungi, plants and animals. The upper limit of the soil is water or air and the lowest limit is difficult to define but is normally thought of as the lower limit of common rooting depth of native perennial plants, a boundary that is shallow in deserts and tundra and deep in humid tropics.

Soil is a dynamic layer of earth's crust. It takes about 200 – 1000 years to form 2.5 cm of soil, depending upon various factors such as climate, soil type, processes involved in decomposition of parental rocks etc. The soil productivity is determined by many soil factors such as soil texture; availability of water; permeability; soil water porosity; soil pH; organic matter and inorganic nutrient contents; cation exchange capacity; microbial population etc. The topography and biotic factors also play an important role in determining the soil conditions. Soil plays a key role in determining the quality and composition of the biosphere. Infact, biosphere develops over the soil and soil provides nutrition to the plants and abodes the microorganisms. The land is vital to our existence due to the following facts:

1. It preserves terrestrial biodiversity and genetic pool.
2. It regulates water and carbon cycles.
3. It acts a store house of basic resources like ground water, minerals and fossil fuels.
4. It becomes a dump of solid and liquid wastes.
5. It forms a basis for human settlement and transport activities.
6. Its top soil horizon, which is only a few cm thick, supports all plant growth and is hence the life support system for all organisms, including man.

Land Degradation and its Causes

Land degradation refers to deterioration or loss of fertility or productive capacity of the soil. All modern growth oriented activities are directly or indirectly causing impacts on land. According to UN studies, 23 per cent of all usable land, excluding mountains, deserts, polar regions etc. has been degraded to such an extent that its productivity has been affected. The main causes of this are deforestation, water logging, excessive use of chemicals, agricultural mismanagement, soil salinity, fuel wood consumption, overgrazing, planting unsuitable crops, poor crop rotation, poor soil and water management, frequent use of heavy machinery such as tractors, heavy industrialisation and urbanisation.

Rapid rate of increase of human population coupled with man's increased demand of food and other needs, has brought the land under pressure. The per capita land resource is showing marked decrease in many thickly populated countries. In India, it may go down from 0.33 hectare to 0.25 hectare in near future. Equally responsible is heavy industrialisation, demanding space and raw materials and causing pollution. In Asian tropics, land degradation is a major problem. It is estimated that about one-third of the irrigated agricultural land, about half of rainfed crop land and about three-fourth of the pastoral land are degraded. In India alone about 170 million hectares of productive land is degraded to a greater or lesser extent. The cause being decimation of much of the forests in the lowland, forest conversion in uplands for timber to support the industries else where, mining, energy generation through large hydro-electric projects etc. Ecological problems and land degradation have accelerated rapid marginalisation of a large section of the rural communities and needs to be viewed upon seriously.

In India, land degradation is a key issue, especially in rural environment. Some of the important issues include improving fertility of degraded infertile soils, improving the sub-soil water balance depleted due to over-exploitation of ground water for agriculture, saline and alkali soil (*usar* land) formation in areas under green revolution agriculture especially in Haryana, Punjab, Western Uttar Pradesh etc. The changing land use patterns such as utilisation of agricultural land for urbanisation and growing soil degrading plants for industrial requirements etc. also need to be paid great attention. Soil erosion and contamination of soil with industrial wastes like fly ash, press mud or heavy metals, all causing soil degradation, also need to be looked upon.

Man-Induced Landslides

Though land resources are very much related to natural disasters but man induced landslides are due to various anthropogenic activities, such as construction of big dams, river valley projects, hydro-electric projects, water storage reservoirs, roads and railway lines, raising buildings, mining activities in the hilly areas, disturbing the balance of nature, are making such areas vulnerable to landslides. Landslides have affected the stability of hill slopes and damaged vegetation cover around roads. They have also affected many development works in the hilly regions. Landslide is a common feature in the hill areas of Darjeeling (West Bengal) due to rainfed moist hill land coupled with road and rail road construction and repairing activities. Earlier there were few reports of landslides between Rishikesh and Byasi on Badrinath highway area in Uttarakhand but after the highway was constructed, 15 landslides occurred in one year. During construction, broadening and extension of roads; mining operations; and other such activities huge portions of fragile mountainous blocks are cut or destroyed by dynamite and thrown into adjacent valleys and streams. These landmasses weaken the already fragile mountain slopes and result landslides. Landslides increase turbidity in neighboring water bodies such as rivers, streams, springs etc., thereby reducing productivity of these water bodies. Blockage of roads, in hilly areas of tourist importance compel many tourists to be stranded in these places or on the way causing many problems and require measures for relief. The landslides can be checked by the construction of drainage ditches and walls along the hill slopes.

Soil Erosion

Soil erosion refers to the loss or removal of the top fertile layer of the soil by action of wind, water, ocean waves, glaciers and human activities. It results in the loss of soil fertility. Causes of soil erosion are many but deforestation without reforestation, overgrazing by cattle, surface mining without land reclamation, irrigation techniques leading to salt build-up, waterlogging of soil, farming on land with unsuitable terrain, soil compaction by crampling of cattle and heavy machinery such as tractors, threshers etc., cause the top soil vulnerable to soil erosion. Moreover, loss of green cover, strong winds and chemical pollution also severely affect the fertility of soils and the soil's ability to act as buffer and filter for pollutants, regulator for water and nitrogen cycle and habitats for biodiversity. Other factors affecting soil erosion are cultivation on hill slopes lack of proper irrigation; duration, intensity and distribution of rainfall and floods; social factors; topography; drainage area; kind of vegetation cover type and extent of covering; presence or absence of well defined channels; hydraulic properties and permeability etc. of subsoil. It is estimated that about one-third of world's cropland is getting eroded. World's two-third of the seriously eroded and degraded land lies in South Africa.

The soil erosion is basically of two types:

1. **Geological Erosion** or **Natural Erosion.** Geological erosion is caused by gradual removal of the top soil by natural processes which bring about equilibrium between physical, hydrological and biological activities and maintain a natural balance between erosion and renewal.
2. **Accelerated Erosion.** Accelerated erosion is caused due to man-made activities. The rate of accelerated erosion is much faster than the rate of soil formation. It is caused chiefly by population pressure, deforestation, change in land use pattern, mismanaged agricultural practices, overgrazing, intensive cropping, mining and constructions leading to developments etc. The soil erosion accounted by overgrazing is 35 per cent and that of the deforestation is 30 per cent. Unsustainable methods of farming cause 28 per cent of the soil to become eroded.

Soil erosion is very extensive in Australia, Central Africa, USA, Spain, Russia, Nepal, India etc. In India, extreme soil erosion by wind and water, is resulting 48,000 hectare of land to become eroded every year.

Soil erosion has resulted in greatest menace to agriculture. Water erosion is common in Maharashtra, Karnataka, Gujarat, Andhra Pradesh, Punjab, Uttar Pradesh, Orissa and Bihar. Wind erosion is common in Rajasthan, Punjab, Madhya Pradesh, Uttar Pradesh etc. Huge sandy areas in Punjab and Haryana are the example of wind erosion. The 700 year old Thar of Rajasthan, which was once a sea coast rich in vegetation and civilisation, has been converted into India's largest desert. In India overgrazing alone is responsible for 18.5 per cent of global loss of soil fertility. It is estimated that the fertile top soil of Shivalic hills, which took about 24,000 years for its ecological construction of fertile top soil, is being lost by flooding at the rate of 6 cm per year. In North East Hill States of India, shifting cultivation, *i.e.*, slash and burn or *jhum* cultivation is the main cause of soil erosion.

Prevention of Soil Erosion

Soil erosion can be prevented by various methods of soil conservation. The best course is large-scale tree plantation of soil binding shrubs and development of grasslands. Plant roots hold the soil and prevent erosion caused by flood, wind and water. Leaves of the plants, especially of trees, intercept the rain and break the impact of rainfall water. Vegetation, as a whole, protects the soil surface from action of wind, run off water and solarization and reduces run off of organic matter. Planting trees, shrubs or grasses stabilises the soil and checks soil erosion. Grasses, in particular, can be grown on soils which are otherwise not suitable for the cultivation of crops. Grasses are also useful for protecting bunds, water ways, highway alignments, badly eroded areas, gully control and for stabilising sand dunes. Farm forestry is another important effort in soil conservation. Planting of trees as wind breaks in the shelter belts is necessary to check wind erosion. Erosion caused by floods can be checked by small dams. Another method to check soil erosion caused by water is to develop a well controlled drainage system which could prevent the uncontrolled water flow. Application of various agronomic methods, forestry management measures or change in land use patterns and mechanical or engineering methods, to check accelerated soil erosion, have also been applied.

Erosion by the sea waves and currents has been checked to a large extent in Orissa, Andhra Pradesh, Kerala and some other coastal areas by the construction of broad walls of stone along the coasts with heavy plantation of *Casuarina* tree. In hill slopes or terraced fields, the alternation of crop beds with erosion resistent vegetation is practiced so that the soil gets stabilised. In cyclone-prone and desert areas, the effect of strong winds is checked by putting barriers of shrubs and trees.

Desertification

Desertification refers to land degradation in arid and semi-arid areas caused by climatic changes and human activities. It is the combined effect of accelerated erosion by wind and water, woodland destruction, soil waterlogging, salinisation and overgrazing in dry land environment etc. Desertification occurs slowly but covers a large area due to merger of different degraded land areas in close vicitnity. Desertification is progressing slowly but appearing over the planet like a skin disease wherein patches of degraded land, erupting separately, gradually join together. As a result of desertification, productive potential of the arid and semi-arid lands falls by 10 per cent or more. Moderate desertification brings about 10-15 per cent drop in productivity, severe desertification causes 25-50 per cent drop while severe desertification results in more than 50 per cent drop in productivity and usually creates huge gullies and sand dunes. Desertification leads to the conversion of rangelands and irrigated croplands to become deserted, so that the agricultural productivity falls. It results loss of vegetation cover, depletion of ground water, salinisation, severe soil erosion, degradation of ecosystem etc. The deserts are the places where water is not even in traces for miles together and temperature reaches 134°F and is hard to find human life in such an environments. Causes of desertification are many. In most deserts, the amount of evaporation is greater than the amount of rainfall which in these areas never exceeds 10-15 inches. Water is hardly retained as it is not soaked into the earth and rushes off in torrents, causing gully erosion. That is why water is not seen even in traces for miles together except for short oasis. Many of the deserts do not receive rains for years together. The moisture in these areas is insufficient to support normal life. The low humidity permits upto 90 per cent of solar radiation to penetrate the atmosphere and heat the earth resulting in high temperatures. The nights are very cold due to loss of heat into the atmosphere through radiation. Scarcity of water coupled with extremes of temperature lead to dust storms, that erode the soil which is unprotected by vegetation. Violent dust storms sometimes carry sand dunes to large distances.

Harsh extremes of heat make deserts a unique and most fragile ecosystem on the earth limiting plants growth and animal survival. Despite the fact we can not live in such torturous environment, many plant and animals thrive in these deserts. But envading drought resistant plants have mastered to survive in deserts by developing various adaptations. Animals survive due to their water retention potential and dependence on succulent plants.

Among other causes affecting desertification are deforestation affecting run off water holding capacity, overgrazing affecting microclimate near the ground, climatic factors, over-exploitation of the land for short term gain, war etc.

Impact of global desertification is threatening every fifth of men. Its effects are visible all over the world, especially in Asia, African Sahel-just South of Sahara, parts of Central and South America, Western half of the United States, Australia and along the Mediterranean. In UNEP's *World Atlas of Desertification*, about 1475 million hectare of land is shown to be in Asia, among which India figures prominently. It is expected that in last 50 years, human activities have been so, that they are much responsible for desertification of land area of about 900 million hectare, equal to the size of Brazil. The UNEP estimate suggests if sincere efforts are not made in the protection of deserts then 63 per cent of rangeland, 60 per cent of rainfed cropland and 30 per cent of irrigated cropland will suffer from desertification. It is further estimated that if present rate of desertification continues then by 2010, it will affect such lands which are presently occupied by 20 per cent of human population. Recognising the global need to fight against desertification, the United Nations declared the year 2006 as the 'International Year of Desert and Desertification (box 8.1). Although Earth Summit, 1992 had earlier supported a new integrated approach to the problem and emphasised action to promote sustainable development at community level. UN Convention on Desertification adopted in 1994, came into force in 1996.

Box 8.1 2006: International Year of Deserts and Desertification

United Nations proclamation of 2006 as the 'International Year of Deserts and Desertification' was aimed at global public awareness of the advancing deserts. It also centreed round the ways to safeguard the biodiversity of the arid lands. As a matter of fact desertification is a growing concern today especially in view of land degradation, loss of biological productivity caused by man-induced factors and climate change. It has affected about 33 per cent of earth's surface and over a billion people. It has a divesting consequence in term of socio-economic costs. According to Kofi Annan, former UN Secretary General, it is one of the world's most alarming processes of environmental degradation. Ignored at our peril, desertification is a global issue as dry lands are also home of the some of the magnificent ecosystems-the deserts of this world. These unique natural habitats with their incredibly diverse fauna have been home to some of the world's oldest civilization. Deserts stand like open-air museums bearing witness to bygone periods. Rightly the celebrations were made to respect the fragile beauty and unique heritage of the world's deserts, deserving protection.

In India, large parts of the hot arid regions of Western Rajasthan are seriously facing the problem of desertification mainly due to high speed winds, water logging, secondary salinisation, overgrazing etc. The Thar desert in India was formed by degradation of thousands of hectares of productive land due to natural phenomenon linked to climate change and abusive use of the land.

Among famous hot deserts of the world are Sahara Deserts-biggest of all deserts located in North Africa, Arabian Peninsula or Arabian Desert almost entirely sandy with most extensive stretches of sand dunes, Kalhari Desert in south western Africa with sand dunes and gravel plains, Thar Desert located partly in India and partly in Pakistan, Sonoran in south western United States, Monte in Argentina, Chihuahuan in north central Mexico and Australian desert.

Among practical measure to prevent desertification and restore deserted land it would be necessary to prevent soil erosion, improve water resource management, sustainable vegetation cover management, aero-seeding over shifting sand dunes, agroforestresty ecosystem, afforestation and reforestation, narrow strip planting, wind breaks and shelter belts of live plants, introduction of new varieties and species with a capacity to tolerate aridity and salinity and environmentally sound human settlements.

Cold Deserts: These deserts are the places which are extremely cold because of very low temperature (-20°C to 4°C), water in any utilizable form is scare, precipitation is in the form of snow and that too only 10 inches or less per year, every available water drop is in frozen form and it is very difficult for life to survive on this frozen and barren land. Such environment is found on the peaks of high mountains like Ladakh and near the poles such as parts of Antarctica. The less cold deserts are inhabited by some bacteria, algae, lichens, ephemeral flowering plants (which complete their life cycle in few weeks only), migratory birds and warm blooded animals. Some common cold deserts of the world are Gobi Desert which extends from Mongolia to the North China and is covered mainly by sandy soil and areas of small stones called 'Gobi'; Great Basin-covering parts of Idaho Nevada, Oregon and Utah (Great Salt Lake is located in this area); Atacama Desert on the coasts of Peru and Chile (the driest area on the earth); Iranian Desert in Iran, Afghanistan and Pakistan; Namib Desert in coast of south western Africa; Takla Makan in western China; and the Turkistan desert in parts of the middle East and south eastern Russia.

A cold desert in India is Ladakh. It has a fragile ecosystem with high speed winds, dry atmosphere, rocky and uneven terrain, very low temperature, with hardly any tree and vegetation and no visibility of wild life. In less cold areas occur wild roses, willow grooves, some herbs, brown bear, marmots yak (a wild ox), nyan (largest sheep in the world) and bharal-

the blue sheep (smallest sheep in the world). Around 170 species of birds (many of them being migratory) are found in Ladakh. About 425 plant species, mostly of medicinal value, have also been reported from this area.

Chapter Summary

Land, the major component of lithosphere, is one of the chief components of biosphere. Land forms about 29.1 per cent of earth surface and covers an area of 148, 950,200 sq. km. It is the most precious resource because of its diverse uses by mankind and as the source of many materials essential to man and other organisms. Many biogeochemical cycles, which sustain life, take place in and on the land. Land supports large forest areas, several habitats of living organisms and many water bodies. Man uses land for agriculture and horticulture; food; energy production; extraction of minerals, oil, coal, petroleum etc.; dwellings; commercial and industrial purposes; waste disposal etc. India with a global land area of only 2.4 per cent supports world's 15 per cent population. Land use patterns vary from country to country but global major part of the land (about 30 per cent) is occupied by forests. In India about 43.6 per cent of land is under agriculture. The soil, uppermost part of the land, supports vegetation and terrestrial biodiversity. The land is being degraded due to modern growth and production oriented activities directly or indirectly. Development activities are resulting landslides in the hilly regions. Improved but mismanaged agricultural practices are causing soil erosion which can be prevented by raising vegetation cover and using other mechanical devices. The most serious land degradation is desertification which is taking place due to the combined effects of climate changes and accelerated erosion by wind and water. Woodland destruction, soil waterlogging, salinisation, overgrazing etc. also cause desertification. Desertification is slowly spreading like a skin disease and taking into its grip a considerable part of the planet. Keeping in view, the year 2006 has been declared as the International Year of Deserts and Desertification by United Nations to bring global awareness to safeguard the land and the biodiversity of deserts.

Study Questions

1. What makes to call the land as a treasure of resources? Explain.
2. What is land degradation? What are various causes of land degradation? Describe.
3. Write an account of deserts and desertification.
4. Briefly describe-
   (i) Landslides
   (ii) Land Use Pattern in India
   (iii) Soil Erosion
   (iv) Cold Deserts

Objective Question. *Select the correct answer.*

1. The biggest Indian desert is-

(1) Gobi (2) Sahara

(3) Thar (4) Takla Makan

Answer

1. (3)

9

CHAPTER

Role of an Individual in Conservation of Natural Resources

LEARNING OBJECTIVES
Introduction • Population Increase and Industrial Revolution v/s Rapid Rate of Consumption of Natural Resources • Resource Use Patterns • Role of an Individual in Natural Resource Conservation

Introduction

A natural resource can be defined as any component of the natural environment that can be utilised by man to sustain his life and promote his welfare. It can be a substance, a natural process, a phenomenon or land, soil, water, air, energy, minerals, food, forest etc. Some of the resources such as air, water, food and land, are important because without them life can not be sustained. They are called life supporting systems. Apart from them, there are many other resources which fulfill other needs of human beings. Even before the beginning of civilisation man had been engaged in exploitation of natural resources. During the last millennium, amount of use of natural resources per year began spiking. The rise in the consumption of resources is visible but its effects are only partially noticeable. But one thing is very clear that with this rapid rate of consumption, the natural resources are bound to be depleted or lost, firstly the non-renewable and then the renewable. Steady economic growth has led to an extraordinary level of unsustainable consumption. In mad race to consume more and more, man is using up earth's finite resources far faster than their regeneration by natural process. Thus it becomes responsibility of all of us to contribute towards their conservation to keep the mankind flourish.

Population Increase and Industrial Revolution v/s Rapid Rate of Consumption of Natural Resources

The rapid rate of growth in human population began since 1960s and is continuing till today. Every year, the size of the world's population increases by as much as it did in the whole of the century throughout most of the human existence. Since equivalent of seven Kolkatas, *i.e.,* about 80 million people are added every year to the world's population, it is expected that the global population may double every 30 years or so. Of these 80 million, most are poor and need water, food, shelter, health care, education and many more things. This exponential growth of human population is causing a great pressure on our resources and resulting a steady impoverishment of our biological systems. The obvious result is deterioration in the quality of the environment and depletion or loss of natural resources. At any point of time in coming days, the system could collapse, leaving the humanity with acute shortage of adequate food, pure water, fresh air and even scarcities of all types.

Within last two or three centuries, the fundamental change in man's relationship with nature began with scientific and industrial revolution that started from Europe in sixteenth and seventeenth century and spread all over the world. The new attitude towards nature came with the idea of progress that advocated superior role of man as master of nature. Colonialism and spread of western education and culture played a key role in conveying this idea of progress to large parts of the world and today most countries swear by this notion of growth and development through science, technology and industrial expansion. And all this marked the beginning of the exponential rate of consumption of natural resources beyond ecological foot print, *i.e.,* the extent of the environmental sphere to sustain life.

Resource Use Patterns

Resource use patterns of people find a great difference in the developed and developing countries of the world. In economically developed countries, people having aspirations for better quality of life demand more resources than necessary for a reasonable living, thereby exhaust resources and degrade the global environment considerably. The people in developing countries, on the other hand, have small resource need due to their simpler way of subsistance for existence. But their population outbreak coupled with less environmental awareness and arousing desire to rapidly upgrade their living conditions, leads to reckless destruction of natural resources.

Indians are fortunate in that nature has gifted them abundance of natural resources to sustain their lives and fulfil other needs. The human societies in our country have developed and flourished within the magnificent environment. The roots of environmental values are deep in our ancient literature. Our classical literature abounds with the message that the resources should not be used wastefully but should be conserved. We believe in *care for nature and share with others*. But overuse of natural resources in our present day society is resulting rapid depletion of resources and causing several environmental problems.

Role of an Individual in Natural Resource Conservation

Natural resources belong to each one of us. Thus it becomes the responsibility of all of us to contribute towards their sustainable use and conservation to keep the mankind flourish. Small droplets of water together form a big ocean. Why not we? Although efforts are being made to conserve natural resources at international, national and local level, the efforts of an individual can go long way. Individual's efforts, combined together, can play a vital role in the wise use and conservation of natural resources and develop a better tomorrow.

While planning for wise use and conservation of natural resources, the things to be given priority include awareness of the consequences of resource loss, planning for resources, use of alternate resources and minimize resource use.

1. Individuals Role in Water Economy. The measures to conserve water resources include-

1. Not keeping water taps running even while brushing, shaving, washing etc.
2. Checking for water leak in pipes at your home and in your area and arrange for any leak to be repaired or plugged. Small pin-hole sized leak will lead to a wastage of huge amount of water.
3. Adopting minimum water use patterns.
4. Installing water saving toilets that use optimum water per flush.
5. Adopting rainwater harvesting devices in your house to conserve water for future use.
6. Installing a small system for collecting normally wasted water in your home and using it for watering lawn and kitchen garden.
7. Filling water in washing machine to the level required for the clothes to be washed.
8. Watering lawn and kitchen garden plants in the evening to minimise evaporation losses and not watering them in the mid day.
9. Cooperate beach clean up activities in coastal areas.
10. Save wetlands, lakes, pods, wells etc. in your locality.
11. Observe March 22 as World Water Day.

Box 9.1 March 22: World Water Resource Day

The decision to observe March 22 as World Water Resource Day every year was taken by UN General Assembly in 1992 at UN Conference on Environment and Development (Rio Summit). The Day marks a key date for mobilizing political will and encouraging society involvement in reminding the impending water crisis and to increase public awareness about the conservation, preservation and protection of water resources. The day attempts to increase participation and cooperation from governments, international agencies, NGOs, and the private sector. One of the nominated UN agencies is incharge of celebrating and promoting a new theme each year guided by Administrative Committee on Coordination (ACC) of Water Resource. The theme for 2002 was "Water for Development". For 2004 it was "Efficiency of Water Resource System" with sub-themes on critical review of current level of efficiency and measures for improving efficiency of water resource systems. The 2004 World Water Resource Day function was organized by the Central Water Commission (CWC) highlighting the judicious utilization of the available water resources in India and the need to conserve these resources to the optimum. For 2005 the theme was "Water for Life". March 22 is, therefore, a unique occasion to remind everybody of the extreme importance of water for maintaining the environment.

12. Observe February 2 as World Wetland Day.
13. Join Youth Water Team or any such NGO engaged in water conservation.

2. Individuals Role in Forest Conservation. The measures to conserve forests, save trees and planting new trees include-

1. Not felling the trees in forests, farms, roads or houses if they are green.
2. Not uprooting the existing trees while constructing a house but planting fast growing plant species in open area of the house.
3. Planting ornamentals, herbs, shrubs or suitable trees in and around the house.
4. Maintaining lawn and garden in open place in your house, if possible.
5. Participating in community plantation programmes.
6. Encouraging mass-scale tree plantation programmes through students, NCC cadets and NSS volunteers on barren culturable lands.
7. Cooperate *Chipko* or *Appiko* movements or other such NGOs engaged in saving trees and planting new ones.
8. Plant trees generously in barren lands.
9. Tag tree plantation with year ceremonies such as birth day, marriage anniversary etc.
10. Observe July 1-7 as *Van Mahotsava* Week.

11. Observe December 11 as Mountain Day- mountains being source of fresh water.
12. Encourage 'adopt a tree' programme, *i.e.*, 'each one tree one'.
13. Observe March 21 as World Forest Day
14. Discourage using paper for correspondence and promote use of e-mail

3. Individuals Role in Conservation of Minerals. Some of the measures to conserve minerals are-

1. Minimise use of minerals which are likely to be depleted or exhausted.
2. Minimise use of jewellery to conserve scarce minerals.
3. Recycle and reuse metals and glass.
4. Stop buying soft drinks in metal containers.
5. Buy durable products that last long.
6. If buying a car, buy a small and efficient one.
7. Repair and reuse bicycles.
8. Use recyclable utensiles.

4. Individuals Role in Food Security. Some of the measures to achieve food security are-

1. Sustainable use of food and not wasting it.
2. Eating only as much as required for sustenance of life.
3. Consuming local and seasonal vegetables and fruits so as to save energy on their transportation, storage and preservation.
4. Grow vegetables by organic methods in open area of your house or an establishment. Use indigenous species, adopt polyculture growing different varieties and practice composting.
5. Buy food from local market and small traders and not from super-markets.
6. Buy only organically grown food.
7. Discourage packed, canned and preserved food.
8. Shift from non-vegetarian to vegetarian.
9. Observe October 16 as World Food Day and November 21 as World Fishery Day.

5. Individuals Role in Energy Conservation and Saving Energy. Some of the measures are-

1. Turning off lights, fans or other electric appliances when not in use.
2. Replacing bulbs and tube lights with Compact Fluorescent Lamps (CFLs).
3. Obtaining as much of light and heat as possible from natural source such as sun.
4. Drying the clothes in sun instead of drier of washing machine, if it is a sunny day.

5. Using solar cookers for cooking food which will be more nutritious and will cut down LPG/CNG expenses.
6. Building the house with provision for sunspace which will keep the house warmer and will provide more light.
7. Buying energy efficient appliances when going for a new or replace old ones. Always checking energy consumption figure. Installing electronic regulators for fans and auto-switch off device for area of staircases
8. If building new house, minimising use of fired bricks that use considerable coal energy during their making.
9. Minimizing use of automobiles by using bicycle, public transport, car pool etc.
10. Trying to reside near place of work, if possible.
11. Keeping vehicles tuned for low consumption of fuel.
12. Checking fuel consumption data while buying a new vehicle.
13. Following the advice given by Petroleum Conservation Research Association (PCRA) with regard to energy conservation.
14. Making shut off personal computers, T.V. sets, music system etc. when not in use. Replace bulky CRT monitor with flat and thin LCD monitor that consumes much less energy and is easier on eyes.
15. Wearing adequate woolen clothes during winter instead of using heat convector.
16. Avoiding use of air conditioners. If using them lowering the load by thermostat setting, as for every one degree rise in temperature setting, saves 3-5 per cent of electricity.
17. Growing deciduous trees at proper place outside the house. They will cut off intense heat of summers, cut off electricity consumption and will provide cool breeze.
18. Observe December 14 as World Energy Conservation Day.

6. Individuals Role in Sustainable Agriculture and Soil Protection. Some of the measures are-

1. Discouraging monoculture practice in agriculture to save plant diversity.
2. Adopting mix cropping to check depletion of some specific soil nutrients.
3. Adopting drip irrigation to avoid washing out of soil nutrients.
4. Reducing use of chemicals such as fertilizers and pesticides to check soil pollution.
5. Using organic fertilizers or bio-fertilizers to retain the soil potential.
6. Controlling pests and pathogens by manipulating cropping patterns and using biological control methods.

7. Avoiding over-irrigation without proper drainage to prevent water logging.
8. Observing December 23 as Farmers Day.
9. Observing June 17 as World Day to Combat Desertification and Deserts.
10. Observe November 21-27 as National Land Resources Conservation Week.

Chapter Summary

A natural resource is any component of the natural environment that can be used by man to sustain his life and promote his welfare. The land, air, water, food, energy, minerals, forests-life supporting systems, are important because life can not be sustained without them. There are other resources which fulfill other needs of humans. Exploitation of resources has been a long practice but during the last millenium use of resources per year began spiking. The rise in consumption of resources is visible but its effects are only partially noticeable. But with this rapid rate of consumption, the resources are bound to be depleted or lost, firstly the non-renewable and then the renewable. Steady industrial and economic growth and rapid increase in human population has led to an extraordinary levels of unsustainable consumption. In mad race to consume more and more, man is using up finite resources of the earth far faster than their regeneration by natural processes. Thus it becomes responsibility of all of us to contribute towards their conservation to keep the mankind flourish. The developed countries with lesser population consume more resources whereas developing countries, with plenty of resources and population outbreak, recklessly destroy the natural resources. Since natural resources belong to each of us, all of us should contribute towards sustainable use of water, food, forest, energy and other resources economically and carefully.

Study Questions

1. Write an essay on the role of an individual in conservation of natural resources.
2. Describe Resource Use Patterns.
3. Briefly explain the Population Increase and Resource Use

Objective Question: *Select the correct answer*

1. Which of them is not directly proportional to resource use?
   (1) Population Growth (2) Industrialisation
   (3) Education (4) Tribal Density

2. World Water Resource Day is celebrated on:
   (1) January 26 (2) April 26
   (3) February 22 (4) March 22

Answer

1. (3) *2.* (4)

10

CHAPTER

Equitable Use of Resources for Sustainable Life Styles

LEARNING OBJECTIVES
Introduction • Equitable Use of Resources and Sustainable Development • Sharing and Caring of Resources • Safe Water for All • Food for all • Fuel for All • Carrying Capacity • Green Accounting.

Introduction

The concept of equitable use of resources for sustainable lifestyles has been well recognized in Vedic literature and Upanishads in India. A saga in Isha-Upanishad states *The whole universe together with its creatures belongs to the Lord (Nature). Let no one encroach over the rights and privileges of other. One can enjoy nature by giving up greeds*. Emphasis has always been given on the philosophy to take from nature only what we actually need and not more. Our environmental ethics recognises *Panchtatwa: Kshiti* (soil), *Jal* (water), *Pawak* (energy), *Gagan* (space) *Samira* (air) as the basic resources of the earth. The concept has been recognised in 1987 when the World Commission on Environment and Development (the Brundtland Commission) introduced the term *Sustainable Development* in its report *Our Common Future* for *the development that meets the needs of the present without compromising on the ability of future generation to meet their own needs*. Agenda 21, adopted during the United Nations Conference on Environment and Development (Earth Summit) held in Rio de Janeiro in Brazil in 1992 is a blue print for achieving sustainable development. Later World Summit on Sustainable Development, held at Johannesburg in 2002 reviewed the progress on the Earth Summit and emphasised largely on *new development models*.

Equitable Use and Sustainable Development of Resources

The concept of sustainable development provides a framework for the integration of developmental strategies with environmental protection. This concept includes reducing excessive resource use and enhancing resource conservation, recycling and reuse of materials, waste minimisation with proper technological input and scientific management of renewable resources specially bioresources which has a life cycle and an inherent sustainable qualities.

The more developed countries such as USA, Canada, Japan, New Zealand and Western European countries have only 22 per cent world's population, but they use 88 per cent of its natural resources, 73 per cent of energy and command 85 per cent of income and in turn contribute a major proportion of pollution. On the other hand, less developed countries including India, have very low or moderate industrial growth, have 78 per cent of the world's population and use only about 12 per cent of natural resources and 27 per cent of energy and have 15 per cent of income. Further the gap between the two is increasing due to a sharp increase in the population in the underdeveloped countries. Thus rich have grown richer while the poor have stayed poor or gone even poorer. Since rich nations are using more resources and developing rapidly and are causing more pollution, they are threatening sustainability of earth's life support system. On the other hand, poor nations are still struggling hard with their large population and poverty problems and their share of use of resources is too little leading to unsustainability. The rich nations can not continue to develop indefinitely because many of our earth's resources are limited and even the renewable resources will become unsustainable if their use exceeds their regeneration. Thus the rich countries will have to lower down their resource consumption levels while the bare minimum needs of the poor have to be fulfilled by providing them resources. A fairer and equitable use of resources will narrow down the gap between the rich and the poor and will lead to sustainable development for all.

The concept of sustainability, applied to urban areas, is a paradox as cities are consuming more resources and are creating a native consumer culture which lacks a necessary respect for other cultures of today and tomorrow. It is feared that the future generations may have to bear the costs for unsustainable use of resources by present generations.

Sharing and Caring of Resources

A way out for equitable use of resources for sustainable life style can only be sharing and caring of resources not only among the societies but also with nature. Thus sustainable consumption of resources is a way of adopting life styles to match the supporting capacity of the earth in just and sustainable manner. The viability and continued existence of many civilisations through the centuries depended on their organised material production and consumption patterns according to the perceptions and values of traditional wisdom. The Oslo Symposium in 1994 proposed a working definition of sustainable consumption as *the use of goods and services that respond to basic needs and bring a better quality of life, while minimizing the use of natural resources so as not to jeopardize the needs of the future generations*. Sustainable consumption with reference to resources brings together a number of key issues such as:

1. Meeting the resource needs of the poor and socially backward people of the society.
2. Enhancing the quality of life through better resources and cheaper goods, services and facilities.
3. Improving resource efficiency through skilled manpower, modern and efficient technology and methods.
4. Increasing the use of cost effective renewable resources which enhance efficiency and prevent pollution.
5. Minimising waste by using such methods and technology which may convert it into wealth, resources or energy.
6. Taking a lifestyle prospective through methods of recycling and thus saving virgin resources for future.
7. Taking into account equity dimensions by promoting and implementing schemes, education, training programmes and various employment opportunities to reduce the gap between the rich and the poor.
8. Helping the economically weaker sections of the society towards the self sufficiently.
9. Reducing the use of fossil fuels especially of petroleum.
10. Reducing transportation by providing goods and services as close as possible to the consumers.

A global consensus has to be reached for more balanced distribution of the basic resources such as drinking water, food, fuel etc. so that the poor, in less developed countries, are atleast able to sustain their life.

Safe Water for All

In developing countries less than 75 per cent of the urban and less than 20 per cent of rural population have convinient access to safe water. About 10 million people die annually due to water related diseases. Water related diseases are the leading killers of infants and children. About 5 million infants die every year from intestinal diseases before the complete one year's of life. Every day about 25,000 people die from unsafe water as two-thirds of the people on the earth have no other choice except to drink it, cook with it and bath in it.

According to the World Health Organisation (WHO) 1 hospital bed out of 4 in the world is occupied by a patient who is ill because of polluted water. Provision of a safe and convenient water supply is the single most important activity that could be undertaken to improve the health of people living in rural areas of the developing world.

The problem of providing a convenient supply of safe water to all remains unsolved. In some parts of the world, women and children spend half of their time for fetching water. Some have to walk as far as 25 kilometres or so to reach the source of water. In India, the people of villages in the hilly and desert regions, where nearest water sources are about 1.6 kilometres away, have difficult access to water. Besides, these water sources contain toxic elements which are dangerous to health and are carrier to epidemic diseases like cholera and guinea pig infestation. Many other villages also have inadequate and unprotected drinking water sources. Various schemes and projects undertaken to supply the safe water have met with several difficulties. According to Evaluation Report on the Accessibility of the Poor to the Rural Water Supply, 1980 of the Planning Commission of India, there are several weak points in the programme. Remedial measures need to be undertaken to ensure safe water supply conveniently through bore, tube or drill wells and pipes, require regular and timely supply, adequate number of supply points, better management for breakdown and out-of-order conditions, provision for separate public points for the poor and increase of supply hours.

Besides, measures should also be taken to educate the community to make them aware that water from open dug wells without parapets and by individual collections from ponds, tanks, lakes etc., is basically unsafe.

Box 10.1 **Water Notes**

- India, considered as a whole, is not a water scarce country. It ranks among the top 10 water rich countries in the world (World Resource Institute, 2000-2001)
- Presently, as many as 20 countries are classified as 'Water scarce, based on their current sources of renewable fresh water *vis-a-vis* demand. The number is further rising.
- Worldwide, 1.1 billion people lack access to basic need of water supply and 2.6 billion lack access to basic sanitation. About 4 billion cases of diarrhoea occur every year and claim nearly 2.2 million lives. About 90 per cent victims are the children under five. Thus diarrhoea kills 4,500 children a day, *i.e.* one child every 20 seconds.
- About half the people in developing nations are exposed to contaminated water. The poor people of these countries spend upto 25 per cent of their real income to have access to clean water.

People should also be made aware to keep the traditional sources of water, such as open wells, water-holes, ponds, rivers, lakes etc., safe and to convert open dug wells into sanitary wells by making provision for elimination of contamination. Since fresh water is finite and essential to sustain life, development and environment effective management of water resources is necessary.

All governments of the developing world were signatories to the Mar Del Plata Action Plan adopted by the United Nations Water Conference held in Argentina in 1977. UNICEF provided assistance to governments of more than 80 developing nations to improve and increase water supplies to make available enough water for drinking, cooking, food production and washing to everyone. A decade long Government of India's plan with the ambitious target of providing safe drinking water to 1,90,000 problem villages, still could not be achieved to provide a source of safe water. To tackle the problem of providing rural areas with safe drinking water, National Drinking Water Mission was set up to identify new sources where there are none, better management of existing water resources, cost effective and practical technologies through existing scientific know how, and ensuring sustained supply of safe drinking water to the community.

Food for All

The global average daily per capita energy consumption from food is about 22,00 kilocalories (kcal). For meeting this entire energy requirement from herbivores-the second tropic level, it would be equivalent to 22,000 kcal of herbivore which, in turn, would require 2,20,000 kcal per day of the producer-first trophic level. Assuming the producer-the green plants one per cent efficient, it would need 22,00,000 kcal per day of solar radiation. With the availability of 500 calories of solar radiation per cm^2 per day, 44,000,000 cm^2 area on 0.44 hectares of land will be required to feed one person. Thus 1.4×10^9 people, *i.e.*, 2.5 times the present world population, which means 4×10^9, would require much more energy. Within the coming five decades, the population will cross the limit. Presently 3×10^9 metric tonnes of food is produced annually in the world of which only 2 per cent comes from water in the form of fish and the rest comes from 1.4×10^9 hectares of cropland and 3×10^5 hectares of grassland. Animal food, which mostly comes from grassland, comprises 20 per cent of the total food production.

The food items coming from different trophic levels, enter into the composition of the diet of human population which shows a wide range of variation due to ecological, social, cultural, religious and personal preferences. The greatest advantage can be derived from taking food items at the producer level and in the long run from ecological point of view it may become necessary to include a higher percentage of vegetable foods in the diet or even to consume a vegetarian diet only.

Although agricultural food production can be increased by the use of fertilizers, pesticides and irrigation but it will increase the pressure on reserve stock of fossil fuels especially oil and will increase environmental problems and encroachment for croplands and other agriculture oriented problems. To be able to provide adequate food to all in future in a sustainable and ecologically sound manner, it would become essential to popularize organic foods (grown by organic farming) using organic manure in place of fertilizers and adopting biological pest control. Sustainable farming; a farming system closer to a natural ecosystem, using optimum land without losing soil fertility and optimum available natural resources, minimising the needs for external chemical inputs, use of crop rotation methods, multiple farming, green manuring etc. can also bring about sustainable food production. Besides, tissue culture also offers new possibilities for mass production, particularly for crops which are normally difficult to multiply.

Aquaculture , *i.e.,* intensive propagation of aquatic organisms, when fully integrated into rural development schemes, is expected provide plentiful, reasonably priced, fresh and locally available protein rich food and also generate much needed employment. In aquaculture fish and other aquatic organisms should be regarded as renewable resource that if properly managed can be sustained perpetually. Role of dairy for production of milk and dairy products is considered significant. The milk is protein and calcium rich. Besides cows and buffaloes, bulls can also be used as draft animals.

Fuel for All

The twin problems of scarcity of fuel and environmental degradation in developing countries can be solved by social forestry, utilisation of biogas, use of solar cookers, LPG, CNG etc. Consuming 25 per cent of uncooked food such as raw vegetables, dry fruits, germinated grains, ground nuts and green leaves, can also reduce the amount of fuel consumption. Alley cropping, raising fast growing leguminous trees such as *Leucaena* or shrubs between food crops, can also solve the fuel problem to some extent.

Carrying Capacity

Carrying capacity of the region or a system can be described as number of individuals of a species that it can sustain. In case of human beings, it presents rather a complex situation where the region or a system has not only to bear the load of his basic needs but also all other associated activities including industrial or development projects which have direct impact on limited natural resources and environmental quality. The carrying capacity may be supportive capacity or assimilative capacity. The supportive carrying capacity of a region or a system provides an assessment of the stock of the available resources with their regenerative capacity on natural or sustainable basis. The assimilative carrying capacity is an assessment of the maximum amount of pollution load that can be discharged without violating the best designed use of these basic components of the environment. Thus carrying capacity of a region or system gets affected by the use of resource base beyond regenerative capability of their supportive capacity or if discharge or generation of pollutants or wastes products is not within the assimilative capacity.

Green Accounting

Green accounting is a widely prevalent concept both in developed and underdeveloped countries. Basically green accounting underlines the same principle as enumerated in concept of sustainable development and carrying capacity, *i.e.,* use of natural resource base in planned and judicious manner causing minimum or no impact on the quality of the environment. However, it conveys by providing us an economic interpretation of both resource base and the quality of the environment as against the conventional accounting in terms of Gross Domestic Production (GDP). Precisely, it gives us a uniform level by converting both natural resource base and the environment quality in monetary terms, therefore, making the task easier for planners and policy makers to formulate further programmes or strategies of developments.

Chapter Summary

The concept of equitable use of resources for sustainable living has been well recognised in Indian Vedic literature and Upanishads. The concept was recognized in 1987 when World Commission on Environment and Development (the Brundtland Commission) introduced the term *Sustainable Development* in its report *Our Common Future* for *the' development that meets the needs of the present without compromising on the ability of future generations to meet their own needs*. Agenda 21 adopted during Earth Summit, Rio, 1992, is the blue print of achieving sustainable development. Later World Summit on Sustainable Development, Johannesburg, 2002, reviewed the progress on the Earth Summit. The equitable use of resources and sustainable development aims at reducing excessive resource use and enhance resource conservation. A way out for equitable use of resources for sustainable life styles can only be sharing and caring of resources so that each of us can get safe water, hygienic food, proper health care and other things needed to sustain life. In other worlds we should use the resources within the limit of carrying capacity of the earth.

Study Questions

1. Briefly discuss the equitable use of resources for sustainable development.
2. Briefly describe:
   (i) Sustainable Consumption
   (ii) Carrying Capacity
   (iii) Green Accounting

Objective Question: *Select the correct answer*

1. Use of natural resources is excessive in:
   (1) India (2) Asia
   (3) Brazil (4) United States

2. Agenda 21 adopted during U.N. Conference on Environment and Development (Earth Summit) is a blue print for:
   (1) Controlling environmental pollution
   (2) Conserving wildlife
   (3) Achieving sustainable development
   (4) Conserving forests

3. Developed countries have:
   (1) High population (2) More use of natural resources
   (3) Less use of energy (4) Less use of natural resources

4. U.N. Water conference was held in the year:
   (1) 1965 (2) 1973
   (3) 1977 (4) 1997

Answer

1. (4) *2.* (3) *3.* (2) *4.* (3)

Ecosystems

- *Concept of an Ecosystem*
- *Structure and Functions of an Ecosystem*
- *Producers, Consumers and Decomposers*
- *Energy Flow in an Ecosystem*
- *Food Chains, Food Webs and Ecological Pyramids*
- *Various Ecosystems*
- *Ecological Succession*

11

CHAPTER

Concept of an Ecosystem

Learning Objectives

Definition of Ecosystem • Characteristics of an Ecosystem • Complex Nature of Ecosystem • Types of Ecosystems • Terrestrial Ecosystems • Aquatic Ecosystems • Natural and Artificial Ecosystems • Components of Ecosystems • Biotic Components • Abiotic Components • Interrelationships • Concept of Homeostasis.

Definition

Sir Arthur G Tansley (1935) coined the term Ecosystem and defined it as '*the system resulting from the integration of all living and non-living factors of the environment*'. Several parallel terms are available in the literature, *e.g.* biocoenosis, microcosm, holocoen, geobiocoenosis, bio-system, ecosom etc.

The word ecosystem is not a synonym of habitat, community or some other similar descriptive terms. Rather, it is a technical term in ecology and refers to *a system of living and non-living components interacting as a whole*.

Characteristics of an Ecosystem

Smith (1966) proposed following general characteristics of most ecosystems-

(1) The ecosystem is a major structural and functional unit (of ecology).
(2) The structure of an ecosystem is related to its species diversity. The more complex systems have high species diversity.
(3) The function of an ecosystem is related to energy flow and material cycling through and within the system.
(4) The relative amount of energy needed to maintain an ecosystem depends on its structure. The more complex the structure, the lesser the energy it needs to maintain itself.

(5) Ecosystems mature by passing from less complex to more complex states. Early stages of such succession (low biodiversity) have an excess of potential energy and a relatively high-energy flow per unit biomass. Later stages (high biodiversity) have less energy accumulation and its flow through more diverse components.

(6) Both, the environment and the energy fixation in any given ecosystem, are limited and cannot be exceeded without causing serious undesirable effects.

(7) Alterations in the environments represent selective pressure upon the population to which it must adjust. Organisms, which are unable to adjust to the changed environment, must need vanish.

Complex Nature of Ecosystem

Ecosystem is thus a bioenergetic approach in which both living (biotic) and non-living (abiotic) components are interacting, interdependent

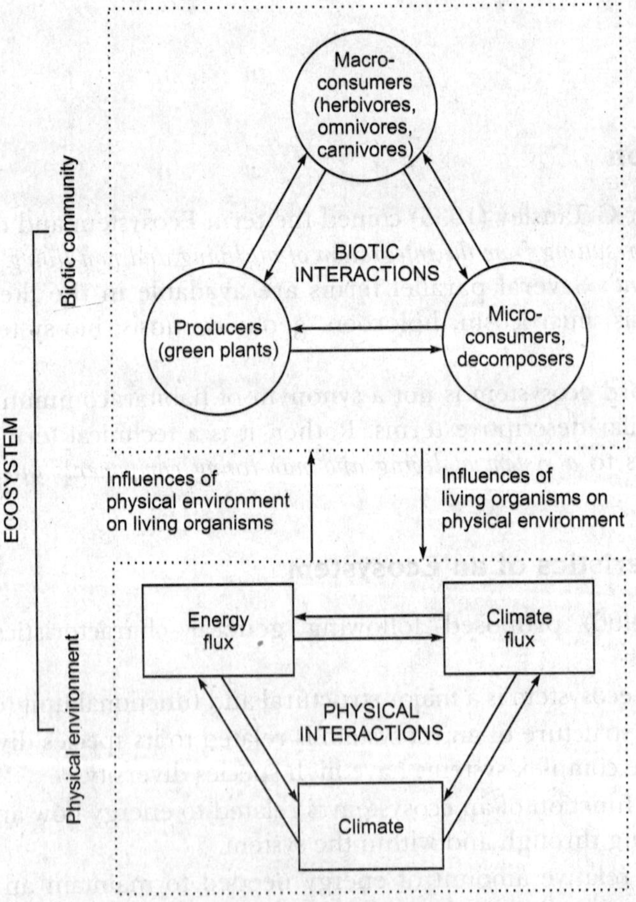

Fig. 11.1. Some important interactions found in and between biotic and abiotic components (after Cox and Atkins, 1979).

and inseparable. Cox and Atkins (1979) explained some important interactions between the biotic and abiotic components of a typical terrestrial ecosystem (fig. 11.1).

The ecosystem is an integrated unit (or zone) of variable size comprising vegetation, fauna, microbes and the environment. Most ecosystems possess a well-defined soil, climate, flora and fauna (or communities) and have their own potential for adaptation, change and tolerance. Ecosystem's function involves a series of cycles (such as water cycle, bio-geochemical cycle etc.). These cycles are driven by energy flow, the energy being the solar energy. Continuation of life requires a constant exchange and return of nutrients to and from (amongst) the different components of the ecosystem.

In simple words, ecosystem is a dynamic system which includes both organisms (biotic components) and abiotic environment influencing the properties of each other and both are necessary for the maintenance of life. It can be recognised as self-regulating and self-sustaining units of landscape.

TYPES OF ECOSYSTEMS

Infact, whole of the earth constitute the giant ecosystem, called biosphere but it has been further divided either on the basis of habitats or on basis of ecoclines, spatial scales, uses, source and level of energy, stages of ecosystem development, stability or instability etc.

Types of ecosystems, recognised on the basis of habitats, are simple and most convenient for study purpose. The habitats exhibit physical environmental conditions of a particular spiritual unit of the biosphere. These physical conditions determine the nature and characteristics of the biotic communities and, therefore, there are spatial variations in the biotic communities. Based on this consideration, the ecosystems are divided into two major categories-

(i) Terrestrial ecosystems

(ii) Aquatic ecosystems

These have been further divided (table 11.1).

COMPONENTS OF ECOSYSTEMS

An ecosystem comprises two very closely interconnected parts, called components. These are the living parts called **biotic components** and the non-living parts called **abiotic components**. *Biotic component* is an important part of ecosystem in which energy flows in the form of food from one organism to another organism within a community. *Abiotic components* are another part of ecosystem comprising inorganic, organic,

TABLE 11.1. Kinds of Ecosystems

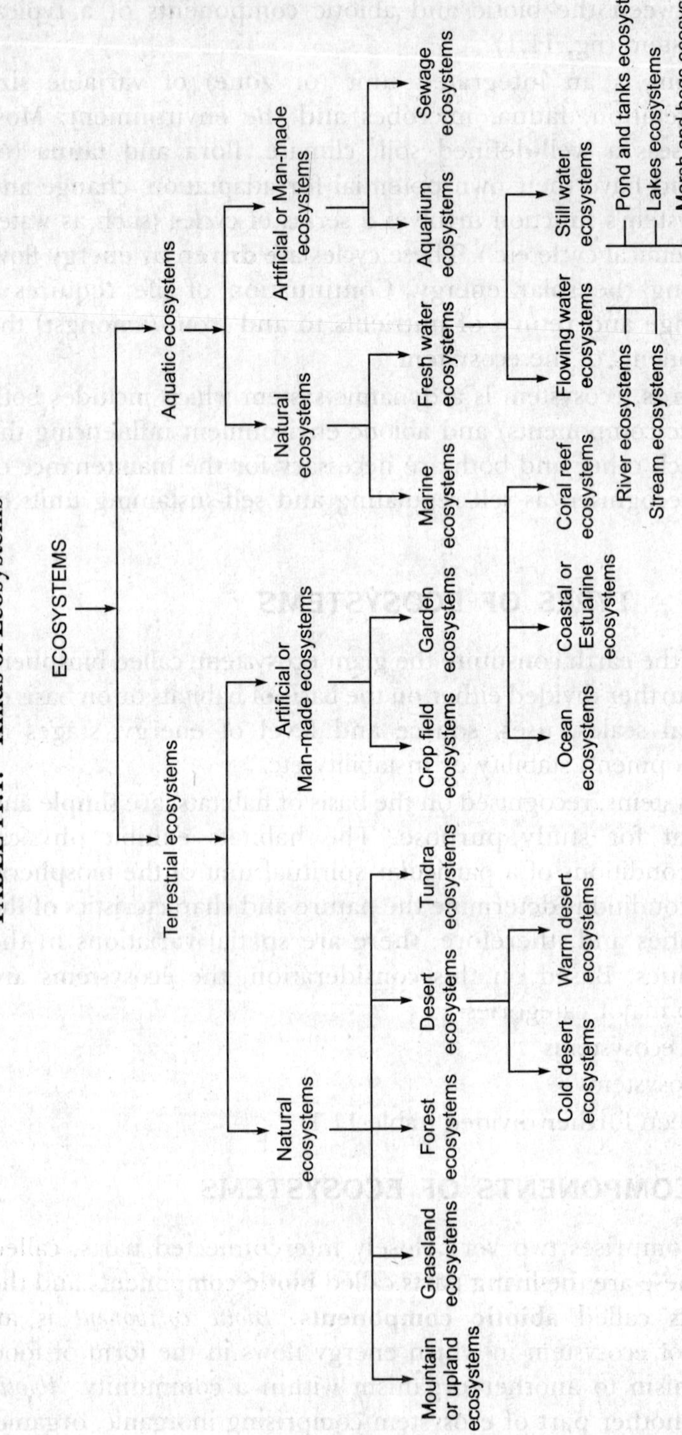

and physical or climatic components. Together, these two components determine the characteristics of the ecosystem.

(I) Biotic Components

Biotic components are basically divided into following categories–

(i) Producers or Autotrophs
(ii) Consumers or Heterotrophs
  (*a*) Primary consumers - herbivores
  (*b*) Secondary consumers - carnivores
  (*c*) Tertiary or top consumers - carnivores
(iii) Parasites and Detrivores
(iv) Decomposers

Producers are green plants or **autotrophs** which synthesise their own food by the process called photosynthesis.

Consumers are heterotrophic organisms which directly or indirectly depend on producers for energy. Energy is procured in the form of food. Consumers have been divided into three categories-

(a) **Primary consumers**. These organisms are herbivores which depend upon green plants for their food, *e.g.* cow, goat, rabbit, some insects etc.

(b) **Secondary consumers**. These are those carnivores that use herbivores as their source of food, *e.g.* snake, frog etc.

(c) **Tertiary consumers**. These are also called top consumers and these kill and eat the secondary or primary consumers. Example of top carnivores are lion, tiger, hawk etc.

Parasites are those organisms, which obtain their food directly from living organisms (hosts). **Detrivores** or **scavengers** are small animals, which feed on dead bodies of other organisms, such as termites

Decomposers are generally saprophytic microbes which decompose the dead organisms, such as bacteria and fungi. These decompose organic complexes into simple substances, which are either mixed within the soil or evaporated to the atmosphere. These are also called **microconsumers**.

Producers, consumers and decomposers remain linked together by **food chain** wherein each category forms a **trophic level**.

(II) Abiotic Components

These are non-living components of an ecosystem and include-

(i) **Inorganic substances**. These include elements (minerals), water, gases etc., which are required for the synthesis of organic substances by the producers. These keep on circulating within the ecosystem.

(ii) **Organic substances**. These substances are derived from dead plants and animals as well as from their excreta. Such substances are decomposed to release the minerals.

(iii) **Climate**. It includes factors like rain, heat, light, temperature, humidity, wind etc. The interaction of various climatic factors determines the nature of the ecosystem.

INTERRELATIONSHIPS BETWEEN BIOTIC AND ABIOTIC COMPONENTS

The various factors of the biotic and abiotic components are inter-related to form a functional and balanced ecosystem. In all ecosystems, solar energy is trapped by **producers** (green plants) and converted into chemical energy (food) by the photosynthetic process. This food is the source of energy for **primary consumers** (herbivores) for their life activities. The primary consumers are the source of food of the **secondary** or **tertiary consumers** (carnivores). On death of producers and primary, secondary or tertiary consumers their dead bodies (organic matter) pass into the soil where they are decomposed by decomposers. By decomposition process, the organic matter is broken down to release the inorganic elements, which are again absorbed by the plants (producers) for metabolic activities. In an ecosystem, the minerals move in cyclic manner but energy movement is always unidirectional (fig. 11.2).

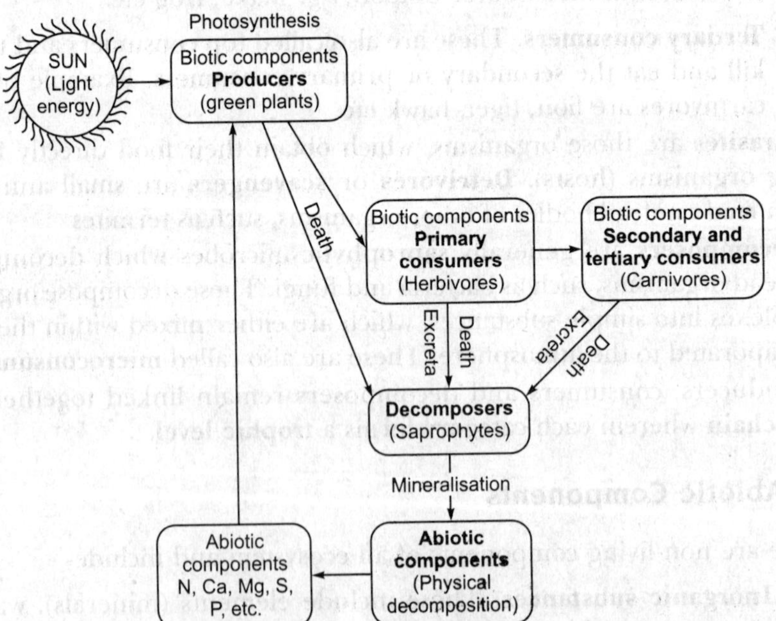

Fig. 11.2. Different components of an ecosystem showing interrelationships.

Concept of Homeostasis

Natural ecosystems are capable of self-regulation and self-maintenance, as they are able to maintain a stable steady state. The proposition refers to the concept of homeostasis (*homeo* = same, *stasis* = standing). Odum (1971) defined homeostasis as the tendency of natural ecosystems to resist change and to remain in a state of equilibrium. A homeostatic condition within an ecosystem implies that there would be a balance between production, consumption and decomposition, as well as between all species within the system.

Chapter Summary

The term **ecosystem** coined by Tansely (1935) refers to the bioenergetic approach of ecological studies in which both living (biotic) and non-living (abiotic) components are interacting, interdependent and inseparable. Several interactions are found between the components. Ecosystem is a major structural and functional unit. The functions are related to energy flow and material cycling. Ecosystem matures by passing from less to more complex states. Complex stages have high biodiversity and less energy accumulation. Various biogeochemical (mineral) cycles are driven by energy flow, the energy being the solar energy. Continuation of life requires a constant exchange and return of nutrients to and from (amongst) the different components of the ecosystem.

Ecosystems have been classified variously but classification on the basis of habitat is simple and popular. Two major categories of ecosystem are the terrestrial ecosystems and the aquatic ecosystems, each one has both natural and man-engineered (man-made) type. Each ecosystem has two types of components, the biotic and abiotic components. Biotic components are further divided into:(*i*) Producers (autotrophs), (*ii*) consumers (heterotrophs) and (*iii*) decomposers (saprotrophs). Consumers comprise primary (herbivores), secondary (carnivores) and tertiary (top carnivores) types. Abiotic components include inorganic and organic substances and climate. The two types of components are interacting and inseparable.

Ecosystems are self-regulating and self-sustaining units of landscape.

Study Questions

1. Define what is an ecosystem? Describe its characteristic features with suitable illustrations.
2. What are different types of ecosystems? Give a brief account of their classification.
3. What are the various components of an ecosystem? Explain them with suitable examples.
4. Describe the general characteristics of an ecosystem.
5. Explain the various types of biotic components found in an ecosystem.
6. What are non-living components of an ecosystem? Explain briefly.
7. Enumerate briefly the interrelationships between biotic and abiotic components of an ecosystem.

8. Who coined the term 'ecosystem'? Explain.
9. Define ecosystem.
10. What are producers? Describe.
11. Name various types of consumers.
12. Write two important characteristics of ecosystem.
13. Name two components of ecosystem.
14. Define homeostasis.

Objective Questions: *Select the correct answers:*

1. The term 'ecosystem' was proposed by:

(1) Carl Mobius (2) A. Tansely
(3) E. Odum (4) E. Clement

2. The most appropriate definition of an ecosystem is:

(1) The biotic community of organisms together with the environment in which they live.
(2) The abiotic components of the habitat.
(3) The community of organisms interacting with one another.
(4) The part of earth which inhabits living organisms.

3. Which of the following occurs in abiotic components of ecosystem?

(1) Cycling of Material (2) Flow of energy
(3) Consumers (4) Both (1) and (2)

4. Energy is the driving source of an ecosystem. It primarily comes form:

(1) Reserve food material (2) Sun Light
(3) Heat liberated during respiration (4) Photosynthesis

5. Photosynthesis is found in-

(1) Producers (2) Consumers
(3) Decomposers (4) None of these

Answers

1. (2) *2.* (1) *3.* (4) *4.* (2) *5.* (1)

12

CHAPTER

Structure and Functions of Ecosystem

LEARNING OBJECTIVES

Introduction • Structure of an Ecosystem • Biotic Components • Abiotic Components • Functional Aspects of an Ecosystem • Productivity • Bio-geochemical Cycles • Carbon Cycle • Nitrogen Cycle • Phosphorus Cycle • Sulphur Cycle.

Introduction

The study of an ecosystem is not merely a description of the biotic community and its abiotic environment. It involves the understanding of a whole network of relationships between the living and non-living components. Two major aspects of the ecosystem are, the **structure** and the **function**. Under the **structure,** studies are made on-

(i) the composition of biological community including species numbers, biomass, life history and distribution in space etc.,

(ii) the quality and distribution of the non-living materials, such as nutrients, water etc., and

(iii) the range or gradient of conditions of existence such as light, temperature etc.

Under the **function** of an ecosystem, studies are made on-

(i) the rate of biological energy flow,

(ii) the rate of materials or nutrient cycles,

(iii) the biological or ecological regulation of organisms by the environment or vice versa, and

(iv) the productivity.

STRUCTURE OF AN ECOSYSTEM

An ecosystem has two major components, the biotic and the abiotic components.

I. Biotic Components

Biotic components are the living organisms of an ecosystem. These have various modes of nutrition and are accordingly classified into:

(1) Producers,
(2) Consumers and
(3) Decomposers.

(1) Producers

Producers are green plants that manufacture carbohydrates by the process called photosynthesis. Photosynthesis is the first and the single most process in which solar energy (radiant energy) is converted into chemical energy (potential energy) and since it is carried out only by the green plants, the green plants in an ecosystem are called **producers**. Term producer is infact incorrect as green plants do not produce energy, the more correct would be to call **transducers**, as these transduce the solar energy in the form of food (potential energy).

Producers are autotrophic, *i.e.*, self nourishing. Besides the green plants, a variety of photosynthetic bacteria and protozoans also produce organic substances in different habitats, but in a very insignificant quantity for any ecosystem.

In terrestrial ecosystems, the main producers are usually the rooted plants, such as herbs, shrubs and trees but in aquatic ecosystems, these are either floating, submerged or rooted submerged or rooted emergent plants.

(2) Consumers

The animals that consume directly or indirectly the energy of food found in green plants are called consumers. Consumers are also known as phagotrophs as these generally ingest or swallow the food. These are heterotrophic organisms and are of different categories such as-

(i) **Primary consumers**. These are purely **herbivores** (*i.e.*, herbivorous animals) that are dependent for their food on producers (*i.e.*, green plants). Some common examples of the primary producers are cattle, deer, goat, rabbit, insects etc. These are also called **first order consumers** and are the chief food source for carnivores.

(ii) **Secondary consumers**. These are the **second order consumers** and are carnivores. Generally, they are flesh eating animals that are adapted to consume herbivorous animals, *e.g.* fox, wolves, cats, snakes etc.

(iii) **Tertiary consumers**. They are popularly called **top consumers** as these are top carnivores, which prey upon other carnivores and herbivores, *e.g.* lion, tiger, hawk, vulture etc.

(3) Decomposers

Decomposers are micro-organisms, like fungi and some bacteria, which are incapable of manufacturing their food and live on dead and decaying plants or animals parts. As such they are special kind of consumers and are commonly called decomposers, or **saprotrophs** (or detrivores). These decompose complex organic substances of the dead parts into simpler forms to make them available to other organisms, the producers. By releasing enzymes, they carry out extracellular digestion of the dead remains and convert them into simpler inorganic substances. The decomposers play an important role in maintaining the dynamic nature of ecosystems.

The path of food in an ecosystem begins with producers (green plants), being eaten by herbivores, followed by carnivores. The food relation in its simplest form expressed as grass → deer → tiger → micro-organisms representing a producer, a primary consumer, a top consumer and a decomposer respectively, is called a **food chain** and food getting interactions made up of several feedings are called **trophic levels**. The first trophic level is made up of green plants which use energy from sun light to make food through the process of photosynthesis. Organisms of this trophic level are called **primary producers**. Organisms depending for their food on primary producers are called **consumers of first order** or **primary consumers**, or herbivores. Animal populations that feed upon other consumers are **second order consumers** or **secondary consumers** and these are called **carnivores**.

A model representation of ecosystem structure is given in fig 12.1.

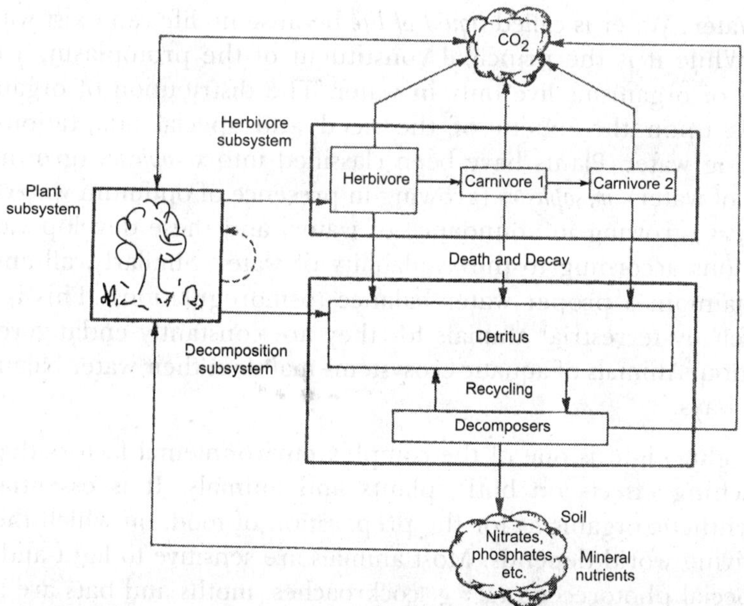

Fig. 12.1. A model representation of an ecosystem structure.

II. Abiotic Components

The non-living parts of an ecosystem are called abiotic components. Among the abiotic factors the most important are **climatic** and **edaphic factors**. The climatic factors comprise temperature, humidity, rain etc., while the edaphic factors comprise soil and substratum. The ability of organisms to utilise and combat various abiotic factors are different and it may limit their distribution, behaviour and relationship with other organisms. Some important factors are as follows:

(1) **Temperature**. Temperature influences all forms of life by exerting its action through increasing or decreasing the vital activities of organisms, like behaviour, metabolism, reproduction, development and death. Normal life activities continue in a specific range of temperature, called *optimum temperature range,* beyond that there is a minimum or maximum temperature that organisms can tolerate. Temperature is a universal influence and is frequently a limiting factor for the growth or distribution of plants and animals. Many animals and plants develop various adaptations to withstand extremes of temperature to a limited extent. Some of the plants and animals produce heat resistant spores, seeds, cysts, eggs etc. to tolerate extremes of temperature. Dried seeds and cysts escape from freezing. A polar bear can live in very cold regions and can hibernate during winter. Some birds and mammals migrate to warmer places to avoid extremes of cold. Some desert animals live inside burrows to avoid the intense heat of the desert. Plants below 0°C and above 45°C are killed.

(2) **Water**. Water is called *liquid of life* because no life can exist without water. While it is the principal constituent of the protoplasm, a large number of organisms live only in water. The distribution of organisms depends upon the extent of the need and special adaptations for conserving water. Plants have been classified into *xerophytes* (growing in scarcity of water), *mesophytes* (growing in presence of optimum water) and *hydrophytes* (growing in abundance of water) and these develop various adaptations according to the availability of water. Similarly, all animals must maintain a proper water balance (osmoregulation). This is very much felt by terrestrial animals for they are constantly endangered by desiccation. Animals of aquatic ecosystems maintain their water balance in various ways.

(3) **Light**. Light is one of the complex environmental factors that has far reaching effects on both, plants and animals. It is essential for photosynthetic organisms for the preparation of food, on which the rest of the living world depends. Most animals are sensitive to light and they have special photoreceptors, *e.g.* cockroaches, moths and bats are active

during the night. Light acts as an important limiting factor in the manifold activities of organisms. Ecologically, the quality, quantity, intensity and duration of light are quite important and organisms react vividly in their response. Some well known phenomena, related to response of light, are seed germination, photoperiodism, phototropism, circadian rhythms, photokinesis, phototaxis etc.

(4) **Humidity**. Humidity, *i.e.,* moisture content of the atmosphere is also an important factor. It directly regulates the rate at which water evaporates from the body surface of land organisms by transpiration, perspiration and other means. Different plants and animals show different adaptations to withstand dry conditions.

(5) **Wind**. Blowing air largely determines weather conditions. It affects the plants in various ways, particularly their rate of transpiration. In high wind velocity, only the plants with strong root system and tough stem and leaves, can survive, *e.g.* coconut-plants growing in sea shore regions. Wind also helps in dispersal of seeds and fruits.

(6) **Soil pH**. Soil pH determines acidity or alkalinity of the soil which makes the soil habitable to the plants and other organisms. Most organisms thrive in an optimal pH, *e.g.* a soil with 6.5 to 7.5 pH is commonly suitable to plant growth. However, some plants and aquatic animals require acidic conditions, (pH below 7), others need neutral or alkaline, (pH above 7), conditions. pH of the soil and water has a strong influence on the distribution of organisms.

(7) **Mineral elements**. Soil constitutes the house of minerals. Availability and concentration of essential mineral elements control the distribution of microbes, plants and animals. Some elements are essential for plant growth, those required in a large quantity are called *macro-elements* (*e.g.* Calcium, Magnesium, Nitrogen, Phosphorus, Sulphur, Potassium, Iron etc.) and those required in traces are called *micro-elements* or micro-nutrients (*e.g.* Zinc, Manganese, Boron, Copper, Molybdenum etc.). Without these elements, normal growth of plants is not possible. Salinity in soil or water greatly affects the distribution of organisms. Special salt secreting glands occur in plants and animals to eliminate excessive salt. Most mangrove plants, growing in saline water, develop vivipary and negatively geotropic breathing or respiratory roots called pneumatophores as special adaptations.

(8) **Topography**. Surface configuration or physical feature of an area is called topography. It influences the distribution of organisms as much as wide geographic separations. Indirectly, it also affects other factors like wind, water current, light or wave action. The distribution of plants in north and south facing hills of Himalayas is remarkably distinct. The

centre and edge of a pond or a stream, top side and under side of a rock, north and south faces of a ridge are generally inhabited by different organisms.

(9) **Background**. The background of the habitat also determines the distribution of animals by enabling them to camouflage against the colour, texture and pattern. Desert animals like lion and the camel are sand-coloured. Most of the jelly fishes and sea cucumbers are grassy in appearance. The chameleon can change its colour according to its background.

FUNCTIONAL ASPECTS OF AN ECOSYSTEM

Functional aspects of an ecosystem broadly include processes such as energy flow, nutrient cycling, productivity etc. Without adequate knowledge of functional aspects, the concept of ecosystem can never be complete, and, therefore, both, structure and function are best studied together. Function reveals working of an ecosystem under natural conditions which shows that the living (biotic) and non-living (abiotic) components are so intricate that their separation from each other becomes practically impossible. The functional aspects can be better explained as follows:

(1) In the initial stage, green plants, also called producers, fix radiant energy (solar energy) in the presence of chlorophyll (green pigments), carbon dioxide and water. The process is called **photosynthesis** and through this process carbohydrates (food materials) are synthesized in the plants. Carbohydrates comprise chemical energy which is being used by plants themselves for their own metabolic activities as well as by herbivorous animals like cattle and insects for their life activities. Several other kinds of food materials are produced by producers as well as by herbivores such as proteins and fats utilizing carbohydrates as a source of energy and the substrate. These processes require several minerals that are absorbed by the plants from the soil, and as such these are called essential nutrients. Mineral nutrients, which are required in a large quantity are called **macronutrients** (such as nitrogen, phosphorus, sulphur, magnesium, potassium, iron, calcium, besides carbon, hydrogen and oxygen) and those required in trace or small quantity are called **micronutrients** or trace elements (such as copper, zinc, manganese, molybdenum, boron etc.). Likewise several other elements are essentially required by plants and animals in specific manners.

(2) The food that comprises energy bound organic matter in the organisms is the productivity of an ecosystem and constitutes the biomass. The rate of biomass production is called **productivity**. The **productivity** is defined as *the amount of organic matter produced over a specified period of time*

irrespective of whether or not it all survives to the end of that time. Thus, the production of an ecosystem per unit time is commonly proportional to the mass or quantity of available material and also to the amount of energy for the transformation and turnover of matter involved in production process. The *autotrophs* determine the available energy whereas the *heterotrophs* affect the turnover and cycling of matter within an ecosystem. The efficiency of productivity and cycling of matter is affected by physico-chemical and climatic factors etc. The two ecological processes, the energy flow and the mineral cycling, may be considered as the *heart* of the ecosystem dynamics.

(3) The food synthesised (productivity) by green plants is either taken away by herbivorous organisms by eating them or is lost by the death and decay of plants or plant parts. The herbivorous organisms are eaten by carnivorous animals. The transfer of food energy from the producers through a series of organisms (herbivores to carnivores to decomposers) with repeated eating and being eaten, is known as a **food chain**, and each constitutes the **trophic level**.

In any food chain, green plants occupy the *first trophic level* (*i.e.*, producer level) and are called **primary producers**. Herbivores which eat the plants and utilize their energy constitute *the second trophic level* and are called **primary consumers**. Herbivores in turn are eaten by carnivores and such carnivores form the *third trophic level* and are called **secondary consumers**. Secondary consumers are strictly carnivorous. These in turn may be eaten still by other carnivores which constitute the *third trophic level* and such organisms are called **tertiary consumers**. This simple chain of eating and being eaten away is known as food chain, *i.e.*,

Primary producers (Green plants)	→	Primary consumers (Herbivores)	→	Secondary consumers (Carnivores)
				↓
				Tertiary consumers (Top carnivores)

(4) Organisms of all trophic levels are subjected to death and decay. So either through a food chain or due to death and decay, the productivity is ultimately decomposed step by step releasing minerals in the soil and energy in the atmosphere. Since these minerals are again absorbed by the plants and utilised again and again, their movement in the ecosystem is called **cyclic**, and since energy is ultimately lost in the atmosphere and not utilised again, its movement is **unidirectional**. The movement of materials in cyclic manner and that of energy in non-cyclic (unidirectional) manner can be best explained with **biogeochemical cycle** (fig. 12.2).

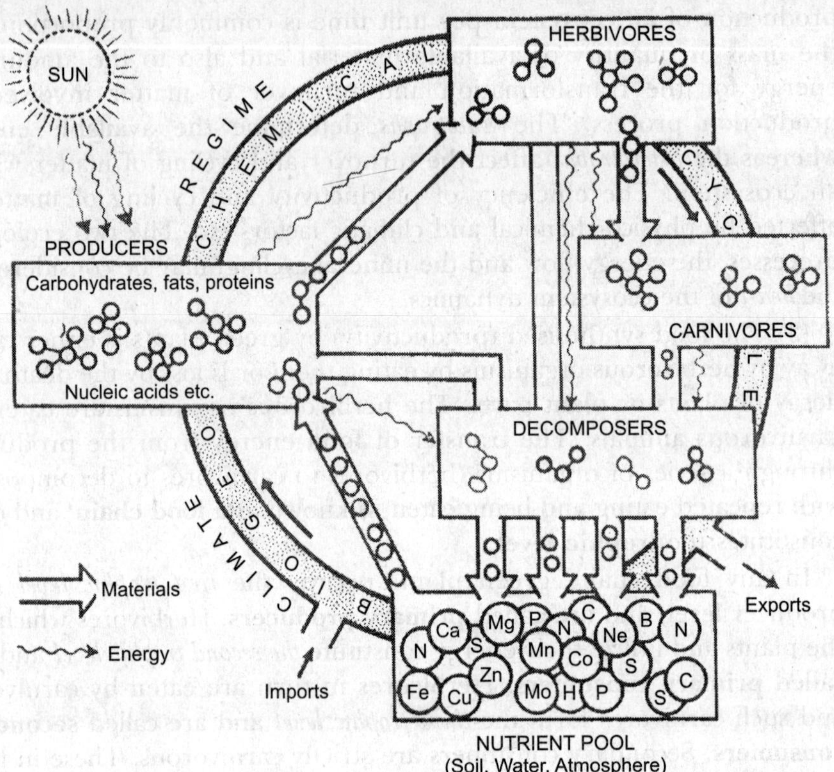

Fig. 12.2. Structure and function of ecosystem consolidated as biogeochemical cycle. The cycle shows cyclic movement of minerals and unidirectional movement of energy (wavy line).

Thus, functional aspects of the ecosystem in its two components are broadly the *productivity* (in biotic components) and the *mineral cycles*.

Productivity

The rate of biomass production is called **productivity**. Productivity in ecosystems, is expressed in terms of dry matter (biomass) produced or energy captured per unit area of land per unit time. Energy is expressed in calories /cm^2/year and dry organic matter in g/m^2/year. By these values, the productivity of different ecosystems can be easily compared.

Productivity is of the following types-

(1) Primary productivity: Green plants fix solar energy and accumulate it in the organic forms as chemical energy. It is the first and the basic form of energy storage. The rate at which the energy accumulates in the green plants (producers) is known as *primary productivity*. It is further distinguished as-

(i) ***Gross primary productivity***: The total energy trapped by the green plants is referred to as gross primary productivity. It is the **total photosynthesis** or **total assimilation.**

(ii) ***Net primary productivity***. A good fraction of gross primary productivity is utilized in respiration of green plants and other plant losses as death etc. The amount of energy left in green plants after these losses forms the net primary productivity, *i.e.*,

Gross productivity	–	Energy lost in respiration	=	Net primary productivity

It is the net primary productivity which is available to the consumers.

(2) Secondary Productivity: The net primary productivity results in the accumulation of plant biomasss and serves as food for herbivores and decomposers. Now the rate at which the heterotrophic organisms resynthesise the energy yielding substances is called secondary productivity, *i.e.*, it is the productivity of animals and saprobes in the ecosystem. Secondary productivity actually remains changing, *i.e.*, keeps on moving from one organism to another organism.

(3) Net Productivity: It refers to the rate of storage of organic matter not used by consumers or heterotrophs and is thus the rate of increase of biomass of the primary producers which has been left over by the consumers. It may then be consolidated on month, season or year basis.

Biogeochemical Cycles

Some forty elements are known to be required by living organisms. These elements tend to be used over and over again, or to be recycled from environment to organisms and back to the environment again. Thus, *biogeochemical cycles* involve pathways of elements moving repeatedly between inorganic forms and organic molecules (fig. 12.2).

Broadly, these can be divided into **gaseous cycles** and **sedimentary cycles.**

The **gaseous cycles** are the *carbon*, *nitrogen*, *oxygen* and *hydrological* cycles. These elements are moved about in large amounts, with the earth's atmosphere serving as the main inorganic storage reservoir.

The **sedimentary cycles** are said to be 'imperfect' cycles because the constituent elements end up in sedimentary rock, a reservoir from which recycling is very slow. These include *phosphorus cycle*, *sulphur cycle* and cycles of their minerals. These cycles extend across long periods of geologic time (hundred to millions of years).

BIOGEOCHEMICAL CYCLES

Flow of energy in the ecosystem is unidirectional. In contrast to it, chemical elements including all essential elements of protoplasm like water, carbon, phosphorus, nitrogen etc. tend to circulate in the biosphere in characteristic paths. These, more or less circular paths, are known as biogeochemical cycles.

Biogeochemical cycles are basically divided into two types-

(i) **Gaseous cycles** in which the reservoir is in the atmosphere or hydrosphere. It includes carbon, oxygen, hydrogen and nitrogen cycles.

(ii) **Sedimentary mineral cycles** in which the reservoir is in the lithosphere, *i.e.,* earth crust. It includes phosphorus and sulfur cycles etc.

Both these cycles involve biological and non-biological agents and are more or less tied with another cycle. Some important cycles are discussed.

Carbon Cycle

Carbon is contained in every organic compound, *e.g.* carbohydrates, proteins, fats, nucleic acids etc. and it makes up a living being.

(1) ***Carbon dioxide goes into the living being from atmosphere***

The main source of carbon for the living world is the free carbon dioxide (CO_2) available in the atmosphere and the producers are the first living organisms to entrap carbon from CO_2 by the process called photosynthesis. It can be briefly explained as follows:

$$6CO_2 + 12H_2O \xrightarrow[\text{chlorophyll (green plants)}]{\text{sunlight energy}} \underset{\text{(carbohydrate)}}{C_6H_{12}O_6} + 6H_2O + 6O_2$$

CO_2 present in atmosphere is about 0.03 per cent or 300 ppm.

About 4–9 × 10^{13} kg. of carbon is fixed or assimilated annually in the photosynthesis. The simple carbohydrates like glucose are converted into complex carbohydrates (polysaccharides) like starch and cellulose. Starch is commonly stored in plant tissues.

Plants are eaten by herbivores that digest these carbohydrates and resynthesise other carbon compounds of their own needs to obtain energy and carry on their life processes. Herbivores are eaten by carnivores, *i.e.,* carnivores digest the carbon compounds of herbivores and resynthesise them into compounds of their own choice.

(2) ***Carbon gets back into the atmosphere***

Carbon returns back to the atmosphere mainly by the following ways-

(i) Carbon in the form of CO_2 is released by the producers, consumers and decomposers through the process called respiration.

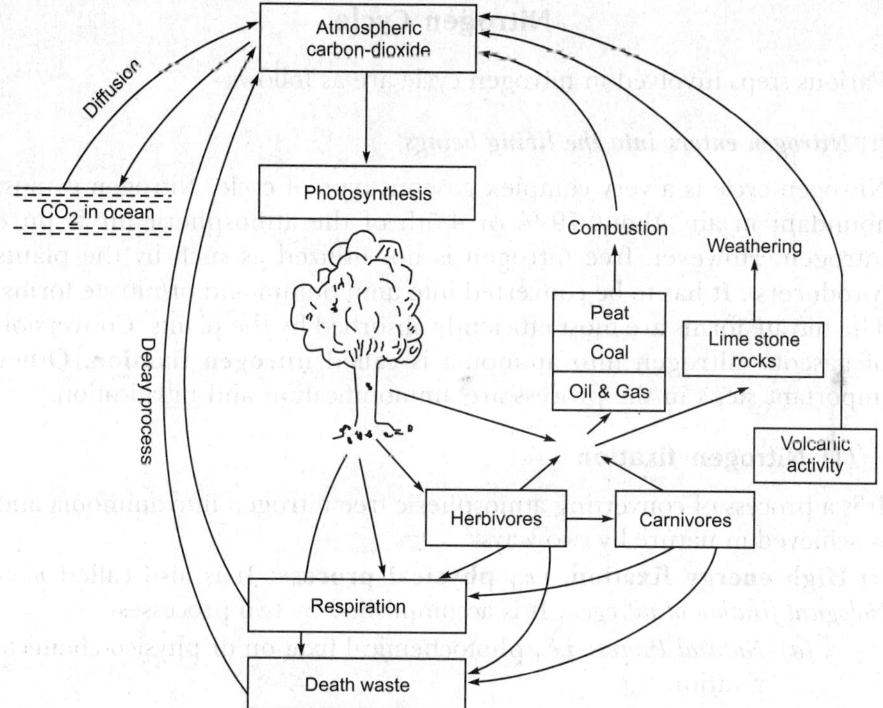

Fig. 12.3. Schematic representation of carbon cycle.

(ii) The dead remains of plants and animals undergo decay process due to the activity of saprobes (decomposers). The decomposers release the locked up CO_2 into the atmosphere.

(iii) Part of organic carbon gets burried in the earth's crust and is converted into fossil fuels (coal, oil, and gas), though this step takes million of years. By burning of this fuel, carbon dioxide is returned to the atmosphere.

(iv) Part of organic carbon gets burried in the earth's crust and in course of time forms lime stone rocks. By weathering of these rocks CO_2 is released back into the atmosphere, though the process is very slow.

(v) Volcanic activities of the earth and hot springs process out about 100 million tons of CO_2 back into the atmosphere, every year.

(vi) Carbon dioxide dissolved in water of the ocean is the second important source of the gas. The interchange between the two phases, *viz.*, atmospheric CO_2 and dissolved CO_2 in oceans, occurs through diffusion. The direction of diffusion is dependent upon relative concentration. Passage into aquatic phase also takes place through perception. A litre of rain water contains about 0.3 ml of CO_2. It can be shown by simple reaction.

$$\underset{\text{(Water)}}{H_2O} + CO_2 \rightleftharpoons \underset{\text{(Carbonic acid)}}{H_2CO_3}$$

Nitrogen Cycle

Various steps involved in nitrogen cycle are as follows -

(I) ***Nitrogen enters into the living beings***

Nitrogen cycle is a very complex gaseous type of cycle. Nitrogen is most abundant in air. About 79 % or 4/5th of the atmospheric air is pure nitrogen. However, free nitrogen is not utilized as such by the plants (producers). It has to be converted into ammonium and or/nitrate forms. The nitrate forms are most efficiently absorbed by the plants. Conversion of gaseous nitrogen into ammonia is called **nitrogen fixation**. Other important steps in the process are ammonification and nitrification.

(1) Nitrogen fixation

It is a process of converting atmospheric free nitrogen into ammonia and is achieved in nature by two ways-

(i) **High energy fixation,** *i.e.,* **physical process**: It is also called *non-biological fixation of nitrogen*. It is accomplished by two processes-

(a) Natural Process, i.e., photochemical fixation or physico-chemical fixation

(b) Man made process, i.e., industrial fixation

(a) In ***photochemical fixation of nitrogen***, the high energy required for dissociation of nitrogen molecule and for its combination with oxygen and hydrogen of water is provided by lightening. The energy can be provided by cosmic radiation and meteorite trails.

(b) The ***industrial fixation of nitrogen*** infact refers to man made nitrogen fertilizers in various industries. Most chemical fertilizers are in the form of nitrates or ammonium compounds and are supplied in agriculture fields to increase the soil fertility, thereby crop yield.

(ii) **Biological fixation of nitrogen**: It is carried out by certain bacteria, cyanobacteria and few fungi. About 95% of the total fixed nitrogen on earth occurs through this process, and this nitrogen is fixed in the form of ammonia that converts into ammonium salts or nitrates by other processes. N_2-fixing organisms are also called *diazotrophs*. Process is basically of two types-

(A) Non-symbiotic nitrogen fixation: It is processed by micro-organisms living free in the soil. A few examples of these nitrogen fixers are:

Bacteria:

- Anaerobic forms- *Bacillus, Clostridium pasteurianum*
- Aerobic forms- *Azotobacter, Azomonas*
- Photosynthetic forms- *Rhodospirillum, Chromatium*

Fungi: *Pullularia,* yeasts

Blue green algae (cyanobacteria): *Nostoc, Anabaena, Aulosira, Tolypothrix, Scytonema* etc.

(B) Symbiotic nitrogen fixation: Nitrogen fixation carried out by micro-organisms living in symbiotic association with higher plants is known as symbiotic nitrogen fixation. In agro-ecosystems, about 200 species of nodule forming legumes are important nitrogen fixers. Inside their root nodules live different species of *Rhizobium* (bacteria) which are adopted to fix atmospheric nitrogen.

By biological nitrogen fixation process about 140-700 mg/m^2/yr N_2 is fixed which is roughly about 1g/m^2/yr.

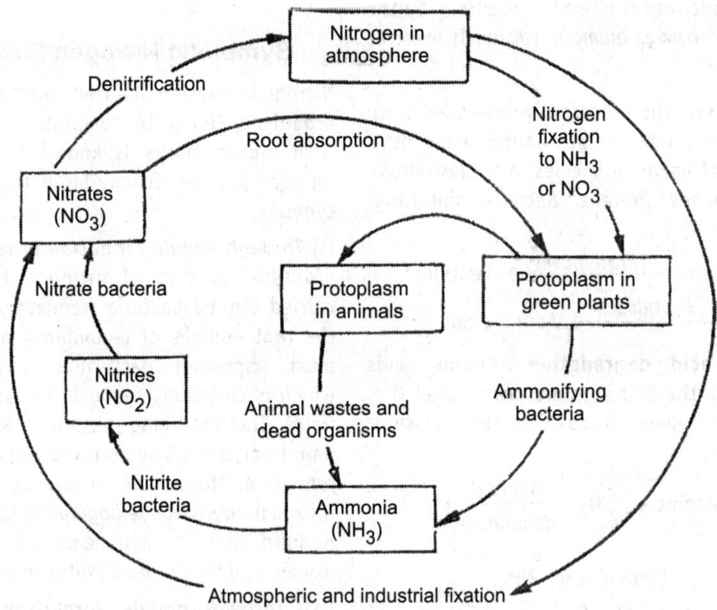

Fig. 12.4. Schematic representation of nitrogen cycle (after Southwick, 1976).

(2) Ammonification

Plants incorporate fixed nitrogen into the protoplasm by amino acid and protein synthesis. In ecosystem, the organic nitrogen compounds may then follow any one of the three routes-

(i) they may be stored or enter in the formation of proteins or nucleic acids within the plant.

(ii) they may be transferred to animal tissues through consumption and assimilation by animals (consumers).

(iii) they may be decomposed to ammonia by death and bacterial action. Some part of animal nitrogen may be decomposed into urea and other excretory products by consumers also.

Box 12.1

Ammonification

On the death and decay of plants and animals and from the excreta of animals, complex organic compounds are released into the soil. These complex organic compounds are acted upon by micro-organisms and broken down into simpler compounds with the release of energy by the process known as *ammonification*. It involves two steps:

Proteolysis: The process involves breakdown of proteins into simple amino acids by a number of bacterial species, *e.g. Clostridium, Pseudomonas, Proteus, Bacillus*, soil fungi etc.

$$\text{Proteins} \xrightarrow{\textit{Proteinases}} \text{Peptides} \xrightarrow{\textit{Peptidases}} \text{Amino acids}$$

Amino acid degradation: Amino acids formed in the first step are deaminated (i.e. removal of amino group) with the release of ammonia.

$$\text{Alanine} + \frac{1}{2}O_2 \xrightarrow[\textit{deaminase}]{\textit{Alanine}} \text{Pyruvic acid} + NH_3$$

•

Nitrification

Ammonia produced by degradation of manures and organic matter is converted into nitrates by the process of nitrification. It involves two steps and is carried out by 2 different groups of micro-organisms.

Oxidation of ammonia into nitrite: Ammonia is oxidised to nitrites by *Nitrosomonas, Nitrosococcus* and *Nitrosolobus* bacteria. The step is also called *nitrosification*.

$$2NH_3 + 3O_2 \xrightarrow[\textit{monas}]{\textit{Nitroso}} 2HNO_2 + 2H_2O$$

Oxidation of nitrite into nitrate: *Nitrobacter, Nitrocystis* and *Bactoderma* group of bacteria are mainly responsible for oxidation of nitrites into nitrates.

$$HNO_2 + \frac{1}{2}O_2 \xrightarrow{\textit{Nitrobacter}} HNO_3$$

•

Symbiotic Nitrogen Fixation

Nitrogen fixation carried out by micro-organisms living in symbiotic association with higher plants is known as *symbiotic nitrogen fixation*. It may be of the following types:

(i) ***Through nodule formation in leguminous plants:*** This kind of nitrogen fixation is carried out by bacteria frequently found in the root nodules of leguminous plants. The most important bacterium is *Rhizobium* which is commonly found in the soil of crop fields. The free-living rhizobia enter the root hair from the tip region and induce *nodule formation*. The nodule so formed contains a pink coloured *leghaemoglobin* pigment and nodulin protein. Leghaemoglobin absorbs oxygen and thus protects *nitrogenase* activity.

(ii) ***Through nodule formation in non-leguminous plants:*** Nodule formation has been found to occur in quite a few non-leguminous plants, *e.g.* root nodules of *Casuarina* and *Podocarpus* and leaf nodules of *Pavetta* and *Psychotria*.

(iii) ***Through non-nodulation symbiotic associations:*** Non-nodulating, nitrogen fixing, symbiotic associations have been reported to occur between cyanobacteria and bacteria and plants, *e.g. Anabaena azollae* grow symbiotically with *Azolla* (water fern), A. *cycadae* with *Cycas* coralloid root, *Nostoc* with *Gunnera macrophylla* and *Azotobacter paspali* with *Paspalum notatum*.

On death and decay of plants and animals and from the excreta of animals, the complex organic compounds are released into the soil where they are decomposed by micro-organisms and are broken down into simpler compounds with the release of energy by the process known as *ammonification*.

(3) Nitrification

The conversion of ammonia into nitrates is known as nitrification. The process is carried out by soil bacteria in two steps:

(i) Ammonia is converted into nitrite by *Nitrosomonas, Nitrosococcus* etc., *i.e.,* $NH_3 \rightarrow NO_2$. The process is also called *nitrosification*.

(ii) Nitrites are converted into nitrates by bacteria like *Nitrobacter* and *Nitrocystis, i.e.,* $NO_2 \rightarrow NO_3$

Nitrates are directly absorbed by the plants and are incorporated into proteins, nucleic acids or other nitrogenous organic compounds. Some nitrates may be stored in the decomposing humus, immobilized by the bacteria and some may be leached away into the ponds, lakes and streams etc. and finally into the sea.

(II) ***Nitrogen gets back into the atmosphere***

Nitrogen gets back into the atmosphere by the process called *denitrification,* carried out by denitrifying bacteria like *Pseudomonas*. Such bacteria break the nitrate ions into nitrite and then into ammonia or molecular nitrogen which gets back into the atmosphere. The nitrogen cycle is thus completed.

The cyclic flow of nitrogen in the ecosystem requires balanced action of bacteria (many species) so that sufficient levels of plant nutrients are maintained.

Phosphorus Cycle

It is a sedimentary type of cycle as phosphorus has its main reservoir not in the air but in rocks and other natural deposits formed during geological ages. However, in ecosystem, it is more abundantly found in plants and animals as compared to abiotic environment. In living components, it is a constituent of biomembranes, bones and teeth. It is essential for metabolic processes that convert food to usable energy and structure of DNA and certain coenzymes.

Phosphorus cycle is of vital importance for the maintenance of life (fig. 12.5). Crystalline rocks are major natural resource of phosphorus in which it is mainly found in the form of ferric and calcium phosphate. It is released slowly in soluble forms by leaching, erosion and mining.

Phosphorus is also added to the soil by man in the form of artificial fertilizers, made using phosphate rocks.

The soluble forms of phosphorus are absorbed by the plants via the roots and assimilated in the plant body. It is transferred to the animals through food. It returns to the surroundings when plant and animal bodies decompose or animals excrete. Phosphates incorporated in bones and teeth become resistant to decay and may remain outside the natural cycle for a very long time. When phosphates form compounds with metals like aluminium, iron or calcium, it becomes nonavailable to the plants and is thus lost to the cycle.

The run off from the soil and the loss of phosphate to ocean is greater than gain on to the land. Since uplift of sediment being not extensive, only 60,000 tonnes of phosphate is returned to the soil mainly through sea animals (birds, fish etc.) and kelps, whereas two million tonnes of phosphate is lost to sea.

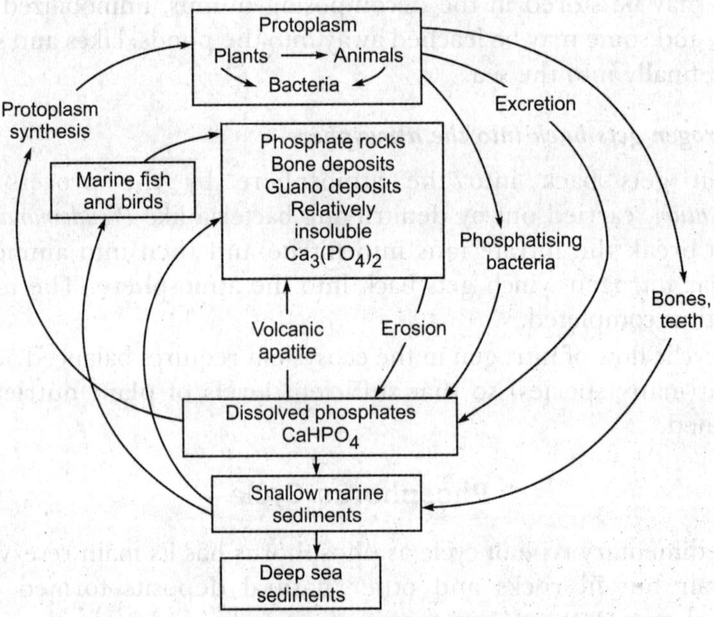

Fig. 12.5. Schematic representation of phosphorus cycle (after Odum, 1983).

Sulphur Cycle

Sulphur cycle is a sedimentary type of cycle as the reserve pool of the element is in soil and the residence time of sulphur in the atmosphere is very short.

In nature, sulphur is found in elemental as well as in combined forms like H_2S (hydrogen sulphide), SO_2 (sulphur dioxide), SO_4 (sulphates) etc.

Plants absorb soluble sulphates from the soil via roots and which thus enters into the living systems. It is used in protein synthesis, vitamins and in some essential oils. Organic sulphur is transferred to animal bodies through food.

Organic sulphur found in plant and animal bodies is aerobically decomposed to sulphate form or H_2S by bacterial action. Anaerobic forms of bacteria like *Beggiatoa* and *Thiobacillus* convert H_2S into elemental sulphur. Such bacteria are chemosynthetic autotrophs. Ultimately, by various decomposition processes, sulphur returns to the soil and thus cycle is completed. Important steps in sulphur cycle are shown in the figure (fig. 12.6). Sulphur is also locked in fossil fuels (coal, petroleum) and is released as sulphur dioxide when these products are burnt.

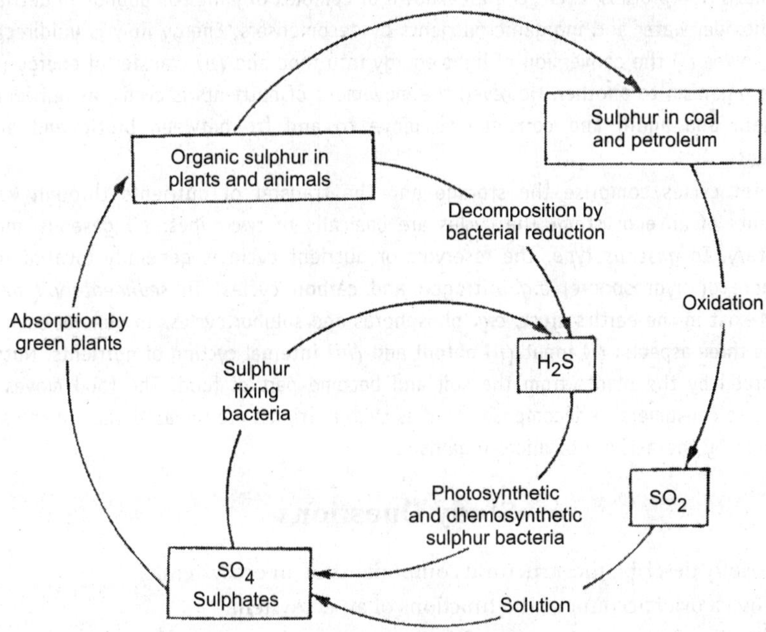

Fig. 12.6. Schematic representation of sulphur cycle (after Southwick, 1976).

Chapter Summary

An ecosystem has a definite organisation comprising structural and functional components. Structural components are divided into abiotic and biotic. The functional components include the volume and rates at which material circulates and energy flows as well as the processes of interaction between the living and non-living components.

Biotic Components are producers, consumers and decomposers. **Producers** are green plants which harvest solar energy by the process called photosynthesis. The products of photosynthesis are stored as food. **Consumers** are heterotrophs. The various levels of consumers consisting of herbivores and carnivores obtain their material and energy requirements by consuming food. **Decomposers** are saprotrophs which survive by decomposing complex organic molecules of dead and decaying organic matter into simpler ones, thus returning the material back to the environment. Energy trapped from sunlight escapes into the surroundings as heat.

Abiotic Components comprise both climatic and edaphic factors, *viz.*, temperature, water, light, humidity, wind, pH of soil, minerals, topography and background of the environment.

Important functions of ecosystem are productivity, decomposition and mineral cycling (bio-geochemical cycles). *Productivity* is involved in the production of food while decomposition leads to the breakdown of organic matter. *Primary productivity* is the rate of capture of solar energy or production of biomass by producers. The rate of total production of organic material or capture of energy is known as *gross primary productivity*. The *net primary productivity* represents the balance after meeting the cost of respiration. *Secondary productivity* is the rate at which food energy is assimilated at the trophic levels of consumers. In decomposition process, there is a breakdown of complex organic compounds in detritus to carbon dioxide, water and inorganic nutrients by decomposers. Energy flow is unidirectional which involves *(i)* the conversion of light energy into food and *(ii)* transfer of energy (food) from one organism to another. However, the movement of nutrients is cyclic, as nutrients are used again and again and continue to move to and fro between biotic and abiotic components.

Nutrient cycles comprise the storage and the transfer of nutrients through various components of an ecosystem. The cycles are basically of two types: *(i)* gaseous and *(ii)* sedimentary. In *gaseous type*, the reservoir of nutrient cycle is generally located in the atmosphere or hydrosphere, *e.g.* nitrogen and carbon cycles. In *sedimentary type*, the reservoir exist in the earth's crust, *e.g.* phosphorus and sulphur cycles. In general each cycle possesses three aspects: *(i)* input *(ii)* output and *(iii)* internal cycling of nutrients. Nutrients are absorbed by the plants from the soil and become part of food. The food moves from producers to consumers to decomposers and as such nutrients are released back to the soil or atmosphere by the activity of micro-organisms.

Study Questions

1. Briefly describe the structural components of an ecosystem.
2. Give a brief account of the functions of an ecosystem.
3. Outline the salient features of the structure and functions of an ecosystem.
4. What is meant by nutrient cycling? Explain by taking example of nitrogen cycle.
5. What are sedimentary cycles? Briefly describe any one of them.
6. List the processes involved in either nitrogen or carbon cycle.
7. Describe the various steps of either Phosphorus or Sulphur cycle.
8. What are various structural components of an ecosystem?
9. Describe productivity types met in ecosystem.
10. What is ecosystem productivity? Explain.
11. What are biogeochemical cycles? Name at least three different types of cycles.
12. Explain nitrogen cycle with the help of diagrams only.
13. Explain phosphorus or carbon or sulphur cycle by line diagrams.
14. What are important functions of an ecosystem? Describe.

15. Describe nitrification and denitrification processes.
16. Name three different kinds of structural components of an ecosystem.
17. What is producer? Give some examples.
18. What are consumers and what are differences between primary and secondary consumers? Describe.
19. Define productivity.
20. Define biogeochemical cycles. Name two types of them.
21. What are decomposers? Explain.
22. Outline the function of decomposers in the ecosystem.
23. What is nitrogen fixation? Describe.
24. Define nitrification and denitrification.

Objective Questions: *Select the correct answers*

1. Biotic components do not include:
(1) Autotrophs (2) Heterotrophs
(3) Saprotrophs (4) Mineral nutrients.

2. Which of the following occurs in abiotic components of an ecosystem?
(1) Cycling of Material (2) Flow of energy
(3) Consumers (4) Both (1) and (2)

3. Food is the source of energy for:
(1) Producers (2) Consumers
(3) Decomposers (4) All of these

4. An example of sedimentary type of nutrient cycle is:
(1) Nitrogen cycle (2) Carbon cycle
(3) Phosphorus cycle (4) Both (1) and (2)

5. The importance of ecosystem lies in:
(1) Cycling of materials (2) Flow of energy
(3) Both (1) and (2) (4) None of these

6. Process of photosynthesis is found in-
(1) Decomposers (2) Producers
(3) Consumers (4) Top consumers

7. Functions of an ecosystem comprises studies on:
(1) Composition of biological community
(2) Distribution of non-living materials
(3) Top consumers
(4) Productivity

8. Process of denitrification is found in:
(1) Nitrogen cycle (2) Carbon cycle
(3) Producers (4) All of these

Answers

1. (4) *2.* (4) *3.* (4) *4.* (3) *5.* (3) *6.* (2)
7. (4) *8.* (1)

13

CHAPTER

Producers, Consumers and Decomposers

LEARNING OBJECTIVES

Introduction • PRODUCERS • Autotrophs • Summary of Photosynthesis • Common Producers of Various Ecosystems • CONSUMERS • Heterotrophs • Oxidation-reduction Process • Common Consumers of Various Ecosystems • DECOMPOSERS • Saprotrophs • Decomposition Process • Common Decomposers of Various Ecosystems.

Introduction

The ecosystems are the basic functional ecological units which include both, the organisms and the environment, each influencing the fuctioning of other and both are necessary for the maintenance of life. The living organisms of various categories of an ecosystem are collectively called biotic components. The biotic components on the basis of mode of nutrition are broadly divided into three categories-

(1) Producers, *i.e.,* autotrophs
(2) Consumers, *i.e.,* heterotrophs
(3) Decomposers, *i.e.,* saprotrophs

PRODUCERS

Producers are those organisms who can manufacture organic compounds from simple inorganic substances. These are the *autotrophic* or self-nourishing components of the ecosystem. Broadly speaking, there are three different types of autotrophs-

(i) **Photosynthetic green plants** that evolve oxygen as byproduct during photosynthesis. Such photosynthesis is called **oxygenic photosynthesis**.

(ii) **Photosynthetic bacteria** that do not evolve oxygen as byproduct during photosynthesis. Such photosynthesis is called **anoxygenic photosynthesis** and is found in bacteria like *Chlorobium* and *Rhodospirillum*.

(iii) **Chemosynthetic bacteria** that do not carry out photosynthesis, rather these use chemical energy in manufacturing of their food. These include bacteria like *Nitrosomonas, Nitrosococcus, Nitrobacter* (*i.e.,* nitrifying bacteria), *Thiobacillus* (sulphur bacteria), *Ferrobacillus,* (iron bacteria), *Methanomonas* (methane bacteria) etc.

The contribution of photosynthetic and chemosysnthetic bacteria, as compared to green plants in ecosystem, is quite negligible and is of only literary interest.

Most important producers in ecosystem are green plants, and their specific property is to carry on photosynthesis. In photosynthesis, solar energy is trapped into chemical energy in presence of chlorophyll and water. A typical **photosynthesis** consists of two processes, *i.e.,* **Light reaction** and **Dark reaction**.

In **light reaction phase** of photosynthesis, light energy is trapped by chlorophyll molecules and is used in the generation of reducing powers (or assimilatory powers). The reducing power remains in the form of energy rich compound, ATP (adenosine triphosphate) and reduced NADP or $NADPH_2$ (nicotinamide adenosine triphosphate). In reduction of NADP, electrons from water molecules are required and as a result water photolyses and oxygen from water is released as byproduct. ATP and $NADPH_2$ are the products of light reaction in photosynthesis.

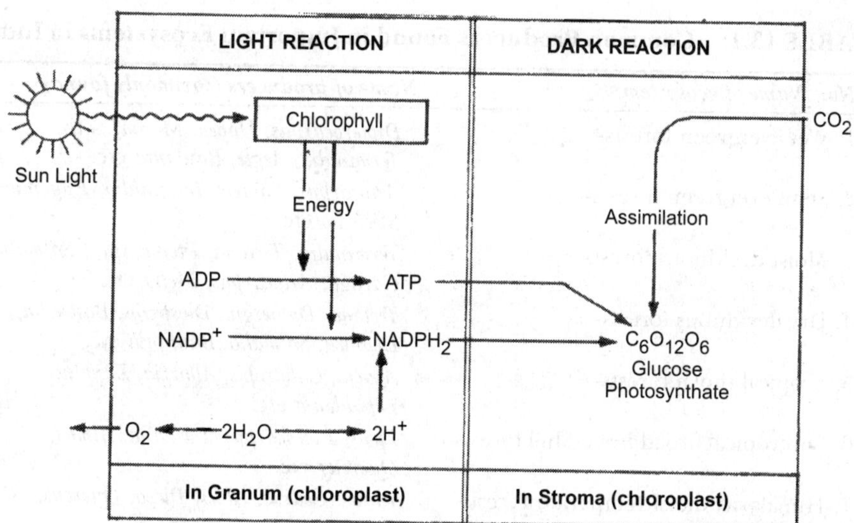

Fig. 13.1. A generalised summary of photosynthesis found within the chloroplasts of green plants (producers).

In **dark reaction phase**, the reducing powers of ATP and $NADPH_2$ are utilised in the reduction of carbon dioxide (or assimilation of CO_2) with the help of a five carbon compound called ribulose 1,5 diphosphate which gets converted into highly unstable six carbon compounds, each splitting into two 3-carbon compounds called phosphoglyceric acid (PGA). PGA molecules are step by step converted into glucose molecules and then into polysaccharides like starch. Starch is the reserve food material of green plants and is used by herbivores as their food. By complicated metabolic processes found in plants, other food materials like proteins and fats are also formed which are used by herbivorous animals.

On the basis of first stable products in the photosynthesis, two types of plants are recognised, *i.e.*,

(a) **C_3 plants**- in which a 3-carbon compound phosphoglyceric acid (PGA) is the first stable product. These are most common and are in majority. Algae, Bryophytes, Pteridophytes, Gymnosperms and majority of Angiosperms belong to this category.

(b) **C_4 plants**- in which 4-carbon compound is the first stable product which may be oxaloacetic acid (OAA), malate or aspartate. Majority of grasses and some dicots are C_4 plants and as they are more efficient photosynthesisers, they are most efficient producers in the ecosystems, *e.g.* maize and sugarcane.

Though producers have basically similar type of photosynthesis, they show a vast form of diversity in different ecosystems. Producers found in terrestrial and aquatic ecosystems and in forests, deserts, grasslands, lakes, rivers and ocean ecosystems are considerably variable. Common producers found in different Indian ecosystems are given in the table 13.1.

TABLE 13.1. Common Producers Found in Important Ecosystems in India

No. Name of ecosystem	*Name of producers (commonly found)*
1. Wet evergreen forests-	*Dipterocarpus, Hopea, Mesua, Salix, Terminalia, Aegle, Bauhinia* etc.
2. Semi evergreen forests-	*Artocarpus, Shorea, Terminalia, Eugenia, Michelia* etc.
3. Moist deciduous forests-	*Terminalia, Tectona, Pterocarpus, Salmalia, Garuga, Artocarpus, Melia* etc.
4. Dry deciduous forests-	*Tectona, Dalbergia, Diospyros, Boswellia, Emblica, Salmalia, Bauhinia* etc.
5. Tropical thorn forests-	*Acacia, Calotropis, Albizzia, Zizyphus, Leptadenia* etc.
6. Subtropical broad leaved hill forests-	*Ficus, Pterocarpus, Lantana, Almus, Quercus* etc.
7. Himalayas moist temperate forests-	*Pinus, Cedrus, Abies, Picea, Quercus, Rhododendron*
8. Deserts-	*Prosopis, Salvadora, Capparis, Calotropis, Casuarina, Opuntia, Euphorbia* etc..

9. Grasslands-	*Dicanthium, Cymbopogan, Arundinella, Phragmitis, Ischaemum* etc
10. Ponds and lakes-	*Trapa, Typha, Sagittaria, Hydrilla, Vallisnaria, Potamogeton, Nymphaea, Eichhornia, Azolla, Salvinia* etc.
11. Ocean or Marine-	Brown algae *(Sargassum, Terbinaria, Ectocarpus)*, red algae *(Gracilaria, Chondrus, Polysiphonia)*, green algae *(Enteromorpha, Ulva, Caulerpa)* etc.
12. Cropland-	Besides crop plants, *Cynodon, Launaea, Euphorbia, Cyperus, Saccharum* etc.

CONSUMERS

These are the **heterotrophic organisms** depending upon autotrophs (green plans or producers) or other organisms or their parts for the food such as animals, birds etc. which ingest other organisms or particulate organic matter. Under this category are included all different kinds of animals that are found in an ecosystem. Consumers are also known as **phagotrophs** (*phago* = to eat). There are different categories of consumers such as primary, secondary, tertiary, scavengers etc.

Primary consumers are exclusively herbivorous animals that are dependent for their food on producers or green plants. The common primary consumers are insects, rodents, rabbit, deer, cow, buffalo, goat, elephants etc. in terrestrial ecosystems and insect larvae, zooplanktons, crustaceans, fish and some molluscs in aquatic ecosystems etc.

Secondary consumers are small carnivores. Generally they are flesh eating animals that are adapted to consume herbivores. The common secondary consumers are fox, wolves, cats, wild dogs, snakes etc.

Tertiary consumers are top carnivores which prey upon other carnivores and herbivores. Common examples are lion, tiger, hawk, vultures etc. in terrestrial ecosystems and cod, halibut, carnivorous fish etc. in aquatic ecosystems.

Besides different categories of consumers, the parasites, scavengers and saprobes are also consumers. The parasitic plants and animals utilize the living tissues for their own life. Scavengers and saprobes utilize dead remains of plants and animals as their food. Since these are smaller in size and difficult to study, these are not accounted in general. Another category of consumers is the **omnivores**, as they derive their energy from both producers and herbivores.

Transfer of material and transformation of energy from one trophic level to another occurs between the different consumers through the process of eating and being eaten. A part of synthesised food is used by organisms at each level for their own maintenance, multiplication and movement. The food is utilized through biological *oxidation-reduction process* which yields energy.

Oxidation-reduction reactions are just opposite to that of photosynthesis, as former is the process of *consumption* and the latter one is the process of *production*. The consumption process in popular terms is called *respiration*, and the three categories predominate in the consumption process.

(a) Aerobic respiration: In the process, oxygen is the electron acceptor and its various steps are just reverse of the photosynthesis. All plants and most animals derive energy for their maintenance, multiplication and movements through this process. In summary, the process can be represented as follows-

$$\underset{\text{(Glucose)}}{C_6H_{12}O_6} + 6O_2 \rightarrow 36ATP + 6CO_2 + 6H_2O + \text{heat energy}$$

Each ATP (adenosine triphosphate) molecule stores about 7.6 kcal energy in single phosphate bond released during oxidation. This energy is utilized by organisms to run their metabolic activities.

(b) Anaerobic respiration: Respiration that occurs in absence of oxygen is called anaerobic respiration, wherein electron acceptors are inorganic compounds. It is commonly found in anaerobic bacteria such as *Desulfovibrio*, methane bacteria and archaebacteria.

(c) Fermentation: It is also an anaerobic process but electron acceptor is an organic compound. Common examples are yeast and fermentation bacteria. They are abundantly found in soil and help in decomposition. The simple representation of the process is as follows -

$$\underset{\text{Glucose}}{C_6H_{12}O_6} \rightarrow \underset{\text{Ethyl alcohol}}{2C_2H_5OH} + 2CO_2 + \text{energy}$$

Some bacteria are capable of both aerobic and anaerobic respiration and are called facultative.

Of the three categories of consumption of food, the aerobic respiration is most common and most efficient.

The common consumers of different ecosystems are given in the table 13.2:

TABLE 13.2. Common Consumers of Different Ecosystems

No. Ecosystem	*Primary consumers*	*Secondary consumers*	*Tertiary consumers*
1. Forests	Ants, flies, beetles, bugs, leaf hoppers, elephant, nilgai, deer, moles, squirrels, shrews, fruit bats, flying foxes, mongooses etc.	Snakes, birds, fox, owls, wolfs, lizards etc.	Lion, tiger

2. Grasslands	Cows, buffaloes, deers, sheeps, rabbits, mouse, insects etc.	Fox, jackals snakes, frogs, lizards, birds	Hawks, tiger
3. Ponds and lakes	Fish, insect larvae, beetles, molluscs, crustaceans, zooplanktons etc.	Insects, fish	Large fish
4. Ocean (Marine)	Fish, crustaceans, molluscs etc.	carnivorous fish *e.g.* Herring, shad, Mackerel etc.	Cod, haddock, halibut (giant carnivorous fish) etc.
5. Cropland	Cattles, men feeding on fruits, smaller animals like rabbits, rats, birds and insects like aphids, thrips, beetles etc.	Frog, birds	Hawks, snakes

DECOMPOSERS

The non-green organisms like fungi and some bacteria, which are incapable of producing their food, live on dead and decaying plants or animals parts (*detritus*) and are consumers of special kinds. They are called **decomposers**. These are also known as **saprotrophs**. They play an important role in maintaining the dynamic nature of ecosystems.

Decomposers carry out decomposition process by which complex organic materials of dead remains are broken into simpler compounds that can again be utilized by green plants (producers). By this process, decomposers obtain energy and nutrients. Decomposition, in simple term, is a reverse reaction of photosynthesis and can be represented as follows:

$$\underset{\text{(Carbohydrate)}}{C_6H_{12}O_6} + 6O_2 \rightarrow 6CO_2 + 6H_2O + 700 \text{ kcal energy}$$

Some organic material such as simple carbohydrates, fats and proteins are decomposed rapidly, whereas others such as cellulose, lignin, cutin, hairs, bones and chitin are decomposed very slowly. Decomposers release different enzymes from their bodies into the dead and decaying plant and animal remains. The extra-cellular digestion of dead remains leads to the release of simpler inorganic substances. It is essentially a vital function in an ecosystem because had it not occurred, all nutrients would have been tied up in dead bodies and no new life could be produced.

Decomposition process comprises three stages-

(i) Formation of particulate detritus by physical and biological action coupled with the release of dissolved organic matter.

(ii) Rapid production of humus accompanied by the release of additional dissolved organic compounds by saprotrophs (*i.e., humification*)

(iii) Slower mineralization of humus (*i.e., mineralization*)

No single type of organism performs complete decomposition. Bacteria, fungi and moulds may work together in breakdown process. Bacteria are more effective on animal flesh and fungi on plant material. Besides bacteria and fungi, protozoa, mites, snails, earthworms, millipedes, insects, centipedes and nematodes, also play important role in decomposition process.

Decomposer organisms like bacteria, fungi and protozoans are although, microscopic in size but these are incredibly abundant in natural ecosystems. As such one gram of soil may contain some one million bacteria, five million actinomycetes, 500,000 protozoans and 200,000 moulds. Their specific number depends upon soil type, moisture, temperature, nutrient level and other environmental factors. Their total biomass is substantially less as compared to that of producers and consumers.

Decomposition process is complex and specific requiring involvement of various types of decomposers. Some are *zymogenous* (decomposing fresh organic matter), while others are *autochthonous* (decomposing humus) or *coprophagous* (decomposing fecal matter) etc.

Some decomposers perform another important function of producing metabolites with regulatory function on other organisms. Such metabolites or chemicals are named as *ectocrines* or *exocrines* or *environmental regulators*. For example, antibiotic substance penicillin produced by the mould (*Penicillium*) inhibits bacterial growth. Such cases are not rare in the soil.

Composition of decomposers varies in different ecosystems (table 13.3)

TABLE 13.3. Common Decomposers of Different Ecosystems

No. Ecosystem	*Decomposers*
1. Forests-	Fungi (*Aspergillus, Penicillium, Agaricus, Saccharomyces, Coprinus, Polyporus, Fusarium, Alternaria, Trichoderma*) etc., Bacteria (*Bacillus, Pseudomonas, Clostridium, Angiococcus* etc.), Actinomycetes (*Streptomyces*), slime moulds etc.
2. Grasslands-	Fungi (*Mucor, Rhizopus, Aspergillus, Penicillium, Cladosporium, Fusarium* etc.), Some bacteria and actinomycetes.
3. Croplands-	Fungi (*Aspergillus, Penicillium, Agaricus, Mucor, Rhizopus, Fusarium, Alternaria* etc.), Bacteria (*Streptococcus, Staphylococcus* etc.), some actinomycetes etc.
4. Ponds or Lakes-	Fungi (*Saprolegnia, Achlya, Aspergillus, Cephalosporium, Pythium, Cladosporium, Trichoderma, Circinella, Curvularia, Paecilomyces* etc), Saprophytic bacteria, Actinomycetes etc.
5. Marine or ocean-	Mainly bacteria and some fungi.

Chapter Summary

The biotic components of the ecosystem on the basis of mode of nutrition are divided into three types: *(i)* Producers, *(ii)* Consumers and *(iii)* Decomposers.

Producers are autotrophs which manufacture their food by the process called *photosynthesis* or *chemosynthesis*. They are of three different types: *(i)* Oxygenic photosynthesisers which release oxygen in the process as byproduct, *e.g.* green plants and cyanobacteria, *(ii)* Anoxygenic photosynthesisers which do not release oxygen, *e.g.* photosynthetic bacteria (such as *Chlorobium* and *Rhodospirillum*), and *(iii)* Chemosynthesisers which use chemical energy in manufacturing of food, *e.g.* bacteria like *Nitrosomonas, Nitrobacter, Thiobacillus* etc. Photosynthesis consists of Light reaction phase and Dark reaction phase. In first phase reducing power is generated using sunlight and water. This power is used in CO_2 fixation and manufacture of carbohydrates. If first stable product is PGA, the plants are called C_3 plants such as algae, bryophytes, pteridophytes, gymnosperms and most angiosperms and if it is OAA, the plants are called C_4 plants such as maize and sugarcane. Producers have highest varieties in an ecosystem.

Consumers are heterotrophs as these can not manufacture their food rather they depend directly or indirectly upon producers. Three categories are well recognised: *(i) Primary consumers* which are invariably herbivores (animals), *e.g.* rabbit, deer, cow etc., *(ii) Secondary consumers* which are small carnivores or flesh eating animals, *e.g.* snakes, fox, wolves etc. and *(iii) Tertiary consumers* which are top carnivores, *e.g.* lion, tiger, vulture etc. Consumers derive their energy by oxidation-reduction process (*i.e.*, respiration). Of the three types of respiration (*i.e.*, aerobic, anaerobic and fermentation), aerobic respiration is most common in which 36 ATP (adenosine triphosphate) per glucose molecule are produced. Type of consumers varies in different ecosystems.

Decomposers are saprophytic *heterotrophs* which feed on dead and decaying plant or animal parts (detritus). They are commonly fungi and bacteria or popularly called microbes. Process of decomposition is essentially a respiration (oxidation-reduction process) by which these decompose complex organic matter into simpler type and release the nutrients in the soil or atmosphere. Decomposition is a vital function in an ecosystem because had it not occurred, all nutrients would have been tied up in dead bodies and no new life could be produced. These are found abundantly in the soil, *e.g.* one gram of soil may contain some one million bacteria. Composition of decomposers varies in different ecosystems.

Study Questions

1. Describe the structural and functional aspects of producers in the ecosystem.
2. Explain the various types of consumers and their role in the ecosystem.
3. What are decomposers? Describe the process involved in the decomposition of detritus.
4. Briefly enumerate the structure and function of various biotic components of the ecosystem.
5. 'The consumers of any ecosystem is a heterogeneous group'. Explain.
6. Give an account of biotic components of ecosystem.
7. Briefly describe the process and products of decomposition.
8. 'Producers constitute the basic component of an ecosystem'. Explain.

9. Decomposers are an essential part of the ecosystem. Justify.
10. What are consumers? Enlist their mode of nutrition.
11. Justify the statement that 'driving force in an ecosystem is formed of producers.'
12. Explain what is detritus?
13. Name two most common trees found in Indian tropical rain forests.
14. Name two producers found in pond ecosystem.
15. Name two dominant deciduous tree species.
16. Give two differences between consumers and decomposers.
17. Give two differences between autotrophs and heterotrophs.
18. What is the main function of decomposers? Explain.
19. What are primary consumers? Give two examples.
20. What is oxidation-reduction process? Define.
21. Give one difference between C_3 and C_4 plants and name one example of each.

Objective Questions. *Select the correct answers*

1. In an ecosystem, the function of producer is to:
(1) Convert organic compounds into inorganic compounds
(2) Transduce solar energy
(3) Convert solar energy into radiant energy
(4) Release energy

2. The category of primary consumers includes:
(1) Eagles and tigers (2) Fish and Whales
(3) Snakes and frogs (4) Cattles and insects

3. Top consumers of grassland ecosystem are:
(1) Carnivores (2) Herbivores
(3) Omnivores (4) None of these

4. Which is characteristically found in temperate forest?
(1) Shorea (2) Prosopis
(3) Rhododendron (4) Salmalia

5. *Vallisnaria* is a component of:
(1) Forest ecosystem (2) Grassland ecosystem
(3) Pond ecosystem (4) Cropland ecosystem

6. Phagotrophic mode of nutrition is found in:
(1) Producers (2) Consumers
(3) Decomposers (4) All of these

7. Complex organic matter is converted into nutrient elements by:
(1) Consumers (2) Decomposers
(3) Producers (4) None of these

Answers

1. (2) *2.* (4) *3.* (1) *4.* (3) *5.* (3) *6.* (2)
7. (2)

14

CHAPTER

Energy Flow in Ecosystem

LEARNING OBJECTIVES
Introduction • Energy • Laws of Thermodynamics • Flow of Energy in an Ecosystem • Energy Flow Charts.

Introduction

Energy is considered as the ability to do work. In an ecosystem, solar energy is stored in producers (by photosynthesis) which is transformed by consumers of different trophic levels. The transformation of solar energy from one trophic level to another, mediated by consumers, is called *energy flow*. The behaviour of energy in the ecosystem follows the laws of thermodynamics.

Laws of Thermodynamics

Two basic laws fundamental to the understanding of energy flow in ecosystem are-

(1) **First law of thermodynamics**: The law states that "*Energy may be transformed from one form to another but is neither created nor destroyed.*" Accordingly, light is a form of energy and it can be transformed into heat, work or potential energy of food depending upon the situation.

(2) **Second law of thermodynamics**: The law states that "*No process involving an energy transformation will occur spontaneously unless there is a degradation of energy from a concentrated to a dispersed form*". It explains that transformation of energy is not 100 per cent efficient. Energy flows from a region of high concentration to a lower one and work is performed through degradation of energy. It is this process that the living organisms use in metabolism.

Flow of Energy in an Ecosystem

The ultimate source of energy for the living world is the sun. The solar energy enters in the form of light rays. About 30% of sun light reaching the earth's atmosphere is reflected back into the space, about 20% of solar energy is used in heating atmosphere and about 50% is used up in heating land, vegetation and water. Only about 1% or less than that of sunlight is used in photosynthesis by producers, and it is this small fraction that supports all life on earth.

The amount of energy fixed by the producers (efficiency of fixing the solar energy) varies in different ecosystems. It is about 0.2 per cent in an aquatic ecosystem and 5-7 per cent in most efficient ecosystems, *i.e.*, energy subsidized cropland ecosystems.

The energy fixed by the producers is in the form of potential energy as food. Food means materials containing energy that organisms can use. It is the means of transfer of both matter and energy in the living world. Potential energy in foodstuffs is passed through the ecosystem in a series of steps of eating and being eaten. Thus, the energy flows from the sun to the producers and then to the consumers and at each transfer a large proportion of potential energy is lost as heat produced during life process such as respiration. And as such, very little energy is left for the consumers of the uppermost trophic level in a long food chain. At each stage of energy transfer, a considerable amount of energy is lost from the food chain. The amount of energy transferred to the next higher level gradually decreases. The decomposition of dead organisms also releases

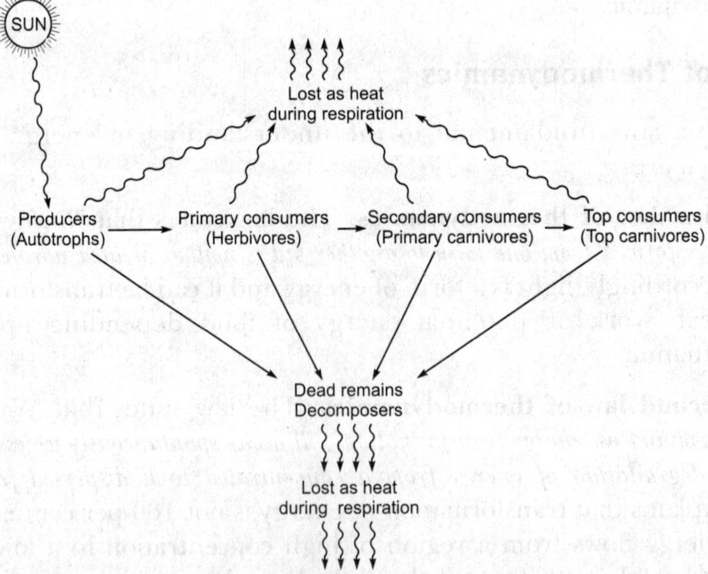

Fig. 14.1. Flow of energy at different levels in ecosystem (wavy lines).

chemical energy. Eventually, all solar energy that entered the living system through the producers goes back into the non-living world, not as light but as heat (fig. 14.1).

It can be better explained taking suitable examples. When a plant is eaten by a herbivore, the total eaten plant material can not be converted into the flesh of the herbivore as much of it remains undigested and is therefore excreted. Similarly when a herbivore is eaten by a carnivore 1kg of the flesh of a herbivore does not make 1 kg of flesh of a carnivore. This is because at each trophic level, only a small part of energy stored in the food can be utilized by the consumer, the major portion being lost as heat. It follows that a herbivore being more active must eat a large number of plants to meet its full requirement of energy. Similarly, a carnivore must eat a large number of herbivores to get the energy it needs. Such inter-conversions of energy in an ecosystem closely follow the laws of thermodynamics.

In general, the efficiency of producers harvesting solar energy into chemical energy is about 10 per cent, while that of herbivores to consumers ranges in between 10-15 per cent and in decomposers it may reach as high as 20 per cent or little more. Efficiency of energy use in

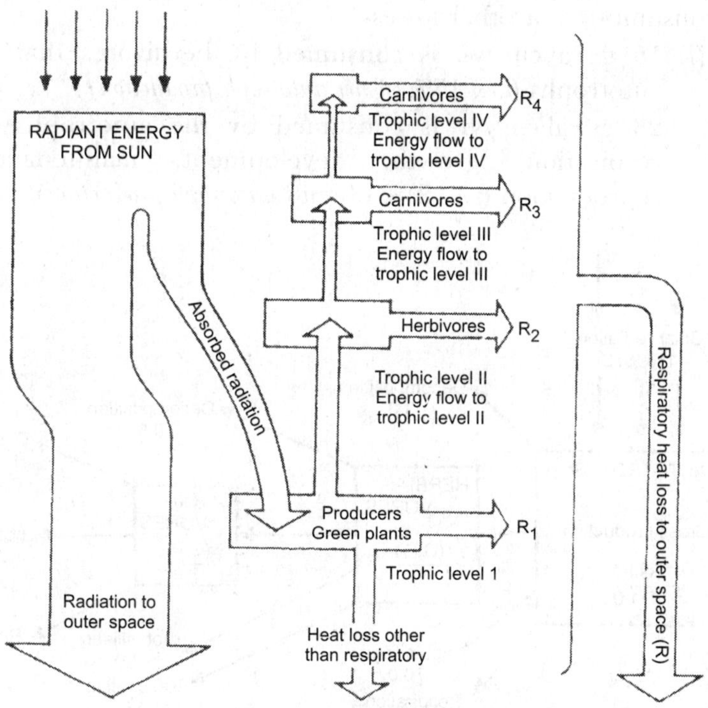

Fig. 14.2. Energy flow in ecosystem. Rectangles represent the stored energy at various trophic levels and R_1, R_2, R_3, and R_4 represent the energy lost in respiration at different trophic levels (redrawn from Lindeman, 1942).

different trophic levels generally increases with its decrease in an ecosystem.

Diagrammatic Representation of Energy Flow

Energy flow in the ecosystem can easily be explained with the help of different models. A few important models are as follows. All these models depict the basic pattern of energy flow in ecosystem.

(i) Boxes and pipes graphic model of energy flow

Lindeman (1942) proposed an energy flow diagram for a lake or fresh water ecosystem in terms of g cal/ cm^2/yr. (gram calories per cm^2 per year). It shows following characteristics (fig. 14.3)-

(1) The total incoming solar radiation is about 118,872 g cal/cm^2/yr. The gross production by autotrophs is about 111.00 g cal/cm^2/yr, *i.e.*, solar radiation equal to 118,761 g cal/cm^2/yr is not utilized by autotrophs.

Thus, the efficiency of capturing energy by autotrophs is only 0.1per cent.

(2) The energy available at the autotroph level shows following consumption or other losses-

(i) 15 g cal/cm^2/yr. is consumed by herbivores that feed on autotrophs (*i.e.*, 17% of *net autotroph production*).

(ii) 23 g cal/cm^2/yr is consumed by the autotrophs in their respiration, growth, development, maintenance and reproduction (*i.e.*, 21% of *gross autotroph production*).

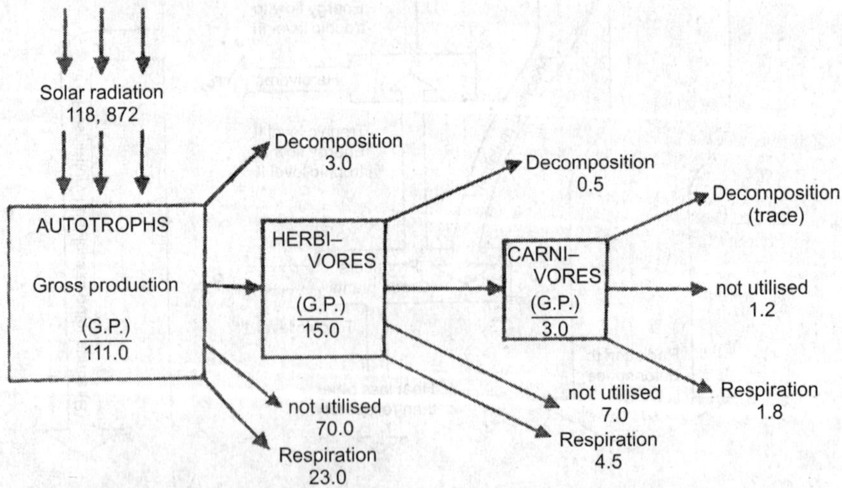

Fig. 14.3. Energy flow diagram for a fresh water ecosystem expressed in g cal/cm^2/yr (redrawn and partly modified from Lindeman, 1942). Abbrev. G.P.= Gross Production.

(*iii*) 70 g cal /cm^2/ yr is not utilized at all and becomes a part of the accumulation sediments (*i.e.*, 79.5% of *net autotroph production*).

(*iv*) 3 g cal /cm^2/yr is used up in the decomposition of autotrophs (*i.e.*, 3.4 % of *net autotroph production*).

It explains that *(a)* at herbivorous stage (first consumers) much more energy is available than is consumed by herbivores, and *(b)* much of energy of autotrophs (G.P.) is lost in various pathways such as by decomposition, respiration and not utilized. It is equivalent to net production, *i.e.*, 3 + 15 + 70 = 88 g cal/cm^2/ yr.

(3) The total amount of energy available at herbivore level is 15 g cal/cm^2/yr. Out of this energy -

(*i*) 3.0 g cal/cm^2/yr (*i.e.*, 28.5%) of net production is consumed by carnivores.

(*ii*) 4.5 g cal/cm^2/yr (*i.e.*, 30%) is used in the respiration of herbivores. It shows the energy loss via respiration by herbivores is more than that of autotrophs where it is only 21%.

(*iii*) 7.0 g cal/cm^2/yr is not utilized.

(*iv*) 0.5 g cal/cm^2/yr is used up in decomposition.

It means that at herbivore-carnivore transfer level a more efficient utilization of resource (or energy) is found as compared to that of autotroph-herbivore transfer level.

(4) Of the total energy available at the carnivore level (*i.e.*, 30 g cal/cm^2/yr)-

(*i*) 1.0 g cal/cm^2/yr (*i.e.*, about 60%) of the total carnivore energy is consumed in respiration. Thus, energy loss via respiration by carnivores is much more than that at the autotroph (21%) and herbivore (30%) levels. It means that there is a progressive loss of energy via respiration from autotroph to herbivore to carnivore level.

(*ii*) The remainder energy (*i.e.*, 1.2 g cal/cm^2/yr) is not utilized. It remains as a part of the accumulating sediments, and only an insignificant amount of energy is subject to decomposition yearly.

Thus, the model shows one way passage of energy flow, *i.e.*, **energy flow is unidirectional**. The energy captured by the producers (autotrophs) does not revert back to the solar input and whichever passes to the herbivores does not pass back to the autotrophs, so is true from herbivores to carnivores. It means without the continuous supply of solar energy the entire ecosystem would collapse.

Odum (1963) proposed a simplified energy flow diagram depicting three trophic levels (boxes) in a linear food chain (fig. 14.4)

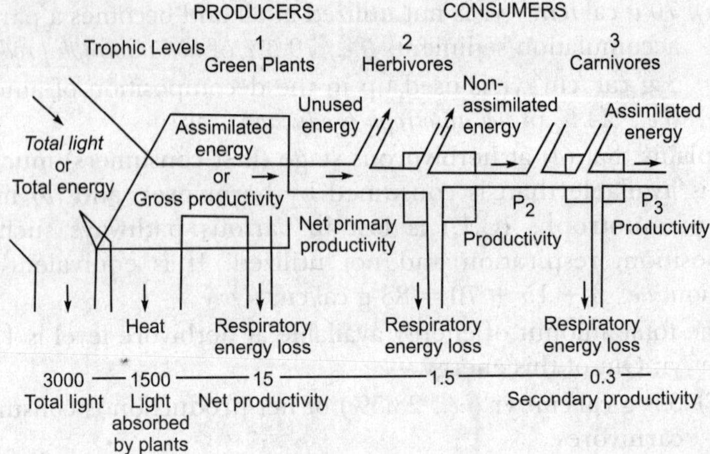

Fig 14.4. Energy flow in three trophic levels (boxes 1,2,3) (after EP Odum 1963).

The model explains that of the 3000 kcal of total light falling upon green plans, approximately 50% (*i.e.*, 1500 kcal) is absorbed, of which only one per cent (15 kcal) is converted at first trophic level. The 15 kcal is the net primary production. Secondary productivity (P_2 and P_3 in fig.) tends to be about 10 per cent at successive consumer trophic levels, *i.e.*, herbivores and carnivores. Percentage of secondary productivity may vary at the carnivore level (P_3 = 0.3 kcal).

The figure shows that there is a successive reduction in energy at successive trophic levels. It means, shorter the food chain, greater would be the available energy as with an increase in the length of food chain, there is a corresponding more loss of energy.

E.P. Odum (1983) proposed a generalised model of Y-shaped or 2-channel energy flow model (Fig. 14.5). This model is applicable to every type of ecosystems. Such models are more realistic than single channel model because-

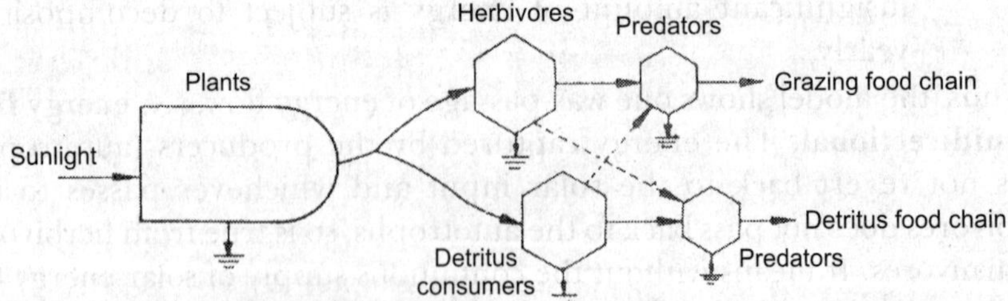

Fig. 14.5. The Y-shaped energy model showing linkage between two types of food chains (after Odum, 1983).

(*i*) It reforms to the basic stratified structure of ecosystems (*i.e.*, all three types of biotic components) (*ii*) It separates the grazing and detritus food chain in both time and space.

(*iii*) It confirms distinction between the macro-consumers (phagotrophic animals) and the micro-consumers (absorptive microbes) which differ greatly in the size-metabolism relations.

(ii) Universal model of energy flow

Odum (1968) proposed the universal model of energy flow applicable to any living component, whether a plant, animal, micro-organism or individual, population or a trophic group. The model can be used to depict a food chain or bioenergetics of an entire ecosystem. The model shows following important features-

(1) Shaded box 'B' represents biomass or living structure of the components.

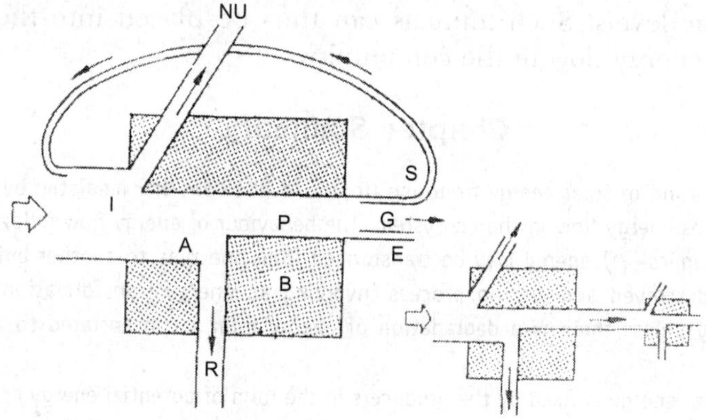

Fig. 14.6. Universal model of energy flow (I = input or ingested energy, NU = not used, A = assimilated energy, P = production, R = respiration, B = biomass, G = growth, S = stored energy, E = excreted energy (after Odum, 1968).

(2) Total energy intake, *i.e.*, input = I which represents light for autotrophs and food for heterotrophs.

(3) 'A' = assimilated energy, is the energy utilized which means gross production or gross photosynthesis in autotrophs and metabolized energy in heterotrophs.

(4) In figure, 'A' separates into 'P' (production) and 'R' (loss of energy due to respiration). 'P' components of energy is available to the next trophic level.

(5) 'P' or production may take a number of forms. It is subdivided into 'G' (growth or addition to biomass), 'E' (assimilated organic matter excreted and secreted) and 'S' (storage or accumulation of food).

(6) 'NU' is energy which is not utilized. The ratio between 'A' and 'NU' is the efficiency of energy assimilation which varies widely. The ratio is

very low in light fixation by plants or food assimilation in detritus feeding animals and is very high in animals and bacteria which feed on high energy food like sugars.

The universal model of energy flow can be used in two ways-

(i) It can represent a species population in which case the appropriate energy inputs and links with other species would be shown as a conventional species oriented food web diagram or

(ii) The model can represent a descrete energy level in which case the biomass and energy channels represent all or parts of many populations supported by the same energy source.

For example foxes feed partly on fruits and partly on small herbivores like rabbits. Here a single box can be used to represent the whole population of foxes to stress on intrapopulation energetics. However, two or more boxes will be used to represent the metabolism of fox population into trophic levels. Such animals can thus be placed into the overall pattern of energy flow in the community.

Chapter Summary

The transformation of solar energy from one trophic level to another mediated by consumers is referred to as energy flow in the ecosystem. The behaviour of energy flow follows two laws of thermodynamics- *(i)* energy may be transformed from one type to another but is neither created nor destroyed and *(ii)* no process involving an energy transformation will occur spontaneously unless there is a degradation of energy from a concentrated to a dispersed form.

Part of solar energy is fixed by the producers in the form of potential energy or food. Food is the medium of transfer of both matter and energy in the living world. Thus, energy flows from sun to the producers and then to the consumers and decomposers and at each transfer a large proportion of potential energy is lost as heat produced during the life processes of the biotic components. Eventually at last, all solar energy that entered the living system through the producers goes back into the non-living world, not as light but as heat.

Efficiency of energy use in different trophic levels is about ten per cent. Energy flow has been explained variously with the help of various models. In all cases, the amount of energy transferred through food to successive higher levels become less and less. Energy flow is always unidirectional.

Study Questions

1. Explain the flow of energy in an ecosystem with the help of suitable diagrams.
2. Describe how energy flow in ecosystem obeys the laws of thermodynamics
3. Discuss the role of various biotic components in the energy flow in an ecosystem.
4. Describe the law of thermodynamics with reference to energy flow in the ecosystem.
5. Draw a chart of energy flow in an ecosystem.
6. Briefly describe the efficiency of energy flow in the ecosystem.

7. Draw a diagram of energy flow for a fresh water ecosystem.
8. Explain the concept that energy flow is unidirectional.
9. Describe what is Y-shaped energy model?
10. Define energy flow in ecosystem.
11. Write two laws of thermodynamics applicable to the energy flow in the ecosystem.
12. What is assimilated energy? Explain
13. Define the term 'potential energy':

Objective Questions. *Select the correct answers*

1. Radiant energy is converted into potential energy by:
   (1) Consumers (2) Producers
   (3) Decomposers (4) All of these

2. Highest potential energy is found in:
   (1) Decomposers (2) Consumers
   (3) Producers (4) Detrivores

3. The efficiency of energy flow in the ecosystem is about:
   (1) 1 per cent (2) 10 per cent
   (3) 50 per cent (4) 100 per cent

4. Energy is returned to the free atmosphere in the form of:
   (1) Potential energy (2) metabolic energy
   (3) Heat (4) Vapours

5. The direction of energy flow is:
   (1) Unidirectional (2) Bi-directional
   (3) Cyclic (4) Any of these

Answers

1. (2) *2.* (3) *3.* (2) *4.* (3) *5.* (1)

15

CHAPTER

Food Chains, Food Webs and Ecological Pyramids

Learning Objectives

Food Chains • Types of Food Chains • Food Webs • Ecological Pyramids • Pyramids of Number • Biomass and Energy • Comparison.

FOOD CHAINS

In all ecosystems, food is the primary source of energy for all animals and plants, but green plants alone are able to trap solar energy and convert it into chemical energy which is stored in the form of food. The chemical energy is locked up in the various organic compounds, such as carbohydrates, proteins, fats etc. All other organisms are consumers and depend directly or indirectly upon this food of plants.

The transfer of food energy from the source in plants through a series of organisms with repeated eating and being eaten is referred to as **food chain** (Odum, 1971). The successive levels of energy flow constituting the links of the food chain are called **trophic levels**. The *first trophic level* in a food chain is thus the **producers** or green plants. Organisms of *second trophic level* are **primary consumers** or herbivores which feed upon the green plants for their nutrition. Smaller carnivorous organisms which prey upon herbivores to obtain their food from the *third trophic level* and called **secondary consumers**. The large carnivores which prey upon secondary consumers to obtain their food are called **tertiary consumers** or top consumers.

Organisms obtaining food from plants by same number of steps belong to the same trophic level. However, omnivores may belong to two different trophic levels.

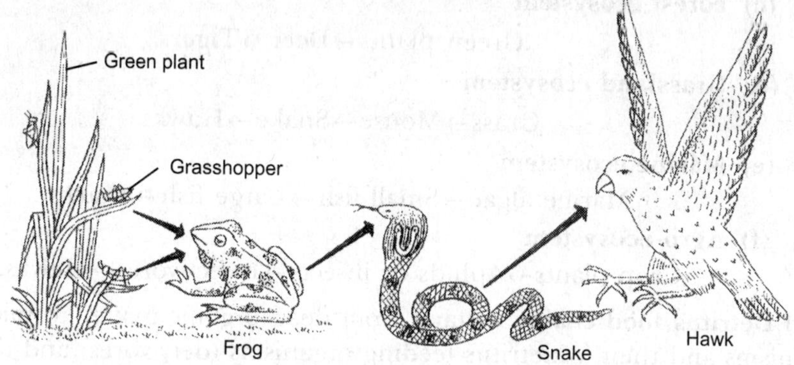

Fig. 15.1. An example of a simple food chain.

Trophic structure may be described in terms of the amount of living material called standing crop, present in different trophic levels at a given time.

Thus, a simple example of a terrestrial food chain may be-

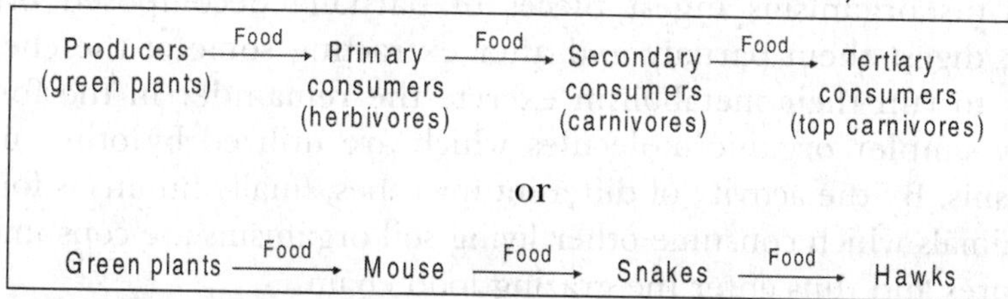

At each transfer, about 80-90 per cent of the potential energy is lost as heat. Therefore, the number of steps in a sequence (food chain) is rarely more than five. The quantity of energy decreases readily in a long food chain, whereas in a short food chain, higher amount of energy is available. Energy is highest at the base and lowest at the top, *i.e.*, the available energy is greater in the organisms nearer to the beginning of the chain.

Types of Food Chains

Mainly food chains are of two types-
(1) Grazing food chain, and (2) Detritus food chain

(1) Grazing food chain: It starts from green plants and operates from herbivores to primary carnivores and so on. Examples of some common grazing food chains are-

(a) Pond ecosystem

Phytoplanktons→Aquatic insects→Small fish→Large fish

(b) Marsh community

Green plants→Butter fly→Dragon fly→Frog→Snake→Hawks

(c) Forest ecosystem

Green plants→Deer→Tigers

(d) Grassland ecosystem

Grass→Mouse→Snake→Hawks

(e) Marine ecosystem

Marine algae→Small fish→Large fish→Sharks

(f) Agro-ecosystem

Crop plants→Aphids or insects→Insectivores→Hawks.

(2) Detritus food chain: It starts from dead organic matter to micro-organisms and then to detritus feeding organisms (detrivores) and their predators. The dead organic matter is decomposed into detritus by micro-organisms like bacteria and fungi. Odum (1970) described a detritus food chain based on mangrove leaves as follows:

Mangrove leaves→Detritus (by microorganisms)→Crabs and shrimps→Small fish→Large fish.

Detritus organisms ingest pieces of partially decomposed organic matter, digest them partialy and after extracting some of the chemical energy to run their metabolism excrete the remainder in the form of slightly simpler organic molecules which are utilized by other micro-organisms. By the activity of different microbes, finally humus is formed. Soil animals which consume other living soil organisms are consumed by carnivores and thus enter the grazing food chain.

Links Between Grazing and Detritus Food Chains

The energy of grazing food chain can enter the detritus food chain and that from detritus food chain can enter the grazing food chain. For example, the dead remains of plants (litter) and animals are decomposed by micro-organisms and converted into detritus and through detritus feeding organisms (detrivores) in soil, energy enters to organisms feeding on detritus feeders. Many of these serve as prey for carnivores in the grazing food chain, *e.g.* a robin eating earthworms. In addition, many insects, (beetles and flies), spend their larval period in the detritus food chain and their adulthood in the grazing food chain. The relationships between the two types of food chains can be illustrated as in fig. 15.2.

However, the energy transfer from detritus food chain to grazing food chain is relatively very low as compared to the amount of energy which flows from grazing food chain to detritus food chain. The detritus food chain differs from the grazing chain in that over a long period of time, it can consume all of the incoming production and can complete the dissipation of biologically fixed energy as heat.

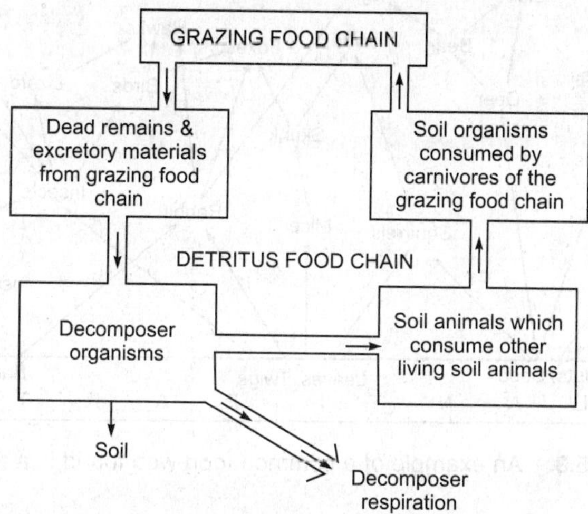

Fig. 15.2. Diagrammatic representation of the detritus food chain showing energy transfers between it and the grazing food chain.

FOOD WEBS

In nature the food relationship can not be explained only in terms of a single food chain, as the same organism may derive food from more than one trophic level and even the same organism may be eaten by several organisms of a higher trophic level. For example, ticks and mites, leeches and blood sucking insects are dependent on herbivores and even on carnivores. Many kinds of animals other than the tiger derive food from various herbivores. Thus, food chains are not isolated linear chains of trophic levels, rather depending upon choice and availability of food, different organisms at each level have food relationship with more than one organisms at the lower levels. In a terrestrial food chain, a rat is eaten by both, the frog and the snake and as such snakes may occupy different positions in the food chain.

Thus in nature, a net work of food chains is found and these interconnected food chains form a structure called **food web**. The food webs become more complicated because of variation in taste and preference, availability and compulsion and several circumstancial factors, *e.g.* tigers normally do not eat fish or crabs but in Sundarbans they are forced to feed on them.

An example of common food web found in a forest can be taken as a good way to understand food web structure (fig. 15.3).

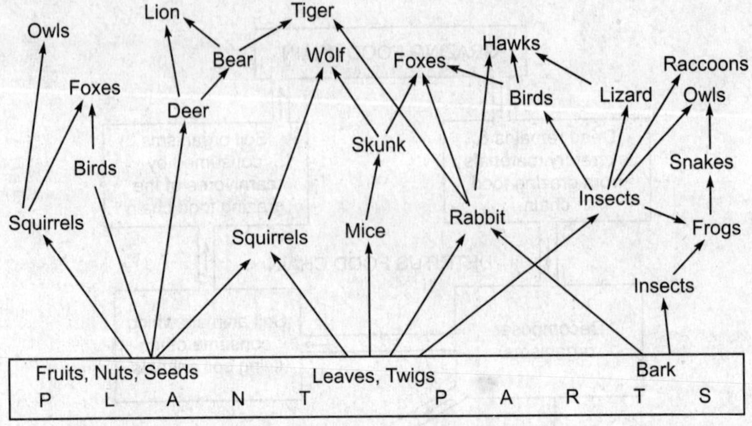

Fig. 15.3. An example of a common food web found in a forest.

Common Features of Food Web

Some of the common features of food web are-

1. The mean proportion of top predators, intermediate species and basal species remains nearly constant in webs of widely differing species diversity.
2. Linkage density is about constant for webs with few species but can increase with increase in species diversity.
3. The lengths of common food chains for top predators are very commonly 2 or 3.
4. Species at higher trophic levels have more prey and fewer predators than those at lower levels.
5. Webs having small numbers of species commonly have internal overlaps in the predators' use of prey species.

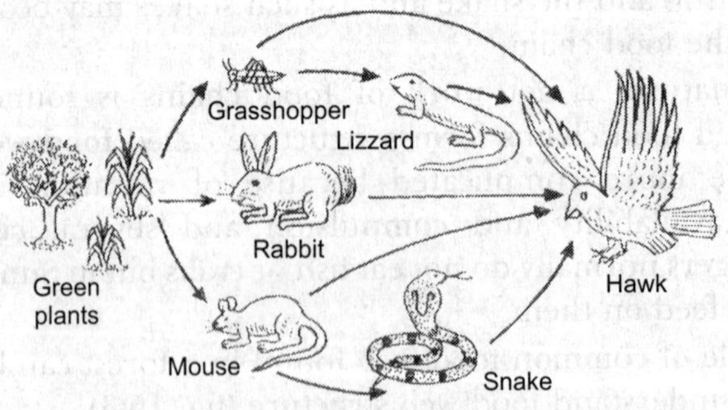

Fig. 15.4. A typical food web of forest and grassland ecosystems.

Importance of Food Webs

Food webs are quite important in maintaining the ecosystem stability. For example, in a grassland if the population of rabbit decreases, the number of alternative host such as mouse will increase. This imbalance may

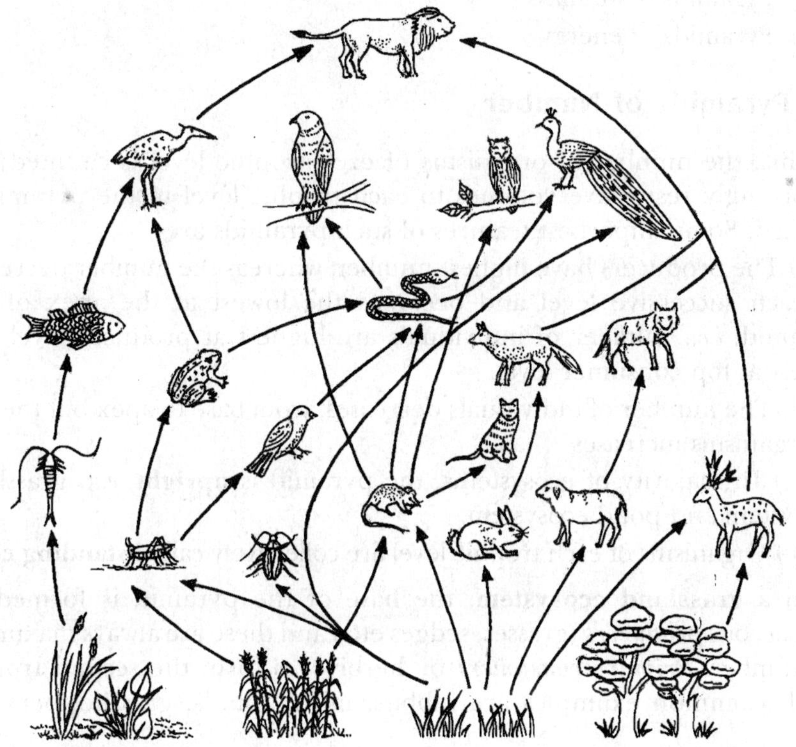

Fig. 15.5. Pictorial representation of the common food webs.

continue further as it may decrease the population of consumers that prefers to eat rabbit. Thus, food webs with alternative food chains do not let the imbalance occur and in this way play a vital role in the maintenance of the stability of the ecosystem.

ECOLOGICAL PYRAMIDS

Ecosystem comprises various trophic levels and these are called producers (autotrophs), primary consumers (herbivores), secondary consumers (primary carnivores) and tertiary consumers (top consumers or top carnivores). Food energy passes from one trophic level to another trophic level, and in each transfer much of its energy is lost as heat and as such each trophic level receives less energy as compared to the previous trophic level. The energy level gradually tapers in the food chain, forming a pyramid like structure. Thus, the graphic representation of relationship between the various trophic levels of a food chain is called **ecological pyramids**. Ecological pyramids were first designed by Charles Elton and as such these are also called **Eltonian pyramids** or **food pyramids**.

Basically three types of pyramids are recognised-

(1) Pyramids of number

(2) Pyramids of biomass

(3) Pyramids of energy

(1) Pyramids of Number

In this, the number of organisms of each trophic level is counted and accordingly respective volumes to each trophic level in the pyramid is allotted. Some important features of such pyramids are-

(i) The producers have highest number, whereas the number decreases in each successive level and becomes the lowest at the apex of the pyramid, *i.e.,* number of individuals are highest at producer level and lowest at top consumer level.

(ii) The number of individuals decreases, from base to apex but the size of organisms increases.

(iii) In majority of ecosystems, the pyramid is **upright**, *e.g.* grassland ecosystem and pond ecosystem.

(iv) Organisms of each trophic level are collectively called **standing crop**.

In a **grassland ecosystem**, the base of the pyramid is formed by herbaceous ***producers*** (grasses, sedges etc.) and these are always maximum in number. ***Primary consumers*** or herbivores form the second trophic level. Common examples are rabbits, mice, aphids, grasshoppers etc.

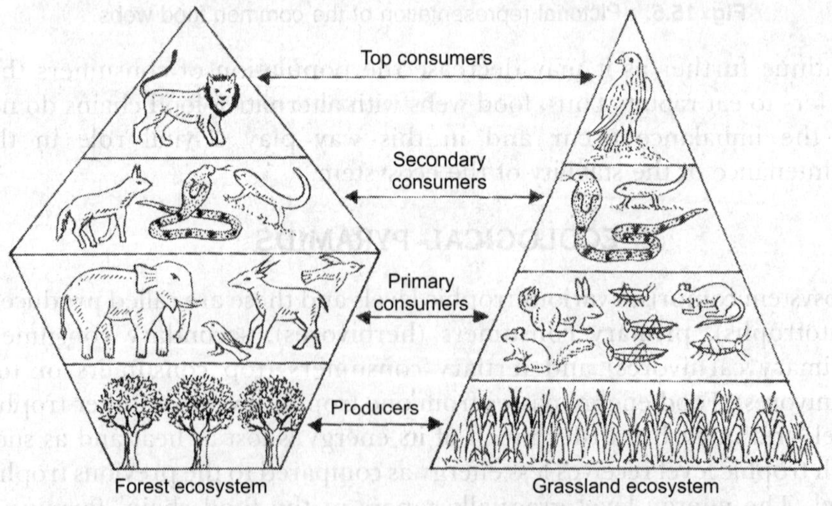

Fig. 15.5. Pyramids of number of forest and grassland ecosystems.

They are lesser in number than the producers. The third trophic level is represented by ***secondary consumers***, also called ***primary carnivores***, and is formed of lizards, snakes, owls etc. which are lesser in number than the primary consumers. The apex of the pyramid is formed of the ***tertiary consumers*** represented by the hawks, eagles and vultures which feed upon the secondary consumers and are the least in number.

In a **pond ecosystem**, the base of pyramids (***producers***) consists of aquatic plants like algae and hydrophytes having highest number. The ***primary consumers*** comprise zooplanktons, larvae and small fish etc. which are fewer in number than producers. ***Secondary consumers*** such as carnivorous fish are still lesser in number and the ***tertiary consumers*** like large fish or carnivorous birds are least in number (fig. 15.6)

(v) In a **tree ecosystem**, the pyramid of numbers is **inverted**. A single tree (***producer***) supports a large number of birds (***primary consumers***), which provide food to a large number of ectozoic and endozoic parasites (***secondary consumers***), which in turn are parasitized by large population of microbes (***tertiary consumers***). In an inverted pyramid the number of organisms increases, but the size decreases (fig 15.6).

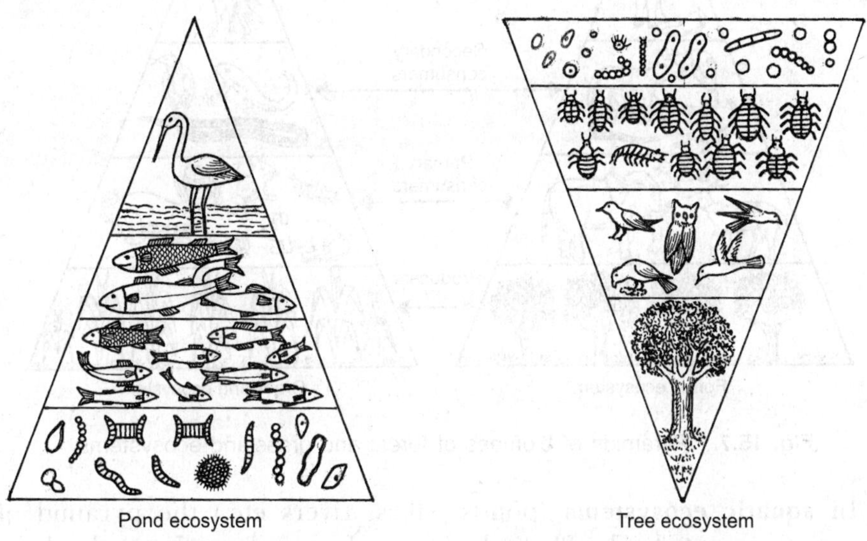

Fig. 15.6. Pyramids of number of pond and tree ecosystems.

(vi) In a **forest ecosystem**, the pyramid is **mixed** or **spindle-shaped**. Trees which are major producers that provide food for birds, deer, squirrels, elephants etc. (herbivores) are fewer in number as compared to primary consumers. So second trophic level is broader than first trophic level. The primary consumers are food for carnivores like fox, jackals, snakes etc. which are fewer in number and as such the third torphic level constricts. The number of top carnivores are still less or they are least in

member, *e.g.* lion, tiger etc. On the whole the pyramid becomes spindle-shaped (fig. 15.5).

(2) Pyramids of Biomass

Biomas is the total weight of dry matter (dry weight) present in the ecosystem at any one time. The pyramid shows the total mass of organisms at each trophic level and provides a rough picture of the overall effect of the food chain relationships for the ecological group as a whole.

In **terrestrial ecosystems** (grasslands, croplands, forests), the pyramid of biomass is **upright**. The **producers** (crop plants, grasses or trees) have the maximum biomass. The **primary consumers** (herbivores) are less in number and have less biomass followed by **secondary** and **tertiary consumers**, respectively (fig. 15.7).

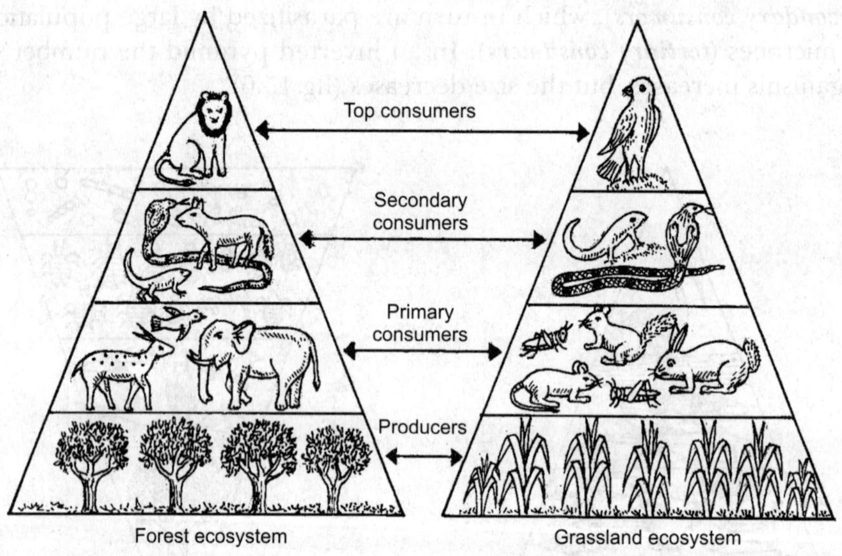

Fig. 15.7. Pyramids of biomass of forest and grassland ecosystems.

In **aquatic ecosystems** (ponds, lakes, rivers etc.) the pyramid of biomass is **inverted**. The Phytoplanktons, algae and small aquatic plants (**producers**) are lighter in biomass and are rapidly eaten by the **primary consumers** (water fleas, insects, larvae etc.) which are somewhat heavier. In ecosystem, the biomass of **primary, secondary** and **tertiary consumers** increases due to their large size and longer life (fig 15.8).

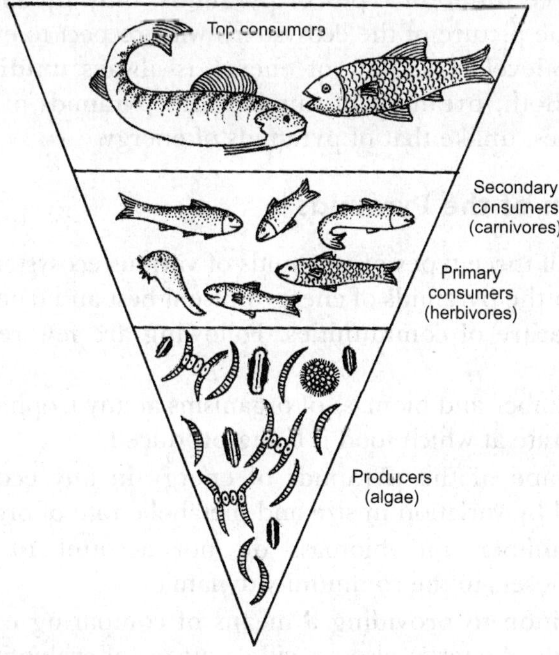

Fig. 15.8. Pyramid of biomass of a pond ecosystem.

(3) Pyramids of Energy

The amount of energy and matter transferred through food to successive higher levels become less and less. Greater amount of energy is available at the **producer** (autotrophs) **level** than at the **primary consumer level** (herbivores). The energy production of primary consumer is greater than that of **secondary consumers** (primary carnivores) and the energy at the **tertiary consumer level** (top carnivores) is produced in minimum amount (fig. 15.9). The deficiency of transfer of energy from one trophic level to another trophic level is more or less about 10 per cent, and as such the general pattern of pyramids of energy remains **upright** in all different ecosystems.

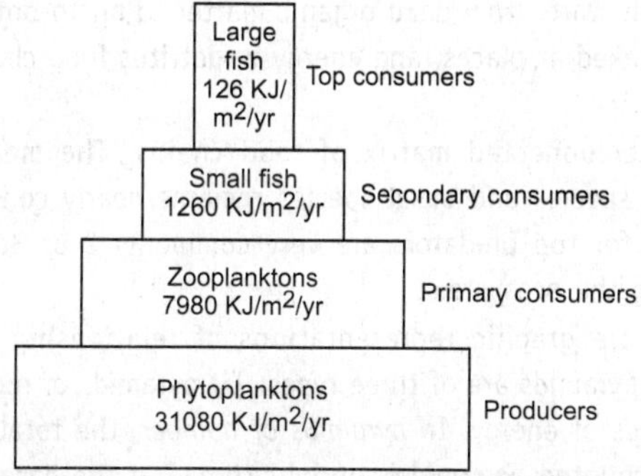

Fig. 15.9. Pyramids of energy in a food chain.

Out of three different types of pyramids, only pyramids of energy presents a true picture of the ecosystems with respect to energy status of each trophic level. The flow of energy is always **unidirectional** and **decreasing**. Both, pyramids of number and pyramids of biomass show variable shapes, unlike that of pyramids of energy.

Comparison of the Pyramids

Comparing all three types of pyramids of various ecosystems, it becomes apparent that the pyramids of energy present best and true picture of the functional nature of communities. Following are few reasons for this conclusion:

(1) The number and biomass of organisms at any trophic level depend on the rate at which food is being produced.
(2) The shape of the pyramids of energy in any ecosystem is not affected by variation in size and metabolic rate of organisms.
(3) The number and biomass do not account for the role of decomposers in the community dynamics.
(4) In addition to providing a means of comparing ecosystems, the pyramids of energy, also provide a means of evaluating the relative importance of populations.
(5) The energy flow in various populations shows that all populations in question are functioning at almost the same trophic level.
(6) The energy flow provides a suitable index for comparative evaluation of all components of an ecosystem.

Chapter Summary

The transfer of food energy from the source in plants through a series of organisms, *i.e.*, from one trophic level to the next trophic levels is called food chain. The trophic levels are producers, primary consumers, secondary consumers and tertiary consumers. Organisms obtaining food from plants by same number of steps belong to the same trophic level and these may be collectively called standing crops. Food chains are of two main types: *(i)* grazing food chain which operates from producers to herbivores to carnivores and *(ii)* Detritus food chain which starts from dead organic matter to micro-organisms (saprotrophs). The two chains remain linked at places, and energy in detritus food chain is transferred from grazing food chain.

Food web is the interconnected matrix of food chains. The mean proportion of top carnivores, intermediate species and basal species remains nearly constant, and the length of common food chains for top predators are very commonly 2 or 3. Food webs are quite important in maintaining the ecosystem.

Ecological pyramids are graphic representations of relationship between the various trophic levels. Basically, pyramids are of three types: *(i)* pyramids of number, *(ii)* pyramids of biomass and *(iii)* pyramids of energy. In *pyramids of number*, the total number of organisms at each trophic level is counted. Pyramid is upright (broad at the base and narrowest at the top) in terrestrial and aquatic ecosystems but spindle-shaped (broad in the middle) in the forest ecosystem and inverted in a tree (or in parasitic food chain). In *pyramids of biomass*, dry weight of standing crop is accounted. It is upright in terrestrial ecosystem but inverted in pond ecosystem. In *pyramids of energy*, amount of total energy at each trophic level is determined. It remains upright in all ecosystems. Thus pyramids of energy present a true picture of the ecosystem and is most accurate.

Study Questions

1. Describe the various types of food chains found in different ecosystems.
2. What are ecological pyramids? Mention their various types.
3. Give a brief account of food chains and food webs.
4. Define food chain and give two examples each in two types of the ecosystems.
5. Differentiate between grazing food chain and detritus food chain.
6. Define ecological pyramids and their types.
7. Describe various types of components found in a food chain.
8. Define food chain and food web.
9. Write a brief account of pyramids of number and pyramids of energy.
10. What is ecological pyramid? define.
11. Give differences between food chain and food web.
12. What is pyramids of biomass? explain.
13. What is standing crop? Define.
14. Draw a diagram of a food chain of a marsh community.
15. Give an example of detritus food chain.

Objective Questions: *Select the correct answers*

1. Which one of the following is a correct food chain?
   (1) Frog → Snake → Grasshopper → Grass
   (2) Grasshopper → Grass → Snake → Frog → Eagle
   (3) Grass → Grasshopper → Frog → Snake
   (4) Eagle → Snake → Frog → Grass
2. Trophic levels are formed by:
   (1) Only plants (2) Only animals
   (3) Only carnivores (4) Food chain linked organisms
3. The highest population is found in:
   (1) Consumers (2) Producers
   (3) Secondary Consumers (4) Predators
4. In grassland ecosystem, the pyramid of energy is:
   (1) Inverted (2) Upright
   (3) Mixed (4) Either (1) or (2)
5. Which is always upright, the pyramid of:
   (1) Biomass (2) Energy
   (3) Number (4) All of these
6. In tree ecosystem, pyramid of number is:
   (1) Upright (2) Inverted

(3) Mixed (4) Spherical

7. In lake ecosystem, pyramid of biomass is:
   (1) Upright (2) Mixed
   (3) Inverted (4) None of these

8. The number of primary producers in a specified area will be highest in:
   (1) Grassland ecosystem (2) Forest ecosystem
   (3) Pond ecosystem (4) Desert ecosystem

9. Which kind of pyramid can never be inverted in any stable ecosystem?
   (1) Biomass (2) Number
   (3) Energy (4) All of these

10. The pyramid of energy is considered as most efficient representation of any ecosystem because it indicates:
   (1) Number of standing crops
   (2) Relationship of plants and animals
   (3) Gross energy of an ecosystem
   (4) The rate of food flow through the food chain

Answers

1. (3)	*2.* (4)	*3.* (2)	*4.* (2)	*5.* (2)	*6.* (3)
7. (3)	*8.* (3)	*9.* (3)	*10.* (4)		

16

CHAPTER

Various Ecosystems

Learning Objectives
Fresh Water Ecosystem (Pond or Lake Ecosystem) • Forest Ecosystem • Grassland Ecosystem • Desert Ecosystem • Cropland or Agroecosystem • Differences Between Natural and Man-made Agroecosystems • Marine ecosystem • Estuarine Ecosystem • Chilika lake • Stream and River Ecosystem.

Enumeration of the complete account of structure and composition of the ecosystems, is not quite possible, so only a few important and specific ecosystems have been accounted.

FRESH WATER ECOSYSTEM (Pond or Lake Ecosystem)

A fresh water or pond or lake ecosytem is a self-sufficient and self-regulating system. The abiotic and biotic components are as follows.

(I) Abiotic Components

The important abiotic components of a fresh water ecosystem are:

(1) *Mineral components* such as calcium (Ca), magnesium (Mg), phosphorus (P), nitrogen (N), sulphur (S) etc., dissolved oxygen (O_2), carbon dioxide (CO_2) contents, solute contents etc.

(2) *Organic components* such as chlorophylls, proteins, carbohydrates, fats etc.

(3) *Climatic factors* such as light, temperature, pH value, hardness and turbidity etc. of water.

(II) Biotic Components

Biotic components of a fresh water ecosytem spread over three categories: the producers, the consumers and the decomposers.

(1) Producers: Two main types of producers are found-

(i) Phytoplanktons: Common autotrophic phytoplanktons (producers) are algae belonging to class Chlorophyceae, Cyanophyceae, Euglenophyceae and Bacillariophyceae, which are popularly called green algae, blue green algae, euglenoids, diatoms, respectively. Important ***green algae*** found in ponds and lakes are *Chlamydomonas, Volvox, Pandorina, Scenedesmus, Chlorella, Spirogyra, Oedogonium, Ulothrix, Pithophora, Cladophora, Chara* etc., ***blue green algae*** are *Oscillatoria, Phormidium, Lyngbya* etc., ***euglenoids*** are *Euglena, Phacus* etc. and ***diatoms*** are *Cyclotella, Melosira* etc.

(ii) Macrophytes: These are hydrophytic higher plants belonging to Angiosperms (flowering plants). Macrophytes are of following types:

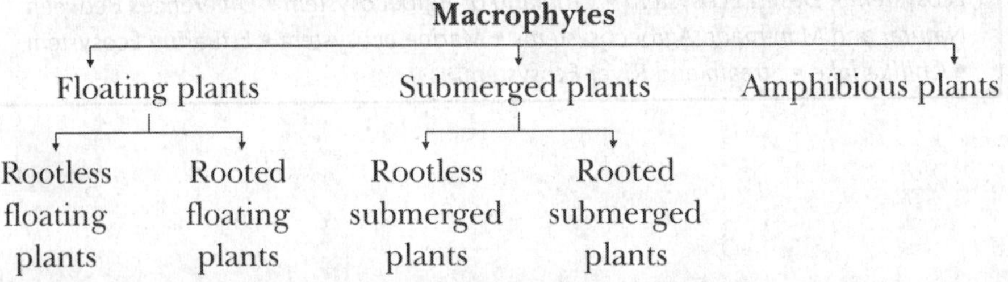

Important and common macrophytes are-

(a) Rootless or floating plants-*Lemna, Wolffia, Spirodela*.

(b) Rooted floating plants-*Eichhornia, Pistia, Trapa,Nymphaea* etc.

(c) Rootless submerged plants-*Ceratophyllum, Utricularia, Najas* etc.

(d) Rooted submerged plants-*Hydrilla, Vallisnaria, Potamogeton* etc.

(e) Amphibious plants (marshy plants)-*Sagittaria, Ranunculus, Typha* etc.

The Phytoplanktons and rooted submerged plants form the main producers of the pond ecosystem.

(2) Consumers: Consumers found in fresh water ecosystem belong to three groups, as follows-

(i) Primary consumers, *i.e.*, herbivores: Common herbivores comprise ciliates, flagellates, other protozoans, small crustaceans like *Daphnia, Copeods* and rotifers, small insects, mites, small fish etc. These feed upon algae and small green plants.

(ii) Secondary consumers: These are primary carnivores that feed upon herbivores and comprise beetles, small fish, frogs etc.

(iii) Tertiary consumers: These are top consumers such as large fish like game fish, water birds etc.

Thus, in a fresh water system, fish may occupy more than one trophic level, *i.e.*, all trophic levels of consumers.

(3) Decomposers or **saprotrophs**: These are aquatic fungi, bacteria and some flagellates that are distributed through out the pond. However, these are commonly found at the bottom where dead plants and animals accumulate. Aquatic fungi commonly found are *Achlya, Saprolegnia,*

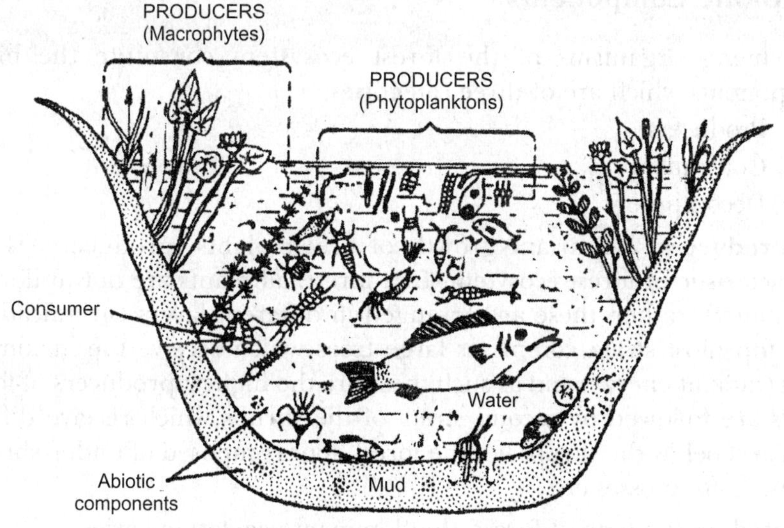

Fig. 16.1. A typical pond ecosystem.

Aspergillus, Penicillium etc. These decompose the organic matter and release the inorganic nutrients for reuse by green plants.

FOREST ECOSYSTEM

Forest consists of a complex biotic community comprising a wide range of producers and consumers. In India, roughly 20.64 per cent of land is covered by forests. In a forest ecosystem, the structural components are as follows.

(I) Abiotic Components

The abiotic or non-living components of forest ecosystem basically include:

(i) ***Inorganic nutrients*** such as Ca, N, P, S, K, Mg etc. are found in the soil in a large quantity. Humus is abundantly present in the forest.

(ii) ***Organic compounds*** in dead remains of plants and animals such as chlorophyll, proteins, carbohydrates, fats etc. are found in the large amount.

(iii) ***Climatic factors*** that affect the forest ecosystem are light, temperature, humidity, rainfall etc. Their interactions largely influence the ecosystem. Light plays a regulating role in the stratification of the vegetation. The annual precipitation (rainfall) ranges between 75 cm and 150 cm.

(II) Biotic Components

The living organisms of the forest ecosystem constitute the biotic components which are of three categories:

(1) Producers
(2) Consumers
(3) Decomposers

(1) Producers: Luxuriant growth of green plants (producers) is the characteristic of forest ecosytem. In a forest, all plants are not uniformly distributed, rather, these are arranged in different layers (stratification). The top most strata comprises large trees which receive the maximum light (radiant energy) and as such these are the highest producers of food. Trees are followed by second strata of the shrubs which receive diffuse light and below the shrubs lies the forest floor composed of under-shrubs, herbs, ferns, mosses etc.

Based on the type of forest, the dominant vegetation varies:

(i) ***Temperate deciduous forests***: The dominant trees found are beech (*Fagus*), birch (*Betula*), elm (*Ulnus*), Maple (*Acer*), oak (*Quercus*), walnut (*Juglans*) etc.

(ii) ***Temperate evergreen forests***: These are dominated by gymnosperms with few scattered angiosperms. Common gymnosperms are *Abies, Juniperus, Cedrus, Picea, Cupressus, Araucaria, Cryptomeria, Pinus* etc. Common flowering trees are *Rhododendrons, Quercus, Betula* etc.

(iii) ***Tropical deciduous forests***: The deciduous trees shed their leaves in autumn. The dominant vegetation of such forests are red silk tree (*Bombax*), amla (*Emblica*), teak (*Tectona*), sal (*Shorea*), *Butea, Terminalia, Syzigium* etc.

(iv) ***Tropical evergreen forests***: A large number of economically important trees are found in these forests. These include cinnamon (*Cinnamomum*), ebony (*Diospyros*), cloves (*Syzigium aromaticum*), mahogany (*Swietenia*), nutmeg (*Myristica fragrans*), rose wood (*Dipterocarpus*), rubber (*Ficus elastica*) etc.

(2) Consumers: Consumers found in forest ecosytem comprise three different categories-

(i) ***Primary consumers***: These are herbivorous animals which feed directly on the producers (green plants). Common examples found are beetles, flies, ants, grasshoppers, spiders, bugs among small animals and deer, elephants, nilgai, mangooses,

rats etc. among large animals which feed on products of trees (leaves, fruits, seeds etc.) Of all the species of insects known, nearly 70-80 per cent occur in the tropical rain forests.

(ii) ***Secondary consumers***: These are primary carnivores which feed on herbivorous animals. These include fox, frogs, snakes, wolf, eagles, vultures, jackals etc.

(iii) ***Tertiary consumers***: These are larger animals or top consumers (or top carnivores) which feed upon the secondary consumers, *e.g.* lion, tiger, leopard etc. Tropical rain forests have highest biome diversity.

Fig. 16.2. Type representation of a forest ecosystem.

(3) Decomposers: These are micro-organisms which feed on dead organic matter. In a forest, a large amount of organic litter (fallen leaves, barks, twigs etc.) is found on the forest floor. The litter is decomposed by a wide group of micro-organisms like-

(a) **Bacteria** *Clostridium, Bacillus, Nitrobacter, Nitrosomonas, Pseudomonas* etc.

(b) **Fungi** *Aspergillus, Polyporus, Fusarium, Agaricus, Morchella* etc.

(c) **Myxomycetes** *Physarum, Dictyostelium* etc.

In a forest ecosystem, the pyramids of biomass and energy are upright but that of number is spindle shaped or mixed.

GRASSLAND ECOSYSTEM

Grasslands are characterized by tall or short grasses, xerophytic herbs and shrubs. These face severe summer drought with only an annual rainfall ranging between 25 and 75 cm. At times, periodic fire devastations are found.

In India, true grasslands are not found as they have been mostly developed in a secondary form by the destruction of forests, prolonged droughts and overgrazing.

Grasslands are an excellent source of fodder for the herbivorous animals and domestic cattles. Common grasses found in Indian grasslands are *Agropyron, Aristida, Cynodon* (Bermuda grass), *Digitaria, Heteropogon* etc.

The various components of grassland ecosystem are

(I) Abiotic Components

(i) *Inorganic* or *nutrient elements* like S, Ca, K, Mg, Fe, N, P etc. are found in the soil.

(ii) *Organic matter* like carbohydrates, proteins, fats etc. are found derived from dead organic matter or plants and animals.

(iii) *Climatic factors* are of extreme nature such as intense light, high temperature fluctuations, severe drought, low rainfall, fire etc.

(II) Biotic Components

Grasslands are not diversity rich. Their biotic components can be classified into three types-

(1) Producers
(2) Consumers
(3) Decomposers

(1) Producers: The primary producers making food are grasses, but herbs and shrubs other than grasses also contribute to the primary production of food. Common grasses found in Indian grasslands are *Aristida, Brachiaria, Cynodon, Setaria* and *Digitaria*. Some grasslands are characterised by the dominance of -

(i) *Schima dicanthium* (South west U.P., M.P., parts of Tamil Nadu and Karnataka)
(ii) *Dicanthium cenchrus* (Haryana, Saurashtra, coastal Maharashtra, T.N.)
(iii) *Phragmites saccharum* (Assam, Bihar, Terai, U.P., Cauvery delta of T.N.)
(iv) *Bothrichloa* (Lonavala area of Maharashtra)
(v) *Cymbopogon* (Aravali, Chhota Nagpur, Satpura)
(vi) *Arundinella* (Nilgiris, Western Ghats)
(vii) *Deyeuxia arundinella* (Upper parts of Himalayas)
(viii) *Deschampsia deyeuxia* (Himalayas above 2500m).

(2) Consumers: Common examples of different consumers are as follows:

(i) ***Primary consumers***: These feed directly on grasses and include herbivorous animals like cows, buffaloes, goats, rabbits, mouse etc. and several other small animals (insects, millipedes etc.)

(ii) ***Secondary consumers***: These are primary carnivorous animals that feed on herbivores. Common examples are snakes, lizards, jackals, foxes etc.

Fig. 16.3. Type representation of grassland ecosystem.

(iii) ***Tertiary consumers***: In grasslands, hawks, eagles vultures, sometimes tigers leopards constitute the panel of tertiary consumers. These are also called top consumers.

(3) Decomposers: The organic matter of the grasslands is decomposed by the microbes like fungi (*Mucor, Rhizopus, Aspergillus, Cladosporium* etc.) and aerobic and anaerobic soil bacteria. By decomposing the organic matter, they release the minerals back into the soil, thus making the soil fertile.

DESERT ECOSYSTEM

Land is deserted in the region where the average annual rainfall is less than 25 cm. In India, about 17 per cent land is desert, of which Thar and Rajasthan deserts are well known. The main components of the desert ecosystem are as follows.

(I) Abiotic Components

In deserts, soil is porous, dry and nutrient poor with scanty organic matter. Climatic factors like temperature and water are found at their extremes. Days are very hot, nights are cold, humidity is extremely low and dust storms are common in deserts.

(II) Biotic Components

Deserts are represented with poor but characteristic biota. Their main biotic components are-

(1) **Producers**: The chief producers of the ecosytem are bushy shrubs and scanty trees, both with long extensive and deep root systems. Xerophytic vegetation is the main feature of deserts. Plants like *Prosopis, Acacia, Calotropis, Opuntia* (cactus), *Euphorbia, Zizyphus*, bunch grass are common. Majority of them are succulents and deep rooted grasses.

(2) **Consumers**: Three types of consumers found are-

- *(i)* ***Primary consumers***: Herbivores like jack and other desert rabbits, rats, kangaroo rats, wood rats, insects, domesticated cattles, seed eating quails and camels are common in deserts. Camel is called *ship of the deserts* as it can live for several days without drinking water.
- *(ii)* ***Secondary consumers***: These are carnivores like desert toads, reptiles, lizards, collar lizards, fox, rattle snakes, insect eating birds etc.
- *(iii)* ***Tertiary consumers***: These are mainly carnivorous birds, turkey vultures but a few wild cats may be found.

Only the organisms having specialised structural, physiological and behavioural adaptations to withstand the extremes of temperature and aridity can survive in deserts.

(3) **Decomposers**: These are thermophilic (loving hot temperature) fungi and bacteria. Due to scarcity of flora and fauna, dead organic matter is less and hence decomposers are also few in number.

CROPLAND ECOSYSTEM (OR AGROECOSYSTEM)

Cropland, also called artificial ecosystem, is a man-engineered ecosystem. The chief components are as follows.

(I) Abiotic Components

Soil remains nutrient rich due to fallen plant debries and substituted chemical fertilizers. It contains relatively a larger amount of elements like Ca, N, Mg, S, P, K, Zn, B etc. Organic matter remains available regularly. Croplands are engineered with variable climatic factors.

(II) Biotic Components

Three types of biotic components are-

(1) Producers

(2) Consumers
(3) Decomposers

(1) Producers: These are various types of crops of economical plants cultivated for human welfare. There may be single species culture such as that of wheat, paddy, maize, mustard, pulses, vegetables etc. or mixed type of them. The harvestable productivity of the crops is always very high, as compared to other ecosystems.

Box 16.1 Differences between natural ecosystems and agroecosystems

No.	*Natural ecosystems*	*Agroecosystems*
1.	They are rich in biotic diversity with less synchronisation of growth.	Biotic diversity is less and growth synchronisation is higher. They are often monocultures.
2.	They have complex trophic structure.	Trophic structure is simple.
3.	They are less vulnerable to catastrophic changes.	Catastrophic changes have severe effects.
4.	They are adapted for weather changes.	They are largely influenced by weather changes.
5.	Incidence of out breaks of insects, pests and diseases are not commonly found.	Such incidences are of common occurrence.
6.	They are more stable and self perpetuating.	They are less stable as being controlled by man through artificial practices.
7.	Vegetation is naturally selected, and are thus continuous in space and time.	Vegetation is selected by man and is thus discontinuous in space and time.
8.	They run for long duration, say for 50 or more years.	They are of short duration sometimes of 3-4 months
9.	They are permanent in nature.	They are of temporary nature.
10.	A kind of natural balance is maintained.	Natural balance is not found.
11.	They are relatively low productive.	They are often highly productive.
12.	Pressure of herbivores on producers is higher.	Herbivore pressure is low.
13.	They are mature from succession point of view.	As succession is disrupted by man, they are not allowed to mature.
14.	They provide wild life habitat.	Wild life habitat is destroyed.
15.	They enrich the soil fertility.	They destroy the soil fertility.

(2) Consumers: Three types of common consumers found in the cropland ecosystem are-

(i) ***Primary consumers***: The primary consumers (herbivores) are small organisms like insects, beetles, aphids etc. and large animals like rabbits, rats, birds, cattles, men etc.

(ii) ***Secondary consumers***: These are insect eating birds, frogs, foxes, jackals etc.

(iii) ***Tertiary consumers***: Top consumers like hawks and carnivorous snakes are found.

(3) Decomposers: These are microbes like fungi and bacteria of saproplytic nature, *e.g. Mucor, Rhizopus, Aspergillus* etc. (fungi) and *Bacillus, Clostridium, Pseudomonas* etc. (bacteria). Soil Microbes are responsible for decay and humification, thus making the minerals available to the producers.

Box 16.2 Man-made Ecosystems

They are also called artificial ecosystems. Common examples of man-engineered aquatic ecosystems are those of larger dams, reservoirs, artificial lakes, ponds, canals, small fishery tanks, Aquaria etc. and those of terrestrial man-made ecosystems are croplands (or agroecosystems), villages, cities, orchards, plantations, gardens, parks etc.

MARINE ECOSYSTEM

It is the largest and most stable system among all ecosystems. The important components of ocean ecosystem are as follows.

(I) Abiotic Components

The abiotic components of the ocean ecosystem do not fluctuate much and hence have little effect on the composition. Chemical components, dissolved oxygen, temperature, light etc. do not change much.

(II) Biotic Components

Different types of biotic components are classified into producers, consumers and decomposers.

(1) Producers: The chief producers are phytoplanktons like diatoms, dinoflagellates, unicellular algae etc. Larger plants are sea weeds belonging to red algae and brown algae. Common red algae are *Gracilaria, Polysiphonia, Chondrus* etc. and brown algae are *Sargassum, Turbenaria, Ectocarpus* etc. Some green algae like *Ulva, Caulerpa* and *Enteromorpha* are also found. Producers in the ocean remain distributed upto the depth to which light penetrates, thus light is the regulating factor.

(2) Consumers: These are as follows:

(i) ***Primary consumers***: Herbivorous organisms like crustaceans, mollusks, and small fish are the common primary consumers.

(ii) ***Secondary consumers***: These are carnivorous fish such as *Herring, Shad,* Mackerel etc. feeding on herbivores.

(iii) ***Tertiary consumers***: These are top carnivores in the food chain such as cod, Haddock, Halibut etc. Carnivorous birds may also be found.

(3) Decomposers: They are commonly marine bacteria which decompose the dead organic matter including excreta. They are frequently distributed in the bottom area. Some marine fungi are also found.

Box 16.3 **Marine Environment**

It is characterised by high concentration of salt and mineral ions. Marine water has about 3.5 per cent salt concentration. Sea is not a single environment. It's vertical zones are determined on basis of light availability- *(i)* **Photic zone** (upper 200 m) with sufficient light for photosynthesis (also called **euphotic zone**), *(ii)* **Aphotic zone** (upto the depth from 200 to 2000 m with insufficient light for photosynthesis and *(iii)* **Abyssal zone** (below 2000 m) with perpetual darkness. As such three major environments are recognised in the ocean basis- *(1)* **The Littoral zone**- comprising the sea floor from coastline to the region with 200 m. depth (also called *continental shelf*), *(2)* **The benthonic zone** comprising continental shelf (Photic zone), aphotic zone and abyssal zone and *(3)* **The pelagic zone** constituting the water of ocean basin. Marine life can be classified into three main categories: *(a)* **Planktons** *i.e.* passively floating or drifting organisms *e.g.* diatoms *(b)* **Nektons**, *i.e.* actively swimming organisms found in the surface or deep waters and *(c)* **Benthonic** organisms, *i.e.* found along the floor of sea bed and include creeping, crawling or sessile organisms. The nekton includes large and small fish, squids and whales and are found in both photic or aphotic zones. In abyssal zones are found predators with sharp teeth and long jaws.

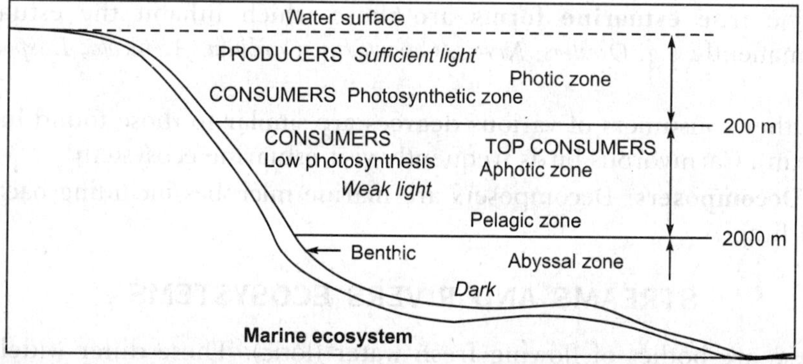

Marine ecosystem

ESTUARINE ECOSYSTEM

Estuary (*aestus* = tide) or brackish water zone is the junction of sea (saline water) with a river (fresh water). Thus, due to mixing of two opposite currents, *i.e.*, unidirectional flow of river and oscillating tidal currents of sea, estuaries develop unique ecological characters. Chemical changes of water is almost a continuous process, as during rains estuaries have larger amount of fresh water while during rest part of the year there is influx of sea water at high tide which results in the increase of salinity. The important components of estuarine ecosystem are as follows-

(I) Abiotic Components

These are almost similar to that of ocean ecosystem with predominating factors like changes in chemical composition of water, turbidity and light.

(II) Biotic Components

Biotic components are as follows-

(1) Producers: These are mainly of three types *(i) macrophytes* such as sea weeds, sea grasses and marsh grasses, *(ii) phytoplanktons*, and *(iii) benthic algae*. Several species of algae grow epiphytically on sea grasses.

(2) Consumers: The estuarine consumers may be divided into three categories: *(i)* stragglers, *(ii)* migratory forms, and *(iii)* true estuarine animals.

The **stragglers** include those animals of fresh water habitat which can tolerate a wide range of salinity changes and those animals of marine habitat which can withstand lower salinity of estuaries. These include wandering animals which may live for shorter or longer periods. Common examples are-*Carcinus* (shore crab), *Gammarus* species etc.

The **migratory forms** are those marine forms which spend their life for shorter periods. These often appear either during larval stage or during breeding season, for example, fishes like salmon and eels.

The **true estuarine forms** are those which inhabit the estuaries permanently, *e.g. Oyesters, Nereis, Gammarus, Cardium, Arenicola, Euspongia* etc.

Other consumers of various degrees are similar to those found in sea (ocean). Carnivorous birds frequently visit estuarine ecosystem.

(3) Decomposers: Decomposers are marine microbes including bacteria and fungi.

STREAMS AND RIVERS ECOSYSTEMS

These are bodies of flowing fresh water (lotic). These differ widely in various abiotic factors, such as volume of water, speed of flow, oxygen

Box 16.4 **Chilika Lake**

Chilika lake is the largest **brackish water lagoon** in India and is located between the 19°28'–19°54' N latitudes × 85°51'–85°38E' longitude being spread in the districts of Puri, Khurda and Ganjam of Orissa on eastern cost. At it peak, it covers 64.3 km × 18.1 km. It is one of the hotspots of the biodiversity in the country and some rare, vulnerable and endangered species listed in the IUCN Red List of threatened animals inhabit the lake area. As per the report of the Asian waterfowl census, it supported upto a million water birds in 1992.

Chilika is presently under threat from both natural and anthropogenic pressures. It was included in the Montreux Record (1993) which highlights priority action to be taken by the contracting party for its conservation. The Ramsar Wetlands Awards for 2002 was conferred to the Chilika Development Authority (CDA) to appreciate their conservation achievements.

Among aquatic vegetation in Chilika, *Hydrilla* and *Vallisnaria* species are predominantly found. Some important animals associated with the lake are White-bellied sea eagle, The Indian false vampire (*Megaderma lyra*), Fishing cat (*Felis viverina*), Tree frog (*Polypedates maculatus*), Green sea turtle (*Chelonia mydas*), Half beak (*Hemirhamphys gaimardi*), Sting ray (*Dasyatis imbricata*), Chilka crab (*Scylla serrata*) etc.

content, and temperature and many other physical and chemical conditions. The nature and composition of flora and fauna depend on the source of the river and the land environment they pass through. Floating population of planktons are generally absent in rapidly flowing zones. The biological activity is found high close to river bed area. Both phytoplanktons and zooplanktons are quite common. Many reptiles, mammals and birds obtain their food from aquatic biota. The general nature and composition of the fauna resemble that of lakes.

Chapter Summary

The important ecosystems found in India are fresh water ecosystem, forest ecosystem, grassland ecosystem, desert ecosystem, cropland ecosystem and streams and rivers ecosystems. They may be classified into natural and artificial (or man-made) ecosystems. Croplands, villages, gardens, artificial lakes etc. are man-made ecosystems. Each kind of ecosystem has its own characteristic features and biota. Various kinds of plants (herbs, shrubs and trees) are found in forest ecosystem, grasses in grassland ecosystems, xerophytic plants in desert ecosystem, hydrophilic plants in aquatic ecosystem etc. as producers. Consumers also vary depending upon climate and habitat, *e.g.* camels are characteristic to deserts; lions and foxes to forests; fishes to aquatic ecosystems. Sharks and whales are found in marine ecosystem.

The most important type of man-made ecosystem is the agroecosystem. It is characterised by low bio-diversity vulnerable to sudden changes. However, agro-ecosystems are simple but highly efficient as regards to productivity.

Study Questions

1. Give an account of the major forest biomes found in India.
2. Give the major characteristics of desert or pond ecosystem.
3. Describe agroecosystems and compare it with natural ecosystems.
4. Write a brief account of marine or estuarine ecosystem.
5. Describe the characteristic biotic components of grassland ecosystems in India.
6. Describe the various types of producers found in pond or lake ecosystem.
7. Explain the differences between natural and man-made ecosystems.

Objective Questions. *Select the correct answers*

1. Largest ecosystem of the world is:
 (1) Grassland (2) Great Lakes
 (3) Ocean (4) Forest

2. Phytoplanktons are:
 (1) Producers of forests (2) Producers of Lakes
 (3) Consumers of sea ecosystem (4) Omnivores

3. A natural ecosystem depends upon:
 (1) Plants (2) Animals
 (3) Man (4) Self-operating system

4. *Vallisnaria* is a hydrophyte of which category?
 (1) Floating (2) Submerged free floating
 (3) Submerged rooted (4) Emergent rooted

5. An example of free floating hydrophyte is:
 (1) *Hydrilla* (2) *Typha*
 (3) *Lemna* (4) *Nelumbo*

6. *Pinus* and *Cedrus* are commonly found in:
 (1) Eastern Ghats (2) Himalayas
 (3) Deserts (4) Ocean

7. A producer found in marine ecosytem is:
 (1) Shark (2) Starfish
 (3) *Gracilaria* (4) All of these

8. Deserts have characteristic presence of:
 (1) Elephants (2) Lion
 (3) Sal tree (4) Camels

9. Succulent plants are likely to be found in:
 (1) Deciduous forests (2) Estuarines
 (3) Deserts (4) Sea ecosystem

10. Highest variety of producers are met in:
 (1) Marine ecosystem (2) Forest ecosystem
 (3) Desert ecosystem (4) Agroecosystem

Answers

1. (3) *2.* (2) *3.* (4) *4.* (3) *5.* (3)
6. (2) *7.* (3) *8.* (4) *9.* (3) *10.* (2)

17

CHAPTER

Ecological Succession

LEARNING OBJECTIVES

Introduction • Characteristics of Succession • Causes of Succession • Types of Succession • General Process of Ecological Succession • Classification of Succession • Hydrosere, Xerosere • Climax Forest Stage.

Introduction

One of the characteristics of the environment is that it is always in a state of constant change due to climatic, biotic and physiographic factors. The vegetation of an area and its related fauna keep on changing over a long period of time. Different communities follow each other in a definite sequence. For example it appears unbelievable that Rajasthan had forests where mughal emperors (some 500 years back) use to go for hunting tigers.

The process of changing communities in a definite sequence is known as **succession**, a term coined by **Hult** (1885). Famous ecologist **Clement** (1916) defined succession as *the natural process by which the same locality become successively colonised by different groups or communities of plants.*

Characteristics of Ecological Succession

The important characteristics of common succession process are-

(1) It is an orderly process of changes in species structure and community and therefore, it is predictable.

(2) It is a completely biological process which is community controlled even though the physical environment determines its pattern, *i.e.* physical environment based biological process.

(3) It ends in a stabilised ecosystem in which maximum biomass and symbiotic function between the organisms are maintained.

(4) Climax community maintains dynamic equilibrium with the environment.

Causes of Ecological Succession

The main causes of succession are-

(1) **Climatic causes**: Plants can adjust with certain range of climatic variation. Acute fluctuations affect them adversely, *e.g.* drought, heavy rainfall, lighting etc. are some important factors for causing destruction of vegetation.

(2) **Topographic causes**: Topographic changes such as erosion or landslides causes destruction of vegetation. Sometimes, soil deposition becomes the cause of destruction and to initiate succession.

(3) **Biotic causes**: Many living agencies including man affect the vegetation, *e.g.* grazing, cutting, deforestation, clearing, jhuming, cultivation etc. are directly responsible for the vegetation changes, and to initiate succession.

Types of Ecological Succession

Succession has been classified in different ways, but following are some important types of succession-

(1) **Primary succession**: This kind of succession begins from the primitive substratum where there was previously no any kind of living matter, *e.g.* rocks, volcanic eruptions or new land exposed from ocean.

(2) **Secondary succession**: It begins from previously built up substrata with already existing living matter. The vegetation of such area has been destroyed due to climatic or biotic factors such as fires, flood etc.

(3) **Autogenic succession**. The developing community brings a change in the environment which becomes unsuitable for them, but opens the way for the growth of a different plant community. Such succession is called autogenic type.

(4) **Allogenic succession**: The habitat is changed due to external factors, like climate, changes in soil composition etc. and becomes unsuitable for the existing community. When such changes favour the growth of a different community, the succession is referred to as allogenic type

(5) **Induced succession**: A succession which is controlled and motivated by man for his own welfare and suitability and maintains a steady community is called induced succession, *e.g.* cultivating crops in a field.

(6) **Autotrophic succession**: A succession which begins on a place rich in inorganic contents but poor in organic contents is called autotrophic type, because it is dominated by autotrophic communities.

(7) **Heterotrophic succession**: A succession that begins in an area rich in organic matter (*e.g.* sewage, forest litter etc.) and dominated by heterotrophic organisms (such as fungi and detrivores) is referred to as heterotrophic succession.

(8) **Retrogressive succession**: Sometimes due to heavy biotic interference, the succession, instead of progressing goes backwards. It is called retrogressive succession or backwardly moving succession, *e.g.* a forest community changes to shrub or grassland community due to deforestation, overgrazing etc.

General Process of Ecological Succession

The process of succession of one community by another of a complex type, till the development of a stable community, is known as **ecological succession**. The whole series of communities which develop in a given area is called **sere**. The first community is called a **pioneer community**, the final community is called a **climax community** and intermediate stages are called **seral stages** or **pioneer stages**. The process of succession can be represented hypothetically as fig. 17.1.

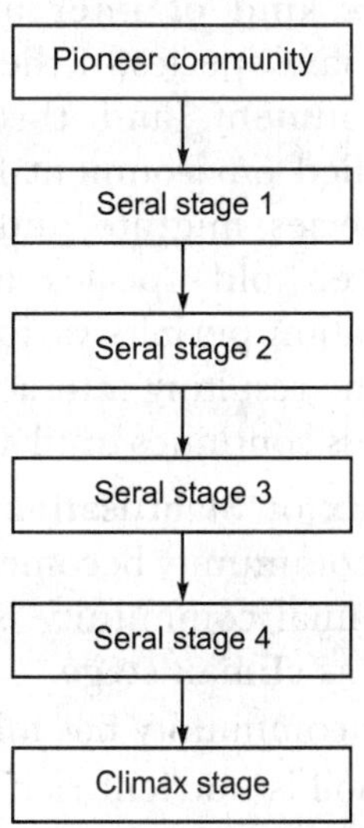

Fig. 17.1. Diagrammatic representation of the process of ecological succession.

The various steps of succession on a sterile habitat are as follows:

(1) **Nudation**: The development of a sterile or virgin area is called **nudation**. It can be done by several catastrophic agencies like landslides, erosion, volcanic eruption, flooding, fire, disease etc. New bare areas can also be created by man, *e.g.* by deforestation, burning, diggings, making reservoirs etc. Thus, nudation may be brought about by topographic, climatic or biotic factors.

(2) **Invasion**: It refers to arrival and settlement of some organisms on the bare areas. The first invaders are called **pioneers**. Invasion comprises three steps:

(i) ***Dispersal*** or ***migration***. Invasion occurs through reproductive units such as spores, seeds, buds etc. The transfer of a species from one area to a new area is called **migration**. It is affected by

dispersal of seeds or spores through the agencies like air, water or animals.

(ii) *Ecesis*: It means establishment of the migrated plant species into the new area. The adjustment followed by the establishment of a species is called **ecesis**. It can take place only after proper germination, vegetative growth and reproduction.

(iii) *Aggregation*: It is the final stage of invasion. As a result of reproduction, the number of species increases and they grow close to one another. This phenomenon is called aggregation.

(3) **Competition and Reaction**: Since the space and nutrition is limited and reproduction continues there ensues a struggle for existence and a kind of inter and intra-specific *competition* sets in. The dominant species of the community begins to modify the physical environment, and this process is called *reaction*. Finding the modified environment favourable for growth more complex type of species migrate and settle in the area. Now, a competition between old species and new species sets in. The changed vegetation permits various kinds of animals to join the community and the resulting interactions further modify the environment. This process continues until a stable and climax community is developed.

(4) **Climax** or **Stabilisation**: After a long process of competition, the final community becomes stable and in equilibrium with the climate. This final community is known as a **climax community** and the stage as **climax stage**.

The climax community has following characters -

(i) Soil is nutrient rich in both inorganic and organic matter.

(ii) Species having long life and resistance to the external disturbances predominate.

(iii) Species diversity of organisms is very high and food chain becomes complex.

(iv) Biomass is maximum and is in steady state.

(v) The community is in perfect equilibrium with the abiotic environment of that area.

In climax community, there is an equilibrium between gross primary production and total respiration, the solar energy captured and energy released by decomposition and between the uptake of nutrients and the return of nutrients to the soil by fall of litter. If undisturbed, such community can maintain itself through the years.

The climax community of all types of terrestrial successions is considered to be a forest or sometimes can be a grassland. It takes about 1000 years to reach climax community in case of primary succession, but only about 200 years for a forest climax in secondary succession. When a climax community which is not a climatic climax for a given area is maintained by man, it may be designated as a **disclimax** (= disturbance climax) or **anthropogenic subsclimax** (= man generated climax).

Classification of Succession

The primary and secondary successions are mainly of following types:

(1) **Hydrosere**-beginning in the aquatic environment.
(2) **Xerosere**-beginning in xeric or dry habitats.
(3) **Lithosere**-beginning on the rock surface.
(4) **Psammosere**-beginning on the sandy habitats.
(5) **Halosere**-beginning on the salty habitat.

HYDROSERE (OR HYDRARCH)

Ecological succession in a pond or lake is called hydrosere. The complete process from the beginning comprises following important stages-

(1) Phytoplankton Stage

It is the initial stage of succession in which pioneer colonisers are planktonic algae which soon becomes associated with aquatic bacteria and protozoans. These algae commonly belong to green algae (Chlorophyceae), blue green algae (Cyanophyceae), and diatoms (Bacillariophyceae). Popular examples of planktonic algae are *Spirogyra, Ulothrix, Oedogonium, Chlamydomonas* and of protozoans are *Amoeba, Paramecium* etc. These forms remain floating in water. These multiply for some times and continuously add a large amount of organic matter and nutrients due to their various activities and their death.

(2) Rooted Submerged Stage

Organic matter built up by phytoplanktons mixes with silt brought with rain water and develops into soft mud at the bottom of the pond. In such condition, especially where light is adequately available, the places become suitable for the plant growth. Slowly with the span of time, rooted submerged hydrophytes like *Elodea, Myriophyllum, Chara, Hydrilla, Vallisnaria, Potamogeton* etc. begin to grow. The seeds and propagates of these plants are brought by birds and other animals which visit the water body for food and other purposes. All these plants form a tangled mass of vegetation. Upon death, the vegetation decomposes only partly due to lack of oxygen. Their remains get deposited at the bottom of the pond and serve three purposes, *(i)* bind the loose bottom sediments into a firmer matrix, *(ii)* increase organic matter at the bottom, and *(iii)* decrease the depth of water and thereby make the pond shallow. Such changes pave the way for other organisms.

(3) Rooted Floating Stage

When the depth of water reaches to 2-5 feet, rooted floating plants colonize the pond bottom. Some common examples of such plants are *Nelumbium, Nymphaea, Trapa, Aponogeton* etc. These are accompanied with other strong floating forms like *Spirodela, Lemna, Pistia, Azolla, Salvinia* etc.

With the changes in vegetation, the animal life changes. Flies like Dragon flies, May flies, Crustaceans like *Asellas, Gammarus, Cyclops, Daphnia* etc. and fishes inhabit the pond.

The decomposing organic matter, formed due to death and decay, further built up substratum making the pond more shallower and environment more conducive for the growth of other forms of life.

(4) Reed Swamp Stage (Amphibious Stage)

In shallow pond water, when conditions for the growth of floating plants become unsuitable, marshy rooted plants like *Scirpus, Typha, Sagittaria, Phragmites* etc. appear. These plants are rooted very firmly in the bottom muck as they have highly branched rhizome and fibrous roots. Their photosynthetic parts grow exposed to air. Such plants build the source by retaining the sediments and accumulation of plant remains. Some vegetation drifts offshore to take roots there and such patches of vegetation finally join the swamp.

Animal life changes due to change in vegetation. Birds like king fisher, heron (*Ardea*), ducks, swamp, sparrow etc. become common, and insects like dragon flies, may fly, giant water bug, scavenger, beetle etc. appear.

(5) Sedge Meadow Stage (Marginal Mats)

When the water table is so shallow as to just cover the surface, the swampy plant stage gives a way to sedge meadow stage. Plants like *Carex, Juncus, Scirpus, Cyperus, Eleocharis, Mentha, Colha, Campanula* etc. make their appearance at the pond edges. These gradually extend towards the centre of pond rapidly and form sod-like mats of vegetation which bind wind-born soil, plant debries and water. This builds up a firmer substratum above the water table which dries in summer due to heavy transpiration by these plants. The dry soil becomes inhabitable by amphibious plants. Plants of this stage form a good quality of humus after their death and decay.

Due to changed vegetation, land animals which feed upon grasses become common at the site.

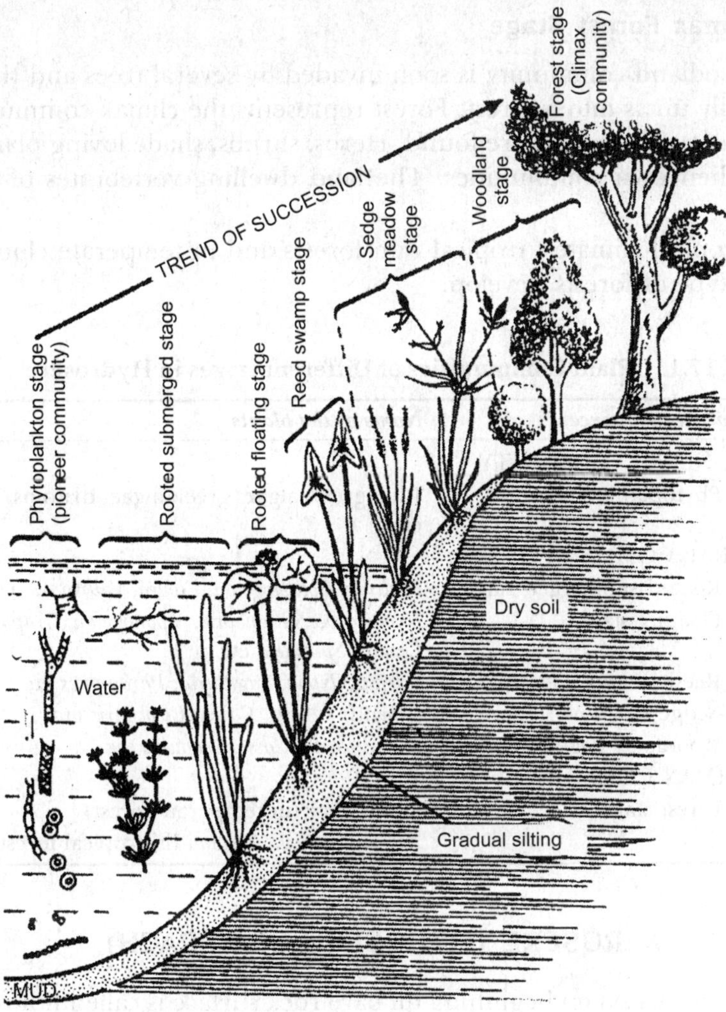

Fig. 17.1. Different stages of hydrosere.

(6) Woodland Stage

When, the land dries, terrestrial herbs, shrubs and then undertrees grow and form open vegetation or woodland. These plants produce more shade and absorb and transpire a large quantity of water. Common plants of woodland stage are: *Salix, Cornus, Cephalanthus, Alnus, Populus, Acacia* etc. In this stage, a large amount of humus, dead organic matter, bacteria, fungi and other micro-organisms grow in the soil. The mineralization is expedited making soil more suitable to a variety of trees. Vertebrates like rabbits and deer gain entry at this stage.

(7) Climax Forest Stage

The woodland community is soon invaded by several trees and the area gradually turns into a forest. Forest represents the climax community in which all type of plants are found. Herbs, shrubs, shade loving plants etc. make their own communities. The land dwelling vertebrates of forests appear.

In tropical climates, tropical rain forests and in temperate climates, a mixed type of forests develop.

TABLE 17.1. Plant Communities of Different stages in Hydrosere

Community (in sequence)	*Name of the plants*
I. PIONEER COMMUNITY	
↓ (1) Phytoplankton stage-	Blue green algae, green algae, diatoms, bacteria etc.
II. SERAL COMMUNITIES	
| (2) Rooted submerged stage-	*Hydrilla, Potamogeton, Elodea, Vallisnaria* etc.
| (3) Floating stage-	*Nymphaea, Nelumbium, Aponogeton, Trapa, Azolla, Lemna, Spirodela* etc.
| (4) Reed-swamp stage-	*Scirpus, Typha, Sagittaria, Phragmites* etc.
| (5) Sedge-meadow stage-	*Cyperus, Juncus, Carex, Eleocharis* etc.
↓ (6) Woodland stage-	*Salix, Acacia, Cassia, Populus* etc.
III. CLIMAX COMMUNITY	
(7) Forest stage -	*Ulmus, Acer* (in temperate forest) *Acacia, Quercus, Tectona* (in tropical forests).

XEROSERE (LITHOSERE OR XERACH)

Ecological succession beginning on bare rock surface is called lithosere or xeric succession or xerosere. It is completed in a series of several orderly steps, each comprising a characteristic plant community. These stages are-

(1) Crustose-lichens Stage

Rock surfaces are dry and hard and are subjected to extremes of temperature. As such these provide the most inhospitable conditions for the growth of plants. Since crustose lichens are able to grow in extreme xeric conditions, they colonise first to rock. The common such lichens are *Rhizocarpon, Lecanora, Rinodina, Lecidia* etc. These lichens produce carbonic acid (lichenic acid) by their metabolic activity which begins to corrode rock surface and initiates its weathering. The dead organic matters of lichens and small particles of rock mix up and develop a thin substratum suitable for other types of lichens. It is very slow and a long time process and may take hundred of years.

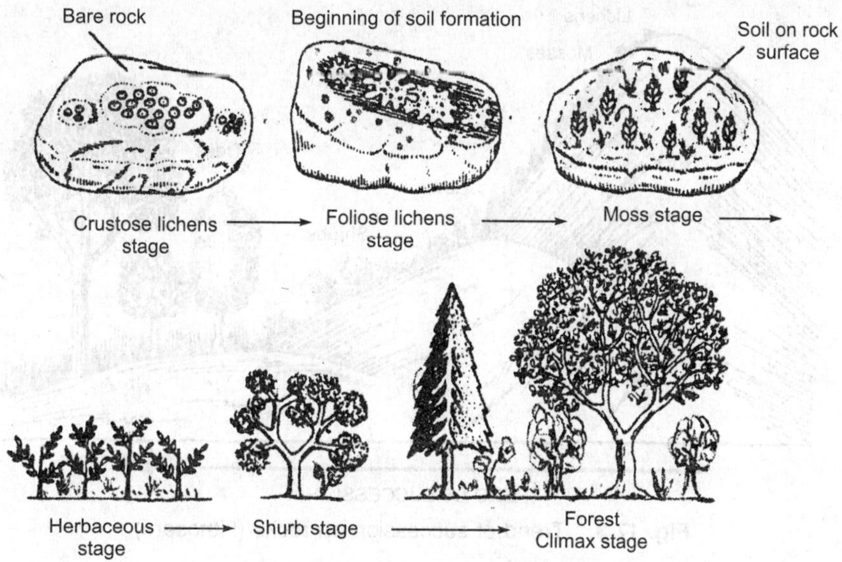

Fig. 17.2. Different stages in Xerosere.

(2) Foliose-Lichens Stage

Foliose types of lichens begin to grow with and after crustose lichens. Foliose lichens like *Dermocarpon, Parmelia, Umbilicaria* etc. are the higher forms of lichens. These have large leaf-like thalli which overlap crustose lichens, thus cutting them off from direct sunlight. This results in the death and decay of crustose lichens enriching the organic matter on the rock surface. Foliose lichens can absorb and retain more water and are able to accumulate dust particles which help in further build up of the substratum accompanied with humus accumulation. As a result, a thin layer of soil develops on rock surface and there is change in habitat. The soil becomes thick enough to hold the next stage of succession. The pioneer animals of lichen stages are a few species of mites, ants and spiders.

(3) Moss Stage

The development of thin soil layer on rock surface favours the growth of some xerophytic mosses like *Polytrichum, Tortula, Grimmia, Bryum, Barbula, Funaria, Hypnum* etc. These migrate with the help of their wind blown spores. The rhizoids of foliose lichens and mosses now compete for water and nutrients and mosses gradually take over. Their death and decay further add the organic matter in the soil. As a result, the thickness of the soil increases and the soil becomes suitable for the growth of other kinds of plants. Animals like mites, spiders, spring tales etc. invade the community.

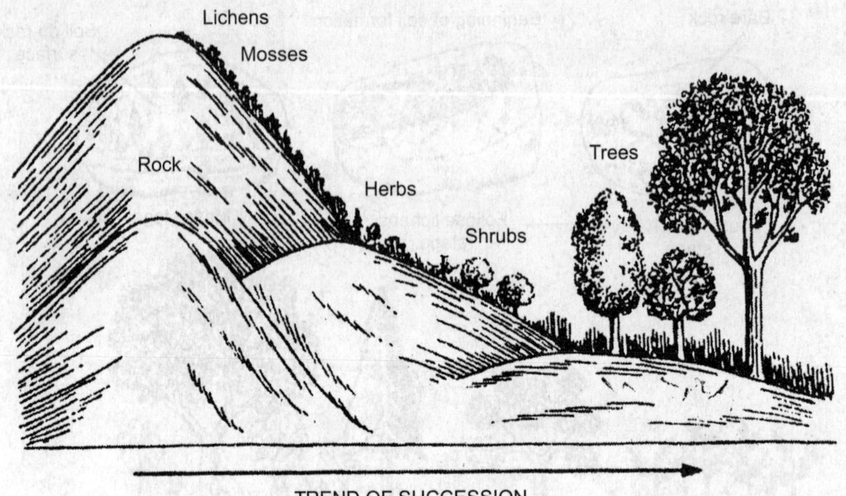

Fig. 17.3. Trend of succession on rocks (Lithosere).

(4) Herbaceous Stage

When organic matter, humus and substratum further increase, more soil is formed. The changed habitat favours the growth of herbaceous plants like *Poa, Sporobolus, Festuca, Solidago* etc. In early stage annual herbs followed by perennial herbs stabilize with the soil. Common perennial herbs may be certain xeric ferns like *Adiantum, Cheilanthus, Asplenium, Actiniopteris* etc. Due to death and decay of short-lived herbs (ephemerals), the humus content of soil continuously increases. Such changes decrease the xeric conditions.

The animal fauna also becomes varied with the complexity of communities. Forms like new species of mites and insects and new forms of insects like *Collembola* and new micro-arthropods invade the area.

(5) Shrub Stage

The changed habitat of the herbaceous stage favours the growth of shrubs which migrate with the help of their seeds or rhizomes from the adjoining area. In moderate climates the pioneer shrubs may be *Zizyphus, Capparis, Zygophyllum, Rhus* etc. and in colder climates, *Rubus, Fragaria, Physocarpus* etc. The herbs, which are now shaded by the overgrowing shrubs, fail to survive and are completely replaced by shrubs with the passage of time. The soil is further enriched by the dense shrubby growth. The animal fauna is further enriched by forms of lizards, centipedes, spiders, birds, shrews etc.

(6) Climax Forest Stage

The shrubs give place to trees and finally a forest results. It leads to stabilise vegetation community. The pioneer trees are xerophytic like *Acacia, Boswellia, Prosopis* etc. These are followed by deciduous trees. A forest has variety of animals including mites, spiders, millipedes, centipedes, slugs, snails, frogs, salamanders, reptiles (turtles, lizards, snakes etc.), squirrels, shrews, foxes, birds etc.

TABLE 17.2. Plant Communities of Different stages in Xerosere

Community	*Name of the plants*
I. PIONEER COMMUNITY	
↓ (1) Crustose -lichens stage-	*Rhizocarpon* and *Lecanora*
II. SERAL COMMUNITIES	
| (2) Foliose-lichens stage-	*Parmelia* and *Dermatocarpon*
| (3) Moss stage-	*Polytrichum* and *Grimmia*
| (4) Herbs stage-	*Poa, Festuca, Aristida* etc.
↓ (5) Shrub stage-	*Rhus, Phytocarpus, Lantana* etc.
III. CLIMAX COMMUNITY	
(6) Forest Stage-	*Quercus, Terminalia, Salmalia, Acacia* etc.

Chapter Summary

Succession is a process of changing communities in a definite sequence in an area. It may occur due to climatic, topographic or biotic causes. There are several types of succession but primary (starting from an area where vegetation never existed) and secondary (starting from an area which earlier had vegetation) are important types. General process of ecological succession comprises nudation, invasion, competition and reaction, climax or stabilisation. Successions are classified on the basis of its place of beginning, *e.g.* hydrosere (in aquatic environment), xerosere (in dry or xeric habitat), lithosere (on the rocks), psammosere (on sandy habitat), halosere (on salty habitat) etc.

Ecological succession in a pond (hydrosere), comprises: *(i)* Phytoplankton stage, *(ii)* rooted submerged stage, *(iii)* rooted floating stage, *(iv)* reed-swamp stage (amphibious stage), *(v)* sedge-meadow stage (marginal mats), *(vi)* woodland stage and *(vii)* climax forest stage.

Ecological succession beginning on rocks undergoes stages like *(i)* crustose-lichens stage, *(ii)* foliose-lichens stage, *(iii)* moss stage, *(iv)* herbaceous stage, *(v)* shrubs stage and *(vi)* climax forest stage.

The biotic community is always dynamic undergoing changes with the passage of time and one community is replaced by the other. The succession begins with the invasion of barren area by the pioneer species followed by other suitable species or serial communities. Ultimately it leads to a climax community which remains stable until there is a vast change in the environment. Primary succession takes longer time as compared to secondary one.

Study Questions

1. What is plant succession? Trace the various stages of succession in a bare rock.
2. Discuss the causes and process of Ecological succession.
3. Describe the process of Ecological succession in an aquatic habitat.
4. Trace the path of changes in vegetation taking place in xeric habitat.
5. Describe the types and general process of succession.
6. Differentiate between.
   (i) Serial and climax community.
   (ii) Primary and secondary succession.
   (iii) Progressive and retrogressive succession.
7. Enlist the different stages found in hydrosere or xerosere.
8. Briefly describe the causes of succession.
9. What are different types of succession? Describe.
10. Define the terms:
    (i) Primary and secondary succession
    (ii) Nudation
    (iii) Invasion
    (iv) Pioneer community
    (v) Seral community
    (vi) Climax community
11. What is succession? Define.
12. Explain the terms:
    (i) Sedge meadow stage
    (ii) Woodland stage
    (iii) Phytoplankton stage
    (iv) Allogenic succession
    (v) Autogenic succession
    (vi) Foliose lichens stage
13. Describe in one sentence:
    (i) Hydrosere
    (ii) Xerosere
    (iii) Lithosere
    (iv) Psammosere
    (v) Halosere

Objective Questions: *Select the correct answers*

1. The term 'succession' was coined by-

(1) Clement (2) Odum
(3) Hult (4) Jefrey

2. The process of changing communities in a definite sequence is known as:

(1) Evolution (2) Variation
(3) Diversity (4) Succession

3. A succession which is controlled and motivated by man for his own welfare is known as:
   (1) Autogenic succession (2) Induced succession
   (3) Allogenic succession (4) Secondary succession
4. The arrival and settlement of some organisms on the bare area is called:
   (1) Migration (2) Invasion
   (3) Ecesis (4) Nudation
5. The community which becomes stable and in equilibrium with the climate is known as:
   (1) Pioneer community (2) Seral community
   (3) Climax community (4) Mixed community
6. A succession beginning on the sandy habitats is called:
   (1) Xerosere (2) Psammosere
   (3) Halosere (4) Lithosere
7. In hydrosere, the first stage is called:
   (1) Rooted submerged stage (2) Reed-swamp stage
   (3) Sedge-meadow stage (4) Phytoplankton stage
8. The amphibious stage in hydrosere is also called:
   (1) Reed swamp stage (2) Sedge meadow stage
   (3) Rooted floating stage (4) Woodland stage
9. The pioneer of Xerosere are:
   (1) Mosses (2) Ephimeral herbs
   (3) Lichens (4) Shrubs
10. Forest stage is the climax community of:
   (1) Hydrosere (2) Xerosere
   (3) Psammosere (4) All of these

Answers

1. (3) 2. (4) 3. (2) 4. (2) 5. (3) 6. (2)
7. (4) 8. (1) 9. (3) 10. (4)

Biodiversity and its Conservation

- *Magnitude, Levels and Gradients of Biodiversity and Biogeographical Classification of India*
- *Value of Biodiversity*
- *Threats to Biodiversity*
- *Endemic and Endangered Species*
- *Conservation of Biodiversity*
- *India as Megadiversity Nation and Hot Spots of Biodiversity.*
- *International and National Efforts for Conserving Biodiversity*

18

CHAPTER

Magnitude, Levels and Gradients of Biodiversity and Biogeographical Classification of India

LEARNING OBJECTIVES

Introduction • Definition of Biodiversity • Global Magnitude of Biodiversity • Levels of Biodiversity: Genetic Diversity • Species Diversity • Community and Ecosystem Diversity • Gradients of Biodiversity • Biogeographical Classification of India.

Introduction

Every organism in this living world, whether it is a plant, an animal or a microorganism, is unique in itself. This uniqueness of individuals is the basis of the diversity that is shown by the living organisms. Different places, such as hills, forests, land or sea side, in different parts of the world have their own typical type of flora (plants) and fauna (animals), together called biota. Biodiversity encompasses microscopic protozoa to about 30 meters long blue whale among animals and *Micromonas* (algae) to about 100 meter tall red wood trees of California among plants. Some conifers live for thousands of years while insects like mosquitos die in few days. Life also ranges from colourful to transparent worms and to brightly coloured birds and flowers. In fact the degree of biodiversity is endless. The environment, which we encounter and which we interact in our day to day life, comprises millions of living organisms. Looking at a patch of forest, one may notice wide variety of plants from tiny grasses to huge trees and animals ranging from a small insects to large mammals, coupled with numerous microorganisms in the soil not seen by naked eye. Looking at a different forest patch at some other place and comparing its biodiversity with that of previous scene, a difference can be seen in the kind of living organisms both plants and animals. This variability among the organisms reflects biodiversity of the forest patch. And such diversity among life differs from place to place. Whether or not a species can occur at one place is largely determined by the environmental conditions of that place. Of course, the range of the tolerance of the species also forms a factor. Considering the total habitats of plants and animals, it may be concluded that the living world abounds with enormous biodiversity which is very much essential for sustainability of the environment. The diversified organisms, living in variety of habitats, possess characteristic features which are time tested and highly evolved to suit them to prevailing environment.

The term 'biodiversity' is often heard in connection with how it is fast disappearing. Loss of biodiversity is a global concern and is being looked upon as one of the world's most pressing crisis. Biologically rich and unique habitats are being degraded, fragmented and destroyed due to the problems caused by exponential growth of human population, over consumption of resources and effects of pollution. The primary reason for the concern is the realisation that biodiversity is being lost even before its dimension is known. Biodiversity loss would check evolutionary capability of flora and fauna to cope up with the changes that have taken place in the environment. How to check the loss of species and erosion of gene pool, is posing a serious challenge before mankind. The way human activities are affecting biodiversity and causing an increased rate of extinction of species, this has to be looked upon with great attention. Indirectly, the loss of biodiversity is resulting in global changes in climate, altered hydrological patterns and other ecological functions. It has now become essential to conserve our biological resources squarally.

Definition of Biodiversity

Biodiversity is a concise form of biological diversity. The term was coined by Harvard scientist EO Wilson in 1985. It refers to variability among living organisms from all sources including terrestrial and aquatic (both fresh water and marine) ecosystems. Biodiversity may be defined as *the variability among millions of species of plants, animals and microorganisms; the gene they contain; and the intricate ecosystem they help build into the living environment*. This bewildering variety of life with uniqueness of individuals on earth, which evolved over millions of years, shows variability among the different life forms, number of species or races within species or gene pools. Biodiversity includes the habitat diversity as well as diversity within the human society. The cultural variability is also covered within the present day concept of biodiversity. In technical parlance, biodiversity refers to the totality of genes, species and ecosystems of a region. Biodiversity includes the rich diversity of forms right from the molecular unit to the individual organisms, to population, community, ecosystem, landscape and biosphere levels. Biodiversity is earth's primary life support system and is a precondition for human survival.

The concept became so popular that it was discussed in the United Nation's Conference on Environment and Development held at Rio de Janeiro in 1992 and a great deal of effort was made by more than hundred countries arriving at an international accord for preservation and conservation of biodiversity. In Conference's Convention on Biodiversity (CBD), biodiversity has been defined as *the variability among living organisms from all sources interalia, terrestrial, marine and other ecosystems and ecological complexes of which they are a part*.

Global Magnitude of Biodiversity

Since last 250 years, there has been a progress in the systematic work on identifying and naming the species present on this planet. But so far it could be possible to collect, describe and name far less number of species than the actual number present on the earth. Biodiversity is not distributed uniform across the globe; greater in some areas than the others. Diversity of species increases from poles towards the tropics. Tropical moist forests cover only 5-7 per cent of land but possess about 50 per cent of world's species. Warm and humid tropic regions of the earth, between the Tropic of Cancer and the Tropic of Capricon, called Megadiversity region, are highly rich and diverse in biodiversity. Of the total biodiversity on this earth, more than 50 per cent is concentrated in Megadiversity Countries namely Australia, Brazil, China, Colombia, Ecuador, India, Indonesia, Peru, Malaysia, Medagascar, Mexico, Zaire, Cameroon, Costa Rica, Ethiopia, South Africa, Venezuela and Philippines. Of these, Brazil is richest in biodiversity, both quantitatively and qualitatively. Vast majority of species are found in the developing countries in the tropical moist forests. Asia pacific region, with its unique identity, is one of the biologically richest regions. It belongs to Indo-Malayan Realm and comprises 27 biogeographic provinces. The region harbours two out of twelve centers of origin and of diversity of crop plans and domesticated animals. Almost all plants consumed today in Europe, originated in the developing countries (table 18.1). Thus genetic diversity, needed to maintain the agricultural system of the world, is mainly found in the developing countries of the world because wild plants still continue to remain in this region. Also variety of wild animals, to be used in developing animal wealth, occur in this region. Most of the medicinal plants and non-agricultural item yielding plants are also found only in these countries.

TABLE 18.1. Plants and their Centers of Origin

No.	*Plant Group/Plant*	*Place of Origin*
1.	Rice, Mustard, Mango, Orange, Radish, Coconut, Banana.	India, Malaysia and Indonesia
2.	Pea, Bean, Okra, Coffee	Ethiopia
3.	Onion, Parsnip, Lettuce, Celery, Turnip, Rubarb	Mediterranean
4.	Tea, Egg plant, Apricot, Cucumber, Sugarcane, Cherry	China
5.	Wheat	Turkey and Afghanistan
6.	Potato	The Andes, South America.

There are about 10 million species of living organisms on this earth, of which only less than 2 million have been identified and described and about 80 per cent have not yet been explored, especially in the tropics and in unapproachable areas of forests, mountains, deserts, depth of the ocean etc. However, with the efforts of projects like Global Information Facility and the Species 2000, new species are being discovered faster than ever before.

Levels of Biodiversity

Nature displays a fascinating variety of organisms with genetic diversity within the species, complex ecological relationships among various species of organisms and great variety of ecological systems. There are three hierarchical levels of biodiversity, interrelated yet distinct enough to be studied separately to understand the interconnections that support life on the earth.

1. Genetic Diversity. Genetic diversity refers to the variation of genes within species. Variation may be in alleles, in the entire gene or in chromosomal structure. When genes, within the same species, show different versions due to new combinations, they result genetic variability. The genetic diversity is perhaps the least obvious part of biodiversity. Each species, from bacteria to higher plants and animals, stores an immense amount of genetic information. This genetic information lies in their genes which number abuot 450-700 in mycoplasma, 4000 in bacterium *Escherichia coli*, 32000-35000 in rice (*Oryza sativa*) and 35000-45000 in man (*Homo sapiens*). The genetic diversity, *i.e.*, variation in genes within species, often increases with the environmental variability and enables a population to adopt to its environment and to respond to natural selection process of evolution. For example, if a species has more genetic diversity, it can adopt better to the changed environmental conditions. The lower the genetic diversity in a species leads to uniformity as in case of large monoculture of genetically similar crops. Taking into this advantage, crop production can be increased but it may prove harmful when the crop gets attacked by insects or pathogenic fungi and thus the whole crop gets threatened. The amount of genetic variation plays a key role in speciation, *i.e.,* evolution of new species, and in the maintenance of the diversity at the species and community level. Many species, in a community, will result greater genetic diversity as compared to a community with few species. Environmental variability often plays a significant role in increase of genetic diversity within a species. Each human being differs widely from all others due to a large number of combinations possible in man's gene that give him specific characteristics. Genetic variability is essential for a healthy breeding population of a species because if the number of breeding individuals is reduced, inbreeding occurs which may lead to extinction of the species. *Oryza sativa* embraces all rice varieties, both wild and cultivated, which show variation at the genetic level and differ in their colour, size, shape, aroma and the nutrient contents of the grain.

2. **Species Diversity.** Species diversity refers to the diversity within a population or between different species of a community or an ecosystem. Species are the distinct units of biodiversity. Each species plays a distinct role in an ecosystem. Therefore, loss of species affects species richness, *i.e.*, number of species per unit area, and this affects the species abundance of the ecosystems as a whole. Richness, abundance and variability of species in an ecosystem represents its species diversity. The species richness also increases with the increase in the area of the site. It is very high in tropical rainforests and coral reefs and low in isolated islands. The species diversity is thus directly proportional to the species richness and species richness to the area occupied by the species. The number of individuals, among species, which may also vary, results into the difference in evenness and equitability and consequently in diversity. In nature, greater diversity is exhibited due to both, the number and kind of species, as well as well as variation in the number of individuals per species. Natural tropical forests have a much greater species diversity than most other regions.

3. **Community and Ecosystem Diversity.** Community and ecosystem diversity refers to the number of ecological niches, various ecological processes and a variety of trophic levels that sustain food chain, food webs and recycling of nutrients. Thus ecosystems exhibit tremendous diversity which has developed over millions of years of evolution. It has a focus on different biotic interactions and role and function of *keystone species*, *i.e.,* the species determining the ability of large number of other species to persist in the community. Generally, the diverse communities, having the component species present in nearly equal abundance, are functionally

more productive and stable even under the environmental stresses such as prolonged droughts, than the uniform communities having extremes of high and low abundance. The number of habitats or ecosystems can vary within a geographical area and act as a measure of biodiversity. Rainforests, deserts, oceans, lakes, estuaries, wetlands etc. are the major ecosystems on this earth where species live and evolve.

The mathematical indices of diversity that have been developed to represent species diversity at different geographical scales are: alfa, beta and gamma diversity

(1) **Alfa Diversity** (*Diversity within a community*). Alfa diversity, also called local diversity, refers to the diversity of organisms which share the same community/habitat. Community combinations of species richness and equitability/ evenness is used to represent the diversity within a community/habitat. A change in species can be observed with the change of community/habitat. In fact alfa diversity refers to popular concept of species richness in an ecosystem and can be used to compare the number of species in different ecosystems.

(2) **Beta Diversity** (*Diversity between communities*). Beta diversity refers to the degree to which species composition changes along an environmental gradient in communities/habitats. There are differences in species composition of communities/habitats along environmental gradients such as altitudinal gradient, moisture gradient etc. Beta diversity increases with higher heterogeneiity in the communities/habitats in a region or greater dissimilarity between communities/habitats. Taking an example of moss community on a mountain, beta diversity is higher if the species composition of moss communities changes at successively higher altitudes on the mountain slope, but it is lower if the same species occupy the whole mountain side.

(3) **Gamma Diversity** (*Diversity of habitat over the total landscape or geographical area*). Gamma diversity refers to the diversity of the habitats over the total land scape or geographical area. It reflects species turnover rate with distance between sites of the similar habitats or with expanding geographical areas. Gamma diversity is applicable on larger geographical scales and is defined as the rate at which additional species are encountered subsequent to geographical replacements made for a given habitat type in different areas.

Degree of diversity of an ecosystem is determined by the following factors:

1. **Habitat Stress**: Diversity is low in habitats under any of the stresses like harsh climate, pollution etc.

2. **Geographical Isolation**: Diversity is less in an isolated region like an island in which if a species disappears due to random events such as *Tsunami*, it can not be easily replaced because the organisms from mainland would find it difficult to reach island and colonise it.
3. **Dominance by One Species**: The dominant species consumes a disappropriate share of the resources and does not allow many species to evolve and flourish.
4. **Availability of Ecological Niches**: A complex community offers a greater variety of niches than a simple community and this promotes greater diversity.
5. **Edge Effect**: The diversity is greater in ecotones or transition areas between ecosystems.
6. **Geological Experience**: Old and established ecosystems such as tropical rainforest, that have not experienced many changes, have high diversity whereas an Arctic ecosystems that have undergone many changes did not allow many species to establish.

Point diversity refers to the diversity on the smallest scale, *i.e.*, the diversity of microhabitat whereas ***epsilon diversity***, also called regional diversity, refers to total diversity of a group of areas of gamma diversities. For example a leaf of a plant may be considered as an unit of point diversity; a single plant as an area of alfa diversity; a group of plants occurring together as an area of gamma diversity; and the forest, within which the plants are located, as an area of epsilon diversity.

Biodiversity is essential for continued survival of species and natural communities. Selection and genetic manipulation cause the loss of biodiversity in both plants and animals leading to decrease in disease resistance, reproductive viability and adaptability to changing environment. The loss of genetic diversity may be considered irreversible, although the potential exists in the conservation of genetic materials of endangered species and indigenous breeds.

Gradients of Biodiversity

Changes in latitude and altitude markedly affect the biodiversity. On moving from high to low latitudes, *i.e.*, from poles to equator, the species diversity increases. Similarly there occurs a decrease in species diversity from lower to higher altitudes on mountains. The two major factors – the latitudinal and altitudinal gradients, largely affect the species diversity, although, regional and taxa-related exceptions do occur. It is also expected that if the physical environment is more complex and heterogenous, the flora and the fauna will also be more complex and diverse. Since one thousand metre increase in altitudes results in a temperature drop of 6.5°C, this temperature drop coupled with greater seasonal variability at higher altitudes, are the major factors for decrease of biodiversity in high altitude regions such as mountain peaks.

The centers of the greatest biodiversity, the megadiversity regions of the world, tend to be in the tropics because of the following reasons:

1. Over long geological times, the tropical areas have had a more stable climate, never been glaciated, therefore, local species continued to thrive and live there itself.
2. There has been more time for tropical communities to evolve, allowing them greater degree of specialisation and local adaptation to occur.
3. Tropical areas receive more solar energy over the year, making tropical communities more productive and, therefore, resulting higher productivity and a greater resource base that can support a wider range of species.
4. Tropical areas, especially tropical rainforests, receive more precipitation which enable the plant species dominated by flowerings trees, their new generations and underlying vegetation to grow round the year, providing rich habitats to many animal species including wildlife.
5. Dense forest canopy in tropical regions prevents light from reaching forest floor, precluding the development of any ground flora.
6. Warm temperature and high humidity of the tropical areas provide favourable environmental conditions for many species of plants, animals and microorganisms.
7. In tropical regions, due to hot and humid climate, the greater pressure from pests, parasites and diseases, does not allow any single species to dominate, giving opportunity to many species to co-exist.
8. Complex physical environment in tropical regions provides many niches for specialist species.
9. In tropical regions, higher rates of outcrossing among plants leads to higher levels of genetic variability, providing more chances of speciation and making it possible for a larger number of species.

In temperate regions, on the other hand, the climate is severe with short growing period for plants. Moreover, the climate in these regions could not be stable over geological time and hence the local species had a tendency to disperse to other areas. Also in these areas, reduced pressure from pests, parasites and diseases due to cold, there are only few dominating species that exclude many other species.

BIOGEOGRAPHICAL CLASSIFICATION OF INDIA

India is the seventh largest country in the world as regards its geographical area. The country lies north of equator between $8^0$4' and $37^0$6' north latitude and $68^0$7' and $97^0$25' east longitude. It measures 3214 km from north to south and 2933 km from east to west with a total land areas of 3,287,263 sq km It is bounded on south-west by Arabian sea and in the south–east by the Bay of Bengal. On the north, in north-east and north-west lies the Himalayan ranges. Kanyakumari constitutes the southern tip of Indian peninsula where it gets narrower, loses itself into the Indian ocean.

India is endowed with rich biodiversity spread over 10 biogeographically distinct zones (Rodgers and Pawar, 1988) due to varying physical conditions and species groupings. Each zone has its own characteristic climate, soil and topography and contains distinct species of flora and fauna. Among these 10 biogeographical zones, Deccan Peninsula has the most extensive coverage (about 42 per cent) of the land mass. The Western Ghats and North–East India account for 4 and 5.2 per cent of the geographical areas, respectively. Each biogeographical zone encompasses several habitats, biotic communities and ecosystems and depicts insight about distribution and environmental interactions of flora and fauna of the country. The country occupies about 2.45 per cent of world's area, has 8.10 per cent of global biodiversity with a species count of about 0.131 million. A sizable percentage of species are endemic, *i.e.,* species exclusive to India. About 33 per cent of flowering plant species (with 140 endemic genera) recorded in India are endemic. Endemics are mainly concentrated in North-East India, North-East and North-West Himalayas, Western Ghats and Andaman and Nicobar Islands. Western Ghat encompasses a very high number of endemic species especially of amphibian animals. The Gangetic plains are generally poor in endemism while Andaman and Nicobar Islands contribute atleast 220 species to endemic flora of India. Among Indian fauna, 396 species of higher vertebrates, 44 species of mammals, 55 species of birds, 187 species of reptiles and 110 species of amphibians are endemic.

The Indian subcontinent has consistently been placed under Palaeotropic Realm (or kingdom) under the Indian Region. Its unique phytogeographical and agro-ecological diversity provides it a wide variety of agro-climatic zones that harbour a rich repository of biological resource. The country represents almost all biogeographical regions of the world. According to a recent classification given by Wild Life Institute of India, the country's biological wealth can be observed as represented by 10 broad biogeographical zones (fig. 18.1). These are: Trans-Himalayan (Upper region), Himalayan (North-West, West, Central and East Himalayas), Desert (Kutch, Thar and Ladakh), Semi-arid (Central

India and Gujarat-Rajwara), Western Ghats (Malabar Coast and Western Ghat Mountains), Deccan Peninsula (Deccan-South Central, Eastern and Chhota Nagpur Plateau; and Central Highlands), Gangetic Plain (Upper and Lower Gangetic Plain), North–East India, (Brahamputra Valley and North-Eastern Hills), Islands (Lakshadweep, Andaman and Nicobar) and Coast (West and East Coasts).

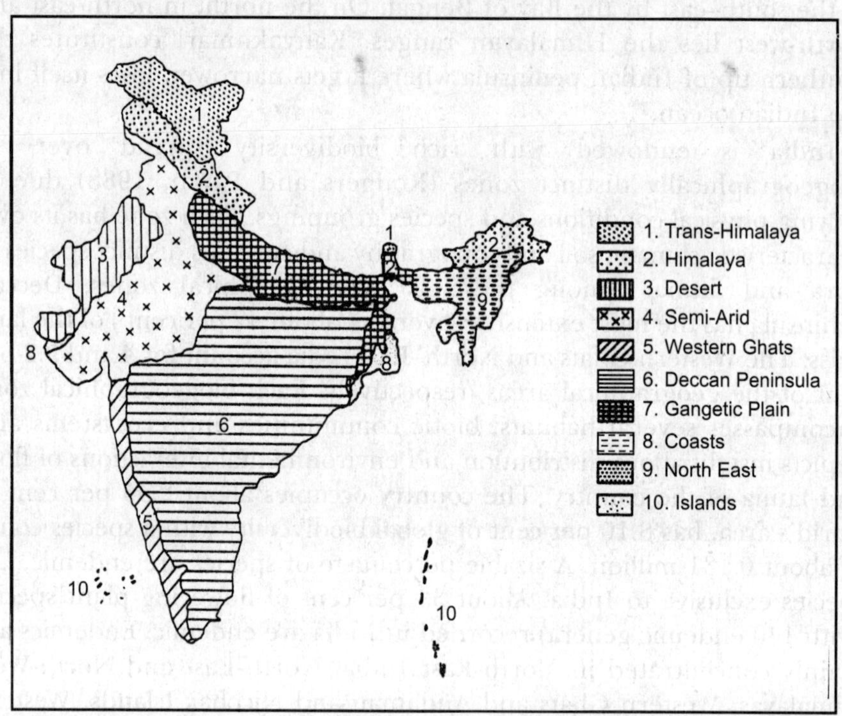

Fig. 18.1. Biogeographical regions of India.

1. Trans-Himalayan Zone (Upper Region) : It is an extension of Tibetan plateau including the high-altitude cold desert in Ladakh and Lahul Spiti with sparse mountain vegetation and possesses the richest wild sheep and goat community in the world. The snow leopard is found here, as is the migratory black-necked crane.

2. Himalayan Zone: Himalayan mountain wall extends over a distance of 2400 km from north-west region of Kashmir to the east upto NEFA. It comprises diverse biotic provinces and biomes. There are four biotic provinces – north–west, west, central and east Himalayas. Based on altitude, there are three zones of vegetation in Himalayan range which correspond to three climatic belts: ***First*** the sub-montane or lower region (tropical and subtropical), that extends from plain to foot of hills upto

5000-6000 ft. altitude, has vegetation dominated by trees of *Shorea robusta* (Sal), *Dalbergia sissoo* (Shisham)–in riverain areas, *Acacia catechu* (Kutch or Kathaa), *Eugenia jambolina* (Jamun), *Butea monosperma* (Dhaak) *Anthocephalous kadamba* (Kadamb), *Albizzia procera, Cedrela toona, Steropermum* sp. etc. ***Second*** the temperate or montane zone, ranging between 5,500-12,000 ft altitude, has vegetation dominated by *Pinus excelsa* (Kail), and *P. longifolia* (Chir), *Cedrus deodora* (Deodar), *Quercus lamellosa, Q. incana* and *Q. lineata, Cedrela, Eugenia* etc. in the lower region and conifers such as *Abies pindrow, Picea morinda, Taxus baccata, Cupressus torulosus, Juniperus* sp. etc. in the upper region. ***Third*** the alpine zone, above 12,000 ft. altitude, is the limit of tree growth (known as *timber* or *tree line*) where shrubby growth of *Betula utilis, Juniperus* and *Rhododendron* is found in grassy areas. At about 15000 ft. altitude and above *snow-line*, plant growth is almost nil. It is worth noting that though all biotic provinces of Himalayan zone are similar in vegetational zones, but eastern Himalayas have more tropical elements, greater variety of oaks and rhododendrons and less of conifers than the western Himalayas due to higher percipitation and warmer conditions in the eastern Himalayas.

3. Desert Zone: This arid area, west of the Aravali hill range, comprises three biotic provinces, *viz.,* Kutch (salt desert of Gujarat), Thar sand desert of Rajasthan and Ladakh. The north-west desert region (Kutch and Thar) incorporates parts of Gujarat, Rajasthan, Haryana and Delhi with climate which is very hot and dry (in summers) and very cold (in winters) and rainfall less than 700 mm. The region has extensive grassland with mostly xerophytic plants such as *Acacia nilotica* (Babool), *Prosopis spicifera, Salvadora oleoides, Tamarix dioica, Tecomella* sp. etc. and the ground vegetation is dominated by the species of *Calotropis, Eleusine, Panicum antidotale, Tribulus terrestris* etc. The Great Indian Bustard, a highly endangered species of animalas, is found in this north–west desert region. The Ladakh region, the cold desert region, on the other, hand has sparse vegetation.

4. Semi-Arid Zone: This area, between the desert and the Deccan plateau including Aravali range, comprises Madhya Pradesh, Chattisgrah, parts of Orissa and Gujarat. On the basis of the amount of rainfall, the forests in this region have developed into thorny, mixed deciduous and 'Sal' type. The vegetation of forests is mostly constituted by *Tectona grandis* (Teak), *Diospyros melanoxylon* (Tendu), *Butea monosperma* (Dhaak), *Terminalia tomentosa, Dalbergia latifola* etc. The thorny vegetation is dominated by *Carissa spiranum, Acacia leucophloea, Acacia catehu, Butea frondosa* etc.

5. Western Ghat Zone: This area of hill ranges and plains, running along western coastline, comprising the Malabar coast and Western Ghat mountains of India, extends from Gujarat in the north to the Cape of

Camorin in the south. This region receives heavy rainfall and has four types of vegetation, *viz.*, tropical north evergreen forests, subtropical or temperate evergreen forests, mixed deciduous forests and the mangrove forests. The tropical moist evergreen forests are very luxuriant and multistoried and have tall trees such as *Dipterocarpus indicus, Tectona grandis, Cedrela toona* and *Sterculia alta* and many species of bamboos. In the temperate evergreen forests of Nilgiri hills, trees such as *Michelia nilgarica, Eurya japonica* and *Gardenia obtusa* are most common.

6. Deccan Peninsula Zone: This area of south and south–central plateau, south of the river Tapti, comprises five biotic provinces, *viz.*, Deccan Plateau (south), Central Plateau, Eastern Plateau, Chota Nagpur Plateau and Central Highland. The zone is semi-arid region and lies in the rain-shadow of Western Ghats. The precipitation hardly exceeds 100 mm. This zone has central hilly plateau with forests of *Boswellia serrata, Hardwickia pinnata* and *Tectona grandis*.

7. The Gangetic Plain: Gangetic plain, along Ganga river system, lies in the north extending upto Himalayan foothills. This region, comprising Uttarkhand,Uttar Pradesh, Bihar and West Bengal, is the most fertile region due to alluvial soil. The major climatic factors, the temperature and precipitiation together, are responsible for the distinctive type of vegetation in the region. The precipitation varies from less than 700 mm in Western Uttar Pradesh to more than 1500 mm in West Bengal. The vegetation is chiefly of tropical moist and dry deciduous forest type. *Dalbergia sissoo* and *Acacia nilotica* are most common near the foothills of the Himalayas in north-western Uttar Pradesh; *Capparis aphylla* (Kareel), *Acacia nilotica* (Babool), *Saccharum munja* (Moonj) etc. are most dominant species in south-western Uttar Pradesh; *Madhuca indica* (Mahua), *Butea monosperma* (Dhaak), *Diospyros melanoxylon* (Tendu), *Acacia catechu* (Kutch or Kathaa), *Azadirachta indica* (Neem) *Mangifera indica* (Aam), *Ficus religiosa* (Peepal), *Cordia myxa* (Lashoda), *Ficus benghalensis* (Bargad) etc. are the characteristic species in the eastern Uttar Pradesh; *Rhizophora mucronata*, *Acanthus ilicifolius*, *Rhizophora conjugata*, *Bruguira gymnorhiza* etc. are most common species of extreme swampy and halophytic vegetation in the Gangetic delta region. Besides, some weeds and grasses like *Xanthium strumarium* (Baghnakhi), *Argemone mexicana* (Pilikateli or Satynaasi), *Cassia tora* (Panwar), *Bothrichloa pertusa* etc. are also found in the Gangetic plain zone.

8. The North-East India: It comprises non-Himalayan hill ranges and is one of the richest floristic regions in the country. This region receives the heaviest precipitation with Cheerapunji that used to have as much as more than 10,000 mm. The temperature and wetness are also very high, resulting in the dense tropical evergreen forests. The common trees are

Mesua ferrea, Michelia champaka (Champa), *Dipterocarpus macrocarpus, Shorea robusta* (Sal), *Alstonia scholaris* (Chitwan), *Sterculia alta* etc. and many bamboo species like *Bambusa pallida, Dendrocalamus hamiltonica* etc. Many grass species including *Imperata cylindria, Saccharum aurandanceum* etc., and insectivorous plants like *Nepenthes khasiana* are also present. In the hilly tracts, *Pinus khasya* (Khasi Pine) is the dominant conifer present. In the northern cooler regions, *Alnus nepalensis, Rhododendron arboreum* etc. are also present. Besides, the region has several wild relatives of cultivated plants such as banana, mango, citrus, peeper etc.

9. The Islands: The Islands of Lakshadeep in the Arabian sea and Andaman and Nicobar in the Bay of Bengal, have a wide range of spreading coastal vegetation like mangroves, beach forests and in the interior some of the best preserved evergreen forests of tall trees such as *Rhizophora, Calophyllum, Mimusops, Largerstroemia* and *Dipterocarpus*.

10. The Coast: India has a coastline of about 7516.6 km. Mangrove vegetation is the characteristic of the estuarine tracts along the coast, such as at Pichavaram near Chennai and Ratna Giri in Maharashtra.

Chapter Summary

Every organism of living world is unique and this uniqueness of individuals is the basis of biodiversity. The term biodiversity, coined by EO Wilson in 1895, refers to variability among living organisms on the planet earth. In technical parlance biodiversity refers to the totality of genes, species and ecosystems of a region. The concept became so popular that United Nation's Conference on Environment and Development held at Rio in 1992 led Convention on Biological Diversity (CBD) was held to protect and preserve the biodiversity on this earth. Of the total biodiversity on this earth, more than 50 per cent is concentrated in megadiversity countries. Vast majority of the species are found in developing countries in tropical moist forests. Plants such as rice, mustard, mango, radish, coconut, banana and orange owe their origin in India. There are three hierachial levels of biodiversity, *viz.*, genetic diversity, species diversity and ecosystem diversity. Genetic diversity refers to variation of genes within a species, species diversity refers to diversity within a population or between different species of a community or an ecosystem whereas community or ecosystem diversity refers to the number of ecological niches, various ecological processes and a variety of trophic levels sustaining food chain, food web and recycling of nutrients. Species diversity is spread over alfa, beta and gamma diversities based on mathematical indices.

India is a seventh largest country in the world as regards its geographical area. The country is endowed with rich biodiversity spread over 10 biogeographical zones due to varying physical conditions, climates, soils, and topography and species groupings of flora and fauna. Each biogeographical zone encompasses several habitats, biotic communities and ecosystems. Endemism centres round mainly in Eastern Himalayas, Western Ghats and Andaman and Nicobar Islands.

Study Questions

1. Write a brief account of the global magnitude of biodiversity.
2. What are the various levels of biodiversity? Describe each of them.
3. What are the various biogeographical regions of India? Discuss their peculiarities.
4. Define Alfa, Beta and Gamma Diversities.
5. Define Biodiversity.
6. Define the Gradients of Biodiversity
7. Discuss the features of Indo-Malayan Realm.

Objective Questions. *Select the correct answers*

1. One of the most biologically richest region of the world is:

(1) Indo-Malayan Region (2) Russia
(3) West Africa (4) South America.

2. Rice is native of:

(1) China (2) Japan
(3) Indonesia (4) India

3. Species diversity within a living community is known as:

(1) Alfa diversity (2) Beta diversity
(3) Gamma diversity (4) Genetic diversity

4. The species determining the ability of large number of other species to persist in the cummunity are called:

(1) Original species (2) Keystone species
(3) Dominant species (4) Competetive species

5. As regards its geographical area in the world, India ranks:

(1) Fourth (2) Fifth
(3) Sixth (4) Seventh

6. One thousand metre increase in altitude results a temperature drop of about:

(1) 5.5°C (2) 6.5°C
(3) 7.5°C (4) 8.5°C

Answers

1. (1) 2. (4) 3. (1) 4. (2) 5. (4) 6. (2)

19

CHAPTER

Value of Biodiversity

LEARNING OBJECTIVES

Introduction • Value of Biodiversity: Direct Values • Consumptive Use Value • Productive Use Value - Timber Value • Fishery Value • Indirect Values -Ecosystem Service Values • Social Services Values • Aesthetic Value • Ethical Value • Option Value.

Introduction

We share with earth with about 10 million other species of living organisms, of these more than fifty per cent occur in the tropics. The services we use from ecosystems, such as clean water, food, fuel, fibre, medicines, climate control, energy, inputs for industries etc. can not be provided without biodiversity. Failure to use biodiversity sustainably will perpetuate inequitable and unsustainable growth, deeper poverty, new and more rampant illness, continued loss of species and a world with ever-more degraded environment which are less healthy for people. Unfortunately, no one knows which species of living beings may prove to be of a particular value to mankind. The biodiversity, *i.e.,* life insurance of our changing world, guarantees the continued provision of ecosystem goods and services and ensures that the world has the capacity to adapt future changes.

As advances in reducing poverty and improving well being for human population are made, we will more clearly understand the need for effectively functioning of ecosystems. A wide range of crop and livestock genetic diversity is essential to ensure that our agro-systems can adapt to new challenges from climatic changes, pests and diseases. The biological wealth in marine environments is needed to feed growing populations and provide livelihoods for coastal communities around the world. Wetlands are needed as water regulators to protect us from floods and storm surges and in moderating climatic change with other ecosystems such as forests and to act as living filters for pollutants and excess of fertilizers. We must not forget that biodiversity is central to many of the world's cultures, the source of legend and myth, the inspiration for art and music, basis of medicinal knowledge and so on. Provision of these services by biodiversity depends on maintaining it with best of our efforts. Importance of sustainable use of biodiversity in achieving the Millennium Development Goals has already been recognized by world leaders in their support for achieving a significant reduction in the rate of biodiversity loss by 2010-the so -called 2010 target. Biodiversity can indeed help alleviate hunger and poverty, can promote good human health and be the basis for ensuring freedom and equity for all. All of us rely on biodiversity, directly or indirectly for our health and welfare. The 2010 target is thus foundation for our well–being and continued sustainable existence.

VALUE OF BIODIVERSITY

Intricately woven biosphere is a life support system to human race. Each species in this system has its own role and own importance. In combination, various species of plants, animals and microorganisms, enable the biosphere to sustain the human race. Biological diversity plays a vital role in biosphere's health, stability and functioning properly. Biodiversity concerns to all human beings and is most directly related to everyday life of the people. It provides them their food, fuel, fodder for livestock, housing material, medicine and spiritual sustenance. Hundreds of millions of small farmers, fisher folk, herders and hunter gathers-a substantial number of them being tribals, that depend upon the diversity of species, genetic varieties and ecosystem services, depend on biodiversity for their livelihoods and cultural lives.

We should care about all this diversity of life because wild species of living organisms have provided essential products since human first set his foot on the earth and still they are continuing to provide a basis for our needs of existence. Apart from our basic needs and other necessities, biological diversity provides ecological services such as soil formation, nutrient cycling, solar energy absorption, management of biogeochemical and hydrological cycles, waste disposal, air and water purification, maintenance of chemical composition of the atmosphere etc. Biodiversity plays a major role in determining world's climate. Diversity of biological communities such as lakes, forests, mangroves, wildlife sanctuaries, coastal areas etc. forms a basis of many recreational activities like picnics and camps, fishing, visiting sea shores, wildlife watching etc. Many mountain societies have cultural practice of maintaing biodiversity and other landforms because of their religious significance.

Value of biodiversity goes far beyond into the future. We know only about few species out of numerous that exist. And every year many species, out of known and unknown, are lost. Infact the rate of their loss far exceeds than their exploration by us. Due to lack of our understanding of the complex interrelationships between organisms, it is difficult to determine whether a specific species is ecologically valuable or essential but when even seemingly insignificant species is removed, it affects the ecology of the system of which it is a component. Keeping in view the above fact, preserving a diversity of life on the earth has come to be an accepted goal. Often this goal comes into conflict with other goals, such as economic development. The question arises between how much diversity and at what cost. To find the answer to this question, it is important to think carefully about value of biodiversity. Environmental economics provides methods of assigning economic values to species, communities and ecosystem. These values include the harvest value of bioresources, the value provided by un-harvested bioresources in their natural habitats, and the future values of bioresources.

The value of earth's biodiversity can broadly be divided into two categories direct use value and indirect use value.

1. Direct Use Value

Direct value, also known as use value and commodity value, is assigned to the benefits derived from biodiversity directly or indirectly from the products that are harvested from it. These values are further sub-divided into two: consumptive and productive use values.

A. *Consumptive Use Value*

Human welfare and availability of bioresources go hand in hand. Economic welfare of mankind centred round consumption and utilization of several hundreds of species of biological diversity of great potential during the course of time as human civilisation grew. For thousands of years, man has been engaged in exploration of plant and animal wealth. Originally, plants were consumed directly from the wild and animals by hunting within man's reach. The biodiversity or their products can be harvested and consumed directly as food, fuel, fibres, drugs, fodder, household accessories, wool, leather, paint, resin, wax, rubber, silk, feathers, thatch, timber, decorative items etc.

(1) Food Value. Human consumes a number of wild and semi-wild species of plants as food. About 8,000 species of edible plants have been reported from wild (Myeres, 1957) but man has used around 5,000 species of plants as food. About 90 per cent of the present day food crops have been domesticated from wild tropical plants. Although there are about 1,000 species of cereals, presently only about 200 species have been domesticated as food crops and of these less than 20, cultivated to produce 85 per cent of world's food, are major international crops of economic importance. Wheat, rice and corn, the three major carbohydrate crops, yield nearly two-third of the food, sustaining world's population. Although relatively few plants contribute to food production at global level, but at local level, particularly in tropics, a very wide range of plants provide food in various forms. For example, villagers in Indonesia are estimated to consume about 4,000 native plant and animal species as food. The food is mostly derived from fruits and seeds or some other parts of the plants such as roots, leaves, flowers, stems or their modifications, such as tubers, rhizomes, bulbs etc. Many species of marine algae and mushrooms also form the source of the food. Livestock and wild fauna also contribute greatly to feeding humans. Staple crops, meat and fish form bulk of our diet while vegetables and fruits add variety to our diet and are indispensable source of vitamins and minerals. Luxury food items such as tea, coffee and food adjuncts such as spices and condiments have no food value, yet they are the world's most valuable agricultural commodities on which the economy of many countries rests. Crops grown to add taste to foods such as sugarcane, sugar beet and sugar mapple also find important place in world's economics. In modern agriculture, bidiversity is used as a source of new crop material for breeding improved varieties and new biodegradable pesticides. Our agricultural scientists, even now, make use of existing wild species of plants, that are closely related to our crop plants, for developing new hardy strains because wild relatives usually possess better tolerance and hardiness. For example, rice grown in Asia is protected from diseases by genes received from a wild rice species *Oryza nivara* from India.

(2) Medicinal Value. Many wild plants have evolved chemical defennce mechanism by developing highly specific toxins which, if deliberated in right way, in right dose or altered chemically, can be used to check disease causing agents or even cancer. Many of these chemicals are derived from plants that had been used in the traditional medicines. The bark of *Cinchona* tree was used to cure malaria by Peruvians and this lead to the discovery and use of quinine in anti-malarial treatment. The pharmaceutical industries largely depend on various plant species furnishing natural product's input. Currently, 25 per cent of all prescription drugs are derived from mere 120 species of plants or are chemically altered versions of plant products and more than 50 per cent of them are modeled on natural compounds. About 12 prescription drugs are derived from higher plants. Some of these natural medicinal products, their source and uses are listed below (table 19.1):

TABLE 19.1. Natural Medicinal Products

No.	*Products*	*Source*	*Use*
1.	Aspirin	Willow bark	Anti-inflammatory
2.	Morphine	Opium poppy fruit	Analgesic
3.	Codeine	" " "	Releaving from cough
4.	Quinine	*Cinchona* bark	Anti-malarial
5.	Atropine	*Atropa belladona*	To dilate pupil of eye
6.	Digitalin	*Digitalis purpurea*	Heart stimulant
7.	Reserpine	*Rauvolfia serpentina*	Against hypertension
8.	Diosgenin	Mexican Yam (*Dioscoria*)	Birth control drug
9.	Cortisone	" " "	Anti-inflammatory
10.	Allantonin	Blowfly larva	Wound healer
11.	Bee venom	Bee	Arthritis relief.
12.	Cytrabine	Sponge	Leukemia cure
13.	Vinblastin	*Vinca rosea*	Anticancer drug
14.	Vincristin	" " "	" "
15.	Taxol	*Taxus brevifolia, T. baccata*	" "
16.	Ephedrine	*Ephedra*	Against asthma
17.	Penicillin	*Penicillium* (fungus)	Antibiotic
18.	Terramycin	*Streptomyces rimosus* (bacteria)	Antibiotic
19.	Bacitracin	*Bacillus subtilis* (bacteria)	Antibiotic
20.	Chlorelline	*Chlorella* (algae)	Antibiotic

Many other living organisms provide several useful drugs and medicines. The United Nations Development Programme (UNDP) estimates the value of pharmaceutical products derived from Third World plants, animals and microbes to be more than $30 billion per year. Many of these, yet unknown and untested such as coral reefs that produce toxins to defend themselves, can find a particularly promising use in pharmaceutical drugs. Some of the plant species native to India, such as 'Neem' and 'Tulsi' etc. also have potential medicinal applications. In India 3,000 plants species are still used as medicinal plants by more than 75 per cent rural populations.

(3) Fuel Value. Communities living near forests obtain fire wood for subsistence and income generation. Firewood collected individually are not marketed but are directly consumed by tribal and local villagers and hence fall under the consumptive value. Use of firewood finds significant role in cottage industries for brick, pottery and bangle firing and fish smoking. In African and Asian countries, where most of the people depend on firewood as their chief source of fuel, the neighbouring forests and woodlands are important source of fuel wood. Till date, the highest use of wood is as fuel.

(4) Other Goods Value. Many other articles of consumptive value include fodder for livestock, variety of natural fibres, thatching material, ornamental plants etc. Seeds from wild species of plants of 'Rudraksh, and Ratti' are used for making necklaces and other ornaments. Banana sap is commonly used as dye and ash for making soap. Besides, marine resource, freshwater and other aquatic animals are also used for solving food problem of local people to provide various types of goods for indigenous use.

B. *Productive Use Value*

Productive use value is assigned to the products that are derived from the wild species and are sold in commercial markets, both at national as well as international levels. Many industries depend on productive use value of biodiversity.

(I) Timber value. Wood harvested from wild species, is the commonest commodity used and traded worldwide. Several wild plants, especially in tropics, produce timber as a source of income generation. Timber resource is one of the most important sources for improving country's economy in many developing countries which adopt it as a national or international trade. Timer export forms a significant part of the export earning of many Third World Countries such as Malaysia, Myanmar, Papua New Guinea, Indonesia etc. These and many other tropical countries are among the major exporters of hardwoods including prized timbers such as teak and mahogany, produced mainly in their natural forests. In many parts of the world, particularly in tropical countries, potentially valuable forests are being degraded through excess harvesting, inadequate management and habitat loss. Many industries, such as pulp and paper industry, saw milling, plywood and charcoal production, transmission, construction poles, railway sleeper and wood craning, thrive on wood resources.

Box 19.1 Bamboo Facts

- Over a billion people in the world live in houses made up of bamboos.
- A sixty feet tree cut for market, takes 60 years to replace but a 60 feet bamboo cut requires only 59 days to replace.
- Some of the species of bamboos grow at the rate of 1.02 cm per hour.
- Majority of bamboo harvested for market are by women and children, most of whom live below subsistance levels in developing nations.
- Currently, the world trade in bamboo is estimated at 14 billion US dollars a year.

(2) Fishery Value. Fish and other fishery products, harvested mainly from wild sources, constitute another class of commodity of great economic importance in global trade. These resources find a significant place in world's food security. There has been five time increase in landing of aquatic resources in past four decades. Marine aquatic resource harvest constitutes about 80 per cent and the remainder from inland fishery and aquaculture. Both inland and marine capture fishery in the last decade has declined but the deficit has been compensated by aquaculture production. Aquaculture along coast, despite being productive in many parts of the world, has severely damaged the coastal environment. Although there are more than 22,000 species of fish, but only about 10 marine fish species make up about 35 per cent of marine culture landing. Commonly harvested marine forms include herrings, sardines and anchovies group.

(3) Other Productive Use Value. Many useful microorganisms have been identified of which some are capable of extra recovery of oil from oil wells, cleaning up oil spills and toxic wastes whereas many others can extract metals from area. A few nitrogen fixing water weeds, such as *Azolla*, are being used as biofertilizers. Several fungi are used to control plant diseases by bio-redemption. Productive use value of biodiversity also incorporates trading of wild genes for use by scientist's for introducing desirable traits in crops and domesticated animals. To productive use value may be included the animal products such as elephant tusks, musk from musk deer, silk from silkworm, wool from sheep, fir from many animals and birds, lac from lac insects. Many industries and trades like textile, silk, leather, ivory work etc. fetch their raw materials from the biological diversity. Many non-wood products of economic value produced by wild species, especially occurring in tropical forests, include medicinal herbs, exudates such as gum and resin, honey beeswax, fibres, essential oils etc., form an important source of income. These goods are used indigenously in various industries or are marketed and exported to many countries. Despite ban on trade in products from endangered species, smuggling for hide, horns, tusk, live organisms etc., worth millions of dollars, are being marketed every year. Developing countries in Africa, South America and Asia including India are the heaven of wildlife trade.

2. Indirect Use Value

Indirect use values of biodiversity are assigned to benefits which do not involve harvesting or destroying of bioresources. Under this category are included ecological benefits such as soil formation, nutrient cycling, waste disposal, air and water purification, education, recreation, future options for human beings and aesthetic, social and cultural values etc.

(1) Ecosystem Service Value. Ecosystem service value of biodiversity refers to the services provided by the ecosystems such as prevention of soil erosion, soil formation, climate regulation, prevention from floods and drought, cycling of nutrients, fixation of nitrogen, waste disposal, cycling of water, role as carbon sink, pollutant absorption, reduction of threat to global warming, maintenance of gaseous composition of the atmosphere, natural pest control, pollination of plants by insects and birds, gene flow etc. It has now been realized that biodiversity is essential for the maintenance and sustainable utilization of goods and services from ecological systems as well as from individual species. The ecosystem services are estimated to be valued in the range of 16 to 54 trillion (10^{12}) US dollars per year.

(2) Social, Cultural and Religious Service Value. Social service value of biodiversity is associated with social life, customs, religion and psycho-spiritual aspects of the people in different societies. Human history reveals that people have related biodiversity to the very existence of human race through cultural and religious beliefs. In India leaves, flowers and fruits or other parts of plants of *Ocimum sanctum* (Tulsi), *Ficus religiosa* (Pipal), *Ficus benghalensis* (Bargad), *Azadirachta indica* (Neem), *Magnifera indica* (Mango), *Prosopis cineraria* (Khejri), *Aegle marmelos* (Bel) etc. are used variously and their trees are planted, which are considered sacred and worshiped by the people. The social life, customs, songs, dances etc. of tribal peoples are intricately woven around wildlife in the forests and are linked with wildlife. Many animals such as several birds, cow, snake, bull, peacock, owl etc. have been considered sacred and find a significant place in our psycho-spiritual arena. Today Indians recognize plants such as lotus and animals such as tiger as symbols of national pride and cultural heritage.

3. Aesthetic Value

Nature makes up our home on the earth and hence all its goodness and beauty is to be valued. The aesthetic value is closely attached with the natural environment and is the source of inspiration for adoption of earth as man's habitat. The diverse flora clothes the earth in amazing beautiful ways. The diverse fauna with innumerous species of animals brings the earth to life. Biodiversity has also a great aesthetic value in ecotourism, bird watching, wildlife viewing, pet keeping, gardening etc. People from far and wide spend a lot of their time and money to visit wilderness areas to enjoy the aesthetic value of biodiversity. Ecotourism is coming up as a new industry and a source of revenue for some countries. The concept of 'willingness to pay' on such ecotourism furnishes a monetary estimate for aesthetic value of biodiversity. Globally, ecotourism is estimated to fetch around 12 billion American dollars of revenue annually.

4. Ethical Value

Ethical value, also known as existence value of biodiversity, involves issues like all life must be preserved based on the concept of '*live and let others live*'. *Ahimsa Purmodharma* is ingrained in Indian culture. The role of biodiversity is to be a mirror on our relationship with the other living species, an ethical view with right duties and awareness. If humans consider that species have a right to exist, they must protect all biodiversity. We may or may not use a species, but we should know the very fact that every species, which exists in nature, gives us benefit directly or indirectly. We are hurt to hear that the particular species such as 'passenger pigeon' or 'dodo' is no more on this earth. Despite we do not derive anything directly from Kangaroo, Zebra or Giraffe, we strongly feel that these species should exist in nature. This is the ethical value or existence value attached to each species. Thus man holds a great responsibility towards preserving and conserving other species. We have a moral obligation to preserve species with lesser powers because extinction of some species will threaten the very existence of man.

Option Value

Holistic perspective on biodiversity suggests that to care diversity values might be based more on what we did not know than on what we do know. The biodiversity can be viewed as primarily capturing the twofold challenge of unknown. The anticipated future uses and values of unknown species are captured in the idea of option values. An option value does not imply just unknown future values of known species but also the unknown values of unknown species or other components of variation. This concept is the core biodiversity because it links variation and values. Thus, option values include the potential of biodiversity that is presently unknown and need to be explored. It is quite possible that we may have some potential cure for diseases like cancer or AIDS, existing within the depths of marine ecosystem or tropical rainforests. Thus option value is the value of biological reserves that may, one day, prove to be an effective option for something important in the future and in future someday any species may prove to be a miracle species. The biodiversity is a precious gift of nature given to us with the guarantee that if we do not harm it, it will continue to sustain mankind in future. Attempts are being made to unwrap the option values of biodiversity by various sectors, such as industries are interested in option values of microorganisms, biomolecules, genes etc.; the agriculture or food sector is busy in exploring option values of cultivated plants; and the pharmaceutical sector is more interested in wild species of plants. Exploration of option values in health care, such as production of recombination proteins, use of genetically modified bacteria to produce insulin for curing diabetes, vaccines against hepatitis B and many other diseases, has given good results.

Chapter Summary

The services we use from ecosystems cannot be provided without biodiversity. Biodiversity is the 'life insurance' of our changing world and guarantees the continued provision of ecosystem goods such as fresh air, clean water, adequate food, fuel, medicines etc. and various ecosystem services. The intricately woven biosphere is a life support system of human race. Value of biodiversity may be direct for consumptive uses such as food, medicines, fuel and other goods; productive for timber, fishes etc.; and indirect value such as ecosystem service value, social service value, aesthetic value, ethical value etc. Biodiversity needs its sustainable use, failing which it will perpetuate inequitable and unsustainable growth, deeper poverty, new and more rampant illness, continued loss of species and a world with more ever degraded environment which are less healthy for people.

Study Questions

1. Write an essay on the value of biodiversity.
2. Define Productive Use Value.
3. Define Consumptive Use Value
4. Define Fcosystem Service Value

Objective Questions. *Select correct answers*

1. Precipitation value of biodiversity is:
   (1) Consumptive Use Value (2) Productive Use Value
   (3) Ecosystem Service Value (4) Essential Value

2. About 85 per cent of world's food is derived from:
   (1) More the 500 species of food crops
   (2) About 200 species of food crops
   (3) About 25 species of food crops
   (4) About 5000 species of food crops

Answers

1. (3), *2.* (2)

20

CHAPTER

Threats to Biodiversity

LEARNING OBJECTIVES

Introduction • Causes of Biodiversity Loss • Habitat loss and Degradation • Habitat Fragmentation • Poaching of Wildlife • Introduction of Exotic Species • Over-exploitation • Pollution and Disturbances • Diseases • Genetic Assimilation • Man-Wildlife Conflict • Causes of Man-Animal Conflict • Remedial Measures to Curb the Conflict, Indian Scenario.

Introduction

The word 'biodiversity' is often noted in connection with how it is fast disappearing on the earth. And if humanity is not careful, we are going to lose the intricately woven web on this planet that makes life possible. Let us take the popular simile about *the ship that is losing one nut at a time, loss of each nut does not seem to be a significant loss in itself but the synergetic effect of loss of several nuts causing the ship to sink.* We are slowly and surely losing the species that keep the earth 'afloat' and if we fail to halt this decline, the ship of biodiversity will sink. The diversity of world's plants and animals is under threat. Indiscriminate devastation of rich biodiversity on our planet is posing a threat not just to plants and animals but to the entire humanity. Although, species may become extinct due to natural selection and other natural causes, but the way species are fast disappearing from the face of the earth due to man's intervention, could cause many more to become extinct even before we know them. When a species is lost, it is lost forever. In Hawaii Island a rare and endemic species of silverword is doomed to extinction because its sole pollinater has become extinct (box. 20.1). Thus a single species can play a crucial role having a kind of domino effect on the community that is disproportionate to its abundance. Global climate change has significantly altered hydrological pattern and other ecological functions triggering or aggravating biodiversity loss. Main causes of threat to the biodiversity are habitat loss and degradation and fragmentation of habitat; introduction of non-native (exotic or alien species), over-exploitation of bioresources; soil, water and atmospheric pollution; intensive agriculture and forestry; diseases; genetic assimilation; poaching; over-population; industrial development etc. (fig. 20.1). If we ignore biodiversity crisis we could lose a sizeable number of species and if we care we might hold the loss to at least ten per cent, but even this amounts conserving millions of species.

Causes Leading to Loss of Biodiversity

In the last half billion years of the history of earth's life, five great catastrophies: the *first* in the Ordovician period (448 million years ago), the *second* in the Devonian (365 mya), the *third* in the Permian (286 mya), the *fourth* in the Triassic (210 mya) and the *fifth* in the Cretaceous period (66 mya), have decimated living organisms. The Cretaceous period saw the decline of dinosaurs and giant pteridophytes. The first four spasms of extinction are believed to have been caused by climate changes. The fifth was due to a giant metorite crash. But the sixth *extinction*, which we are now witnessing, is man's handiwork. In recent times man's practices such as shifting cultivation and many other land use change patterns, spread of urbanisation and industrialisation and rapid pollution growth have all of sudden accelerated disastrous conditions in the natural ecosystems. The root causes of loss of biodiversity are: exponential growth of human population, complicated lifestyle featuring high consumerism, lack of awareness of consequences, narrowing down of the use of a few high yielding varieties of species and ignorance of value of wild traditional varieties.

Box 20.1 **Endangered Pollinators**

The splendor, variety and colour of flowers arises from their need for pollination, critical to fruit and seed production. Pollination is commonly accomplished by insects and other animals. More than 1,000,00 different species of animals play role in pollinating 250,000 kinds of wild flowering plants on our planet. In addition to bees, wasps, moths, butterflies and beetles, more than 15,000 species of birds and mammals serve as pollinator. But the population of the pollinators is declining at alarming rates owing to alteration in their food and nesting habitats and shrinkage in natural ecosystems such as forests and grass lands, pesticide poisoning, diseases and pests, over-collecting, smuggling and trading in certain rare and endangered species etc. There are about 1500 species of butterflies in the Indian subcontinent but their population is dwindling due to pollution caused by indiscriminate use to chemicals in the form of fertilizers and pesticides. The Travancore Evening Brown, the Malabar Tree Nymph, Bhutan Glory and many more butterflies are listed as endangered. The scarcity of natural pollinators has become a critical factor for protecting biodiversity.

The main causes of loss of biodiversity are:

1. *Habitat Loss and Degradation*

In our generation, destruction of habitat is the primary cause for the loss of biodiversity. This single largest factor of loss of biodiversity is the leading killer of the species. According to *Red Data Book* published in 2000 by International Union for Conservation of Nature and Natural Resources (IUCN), habitat destructions is responsible for 73 per cent of the species loss. The rate of loss is so much that about 100 species are lost each day. The loss of habitat and habitat degradation has resulted 89 per cent of threatened bird species, 83 per cent of threatened mammals and 91 per cent of threatened plant species to become extinct. When people cut down trees, fill a wetland, plough a grassland or burn a forest, the natural habitat of a species gets changed or destroyed resulting in killing or forcing out many plants, animals and microorganisms as well as disrupting complex interactions among the species. The main causes of habitat loss are agricultural activities; harvesting or extraction; mining; fishing; logging, deforestation; rapid rate of urbanisations; industrial development; development of associated infrastructure etc. Expansion of human population and human activities are chiefly responsible for habitat destruction. The ever increasing human settlements have been causing destruction of natural ecosystems to meet their requirement of food, shelter and other needs. Millions of hectares of forests and grasslands have been cleared world over for conversion into agricultural lands, pastures, settlement areas or development projects (fig. 20.1). These natural ecosystems, the forests, grasslands and wetlands, are the natural homes of thousands of the species which perished due to loss of their natural habitats. Worst damage has been caused to wetlands

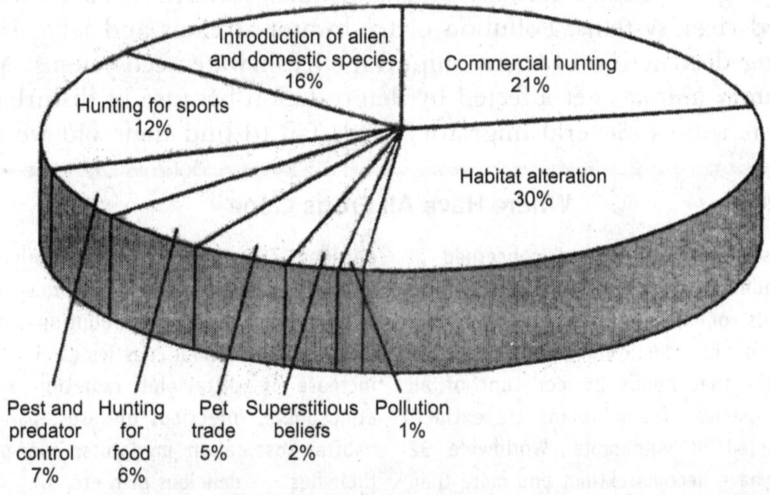

Fig. 20.1. A breakup of human activities that cause extinction of biodiversity.

considering them to be the useless ecosystems. They have been destroyed due to agriculture, drainage filling and pollution. Many significant wetlands have been destroyed due to eutrophication resulted by human activities. Today, the rich biodiversity of wetlands, estuaries and mangroves, with unique species, are under most serious threat.

The greatest destruction to biological communities has occurred within the last 150 years during which human population has increased tremendously. In many highly populated island countries most of the original habitat has already been destroyed. According to IUCN report, more than 50 per cent of the wildlife has been destroyed in 49 out of 61 Old World tropical countries. In India, unwise human deeds, including forest felling without proper understanding of biota, have had devasting effect on species diversity with the rate of man induced extinctions on the rise. Rate of forest destruction in India is 13,000 sq km per year. Fifty years back Kerala had a forest cover of almost 38 per cent of its total land area and it has now reduced to 10 per cent. The areas under threat in India, are the lower slopes of the Western Himalayas and the Western Ghats of peninsular India. The original Himalayan forest areas of about 340,000 sq. km are down by two–thirds and fast disappearing through unregulated logging and conversion of farmland. These forests contain about 9,000 plant species of which 39 per cent are limited to the region as a whole. The Western Ghats, with the forest cover of about 17,000 sq km and home to 4,000 known species of which 40 per cent are endemic, is disappearing at the rate of 2 – 3 per cent per year. As a result of human interventions, including marine pollution, marine biodiversity is also under serious threat due to large scale destruction of fragile breeding and feeding grounds of oceanic fish and other species. Large coral reef areas are being destroyed. Fresh water fish and other aquatic species are becoming threatened because dams and water withdrawls have radically altered river systems. Pollution of freshwater streams and lakes is also causing destruction of many important fresh water ecosystems. Many migratory animals get affected by deforestation because of disturbances in their routes. Several migratory birds fail to find their old wetland,

Box 20.2 **Where Have All Frogs Gone**

Scientists are puzzled and concerned at worldwide decline in frog populations. Hundreds of species of amphibians are vanishing or declining globally. IUCN estimates that about 25 per cent of all known species of amphibians are extinct, endangered or vulnerable. Worldwide 32 species have become extinct and more than 200 species are in decline in the last few decades. The amphibians, especially frog's number is declining not only due to the habitat degradation and pollution but also due to agricultural pesticides in water, increase in ultraviolet radiation in the atmosphere, infectious diseases, parasites, habitat loss, alien predators, over-use of their flesh as delicious dish etc.

which have now been either filled or converted into human settlements. Sometimes human cleanliness destroys the habitat of scavengers such as vultures, kites etc. As a result of habitat loss, the species must adapt to the changes, move elsewhere or succumb to predation, starvation or disease and eventually die. In Western Ghats of India, several rare butterfly species are facing extinction with the swift habitat destruction.

2. *Habitat Fragmentation*

Fragmented habitats are such as a forest patch surrounded by croplands, orchards, plantations or urban areas, *i.e.,* a large continuous habitat is divided into small and scattered patches. Sometimes, the habitats, that formerly occupied wide areas, are often divided into pieces by roads, fields, towns, canals, industries, power line etc. The fragments thus formed are often isolated from one another by highly degraded or modified landscapes. As a result, the population of a species gets separated into isolated groups that not only limit the potential of the species but also the dispersal and colonisation potential of species. These isolated, small and scattered populations are increasingly vulnerable to inbreeding depression, high infant mortality and susceptibity to environmental stresses and finely lead to extinction. With the fragmentation of a forest tract, usually the species occupying deeper parts of forests disappear first. For example many wild life species such as bears and large cats, that require large territories to subsist, get badly threatened as they breed only in the interiors of the forests. Many singing birds are vanishing due to habitat fragmentation. Among other things, for a species to survive, requires a minimum area in an ecosystem. For example large animals like elephants and lions require large areas to move about. Thus habitat fragmentation due to human impact has many advanced effects on biodiversity. Poaching of Indian tigers has risen because of China using its bones in pharmaceutical industry.

3. *Poaching of Wildlife: Commercial Hunting*

Poaching refers to illegal trade of endangered and rare species of animals and plants and products thereof. Illegal trade of wildlife may be for many rare medicinal plants and endangered animal species, including birds and mammals. In recent decades, it has emerged as an insiduous threat to wildlife. Today wildlife is in the crosschairs. Global illegal trade in wildlife is second only to the illegal drugs trade in value. Poaching of wildlife is proving a disaster, threatening biodiversity world wide. It is pushing several species at the verge of extinction. Despite ban on trade in endangered and rare species and their products, smuggling of wildlife and their products such as hide, skin, fur, antlers, meat, horn, tusk, cosmetic, decorative and herbal products continues (table 20.1). The

TABLE 20.1. Animals and Animal Parts Involved in Illegal Trade

No.	*Animals*	*Parts of the body used/purpose of use*
1.	Asian elephant	Tusk (Ivory) used for decoration after carving & medicine. Hair from the tail for magic rings
2.	One-horned rhinoceros	Horn for medicinal purposes.
3.	Bengal tiger	Skin for making hides, bone for medicine
4.	Snow leopard	Fur and other body parts
5.	Sloth bear, Himalayan black bear, Brown bear	Bile, nails, fur for making medicine and nails for black magic
6.	Mongoose	Tail hair for making brushes used in drawings
7.	Civet cat	Meat
8.	Blackbuck	Skin for decoration, antlers for medicine
9.	Fox, Jackal	For hide
10.	Musk deer	Musk for aphrodisiac purposes
11.	Gaur	Skin, head with antlers
12.	Tibetan antelope (Chiru)	Skin/fur for wool, which is known in the international market as "shatoosh" or "king of wool"
13.	Himalayan tahr	For meat
14.	Otters	Skin/pelts
15.	Porcupines	For quills
16.	Hare	For meat
17.	Peacock	Feathers, legs, meat, to make various artifacts, extract oil for aphrodisiac, medicines.
18.	Pheasants (Monal, Cheer, Western fragopan)	For meat and feathers
19.	Falcon (Peregrine and Shaheen)	As pet
20.	Parrots	As pet
21.	Quails, Francolins	For meat
22.	Peafowl	Feathers, meat
23.	Monitor lizards	Hide to make drums,to use as leather
24.	Spiny tailed lizard	Extracting oil for preparation of medicine
25.	Star tortoise	As pet
26.	Snake	For hide, venom
27.	Mugger/Ghariyal	For hide
28.	Frogs	Frog legs for meat
29.	Butterflies	After mounting/ preserving fancy artifacts are being made

poaching pressure is unevenly distributed. The developing countries in Asia, South America and Africa are the richest source of wildlife and the poachers, mostly poor people, depend on this trade for their livelihood. They collect specimens indiscriminately, kill young and old, male and female, often using very cruel methods. The rich countries in Europe, Middle East and South America and some affluent Asian countries like Japan, Taiwan and Hongkong are the major importers and consumers of wildlife species or their products.

Recent studies have revealed that international trade in wildlife is the second biggest threat to species survival after habitat destruction. The trade involves hundreds of millions of species of wild animals and plants. According to various agencies, over 100 million tonnes of fish, 1.5 million live birds and 440,000 tonnes of medicinal plants are in trade in just one year. The global import value, both legal and illegal, of wildlife, is estimated at around US$ 159 billion per year.

Wildlife traders are attracted because of the price they get for a wildlife specimen or its product. In world market a live mountain gorilla fetches US$ 150,000, a panda pelt US$ 100,000, elephant tusk US$ 100 per kg, the leopard skin coat US$ 10,000, a rare hyacinth macaw bird of Brazil US$ 10,000 and so on. Mass killing of some of the birds for sports or for commercial purposes has driven some species like passenger pigeon, whose firs are used for making shuttle cock, to extinction. The worse part of the entire story of poaching is that for every live animal that eventually gets into the market, on an average, 50 additional animals are caught and killed. Moreover, most of the captured live animals die during transit. The tragedy is that poacher finally gets very little, most of the money goes to the middleman. The country of origin also does not benefit in any way, since no taxes or duties are paid. The country merely loses its biodiversity. Each year thousands of seizures are made at ports, air ports and trading organisations, but the trade still continues.

More than 37,000 plant and animals species, including rhinoceros, tiger, leopard, gorilla, butterfly, frog, tortoise, orchid, cactus, mahogany etc are exploited for various purposes. Apart from these, many exotic pets and decorative plants, are sold to collectors. The Convention on International Trade in Endangered Species of Wild Fauna and Flora (CITES) is an attempt to prevent this trade.

In India, rhino is hunted for its horns, tigers for bones and skin, musk deer for musk, elephant for ivory, gharial and crocodile for their skin, whale for combs, spiny tailed lizard for 'sanda' oil etc. Nine Indian animal species, *viz.*, Fin whale, Himalayan musk deer, Green turtle, Hawksbill turtle, Olive Ridley turtle, Salt water crocodile, Desert monitor lizard, Yellow monitor lizard and Bengal monitor lizard, have been severely depleted due to international trade.

In efforts to control wildlife trade, in June 2005, Indian Wildlife Board, headed by Prime Minister gave final clearance to the creation of a Wildlife Crime Control Bureau (WCCB) armed with investigative powers to curb poaching of endangered animals and illegal trade in their parts and products. The greater success recorded in curbing the wildlife trade was made on June 30, 2005 when Delhi Police Crime Branch arrested the most notorious wildlife criminal Sansar Chand who was engaged in the trade of tiger and leopard since last 22 years.

4. *Introduction of* **Exotic** or **Alien Species.**

Exotic or alien species are the non-native newly entering species in a geographical region where they have never been. Introduction of such invasive species in a new area, for which they are foreign guests, may cause disappearance of native species and biological communities through altered biotic interactions. The impact of alien invasive species is now recognized by experts around the world as one of the most significant negative effects on the conservation of natural biodiversity, perhaps second only to habitat loss. Rapid escalation of the establishment of new global trade routes triggered by falling trade barriers, is increasing the risk of new invasions. Water transport, the principal vector for spread and entry of alien specie, is now joined by air transport as a critical agent for such spread. Controls have not kept pace with the spread of organisms and information on known or potential invasions is not readily available in either the host or exporting nations. Global action is required if nations are to retain any semblance of natural biodiversity within their boundaries.

Exotic species are often called '*biological pollutants*, proving to be most damaging agents of habitat alteration and degradation in the world. Exotic species exert large impact, especially in island ecosystems, which harbour world's much of threatened biodiversity. Only some of exotic species are able to establish new area whereas great majority of them do not become established in the new areas. The successful exotic species may kill or eat the native species to the degree of extinction or may so alter the habitat that many native species are no longer able to persist because native species are subjected to competition for food and space. For example, introduction of goats and rabbits in the Pacific and Indian region has resulted in the destruction of habitats of several plants, birds

Box 20.3 **Biodiversity Loss Fact Sheet**

- Habitat destruction is most responsible for biodiversity loss.
- At the moment 1 in 10 of the known plant species are considered threatened globally.
- Nearly half of the global species of animals and microorganisms will be destroyed or severely threatened over next 25 years (Peter Raven is 'Our Diminishing Tropical Forest').
- A section of rainforest, the size of ten city blocks, disappears every minute (Smithsonian Institute).
- Australia has one of the worst records for species destruction.
- More than 75 per cent of the world's rain forests have been destroyed so far.
- Human activities pose serious threat to the biological richness of the oceans which cover 70 per cent of earth's surface and coastal regions which are home to 60 per cent of world's population.

and reptiles. Pathogenic microorganisms, if introduced to new virgin areas, may cause epidemics resulting total elimination of native species. Exotic species, often called *ecological cancer*, have no predators, competitors, parasites or pathogens and flourish well if established in the new geographical regions. They wipe out many local species. A sizeable number of exotic species have been introduced accidentally or intentionally in coastal waters, lakes and wetlands. Unintentional introduction of non-native species takes place as stow aways in aircraft, through the ballest water of oil tankers and cargo ships or as 'hitchhikers' on imported products like wooden packing crafts. In some cases exotic species are delibrately introduced to derivee certain benefits. Such introduction may initially be beneficial, but could cause problems later if the introduced species proliferates at the expense of local ones. Alien invasive species are significant threat affecting 30 per cent of all threatened birds and 15 per cent of all threatened plant species.

In 1959, rabbits were introduced in Australia for sport shooting. Environmental conditions favoured the rabbits and as there were no predators, their population exploded to such an extant that they destroyed the vast area of rangeland, native wildlife and land used for sheep ranching. At last a viral disease was intentionally introduced to check the growth of rabbits.

Introduction of Nile perch, an exotic predatory fish, into lake Victoria in South Africa, threatened the entire ecosystem of the lake by eliminating several native species of the small chichid fish species that were endemic to this freshwater aquatic system. Introduction of water hyacinth ('Jalkumbhi'), a water weed in several tropical countries including India, clogs rivers and lakes and threatens the survival of many aquatic species in lakes and flood plains. In India introduction of *Eucalyptus* spp., *Cryptomeria japonica, Casuarina equisetifolia, Acacia moniliformis, Taxodium distichum, Barringtonia acutangula* has influenced ecological balance of many ecosystems. Introduction of *Lantana camara*, an exotic shrub weed, in many forest land areas of different parts of India, strongly competes with the native species. Introduction of *Opuntia* spp. (Naagphani), *Mikania* spp., *Argemone mexicana* (Pili Katelia or Satyanasi), *Parthenium* sp. ('Congress Grass' or 'Gajar Ghaas'), are the other examples of exotic species found in various parts of India.

5. *Over-Exploitation of Bio-resources*

Over-exploitation of biological resource is one of the chief causes of loss of not only species of economic value but also of biological curiosities. A particular species, if over-exploited, reduces the size of its population to an extent that it becomes vulnerable to extinct. For example, the insectivorous plant, such as *Nepenthes khasiana*. (Pitcher plant), primitive fern species of *Osmunda,Ophiglossum, Psilotum* and gymnosperms such as *Cycas, Gnetum, Ginkgo* etc. and animal species such as frogs, earthworms, cockroaches etc., needed for laboratory teaching work are overused for study purposes. Over use of such bio-wealth coupled with commercial over-exploitation of wild plant and animal species has invariably caused their destruction. Consumpion of wood of Indian wild mago trees in plywood, hunting of whales for tallow and wild medicinal plants for extraction of drugs such as *Podophyllum hexandra, Coptis teeta, Aconitum napellus Rauvolifa serpentina, Lycopodium clavatum, Dioscorea deltoidea* etc. and horticultural plants like several species of rhododendrons and orchids, fall under the over-exploited category. Throughout the world, plant or animal collection for zoos or botanical gardens or for practical work and research is also factor leading towards extinction of many plant and animal species.

6. *Pollution and Disturbances*

Most subtle form of habitat degradation leading to biodiversity loss is environmental pollution. Pollution may alter natural habitats and cause reduction and elimination of populations of sensitive species. Excessive use of pesticides in crop fields is causing the washouts to get mixed in water of adjoining fresh water aquatic ecosystems resulting in decline in the populations of fish eating birds and falcons. Toxic wastes entering the water in fresh water bodies, estuaries and coastal ecosystems are injurious to the biotic components of these systems. They bring about changes in the food chain. Coral reefs along coastal areas are being threatened by

Box 20.4 Dark Face of Wildlife Trade

1. International trade in wildlife is the second biggest threat to species survival, after habitat destruction.
2. International trade in wildlife is the second largest illegal trade in the world after drugs, recently overtaking arms.
3. Every year, an area of the rainforest of the size of the British Isles is destroyed to fulfill the fuel demand, for the trade in wood and its products.
4. Destruction of rainforests results in loss of estimated 50,000 species of plants and animals every year, many of these species have not even been discovered by mankind.
5. The ivory trade in the twentieth century resulted the African elephant population to fall by 10 million between 1900 and 1989.
6. About 9 out of 10 birds, caught from the wild for the pet trade, die before completing their normal age.
7. Cruel methods in trade of wild animals and birds causes a great stress resulting their death during all stages such as capture, transport and keeping.

pollution from industrialization along the sea coasts, oil sleeks during transportation of oil and offshore mining. Cadmium poisoning is causing 'etai etai' disease in fishes and lead poisoning is resulting in the mortality of many aquatic birds like ducks, swans and cranes. These birds offen swallow the spent shotgun pellets that fall into lakes, marshes and wetlands. Water pollution, due to excessive organic loading of water bodies, results eutrophication which drastically reduces biodiversity. Sulphur dioxide, nitrogen oxide, suspended particulate matter, industrial and automobile emissions etc. also cause negative effects on plant and animal species. Acid rain, global warming and ozone layer depletion affect adversely plant and animal species. Noise pollution has been reported to cause wildlife extinction in certain cases.

Natural as well as man-made disturbances such as fire, defoliation by insect's etc. affect biological community adversely. Intensity of man-made disturbances is more. For example, frequent intentional fires in the forests changes the species richness of a community. Massive use of synthetic compounds, vast release of radiations etc. also lead to a change in habitat quality.

7. *Diseases*

Diseases causing organisms, including pathogens, may also be considered as a kind of predators. Human activities may cause spread of disease in wild species. Extent of the disease may further increase when the animals are kept in captivity in zoos, sanctuaries, reserves or even in vast population over a large area. Large number of plant species have become extinct due to disease spread in vast areas. As compared to plants, animals are more prone to infection when under stress.

8. *Genetic Assimilation*

Some rare and endangered wildlife species are threatened by genetic assimilation, as they crossbreed with closely related species that are more numerous and quite vigourous. Opportunistic plants or animals introduced into a new habitat by human may genetically overwhelm local populations. For example, hatchery raised trout introduced into lakes or streams may cause indigenous stock of species to become dilute.

9. *Other Factors*

Among other factors responsible for loss of biodiversity are:

1. ***Distribution Range***: Smaller the range of distributions, the greater is threat for extinction.
2. ***Substitution***: An existing ecological species may be replaced by another one during the course of evolution.

3. ***Reproductive Rate***: Large organisms such as tiger, elephant etc. tend to produce fewer offspring at widely spaced intervals.
4. ***Status in Food Chain***: The higher the status of a species in food chain, the more susceptible it becomes.
5. ***Degree of Specialisation***: The more specialized a species is, more vulnerable it will be to the extinction.

Besides, poverty of people, macroeconomic policies, international trade, policy failures, poor environmental laws, weak enforcement of laws to save wildlife, unsustainable development projects, lack of control over bioresources, are also responsible for loss of biodiversity.

Man-Wildlife Conflicts

Expansion of human population causes shrinkage of natural habitats. As such people and animals are increasingly coming on conflict over living space and food. And under such circumstances, conservation of wildlife becomes a challenging task because the impact of conflict between men and animals is huge. People lose their crops, orchards, livestocks, property and sometimes their own lives. In order to prevent the future conflicts, people often retaliate and kill the animals, many of them being threatened or endangered. Thus man and wildlife conflict has become one of the main threats to the continued survival of many animal species in many parts of the world and is also a significant threat to many local human populations. If an adequate solution to the conflict is not achieved, local support to conservation strategies also declines. There is a continual struggle between policy makers of conservationists of wildlife and anti-conservationists and the people who suffer. The practice of wildlife management is rooted in the intermingling of human ethics, culture perceptions and legal framework.

The man's activities bringin about shrinkage of the forest tracts had an adverse impact on the habitats of large animals. For example, the elephant's habitats has to be very large one because they need a larger home range than any other land inhabiting animals. An elephant herd of 10 to 15 individuals has a daily requirement of huge amount of fodder and water. Since it takes time for plants to regenerate, elephants require a large range of land for their survival. But with the expansion of human settlements, extension of agricultural land and industrial development, man wildlife conflict is bound to occur and adversely affect many species. There are many more examples of wildlife harming man throughout the world such as Baboons in Namibia attacking young cattles, Orangutans destroying oil palm plantations, one-horned Rhino in Nepal damaging crops, European bears and wolves killing livestock etc. The problem affects both rich and poor and the conservation minded people are turning as anti-conservationists.

In India, instances of man-animal conflicts were not uncommon in last few years when in Sambalpur, Orissa, about 200 people were killed by elephants and in retaliation the villagers killed about 100 elephants and badly injured about 30 of them. Frequent incidences of killing of elephants in the border region of Kote Chamarajanagar belt in Mysore have recently been reported. In this region the conflict between man and elephant has arisen because of massive damage caused by the elephants to the cotton and sugarcane crops of the local people. The organized villagers often shoot, electrocute, adopt explosion by hidden explosives in the crops to kill the elephants. In fact more killings are made by local people than by poachers. In early 2004, a man eater tiger was reported to kill 17 people in Nepal inside the Royal Chitwan National Park, 240 km south west of Katmandu. Now the park, which was once renowned for its wildlife, has become a zone of terror to visitors and local peoples. In June 2004, two men were killed by leopards in Powai, Mumbai. A total of 14 persons were killed by the leopards from the Sanjai Gandhi National Park, Mumbai, during 19 attacks creating a panic among local residents. At times reports are coming from Corbett, Dudhwa, Palamu, Ranthamhbore and other National Parks, of conflicting situations. Most unfortunate part of the story lies under the situation when the people who traditionally protected these animals themselves become victim of protected animals. For example, the Bishnoi community around Jodhpur, Rajasthan had always protected the black buck for centuries, as a result black buck community has increased which has led to the destruction of crops of Bishnoi communities. Similarly in Uttar Pradesh, the Nilgais are abundant. These destroy the crops, but since they are considered sacred are left as such without killing them.

Worldwide Fund for Nature (WWF) and its partners have launched a number of projects around the world to curb the conflict between man and wildlife and improve the livelihoods of the people affected. These projects are aimed at deriving simple and creative solutions specific to the specific species or specific areas and to benefit both the animals and local human population by involving local communities actively. The solutions lead to mutually beneficial co-existence.

Causes of Man-Animal Conflicts

Some causes of man-animal conflicts are:

1. Dwindling habitats of elephants, tigers, rhino, bear etc, due to shrinkage of forest covers, compel these animals to move outside the forests and attack the fields or sometimes even humans. Furthermore, human encroachment into the forest areas raises a conflict between man and the animals. This conflict is due to the issue of survival of both, the co-existents.

2. Ill, weak and injured animals have the tendency to attack men. Often the females attack men in view that her new bornes are in danger. Usually the tiger has no intention to kill men, but once tasting human flesh he does not eat any other animal and becomes man eater. Due to difficulty arising in identifying man eater tiger, many innocent tigers are killed.
3. The cash compensation paid by the government in lieu of the crop damaged is not enough. In Mysore it is Rs 400/- per quintal of the expected yield while the market price is Rs 2400/- per quintal. The organised farmers, therefore, prefer to kill the wild animals over compensation.
4. In order to protect the crops, farmers put electric wiring around their ripe crop fields. The elephants trying to enter crop fields get electrocuted, injured, suffer in pain and turn violent.
5. Lack of growing paddy, sugarcane etc. in the forest within sanctuaries to feed elephants, which used to be earlier, the elephants move out of the forest in search of food for which they were used to.
6. Earlier, there used to be wild life coridors through which the wild animals used to migrate seasonally to other areas in groups. Now, due to development of human settlements in these coridors, the routes of wild animals have been destroyed and the animals attack the settlement.

Remedial Measures to Curb the Conflict

Some of the remedial measures to curb the conflict between the man and the animals are:

1. To deal with any major danger, Tiger Conservation Project (TCP) has been provided with vehicles, tranquiliser guns, binoculars and walky talky sets etc.
2. Crop compensation programme has been started by Karnataka government. In many states, apart from crop compensation, cattle compensation and substantial cash compensation, for loss of human life programs, are under consideration.
3. Use of solar powered fencing, along with electric current proof trenches, are being tried to prevent animals from straying into fields.
4. Animals in sanctuaries are given adequate food, fodder, meat, fruits and water, depending on their choice.
5. Cropping pattern is being changed by growing the crops not liked by the animals avoiding damage along the forest borders.

6. Provisions are being made to preserve wildlife corridors for mass migration of larger animals during unfavorable periods. Elephants require about 300 km^2 area for their seasonal migration.
7. Similipal Sanctuary, Orissa observes a ritual of wild animal hunting during the months of April-May during which a part of the forest is burnt to flush out animals. Due to the massive hunting by people, there is decline in population of tigers as they start coming out of the forest in search of prey. With the initiative of WWF-TCP attempts are being made to curb this ritual, the 'Akhad Shikaar' in Orissa.

Now Ministry of Environment and Forests, Govt. of India has taken initative to identify corridors between forest to enable movement of animals. The Wildlife Institute of India, Deharadun has submitted a blueprint for developing corridors in all geographical regions of the country to provide an additional 12,000 sq. km protected area for animals. WWF is currently working to change livestock management to prevent tiger kills and to provide livelihood alternatives to local communities to reduce their dependence on forest resources, thus reducing the likelihood of tiger attacks on humans.

Indian Scenario

To look at larger picture, all is not well within Indian wildlife. Undoubtedly the single biggest threat to wildlife is the exponential rise in population, with an increase from 30 crore to 110 crore, India added about 80 crore people since 1947. This crush of humanity affected all habitats and resulted animals to come increasingly in conflict with humans by eating the crops raised by men, killing the livestock or even people and occasionally fighting for the same recourses. The unprecedented development versus forest debate, has come into being mainly due to the increased needs of the population. These development projects affected wildlife both by diverting the commons, *i.e.*, villages and forest and by the diversion of the local resources such as water, without meeting the needs of local people and animals. Mining has proved to be the other havoc to create larger patches of red moonscaps in Uttarakhand and Jharkhand, thereby driving out many animals including elephants for the time in the recorded history. Moreover, construction of dams in forested river valleys, railway operations, high tension power lines and networking of roads in and around rich biodiversity areas has caused a serious threat to survival of wildlife. There has been sudden rise in incidences of killing of animals due to man animals conflict and poaching. For example the elephant human conflict has risen to such an extent that in India alone at least 200 people are killed by elephants every year and similary about 200 elephants are killed by the people. Other less known creatures suffer worse fate. Habitat destruction of village commons, due to encroachment, goes on. There are problems even for the areas that have been set aside as protected areas for wildlife as more than 4 million people continue to reside in them. About 40 per cent of these protected areas are under pressure of traditional livestock grazing, fodder and timber extraction and collection of non-timber forest produce. Almost 45 per cent have roads running right through these reserve areas. India has 492 sanctuaries and 92 national parks. In India, threat to wild flora and fauna is commonly due to wildlife habitat degradation. Although slow the process is a serious threat to the species conservation. The unsustainable extraction of wild resources affects the immediate survival of species and would lead to a higher category of threat including local or total extinction of species. Indiscriminate poaching and illegal wildlife trade particularly for body parts and derivatives of many wildlife species of both plants and animals such as rhino horn, tiger parts, elephants ivory, musk, bile, sandalwood, sahtoosh, agar wood etc., has caused a serious threat to wildlife. In the last nine years 680 tiger skins were sized. Extraction of sahtoosh, the down wool of endangered Tibetan antelope, causes hunting of this threatened species. In recent years, commercial exploitation of medicinal plants has caused grave threat to the existence of several species of plants. Unless protective measures are taken immediately, several valuable species of animals and plants may disappear altogether.

Chapter Summary

The word 'biodiversity' is often heard in connection with how it is fast disappearing and losing intricately woven web on this planet. Among many causes, the primary reason for threat to biodiversity is habitat loss and habital degradation. According to *Red Data Book* 2000, seventy three per cent loss of biodiversity is due to these two reasons alone. This loss is so much responsible that about 100 species are disappearing each day. Habitat loss is mainly due to human activities. Other reason next to habitat loss is poaching of wildlife and introduction of exotic species. Poaching, *i.e.*, illegal trade of endangered species, second only to the illegal drug trade, is proving disaster and threatening biodiversity world wide. Among other threats to biodiversity are habitat fragmentation, over-exploitation, pollution disturbances of all kinds, diseases, genetic assimilation etc.

Shrinkage of habitats, due to expanding human population, is greatly responsible for man-wildlife conflict. Habitat loss results shortage of desired food of wild animals. As a result such wild animals come out of their habitats and feed on nearby crops and attack people. Every year a large number of people lose their crops, property and even lives. Causes of man animal conflict are many and need remedial measures to curb the conflict. In India, Ministry of Environment and Forests has taken initiatives to identify corridors between forests to enable movement of animals.

Study Questions

1. What are the various threats leading to loss of biodiversity?
2. 'Habitat loss is the primary reason for loss of biodiversity'. Justify the statement.
3. Discuss various reasons for man-wildlife conflict. What are various measures to curb this conflict?
4. Relate the introduction of exotic species and biodiversity.
5. Disscus poaching of wildlife.
6. What is the role of CITES?

Objective Questions. *Select correct answers*

1. Global illegal trade in endangered animal and plant species is more common in:
   (1) Tropical Countries (2) Temperated Countries
   (3) Developed Countries (4) Industrialised Countries

2. The notorious wildlife criminal arrested in India in the year 2005 is:
   (1) Ram Chandra (2) Shyam Chandra
   (3) Sansar Chand (4) Krishna Chand

3. The chief cause of threat to biodiversity is:
   (1) Habit loss and degradation (2) Poaching of wildlife
   (3) Deforestation (4) Air pollution

4. International Union for conservation of Nature and Natural Resources has published a book named:
   (1) Species Plantarum (2) Genera Plantarun
   (3) Red Data Book (4) Black Data Book

5. Corbett National Park is located at:
   (1) Bihar (2) Nepal
   (3) Uttar Pradesh (4) Uttarakhand

6. According to final technical report of National Biodiversity Strategy and Action Plan, Govt. of India, (2005), the country has lost its:
   (1) 50 per cent forest cover (2) 10 per cent mangroves
   (3) 40 per cent wetlands (4) 30 per cent deserts

7. A Ramsar site recognised in 1982 in India is:
   (1) Wooler lake (2) Bharatpur lake
   (3) Chilika lake (4) Sambhar lake

8. I.U.C.N. is now known as:
   (1) WWF (2) CITES
   (3) WCU (4) DST

Answers

1. (1) *2.* (3) *3.* (1) *4.* (3) *5.* (4) *6.* (1)
7. (3) *8.* (3)

21

CHAPTER

Endemic and Endangered Species of India

Learning Objectives
Introduction • Endemic Species of India • Species Extinction • IUCN Red List Categories • Top Ten Most Wanted Species Announced by WWF • Threatened Animal Species of India • Threatened Plant Species of India

Introduction

Endemic species are those species which remain confined only to a particular locality. Such species are very important from the point of view of conservation strategy because their disappearance means extinction of these species as they are not found elsewhere. Thus, they have a value in their uniqueness. The areas with endemism have generally been isolated for a long time, thus enabling the original species to evolve into new genetic entities better adapted to local areas. Isolated mountain tops, valleys and large oceanic islands are usually areas of endemism. Conservation of resources of these areas is very difficult as each area poses its own peculiar problems. Undoubtedly, endemism represents a unique feature in the process of evolution which could be perpetuated and sustained only in the locality concerned by imperfectly understood attributes of environmental quality. This adds to the importance of the habitats in which endemic species thrive. The importance of the habitat is further highlighted by the fact that in most of the cases such localities possess a number of endemic species distributed in several taxonomic categories. The endemic species and the habitats which are likely to be lost for ever should be given top priority for conservation. Once lost, there is no way to recover them. Tropical forests are home to a large number of endemic and native species found only in these locations. The tiny Rio

Plaque Reserve in Equador, with an area of only 0.82 km, has over 250 endemic plant species. Myers (1988), based on degree of endemism in species composition, has identified twelve such localities in tropical regions of world, including one Eastern Himalayas, that need urgent conservation attention.

Endangered species are those species which have very low population and are in considerable danger of extinction. Taxa in danger of extinction and whose survival is unlikely if the causal factors continue operating, fall under the category of endangered taxa. Among endangered category are included those taxa whose number has been reduced to a critical level or whose habitats have been so drastically reduced that they are deemed to be in immediate danger of extinction. International Union for Conservation of Nature and Natural Resources (IUCN) has estimated that at least 25,000 plant species, including lower plants such as mosses, liverworts, fungi, lichens and seaweeds, are endangered due to habitat deterioration, clearence, uprooting and construction. In India nearly 7000 plant species are considered as endangered. IUCN, now called World Conservation Union (WCU) in its Convention on International Trade in Endangered Species of Wild Fauna and Flora (CITES) has proposed measures to conserve endangered species. According to final technical report of National Biodiversity Strategy and Action Plan (NBSAP), India has lost over 50 per cent of its forest cover, 40 per cent of its mangroves and significant part of its wetlands in the past couple of centuries. The report Securing India's Future: The Final Technical Report of NBSAP was released in 2005. At least 40 species of plants and animals have become extinct.

ENDEMIC SPECIES OF INDIA

Global species composition reveals that India is tenth among the plant rich countries in the world, fourth among Asian countries and eleventh according to the number of endemic species of higher vertebrates. Out of 25 hot spots identified in the world, India has two: North-East Himalayas and Western Ghats. Andaman and Nicobar Islands are no less than a hot spot and Chilika has been recognised as a Ramsar site in 1982. Both hot spots-North-East Himalayas and Western Ghats show a high degree of endemism and contain 5,332 endemic species of higher plants, mammals reptiles, amphibians and butterflies. Not only in these hot spots, on the whole, India has a rich element of endemism in its flora. Indian subcontinent has about 49,219 plant species of which about 33 per cent are endemic. Andaman and Nicobar Islands contribute at least 220 endemic species. Gangetic plains are generally poor in endemism. According to Botanical Survey of India report, about 7,000 plant species and 140 genera of the plants, mostly flowering angiosperms, are endemic to India. Of these, the Himalayas and the Khasi Hills account for about 3,000 endemic species and the Deccan peninsula for about 2,000 endemic species. *Sapria himalayana, Uvaria lurida, Alchimandra cathcartii, Magnolia gustavii, M. pealiana, Pachylarnax pleocakpa, Nepenthes khasiana, Dicentra roylei*, several species of *Primula* and *Rhododendron* and the Lady's slipper orchid, *Paphiopodilum insigne* and *P. venustum* are some of the noteworthy endemic species of Himalayas and Khasi Hills. *Antiaris toxicaria, Companula cytinoides, Pedicularis perrotteti* and some species of podostemonaceae are endemic to Western Ghats or Nilgiri Hills in South India. Many deep and semi-isolated valleys are exceptionally rich in endemic plant species such as in Sikkim, with an area of 7,298 sq km, out of 4,250 plant species, 2,550 species are endemic.

The crops, that first grew in India and now spread throughout the world include rice, sugarcane, Asiatic vigna, jute, mango citrus, banana, several species of millets, spices, medicinal, aromatic and ornamental plants. According to Khoshoo (1990), India ranks sixth among the centers of diversity and origin as far as agro-biodiversity is concerned. The origin place of 167 cultivated plants and 320 species of wild relatives of cultivated crops is located in India. India is the center of diversity of jackfruits, cucurbits, disacorias, colocasias, alocasias, black pepper, ginger, turmeric, bamboos, brassicas, tree cotton and secondary center of domestication of tobacco, potato and maize.

Indian subcontinent has about 81,000 species of animals of which a large number of species are endemic. The Western Ghats are particularly rich in endemic species of animals. It is estimated that almost one-third of the animal species found in India have taken refuge in the Western Ghats of Kerala. Western Ghats are particularly rich in amphibians (frogs, toads etc.) and reptiles (lizard, crocodile etc.). About 62 per cent of amphibians are endemic. Perhaps the most notable amphibian genus is the monotypic *Melanobatrachus* which has a single species known only from a few specimens collected in Annamalai Hills in 1870s. It is possibly most closely related to two relict genera found in mountains of Eastern Tanzania. About 50 per cent lizards are endemic to Western Ghats.

The distribution concentration of Indian endemic species of animals is mainly in areas of high rainfall particularly in Eastern India along the mountain chains, South West India (Western Ghats) and Andaman and Nicobar Islands. Endemism in Indian reptilian and amphibian fauna is high. Various species of monitor lizard (*Varanus*), reticulated python, Indian salamander and viviparous toad (*Nectophryne*) are some significant endemic species of the country. Eight amphibian genera, the caesilians *Indotyphlus, Gegeonophis, Uracotyphlus;* the anurans – the toad *Bufoides*; the microphyllid-*Melanobatrachus*; and the frogs *Ranixalus, Nannobatrachus* and *Nyctibatrachus,* are not found outside India. The number of animal species endemic to India can be spread over 16,214 species of insects, 967 species of molluscs 187 species of reptilians 110 species of amphibians 55 species of aves and 44 species of mammalians. The 396 known endemic higher vertebrate species have been recorded from India. Among mammalian species endemic to India, four species: lion tailed macaque (*Macaca silenus*), Nilgiri langur (*Presbyton johnai*), brown-palm civet (*Paradoxurus jerdani*) and Nilgiri thar (*Hemitragus hylocrius*), occurring in Western Ghats, are of conservation significance. India is also a primary center of diversity of many animal species such as of zebu, mithun, water buffalow, chicken, camel etc. and secondary center of domestication of several animals such as horse, goat, sheep, yak, donkey etc.

Whatever may be the cause of endemism, endemic plants and animal species are also considered threatened. Any disturbance in their ecosystem would result in the total extinction of such plant and animal species. Among 150 world wide selected centres of rich plant diversity for conservation, five are in India, these include Agastyamalai Hills, Silent Valley, New Amarambalam Reserve and Periyar National Park (all in Western Ghats) and the Eastern and Western Himalayas.

Species Extinction

Extinction of species is a natural process taking place by natural extinction, mass extinction or anthropogenic extinction. Natural extinctions are related with the change in the environment conditions during which some species disappear and others, which are more adapted to the changed conditions, take their place. This loss of species occurred in the geological past at a very slow rate. In mass extinction, there has been several periods in the earth's geological history where large number of species became extinct because of catastrophies during millions of years. Anthropogenic extinction refers to the disappearance of an increasing number of species from earth's face due to human activities. The man caused huge extinction represents a very severe depletion of biodiversity, particularly because it is taking place within a short period of time.

Over a long geological history many species have disappeared and new species have evolved to take their place. The World Conservation Monitoring Centre has recorded that 533 animal species, mostly vertebrates and 384 plant species, mostly angiosperms, have become extinct since 1960. Humans have caused over 75 per cent of all extinctions since they became hunter and gatherer. More species have gone extinct from the islands than from the oceans or mainlands.

Current rate of extinction of species is 1,000 to 10,000 times higher than the previously occurring rate of extinction. If the current rate of loss of species continues, it could result upto 50 per cent of global loss of the species by the end of 21st century. The tropical forests alone are losing roughly 14,000 to 40,000 species annually at the rate of 2.5 species per hour. Percentage of extinciotn of species in tropical forests, at current rate of deforestation in 2040 would be about 17 per cent and with 50 per cent increase in current rate in the deforestation it may double (fig. 21.1).

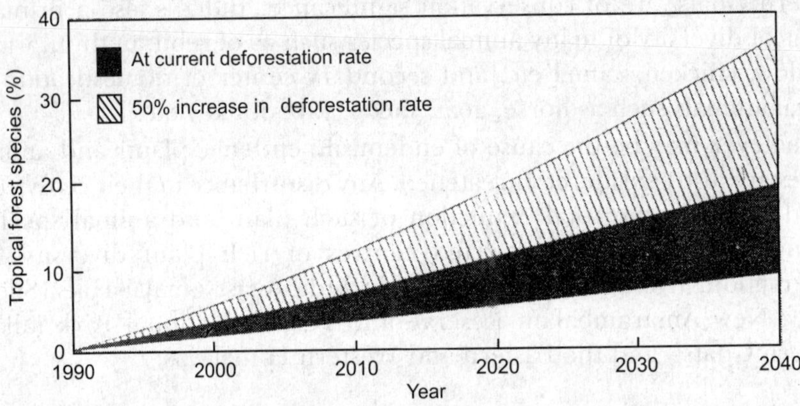

Figure 21.1. Percentage of tropical forest species likely to become extinct.

It is expected that from 10 megadiversity locations in tropical forests, covering an area of 300,000 sq km, some 17,000 endemic plant species and 350,000 endemic animal species could be lost in near future.

Characters of Species Susceptible to Extinction

The species particularly susceptible to extinction are characterised by large body size such as Bengal tiger, lion, elephant; small population size and low rate of reproduction such as blue whale and giant panda; feeding at high trophic levels in the food chain and inability to switch over to the alternate food sources such as Bengal tiger, bald eagle, vulture; fixed migratory routes and habitat specificity such as blue whale and whooping crane; localized and narrow range of distribution such as pitcher plant and many island species; and lack of genetic variability. A small population is more likely to become extinct than a large one, firstly because normal population fluctuation resulting from variations in the environmental conditions such as harsh winter, are more likely to decimate the population, secondly because inbreeding depression will reduce the population's ability to produce viable offspring's. This interaction between demographic and genetic factors acts as an 'extinction vortex' pulling the population numbers low. The species which are threatened with extinction are included in vunlnerable, endangered or critically endangered category.

It is not easy to predict the chance of a species becoming extinct as the data on population numbers are often not available. Even in well studied cases, other factors, such as environmental stochasticity, catastrophic events and the effects of inbreeding, make population size alone an unreliable predictor of extinction. Population Viability Analysis (PVA) is one approach that uses computer modelling to predict a population's chance of extinction. The model incorporates information on the population dynamics of the species, genetic and behavioural factors and environmental effects, such as the state of habitat, interactions with other species and human interference. This type of analysis predicts that there is a 25 per cent chance that the eastern barred bandicoot will be extinct within the next 30 years. This approach also helps in the management of endangered species by allowing the estimation of the minimum reserve area to retain viable population.

The IUCN Red List Categories

International Union for Conservation of Nature and Natural Resources (IUCN), established in 1948 with its headquarters at Switzerland, initiated to evaluate conservation status of species of plants and animals in 1963. Now IUCN is known as World Conservation Union (WCU). It published a Red List for threatened species as a catalogue of taxa that are facing the risk of extinction. The 2000s IUCN Red List is the most comprehensive inventory of the global conservation status of more than 18,000 species of plants and animals of which about 11,046 species are threatened. The Red Data Book aims to impart information about the urgency and degree of need for conservation to the public and policy makers. The making of Red List had a close watch on abundance and quality of natural habitats of a species and potential biological and economic value of the species. The list also provides information to international agreements, such as Convention on Biological Diversity (CBD) and Convention on International Trade in Endangered Species of Wild Fauna and Flora (CITES). To evaluate extinction risk of species, IUCN adopts a set of criteria relevant to all species and regions of the world. The percentage of some major groups of organisms evaluated as critically endangered, endangered, vulnerable and at lower risk are depicted in (fig. 21.2).

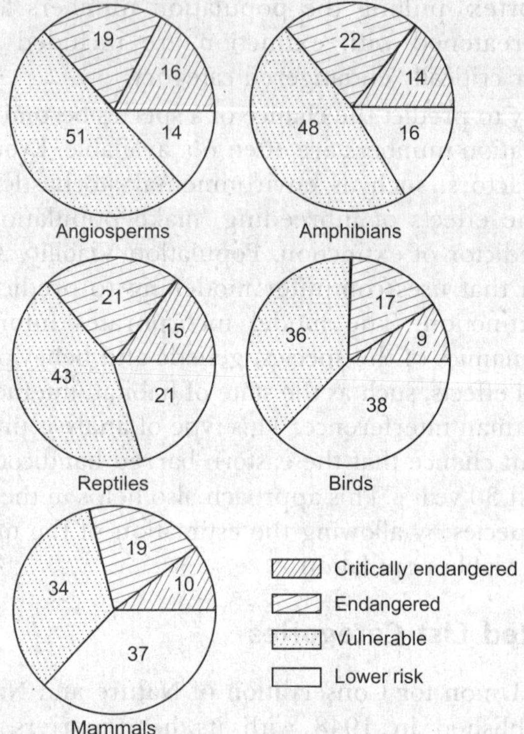

Figure 21.2. The precentage of threatened angiosperms, amphibians, reptiles, birds and mammals categorised as Critically Endangered, Endangered, Vulnerable and at Lower Risk.

According to IUCN Red List, (2000), out of 11,046 threatened species 5,611 are plant species and 5,435 animal species. Out of these threatened species, 1939 (1014 plant species and 925 animal species) have been assessed as critically endangered. The number of plant and animal species of various threat categories in India are shown below (fig. 21.3):

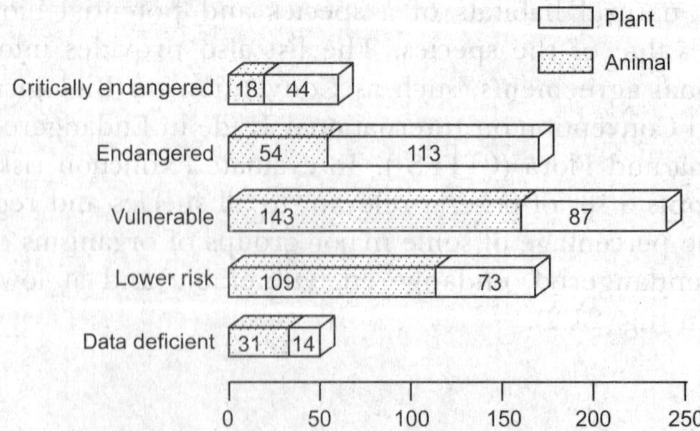

Figure 21.3. The number of plant and animal species of various threat categories in India.

According to volume five of Red Data Book on angiosperm plants started by the Survival Service Commission of the IUCN in 1970, it is estimated that out of the total 0.3 million of plants in the world, over 20,000 were in the category of either endangered or vulnerable or threatened for extinction by the end of the twentieth century. The percentage of threatened angiosperms and four vertebrate groups categorized as critically endangered, endangered, vulnerable and at lower risk are shown in fig. 21.2. The figure indicates that of the species evaluated for the risk in these major groups, 9.6 per cent are critically endangered and 34.50 per cent are vulnerable.

The objectives of the Red List are-

1. Developing awareness about the importance of threatened biodiversity.
2. Identification and documentation of endangered species.
3. Providing global index of the decline of biodiversity.
4. Defining conservation priorities at the local level.
5. Guiding conservation action.

The IUCN has recognised eight Red List categories of species according to the degree to which they face the threat of extinction. These are-

1. **Extinct:** A taxon is extinct when there is no doubt that the last individual has died. It can not be found in the areas it once inhabited nor in other likely habitats.

2. **Extinct in Wild:** A taxon is extinct in wild when exhaustive surveys in known/ or expected habitats have failed to record an individual. It is not seen in wild for 50 years at a stretch, *e.g.* Dodo, Passenger pigeon.

3. **Critically Endangered:** A taxon is critically endangered when it is facing an extremely high risk of extinction in the wild in the immediate future.

4. **Endangered:** A taxon is endangered when it is not critically endangered but is facing a very high risk of extinction in the wild in the near future. Endangered taxon has very low population numbers and its number has reduced to a critical level or its habitats have been drastically reduced. If such a taxon is not protected and conserved, it is in immediate danger of extinction.

5. **Vulnerable:** A taxon is vulnerable when it is not critically endangered or endangered, but is facing a high risk of extinction in the wild in the medium-term future. It is under threat or actually decreasing in number due to over-exploration or habitat destruction. Such a taxon is still abundant but under a serious threat of becoming endangered if causal factors are not checked.

6. **Lower Risk:** A taxon is lower risk when it has been evaluated and does not satisfy the criteria for critically endangered, endangered or vulnerable.
7. **Data Deficient:** A taxon is data deficient when there is inadequate information to make a direct or indirect assessment of its risk of extinction.
8. **Not Evaluated:** A taxon is not evaluated when it has not yet been assessed against the above criteria.

A rare taxon is not endangered or vulnerable at present but is at a risk as it is usually localized within restricted geographical areas *i.e.*, it is usually endemic. Sometimes it is thinly scattered over a more extensive area. It has small population at risk of becoming rarer but not of becoming extinct.

Top Ten Most Wanted Species Announced by WWF

Worldwide Fund for Nature (WWF), previously known as World Wildlife Fund, had announced 10 most wanted species for 2004, based on threats from sustainable trade and consumer demand of:

1. **Tigers** (*Panthera tigris*): There are perhaps less than 5,000 tigers left in wild. Tigers are traded for their skins, bones and various other parts for making medicines in China.
2. **Asian Elephants** (*Elephas maxima*): There are hardly 35,000 to 50,000 Asian elephants in the wild with an additional 150,000 in captivity. In recent years they have suffered from serious habitat loss. Elephants are facing indiscriminate poaching for ivory and meat.
3. **Asian Yew Trees** (*Taxus chinensis, T. cuspidata, T. fauna, T. sumatrana*): Found throughout Asia, the Yew trees are harvested for their bark and needles which contain a chemical *taxol* used in cancer medication.
4. **Pig-nosed Turtle** (*Careltochelys insculpta*): Found only in Papua New Guinea, it is a popular pet worldwide and suffers from high demand of the international pet trade.
5. **Yellow-crested Cockatoos** (*Caatua sulphurea*): Found in Indonesia as exotic-looking birds, cockatoos are highly prized in international pet trade. Their populations is less than 10,000. Convention on International Trade in Endangered Species of Wild Fauna and Flora (CITES) has imposed to put an end to its marketing
6. **Irrawady Dolphins** (*Orcaella bevirostris*): This extremely endangered, Asian dolphin species gets entangled in fishing nets and gets injured from explosives used for dynamite fishing. It is in great demand for displays in zoos and museums.

7. **Great White Sharks** (*Carcharias* sp.): The largest of the sharks poached for jaws, teeth and fins which have great demand worldwide and fetch high prices.
8. **Leaf-tailed Gackos** (*Uroplatus* sp.): All 10 species of this lizard are found in Medagascar. They are sold at a very high price in international market. They are also threatened due to habitat loss and fragmentation.
9. **Humpheal Wrasse** (*Cheilinus undulatus*): It is a bulbous-head coral reef fish. Its reproduction is slow but it has unsustainably been harvested for high demand of its delicacy. Its population is suffering greatly because it is caught and displayed live in tanks for dinners in East Asian restaurants.
10. **Ramin** (*Gonystylus* sp.): This tropical hardwood tree from Indonesia and Malaysia is used to make moulding doors and picture frames. It grows in peat swamp forests. It is illegally logged for valuable wood.

Threatened Animal Species of India

The International Union for Conservation of Nature and Natural Resource (IUCN), now World Conservation Union, symbolises the warning signals for threatened species of animals, which if not protected, are likely to become extinct. In India, about 53 species of mammals, 69 species of birds, 23 species of reptiles, 3 species of amphibians, 3 species of fish and 2 species of molluscs are estimated to be threatened. An unknown number of animal species of insects are endangered. According to IUCN Red Data Book, India ranks second in terms of the number of the species of threatened mammals while sixth in terms of the countries with most threatened birds. According to Red List of threatened animals in India, 18 animal species are critically endangered, 54 endangered and 143 vulnerable while 10 are lower risk and 99 are lower risk near threatened. According to Zoological Survey of India, Asiatic cheetah, pink-headed duck and mountain quail have already become extinct. There are 21,000 species and sub-species of birds known in India, of which large number are endangered. Great Indian bustard, peacock pelican, great Indian hornbill, Siberian white crane, geese, swans, white winged duck, hawks, eagle etc. are endangered. Among reptilian resource, gharial, green sea turtle, Manipur lizard and python and among amphibians viviparous toad is endangered.

About 850 species constitute the mammalian fauna of India of which 81 including 12 species of primates are greatly endangered. Nearly 23 species, including cheetah, are known to have become extinct and many more could have been vanished (Khoshoo, 1996). Those having small and slender bodies are confined to tropical rainforests and open woodland forests of south India and slightly bigger ones inhabit the denser forests of North-East India. Among carnivorous animals Indian wolf, red fox, sloth beer, red panda, tiger, leopard, striped hynea, Indian lion, golden cat, desert cat, dugong, rhino etc. are endangered. Among primates peacock gibbon, lion tailed macaque, Nilgiri languer, golden monkey etc. are endangered. Due to indiscriminate hunting, the killing of tigers has reduced their population to 1827, but under Project Tiger their population is estimated to be 2,287. The Indian lion inhabits Gir forest in Gujarat. Leopard is distributed in wooded forests throughout the country whereas snow leopard is found in the upper Himalayas between an altitude of 2000 to 4000 metres. The great Indian one-horned rhinoceros, once widely distributed in the Gangetic plain, is now at the verge of extinction and is confined to North Bengal and Plains of Assam. It was on the verge of extinction in 1904. Today only few rhinoes have remained in Kaziranga. Indian wild ass is restricted to the saline flats of Rann of Kachch, Gujarat. Presently it is represented by 720 individuals. The musk deer inhabiting high altitude hills of Garhwal, Arunachal Pradesh and Assam has been over-exploited for the musk and has fallen under the category of endangered. The black buck occupies open plains of western and southern India. The wild buffalo is now restricted to certain swampy areas of Assam and Madhya Pradesh. Lion tailed macaque survives only in the forested parts of North-East India. The Nilgiri langur is restricted to the evergreen forests of Western Ghats. The hoolock gibbon is the only ape found in the hilly forest areas of North East India. The Malayan sun beer inhabits the hilly terrain forest of North East India. The Himalayan brown beer is restricted to upper Himalayas form Kashmir to Sikkim. The Manipur brown antlered deer, once distributed in some parts of North East India, is most threatened species. The Kashmir stag has fallen under the category of endangered. The populations of elephants, once found all over the country has restricted to North East and the southern part of the country. Pigmy hog is critically endangered, red panda is endangered and black buck is vulnerable

Threatened Plant Species of India

According to IUCN Red Data List of plant taxa of India, 19 species are extinct and 1149 species are threatened. Of these threatened, 44 species are critically endangered, 113 endangered, 87 valnerable, 215 rare and 690 intermediate. Among endangered 44 are critically endangered and 69 endangered. Botanical Survey of India and Forest Research Institute, Dehradun has published an inventory of rare and endangered species of Indian flora in 1980. According to the inventory in Himalayas and Eastern India 99 plant species; in Rajasthan and Gujarat 4 plant species

Hyphaene dichotoma, Rosa involucrata, Helichrysum cutechium and *Commiphora wightii*; in Gangetic plain the only plant *Aldrovanda*; in peninsular India 15 plant species; and in Andaman and Nicobar Islands 11 plant species are declared as rare or endangered. Some common plants of Himalayas and Eastern India are *Aconitum deinorrhizum* (Aconite), *Atropa acuminata* (Indian Beladona), *Balanophora dioica* (Angiosperm root parasite), *Botrychium virginanum, Colchicum luteum* (Colchicine), *Coptis teeta, Cyathea gigantea, Drosera burmani* (Sundew), *Nepenthes khasiana* (Pitcher plant), *Magnolia griffithii, Osmunda regalis, Picea brachytyla, Abies delavayi* (Chinese fir), *Podophyllum hexandrum, Populus gamblei* (Poplar), *Psilotum nudum, Vanda coerulea, Vanilla pilifera* and several species of rhododendrons; penisular India are *Dioscorea wightii* (Yam), *Entada pursaetha, Gnetum ula, Piper barberi, Podocarpus wallichianus, Santalum album* etc.; and Andaman Nicobar Islands are *Alibanthus cruzii, Dipterocarpus kerrii, Myristica andamanica, Ophioglossum peridulum, Podocarpus neriifolius, Psilotum complanatum, Uvaris nicobarica* etc.

Among well known medicinal plants that are endangered are different species of *Lycopodium, Rauvolfia serpentina* (Serpgandha), *Podophyllum hexandrum* (Hansapadi), *Dioscorea* (Khamalu). Among others reported, *Berberis nilgriensis* is critically endangered, *Benuntinckia nicobarica* is endangered, *Cuppressus kashmiriana* and some species of *Taxus* are vulnerable.

Among other over-exploited economic plants are *Colchicum luteum* (Jafran), *Gloriosa superba* (Ulatchandan) and *Santalum album* (Chandan). Indicriminate collection and senseless destruction, by the students and exploitation by traders, of *Nepenthes khasiana* (Pitcher plant), *Drosera burmani* (Sundew), many pteridophytes such as species of *Psilotum, Botrychium, Ophiglossum, Lycopodium, Isoetes, Helminthostachys* and several gymnosperms such as *Gnetum, Podocarpus, Taxus* etc. is causing them to become rare.

There was a time in the past when India was known for its rich wildlife. The wildlife has gradually declined over the millennia and today it has reached a critical stage when at least 10 per cent of India's plant species and a larger percentage of its animal species are threatened. We seem to be in danger because our wildlife appears to be on the verge of being wiped out. The cheetah and the pink-headed duck are amongst the well known species that have become extinct in recent decade. More than 150 medicinal plant species have disappeared. About 10 per cent of flowering plants, 20 per cent of mammals and 5 per cent of bird species are threatened. Hundreds of crop varieties have disappeared and even their genes have not been preserved.

India had tremendous diversity in wild crop plants. To this, farmers have added a large variety of crop species. There use to be thousands of rice and wheat varieties. The green revolution encouraged farmers to grow the new High Yielding Variety (HYV) seeds, replacing the indigenous ones. Over the years, the traditional species have disappeared and in place of 30,000 varieties of rice Indian farmers now plant just 12 High Yielding Varieties.

According to final technical report of National Biodiversity Strategy and Action Plan (NBSAP), released in 2005, India has lost over 50 per cent of its forests, 40 per cent of its mangroves and a significant part of its wetlands in the past couple of centuries. According to the report 'Securing India's Future' at least 40 species of plants and animals have become extinct, including the cheetah and pink-headed duck while several hundred more are under the threat of extinction. Much of the diversity of crops had been lost or is under threat. All 18 India's poultry breeds are also under threat.

Chapter Summary

Endemic species are those species which are confined only to a particular locality. They are important from conservation point of view because their disappearance means their extinction as they are not found elsewhere. They represent a unique feature of evolution. Once endemic species are lost, there is no way to recover them. Habitat destruction and degradation, especially in the tropical forests and island forest covers, is the main cause of their loss. India ranks eleventh in the world in terms of endemic species of both plants and higher vertebrate animals. India is one of the megadiversity country and has two, out of 25, hot spots of the world. About 33 per cent plant species of India are endemic and are found no where else. Andaman and Nicobar Islands alone have 220 endemic species. About 3,000 plant species are endemic to Himalayas and 2,000 to Deccan peninsula. Many of the animal species are also endemic.

Extinction of species is a natural process but anthropogenic, *i.e.*, man-caused extinction of species is posing a serious threat to biodiversity. IUCN established in 1948, now World Conservation Union (WCU), is taking active interest in evaluation of conservation status of species of plants and animals and has published Red Data Book listing the plants under various categories viz. extinct, critically endangered, endangered, vulnerable, lower risk and data deficient. A large number of Indian plant and animal species fall under various categories of Red List.

Study Questions

1. Discuss the status of endemic species of plants and animals in India.
2. Briefly discuss species extinction and role of IUCN in evaluation of conservation status of various threatened species of plants and animals.
3. Throw light on threatened plant and animal species of India.
4. Discuss role of IUCN.
5. What is Red Data Book? Explain.
6. Discuss Eight Red Data List categories.

Objective questions. *Select correct answers*

1. When there is no doubt that the last individual has died, the species is designated as:
(1) Rare (2) Vulnerable
(3) Threatened (4) Extinct

2. *Ailurus fulgens* (Red Panda), a highly endangered species, is found in which part of India:
(1) Coastal region (2) North Eastern Himalayas
(3) Trans Himalayan region (4) Tarai area

3. Asian elephant is now endangered because it is being killed for:
(1) Skin (2) Ivory
(3) Musk (4) Fur

4. The only ape found in India is:
(1) Gorilla (2) Chiampanzee
(3) Hodock Gibbon (4) Orangutan

5. The full form of CITES is:
(1) Conference on Indian Tropical Endangered Species
(2) Convention on International Trade in Endangered Species
(3) Convention on International Trade in Endangered Species of Wild Fauna and Flora
(4) Convention on Indian Trade in Endangered Species

6. According to final technical report of National Biodiversity Strategy and Action Plan, Govt. of India, (2005), the country has lost its:
(1) 50 per cent forest cover (2) 10 percent mangroves
(3) 40 per cent wetlands (4) 30 per cent deserts

7. A Ramsar Site recognised in 1982 in India is:
(1) Wooler lake (2) Bharatpur lake
(3) Chilika lake (4) Sambhar lake

8. I.U.C.N. is now known as:
(1) W W F (2) C I T E S
(3) W C U (4) D S T

Answers

1. (4) *2.* (2) *3.* (2) *4.* (3) *5.* (3) *6.* (1)
7. (3) *8.* (3)

22
CHAPTER

Conservation of Biodiversity

Learning Objectives
Introduction • IN SITU conservation-Protected Areas: National parks • Wildlife Sanctuaries • Biosphere Reserves • Sacred Forests and Sacred Lakes • Project Tiger • Project Elephant; EX SITU *Conservation-Botanical Gardens • Zoos or Zoological Parks • NBPGR • NBAGR.*

Introduction

The earth is a home of rich and diverse array of living organisms, whose genetic diversity and relationship with each other, constitute planet's biodiversity. It includes species as well as varieties found within the same species, as also the enormous ecosystems and habitats-forests, grasslands oceans and its depths, peats and peat bogs. Worldwide, an area of tropical forest equal to a football field, is disappearing every second. Forests are being cleared for agriculture, cattle ranching and for timber. They get submerged when dams are constructed. When forests are felled we deal a blow to biodiversity, the richness of life. Five great waves of extinction's have decimated our planet's species in the remote past. The sixth wave, which we are witnessing, is largely man made. One-fourth of the earth's total biological diversity, amounting to about 1.6 million species, which might be useful to mankind in one way or the other, is in the serious risk of existence and if we ignore the biodiversity crisis, we would lose these species in the next 25 years. If we respond with the knowledge and technology, we might hold the loss to 10 per cent, but even this amounts to millions of species.

It is a well known fact that the ecosystem are undergoing changes due to pollution, invasive species, over-exploitation by humans, climate change and many other reasons. Most of the people are beginning to recognize that biodiversity is not only for availing the benefits but is also to be conserved in all its forms. The importance of species diversity conservation is usually justified by the direct use of the species by man, as food sources, fibre, fuel, new chemicals and raw materials. When a species is lost, it is lost for ever. Our ethics recall us that we should not deprive our future generation from economic and aesthetic benefits that can be derived from biodiversity. It is our moral duty to look after our planet and pass it on to our future generation in a good health. The decisions taken by us now, as an individual and as a society will determine the diversity of genes, species or ecosystems that remain in future. Top priority should be given to check destruction or degradation of habitats because it is most responsible for the loss of biodiversity. More knowledge is required to conserve biodiversity in perspective of reduced space under increased pressure of human activities. Unless immediate actions are taken to protect the biodiversity, we will lose forever the opportunity of reaping its full potential benefit in the present and in future.

Conservation of biodiversity is concerned with the protection of genes and species and their number in population, ecosystems or habitats. It is important to conserve numerous varieties of plants and animals. Each variety within a species contains unique genes. The diversity of genes within a species increases its adaptability to pollution, diseases, stresses and other environmental changes. When varieties of plants and animals are destroyed the genetic diversity within the species is diminished. Biodiversity conservation refers to the efforts to maintain or inhance biodiversity involving protection, upliftment and scientific management at its optimum level in order to derive sustainable benefits for the present as well as for the future. Biodiversity conservation study is aimed at knowing how human activities affect the diversity of plants and animals and developing ways to protect it. Thus conservation of biodiversity, the earth's biological heritage, is one of today's most pressing environmental issues. The challenge lies before individuals, organisations and

Box 22.1 Thunder Dragon Conserves its Biodiversity

Bhutan, the *Thunder Dragon,* has a large forest cover and enormous biodiversity. The country is also called 'oxygen tank' (or carbon sink) of the world. With a small area of only 40,000 sq km, 72 per cent is under forest cover. This hot spot ecosystem harbours and supports more than 7,000 species of plants, 700 species of birds, and 165 species of mammals and so on. The species richness and variability is due to wide range of altitude, topography and climate. The country, considered as economically weaker in the world, has a great commitment to conserve its rich biodiversity. Till 1993, the country had 4 national parks, 4 wildlife sanctuaries and one protected area covering about 10,513 sq. km. With country's policy of extension of protected areas in 2,000 now 35 per cent of Bhutan's geographical area is protected and the government is determined to maintain a forest cover of at least 60 per cent of its geographical area with the cooperation of local communities. Thus, Bhutan is a rare biodiverse country with its active biodiversity conservation programme.

government agencies for nations and the world as a whole to protect and enhance biological diversity, while continuing to meet people's need for natural resources. If these goals are not met, the future generations will live in biologically impoverished world.

There are two approaches of biodiversity conservation namely *in situ* (on site) conservation which tries to protect the species where they are, *i.e.*, in their natural habitats and *ex situ* (off site) conservation which attempts to protect and preserve a species in place away form its natural habitat.

In situ CONSERVATION

In situ conservation, an age old practice, emphasises the preservation and protection of total ecosystems at their original or natural environment. Kings and courtesans had always encouraged identification and protection of natural areas that had high biodiversity for animal games. Human societies have always taken interest in preserving wildlife areas. The main objective is to recognise a particular biodiversity rich area and to preserve it so that the biodiversity can continue to flourish and evolve. This involves establishment of protected areas, national parks, sanctuaries, biosphere reserves, reserve forests etc. Over past few decades there has been an increase in the number of such areas. Protection of the ecosystem by simply elimating factors detrimental to the existence of species concerned has given good results in conservation of constituent species, known or unknown.

In situ conservation of biodiversity is advantageous in that it is a cheap and convenient method that requires people's our supportive role. It maintains all organisms at different trophic levels from producers to top consumers such as carnivores. In natural environment, organisms not only live and multiply but also evolve and continue to maintain their ability to resist various environmental stresses such as drought, storm, snow, temperature fluctuations, excessive rains, floods, fires, pathogens etc. *In situ* conservation requires only elimination of factors detrimental to the existence of the species and allow the large number of species to grow simultaneously and flourish in their natural environment in which they were growing since a long time. The only disadvantage of *in situ* conservation is that it requires larger areas and minimises the space for inhabiting human population which is increasing tremendously.

Protected Areas

According to World Conservation Union (WCU), previously IUCN, protected areas are the areas of land/ or sea, especially dedicated to the protection and maintenance of biodiversity and natural and cultural resources managed through legal or other effective means. Benefits of protected areas are manifold. They maintain viable populations of all native species and their varieties; maintain the number and distribution of communities and habitats; conserve the genetic diversity of all the species present; prevent human caused introduction of alien (exotic) species; and make it possible for species to shift in response to environmental changes. Many protected areas now protect rare species and wilderness areas throughout the world.

This type of conservation applies mainly to wild flora and fauna. There are different categories of protected areas which are managed with different objectives. Protected areas include National Parks, Wildlife Sanctuaries and several other protected places. World Conservation Monitoring Centre has recognised 37,000 protected areas around the world. Based on National Wildlife Database 2004, India has 584 protected areas including 92 National Parks and 492 Wildlife Sanctuaries covering an area of 4.7 per cent as against 10 per cent internationally. *In situ* conservation of plant and animal species in their natural habitats is achieved through following strategies:

National Parks

Earlier, the national parks, such as Yellowstone Park in USA and the Royal Park near Sydney, Australia, were chosen because of their scenic beauty and recreational value but now the national parks have been given a status of selected areas dedicated to conservation of wildlife along with its scenery *i.e.* natural environment. They are strictly reserved for the betterment of the wildlife and declared as places where activities such as forestry, grazing, hunting, cultivation etc. are not permitted either, imposing legal binding or by certain other ways. No private ownership is allowed. Usually they are hitched to the habitat for specific wild animal species along with others such as tiger, lion, rhinoceros, elephant etc. with their boundaries circumscribed by legislation. No biotic interference is allowed except for the buffer zone where limited human activity for enjoyment through eco-tourism is allowed.

According to the National Wildlife Database 2004, there are 92 existing national parks in India covering an area of 37,921.66 sq km which is 1.15 per cent of the total geographical area of the country. In addition 74 national parks, that would cover an area of 16,630.08 sq km, are proposed in the Protected Area Network Report. In India, first national park named Hailey National Park, which is now known as Corbett National Park, Uttarakhand, came into being in 1936. The Wildlife Protection Act, 1972 empowers the state governments to declare any of their area as national park for protecting, propagating and developing wildlife and its environment. State wise break up of national parks is given in table 22.1.

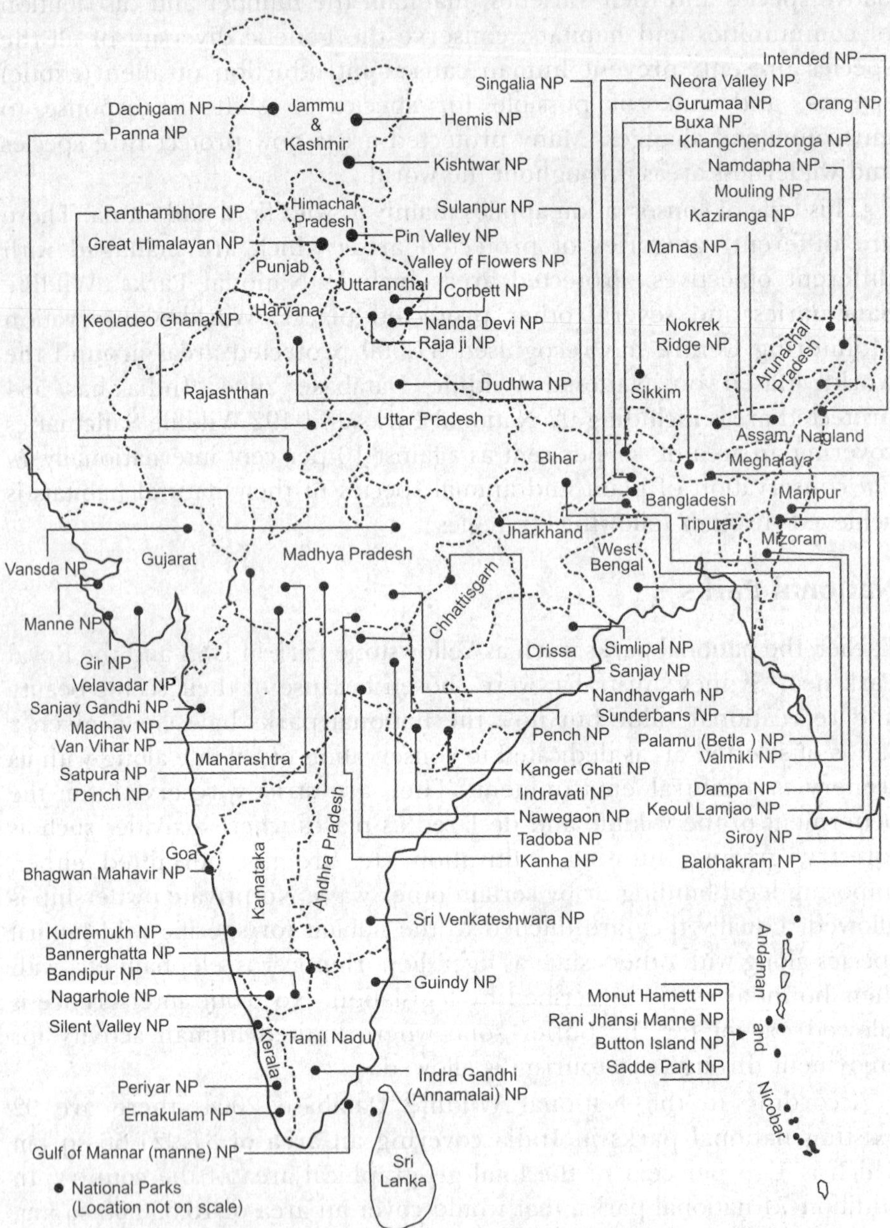

Figure 22.1. Selected National Parks of India.

TABLE 22.1. State-wise Break up of National Parks in India (2004)

State/ UTs	*National Parks*	*Area Covered (km^2)*
Andhra Pradesh	4	373.23
Arunachal Pradesh	2	2290.82
Assam	5	1968.60
Bihar	1	335.65
Chhattisgarh	3	2929.50
Goa	1	107.00
Gujarat	4	480.11
Haryana	2	117.13
Himachal Pradesh	2	1429.40
Jammu & Kashmir	4	4680.25
Jharkhand	1	231.67
Karnataka	5	2435.14
Kerala	4	549.34
Madhya Pradesh	9	3656.36
Maharashtra	5	955.93
Manipur	1	40.00
Meghalaya	2	267.48
Mizoram	2	250.00
Nagaland	1	202.02
Orissa	2	990.70
Punjab	0	0.00
Rajasthan	5	4122.33
Sikkim	1	1784.00
Tamil Nadu	5	307.84
Tripura	0	0.00
Uttar Pradesh	1	490.00
Uttarakhand	6	4077.00
West Bengal	5	1693.25
Union Territories (UTs)		
Andaman & Nicobar	9	1156.91
Chandigarh	0	0.00
Dadar & Nagar Haveli	0	0.00
Daman & Diu	0	0.00
Delhi	0	0.00
Lakshadweep	0	0.00
Pondicherry	0	0.00
India	92	37,921.66

Some important national parks of India, their location and important flora and fauna they contain are given in table 22.2.

TABLE 22.2. Select National Parks (NP) of India

No.	*Name of Park*	*State/UT.*	*Common Flora; Fauna*
1.	Corbett N.P.	Uttarakhand	Moist and dry forest, Khair-sisoo forest; tiger, leopard, fishing cat, Himalayan palm civet.
2.	Rajaji N.P.	Uttarakhand	Moist and dry sal forests, Khair-sisso or riverine forest; tiger, leopard, samber, wild bear, many species of birds.
3.	Dudhwa N.P.	Uttar Pradesh	Moist mixed deciduous forest, terai and bhabar, sal forest; tiger, leopard, elephant, sloth bear.
4.	Guindy N.P.	Tamil Nadu	Evergreen and semi-evergreen forest and grasslands; leopard, elephant, cheetah, gaur.
5.	Khangchendzona N.P.	Sikkim	Wet temperate forest, dry and alpine scrub; musk deer, leopard.
6.	Ranthambhore N.P.	Rajasthan	Dry mixed deciduous forest; tiger, leopard, hyaena.
7.	Keoladeo N.P.	Rajasthan	Dry areas, moist scrap, thorn and mixed deciduous forest; 375 species of resident and migratory aquatic birds.
8.	Similipal N.P.	Orissa	Semi-evergreen forest, sal forest and grasslands; 42 species of mammals, 242 species of birds, 29 species of reptiles.
9.	Intanki N.P.	Nagaland	Semi-evergreen forest, subtropical forest; gibbon, elephant, leopard.
10.	Dampa N.P.	Mizoram	Semi-evergreen forest; tiger, leopard, elephant, swamp deer.
11.	Balaphakram N.P.	Meghalaya	Evergreen and semi-evergreen forest; tiger, leopard, gibbon and many species of birds.

12.	Keibul Lamgo N.P.	Manipur	Swampy forest, hog deer, fishing cat.
13.	Kanha N.P.	Madhya Pradesh	Mixed deciduous and thorn forest, tiger, leopard, gaur, swamp deer, barking deer.
14.	Periyar N.P.	Kerala	Dry mixed deciduous forest, bamboo and cane brakes; tiger, leopard, elephant, wild dog, jackal, wild bear.
15.	Bandipur N.P.	Karnataka	Tropical evergreen forest, southern moist mixed deciduous forest; tiger, leopard, elephant, cheetah.
16.	Dachigam N.P.	Jammu and Kashmir	Moist deodar forest, mixed coniferous forest; hangul, leopard, brown and black beer.
17.	Kaziranga N.P.	Assam	Tropical evergreen forest, cane and bamboo brakes, parkland creeper forest, tropical seasonal swamp forest; one horned rhino, wild buffalo.
18.	Gir N.P.	Gujarat	Dry teak forest, desert thorn forest; lion, panther, spotted deer, chinkara and wild boar.
19.	Great Himalayan N.P.	Himachal pradesh	Oak forest, moist deodar forest; leopard, nyaena, civet cat, brown and Black Bear.
20.	Sanjay N.P.	Chhattisgarh	Sal forests; tiger, leopard, cheetah, wild bear and many species of birds.
21.	Palamu N.P.	Jharkhand	Peninsular sal forest, hilly valley swamp forest; elephant, gaur, cheetah, sambhor, nilgai.
22.	Sri Venkteshwara N.P.	Andhra	Red sandar trees; golden gecko.
23.	Bhagwan Mahavir N.P.	Goa	Semi-evergreen forest, bamboo brakes; leopard, sloth bear, cheetah and many other species

			of animals and birds.
24.	Sultanpur N.P.	Haryana	Swampy vegetation; crustaceans fish and insects thrive during floods which attract a number of birds including king fishers.
25.	Buxa N.P.	West Bengal	Terai and bhabar, sal forest, swamp and rivenine vegetation; tiger, leopard, elephant, clouded leopard, sambhar.

Wildlife Sanctuaries

Like national parks, wildlife sanctuaries are also protected areas dedicated to protect wildlife, but they are aimed at conserving animal species only. Also the boundary of a sanctuary is not limited by state legislation. In these protected areas, killing, shooting, hunting or capturing of wildlife is prohibited except at the discretion of highest authority. Limited biotic interference is allowed in them. Private ownership rights and forestry operations are permissible to an extent that they do not affect the wildlife adversely. Local residents are given responsibility of controlling fire damage, reporting about dead animals and keep an eye on offenders. Visitors are not allowed to carry weapons without permission, set fire, leave any fire burning or its remnant, feed animals and use the chemicals or explosives injurious to any wildlife. The Chief Wildlife Warden of the state is authorised to grant permission to enter or stay in a sanctuary for study, research, photography, tourism, transaction of lawful business etc.

According to National Database Resource, 2004, there are 492 existing wildlife sanctuaries in India, covering an area of 1,16.992 sq km which is 3.56 per cent of the geographical area of the country. Another 217 sanctuaries are proposed in the Protected Area Network Report covering an area of 16,689.44 sq km. State-wise break up of wildlife sanctuaries in India is given in table 22.3.

TABLE 22.3. State-wise Break up of Wildlife Sanctuaries in India (2004)

State/ UTs	*No. of WLS*	*Area Covered (km^2)*
Andhra Pradesh	22	12600.09
Arunachal Pradesh	11	7606.37
Assam	15	943.31
Bihar	11	2949.17
Chhattisgarh	10	3409.13

Goa	6	647.96
Gujarat	21	16422.72
Haryana	8	179.04
Himachal Pradesh	32	5770.85
Jammu & Kashmir	15	10312.25
Jharkhand	10	1862.72
Karnataka	21	3888.22
Kerala	12	2143.36
Madhya Pradesh	25	1758.40
Maharashtra	36	14376.66
Manipur	3	393.30
Meghalaya	3	32.20
Mizoram	4	71.00
Nagaland	3	20.34
Orissa	18	6969.15
Punjab	10	316.73
Rajasthan	23	5447.03
Sikkim	5	265.10
Tamil Nadu	19	2539.82
Tripura	4	603.62
Uttar Pradesh	23	5222.47
Uttrakhand	6	2413.76
West Bengal	15	1203.28
Union Territories		
Andaman & Nicobar	96	389.39
Chandigarh	1	25.42
Dadar & Nagar Haveli	1	92.16
Daman & Diu	1	2.18
Delhi	1	13.20
Lakshadweep	1	0.01
Pondicherry	0	0.00
India	492	1,16,992.41

Selected wildlife sanctuaries, states in which they are located and important fauna they contains are given in table 22.4.

TABLE 22.4. Select Wildlife Sanctuaries of India

No.	*Name of Wildlife Sanctury*	*State/UT*	*Important Fauna*
1.	Anamalai Sanctuary, Coimbatore	Tamil Nadu	Elephant, Tiger, Panther, Gaur, Sambhar, Spotted deer, Sloth bear, Wild dog, Barking deer.
2.	Jaldapara Sanctuary, Madarihat	West Bengal	Rhino, Elephant, Tiger, Leopard, Gaur, Deer, Sambhar, Different kinds of birds.
3.	Keoladeo Ghana Bird Sanctuary, Bharatpur	Rajasthan	Birds such as Siberian crane, Strokes, Egrets, Herons, Spoon bill etc. and Spotted deer, Black

			buck, Sambhar, Wild boar, Blue bull, Python.
4.	Sultanpur Lake Birds sanctuary, Gurgaon	Haryana	Crane, Sarus, Spot bill, Duck, Drake, Green pigeon, Wild boar, Crocodile, Python.
5.	Bir Moti Bagh Wildlife Sanctuary, Patiala	Punjab	Nilgai, Wild boar, Hog deer, Black buck, Blue jackal, Peafowl, Partidge, Sparrow, Mayna, Pigeon, Dove.
6.	Shikari Devi Sanctuary, Mandi	Himachal Pradesh	Black bear, Snow leopard, Flying fox, Barking deer, Musk deer, Chaker partridge.
7.	Dachigam Sanctuary Srinagar	Jammu and Kashmir	Hangul or Kashmir stag, Musk deer, Snow leopard, Black bear, Brown bear.
8.	Mudumalai Wildlife Sanctuary Nilgiri	Tamil Nadu	Elephant, Gaur, Sambhar, Chital, Barking deer, Mouse deer, Four horned antelope, Langur, Giant squirrel, Wild dog, Wild cat, Civet sloth beer, Porcupine, Python, Rat, Snake, Monitor lizard, Flying lizard.
9.	Nagarjuna Sagar Sanctuary, Guntur Karnool and Nalgonda	Andhra Pradesh	Tiger, Panther, Wild beer, Chital, Nilgai, Sambhar, Black buek, Fox, Jackal, Wolf, Crocodile.
10.	Periyar Sanctuary	Kerala	Elephant, Gaur, Leopard, Sloth beer, Sambhar, Bison, Black langur, Egret, Hornbill, Famous for elephants.
11.	Chilika Lake Bird Sanctuary, Balagaon	Orissa	Water fowls, Duck, Crane, Osprey, Golden plover, Sand piper, flaminge.
12.	Manas Wildlife Sanctuary, Kamrup	Assam	Tiger, Panther, Rhino, Gaur, Wild buffalo, Sambhar, Swamp deer, Golden langur, Wild dog, Wild beer.

Biosphere Reserves

Biosphere reserves are special category of protected areas of land and/or coastal environment, representing natural ecosystems or biomes with unique biological communities selected for long term *in situ* conservation of species of plants and animals. People are an integral component of biosphere reserves. UNESCO's Man And Biosphere programme (MAB), dealing with conservation of ecosystem and genetic resource contained therein is credited for launching the concept of biosphere reserve in 1975. Till May 2002, there were 408 biosphere reserves spread over in 94

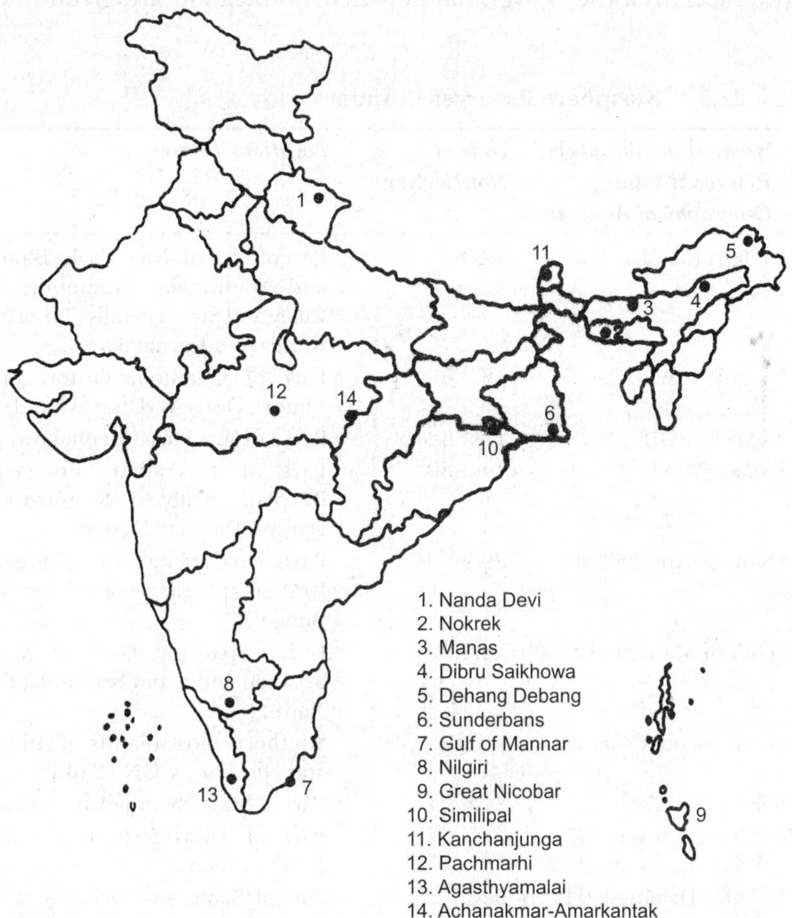

Fig. 22.2. The Biosphere Reserves in India.

countries. In India, the first biosphere reserve, Nilgiri Biosphere Reserve spread over Kerala, Tamil Nadu and Karnataka, came into being on August 1, 1986. There are 13 biosphere reserves in India and the fourteenth Achnakamar-Amarkantak was declared in 2005 (fig. 22.2). Of the 14 biosphere reserves, 4 namely Nanda Devi (Uttarakhand), Sundarbans (West Bengal), Gulf of Mannar (Tamil Nadu) and Nilgiri (Tamil Nadu, Kerala and Karnataka) have been included in the World Network of Biosphere Reserves. A stake holder's consulation was held to review the proposal received in respect of Gulf of Kutch (Gujarat) and a notification was issued by the Ministry of Environment and Forests after finalisation of guidelines for regulatory regime.

Various biosphere reserves, the state/states of their location, geographical area they cover, date of their notification are given in table 22.5.

TABLE 22.5. Biosphere Reserves in India

No.	*Name of the Biosphere Reserve & total Geographical Area (km^2)*	*Date of Notification*	*Locations (States)*
1.	*Nilgiri (5520)	1.8.86	Part of Wynad, Nagarhole, Bandipur and Madumalai, Nilambur, Silent valley and Siruvani hills (Tamil Nadu, Kerala and Karnataka)
2.	*Nanda Devi (2236.74)	18.1.88	Part of Chamoli, Pithoragarh & Almora Districts (Uttarakhand)
3.	Nokrek (820)	1.9.88	Part of Garo Hills (Meghalaya)
4.	Manas (2837)	14.3.89	Part of Kokrajhar, Bongaigaon, Barpeta, Nalbari, Kamprup and Darang Districts (Assam)
5.	*Sunderbans (9630)	29.3.89	Parts of delta of Ganges & Brahamaputra river system (West Bengal)
6.	*Gulf of Mannar (10500)	18.2.89	Indian part of Gulf of Mannar between India and Sri Lanka (Tamil Nadu)
7.	Great Nicobar (885)	6.1.89	Southern most islands of Andaman and Nicobar (A & N Island)
8.	Similipal (4374)	21.6.94	Part of Mayurbhanj district (Orissa)
9.	Dibru-Saikhowa (765)	28.7.97	Part of Dibrugarh and Tinsukia district (Assam)
10.	Dehang Debang (5111.5)	02.09.98	Part of Siang and Debang valley in Arunachal Pradesh
11.	Kanchanjunga (2619.92)	07.02.2000	Parts of North and West Sikkim
12.	Pachmarhi (4926.28)	03.03.99	Parts of Betur, Hoshangabad and Chindwara, Distt. of Madhya Pradesh
13.	Agasthyamalai (1701) (area expanded on 30.3.2005)	12.11.2001	Parts of Thirunelveli and Kenya Kumari Districts in Tamil Nadu and Thiruvananthapuram, Kollam and Pathanmthitta
14.	Achanakmar – Amarkantak (3835.51)	30.03.2005	Parts of Anupura and Dindori distt. of MP and parts of Bilaspur distt. Of Chattisgarh State.

Sites with * have been recognised by UNESCO on World Network of Biosphere Reserves

Biosphere Reserves may have one or more National Parks. For example Nilgiri Biosphere Reserve has two National Parks namely

Bandipur National park and Nagarhole National Park. In biosphere, multiple land use is permitted. A biosphere is not hitched to any one, two or three species but to the ecosystem as a whole *i.e.* totally of all life forms that live in it. In a biosphere reserve wild populations as well as traditional life styles of tribals and various domesticated plants and animal genetic resources are also protected.

There are three district zones viz. *core*, *buffer* and *transition* in a biosphere reserve. The core zone, also known as natural zone, comprises an undisturbed and legally protected ecosystem. The buffer zone, surrounding core zone, is managed to accommodate a greater variety of resource use strategies, research and educational activities. The transition zone, the outermost part of the biosphere reserve, is meant for active cooperation between reserve management and the local people, wherein activities such as settlements, cropping, forestry, recreation and other economic uses are allowed in harmony with conservation goals.

The biosphere reserves are aimed at performing the following functions:

1. ***Conservation***: Biosphere reserves ensure the conservation of land-scapes, ecosystems, species and genetic resources for ecological evidence for the process of evolution to act upon and for traditional resource use.
2. ***Development***: Biosphere reserves promote sustainable economic development culturally, socially and ecologically.
3. ***Monitoring, Scientific Research and Education***: Biosphere reserves render support for monitoring scientific researches and education and furnish information exchange related to local, national and international issues of conservation and development.

Preservation Plots

Preservation plots are specific forests areas identified for preservation and conservation of biological diversity contained in them. Presently, there are more than 309 preservation plots all over the country, of these 287 are in natural forests and 22 in plantation forests.

In situ conservation has proved to be an effective strategy for preservation of biodiversity in India. The country provides the world's largest network of protected areas in *in situ* conservation. The populations of great Indian bustard, tiger, elephant, hungal, crocodile etc. have gone up. It indicates that if proper implementation of conservation measures is done, the loss of biodiversity can be compensated. However, the conservation efforts towards plant species have not been given adequate attention, particularly to those which carry scientific value. Efforts made in the direction of conservation of plant species of potential economic value are insufficient. One sanctuary, each for in *Citrus* and *Nepenthes* (Pitcher plant) has been set up in Meghalaya in North East India and rhododendrons and orchids in Sikkim and Darjeelings, Himalayas. The *in situ* conservation strategy needs holistic view of saving much of the biodiversity and not restricting on protection and conservation of animals alone or limited to only larger animals such as tiger and elephant. The other significant aspect which is considered is proper rehabilitation strategy for rare and threatened species of plants and animals. Equally important is to prepare a comprehensive National Biodiversity Inventory. The most important aspect of *in situ* conservation is to enrich the ecosystem as a whole by involving tribals and local communities by integrating rural development programmes linked with ecosystem conservation. For protection and conservation of specific animal species, certain projects such as Project Tiger, Project Elephant etc. have been launched by Ministry of Environment and Forests, Government of India.

Project Tiger

In India, Project Tiger was lunched in 1973 with an objective "to ensure maintenance of a viable population of tigers in India for scientific, economic, aesthetic, cultural and ecological values and to preserve for all times areas of biological importance as a national heritage for benefit and enjoyment of the people". The Project has been successfully implemented, and at present there are 28 Tiger Reserves in 17 states covering an area of 37,761 sq km. According an estimate the number of tigers which was about 4026 in 1989 went down to about 1233 in 2000. Surprisingly no tiger in Sariska is seen since 2004. A survey of number of tigers in 2008 revealed that there are about 1411 tigers in India. The selection of reserves was guided by the need to conserve unique ecosystem habitat types across the geographic distribution of tigers in the country.

The network of Tiger Reserves include high mountainous terrains of Arunachal Pradesh; the heavy rainfall areas of Assam and West Bengal; the estuarine mangroves of sundarbans; the dry forests of Rajasthan; the foothills of Himalayas in Uttarakhand, Uttar Pradesh and Bihar; the Central Indian Highlands of Madhya Pradesh, Chhattisgarh and Maharashtra; the plateau of Chhota Nagpur (Jharkhand); the hilly tropical and evergreen forest of Orissa; the evergreen forests of Western Ghats in Kerala and Karnataka; the dry deciduous forests of Andhra Pradesh; and the southern moist deciduous forests of Tamil Nadu.

The Project Tiger is undisputedly the custodian of major gene pool of the country and a repository of some of the most valuable ecosystems and habitats for wildlife. The Tiger Reserves are constituted for the purpose of management on a *core-buffer* strategy. In the core area, forestry operations, collection of non-forest timber produce, grazing, human settlement and other biotic disturbances are not allowed and is singularly oriented towards conservation. The buffer zone is managed as a *multiple use area*, with conservation oriented land use, having the twin objectives of ensuring habitat supplement to the spillover population of wild animals from the core, apart from providing site specific eco-development inputs to stake holder communities. The main thrust of the project is protection and mitigation of deleterious human inputs with a view to comprehensively revive the natural ecosystems in the reserves. During the Tenth Plan the major thrust is to further enlarge and diversify the activity and consolidate the progress made under the scheme hetherto.

Project Elephant

In India, Project Elephant was launched in February 1992 to assist states having free ranging populations of wild elephants to ensure long term survival of identified viable populations of elephants in their natural habitats. There are 11 Elephant Reserves spread over 13 states, *viz.*, Andhra Pradesh, Arunachal Pradesh, Assam, Jharkhand, Karnataka, Kerala, Meghalaya, Nagaland, Orissa, Tamil Nadu, Uttarakhand, Uttar Pradesh and West Bengal. States are being given financial as well as technical assistance by the Ministry of Environment and Forests in achieving the objectives of the Project. Help is also provided to other states with small population of elephants for the purpose of census, training of field staff and mitigation of human elephant conflict.

Sacred Forests and Sacred Lakes

There has been a tradition strategy for the preservation of biodiversity in the form of sacred forests in India and many other Asian countries. Sacred forests are the forest patches of varying dimensions protected by the tribal communities on account of their religious sanctity accorded to them. These represent islands of pristine forests *i.e.* most undisturbed forests with no human impact and have been free from all disturbances despite they are frequently surrounded by highly degraded lands. Many states in our country such as Maharashtra, Karnataka, Meghalaya and Kerala have sacred forests which are serving as a refugia for many endemic, rare and endangered taxa. Similarly, some fresh water lakes are also serving the purpose of protection of aquatic flora and fauna. For example Kecheopalri lake in Sikkim has been declared sacred by the people to save aquatic life from being degraded.

Ex situ CONSERVATION

Ex situ conservation involves cultivation of plants and rearing of animals outside their natural habitats. In this strategy, conservation of specific species is done as a sample of genetic diversity, particularly of endangered species of plants and animals, under human care. The *ex situ* biodiversity conservation is done through botanical gardens, zoos, conservation stands and seed, seedling, tissue culture, pollen, gene and DNA banks.

Botanical Gardens

World over, there are more than 1,600 botanical gardens and arboreta. The botanical gardens where specific tree and shrubs species are cultivated, contain more than 4 million plants. The Royal Botanic Gardens, Kew, England, a monumental center that alone contains about 80,000 plant species *i.e.* about 30 per cent of all known species of plants. In India, there are more than 8 botanical gardens. Plant Resources Center, Bhubneshwar is the largest botanical garden. The role of Indian Botanical Garden, Howrah (Kolkata); Lloyd Botanical Garden, Darjeeling; National Botanical Research Institute, Lucknow; and Tropical Botanic Garden and Research Institute, Thiruvanathpuram in this direction are note worthy. The Indian Botanic Garden, Howrah-Kolkata, previously known as Royal Botanic Garden, is the living repository of 15,000 trees and shrubs. It is aimed at conservation of rare flora and multiplication of the selected endangered plant species. The Tropical Botanic Garden and Research Institute, Thiruvanadpuram, with its mandate for conservation of tropical plant diversity, is also recognised as National Centre of Excellence in *ex situ* conservation of tropical plants. It also functions as the National Gene Bank for medicinal and aromatic plants of peninsular India. The Centre is set up to carry out *ex situ* conservation of country's biodiversity and its sustainable utilisation. Many of the botanical gardens now have seed banks, tissue culture facilities and other *ex situ* technologies. Botanical gardens play a significant role in conservation of plant species beyond preservation in the wild and in supply of plants for research and education, thereby taking pressure off of wild population of species.

Zoos or Zoological parks

Zoos contribute to the conservation of animal diversity by propagating and reintroducing endangered species. They also act as centres for research to improve management of captive and wild populations. There are more than 800 zoos around the world with about 3,000 species of mammals, birds, reptiles, amphibians etc. Many of them have well developed captive breeding programmes for restoring those species of animals whose number of surviving species is so small that there is no realistic chance of *in situ* survival. As the number of individuals increases in the captive breeding, they are selectively released in the wild. Zoos also function for off site collection to be used to restore depleted populations, reintroduction of species in the wild and bringing the habitats to their natural state. In India, zoos have been categorised as large, medium, small and mini based on the number of animals and kind and number of endangered species exhibited. Till March, 2006 there were 159 recognized zoos in India, spread over 19 large, 12 medium, 27 small and 101 mini. Government of India adopted a National Zoo policy in 1994, with objectives to complement and strengthen the national efforts in conservation of rich biodiversity of the country, particularly wild fauna.

Apart from conservation, zoos are also aimed to inspire among visitors sympathy for wild animals and understanding and awareness about the wildlife conservation. Zoos are also assigned as rescue centres for rescued orphaned wild animals, subject to the availability of appropriate housing and upkeep infrastructure.

After the amendment of Wildlife (Protection) Act 2003, Rescue Centres and Circuses, *i.e.*, Mobile Mini Zoos, have also been brought under the definition of zoo. The Government has banned use of lions, tigers, leopards, bears and monkeys for performing purposes in circuses in the country and more than 400 lions and tigers in addition to many leopards, bears and monkeys from the circuses have already been rehabilitated in the Rescue Centres. Government of India has set up a National Zoological Park in New Delhi which looks after programs on awareness

Box 22.2 Efforts of an Individual to Conserve Snakes and Crocodiles

With the ban an snakeskin's trade, Irulas, expert snake catchers in Tamil Nadu, lost their livelihood. At that time Romulus Whitaker, *snake* and *croman*, entered their lives and helped them to set up Irula Cooperative Society for extracting snake venom and selling it to the institutes and organizations that make lifesaving anti-venom, a win-win formula stopping killing of the snakes and deriving profit. Romulus Whitaker came to India at the age of 7 bestowed with his natural affinity for snakes and in fact for all wildlife. During his school days at Kodaikanal he used to wander Palani Hills to pick up skill to deal with wildlife. Later he learnt snake catching from Irulas and crocodile catching from Papua New Guinea. In 1972, he along with his friends set up Madras Snake Park which is today on the must see list of every tourist to Chennai. Park has 31 species of Indian snakes, all three species of Indian crocodiles, four species of exotic crocodiles, three species of Indian turtles and five species of lizards. Many species of reptiles, including endangered species Indian python is subjected to captive breeding. He has also established a Crocodile Bank – the gene bank for crocodiles. Snake park is of great educational, scientific and conservation value. Whitaker's life and work throws light on that a single individual can make a significant contribution to the conservation of biodiversity through passion and dedication.

conservation, breeding, research and preparation of literature on animal wildlife.

In India the zoos also function as centres of conservation of animals, education and conservation breeding of endangered species, which have no chance of survival in wild, through coordinated breeding *ex situ* conditions and raising stock for rehabilitating them in wild. There are many ongoing conservation breeding programs in off display conservation breeding centres of the zoos or specially created facilities for the purposes in the country. Conservation breeding programmes for snow leopard and red panda in Darjeeling, Asiatic lion in Junagarh, lion tailed macaque in Chennai, pygmy hog in Guwahati and western ragopan in Himachal Pradesh are some of the oncoming programs. A Central Zoo Authority, created by Government of India, through an amendment in the Wildlife (Protection) Act 1992, has set up a sub-committee in 2005 to monitor and give suggestions/recommendations on various ongoing conservation breeding programs.

Aquaria

Aquaria, like botanical gardens and zoos, also play a prominent role in restoring degraded ecosystems, reintroducing species in wild and restocking of depleted populations of fishes and other aquate animals. Role of aquaria in captive propagation of threatened fresh water species is significant. The World Conservation Union (previously IUCN) is mounting a major effort to develop captive breeding programs for endangered fish species, starting from lake Victoria, the desert fishes of North America and Appalachian fishes.

National Bureau of Plant Genetic Resources (NBPGR)

NBPGR is a special gene bank/seed bank garden established in New Delhi, India to cryoperserve pollen, seeds etc. of agricultural and horticultural vegetatively propagated crops like potato and tissue culture raised forest trees and their wild relatives by using liquid nitrogen at a temperature of –196°C. Cryopreservation is done by storing the desired material at ultra-low temperature either by very rapid cooling used for storing seeds or by gradual cooling and simultaneous dehydration at low temperature used for tissue culture. Thus, preserved materials can be stored for long period of time in compact and low maintenance refrigeration units. The genetic variability of many plant species and their varieties such as of rice, pearl millet, turnip, radish, potato, tomato, onion, carrot, chilli, tobacco, poppy, mustard etc. has been preserved successfully by cryopreservation, without losing their seed viability for future crop improvement and afforestation programs. There are more than 100 seed banks in the world. They hold more than four million kinds of seeds maintained in low temperature and low humidity levels by cryopreservation. Majority of seed banks are in developed countries or are indirectly controlled by them. Most of the seed banks concentrate on about 100 plant species that give 90 per cent of our food. National Facility for Plant Tissue Culture Repository (NFPTCR) is the Center created by NBPGR for the development of a facility for the *in vitro* conservation and storage of varieties of crop plants/trees.

National Bureau of Animal Genetic Resource (NBAGR)

NBAGR is a special gene bank center established in Karnal, Haryana to preserve seamen of domesticated bovine animals. *Ex situe* or off-site conservation is done artificially under human care and is most suited for rare species or species having small remaining population. Advantages of *ex situ* conservation lie in that the organism is assured of food, shelter and security and hence can have longer lifespan and breeding activity so as to produce more offsprings. Moreover, captive breeding and use of genetic practices provide possible reintroduction of animals to the wild at later stage or supplementing the current populations with new stock. Disadvantages of *ex situ* conservation lie in that it can be adopted only for new selected species because of limited space, facilities and finances in the centers. Since *ex situ* conservation is done under a set of favourable environmental conditions, it deprives organisms the opportunity to adopt to the ever-changing natural environment. As such new life forms can not evolve and the gene pool gets constant. The developing countries have the maximum biodiversity but they do not have enough funds and in many cases the expertise needed to manage protected areas well.

Chapter Summary

Biodiversity refers to rich and diverse array of living organisms on the earth. Worldwide biodiversity is fast disappearing and needs conservation. Conservation of biodiversity is concerned with the protection of genetic, species and ecosystem diversity so that our lives sustain presently and in future also. There are two approaches of conservation of biodiversity. *In situ* (on site) conservation is the conservation of biodiversity where they are, *i.e.* in their natural habitats and *ex situ* (off site) which attempts to protect the species in place away from their natural habitats. *In situ* conservation is an age old practice involving protection and preservation of total ecosystems and their original and natural environment by establishing protected areas like national parks, wildlife sanctuaries, biosphere reserves etc. According to NRD (2004), India has existing 92 national parks and 492 wildlife sanctuaries. Corbett National Park is the first national park, which was established in 1936. Both national parks and wildlife sanctuaries are spread over various states and UTs of the country. Biosphere Reserve is a special category of protected area, which includes distinct zones for degree of human interaction with flora and fauna. There are 14 biosphere reserves in India, of which Nilgiri Biosphere Reserve is the oldest. There are also many sacred forests and some sacred lakes in the country protected by local people. India has established various projects such as Project Tiger, Project Elephant etc. The *ex situ* conservation involves establishing Botanical Gardens and Zoos which act as repository of flora and fauna and perform conservation of threatened species of plants and animals, respectively. Besides, they also undergo research, education, awareness and such other functions. National Bureau of Plant Genetic Resource (NBPGR) and National Bureau of Animal Genetic Resource (NBAGR) are seed banks/gene banks which cryopreserve reproductive stocks of indigenous and wild and other plant and animal species.

Study Questions

1. Discuss the various strategies of *in situ* conservation of biodiversity in India.
2. Describe the various methods of *ex situ* conservation of biodiversity in India
3. Discuss Biosphere Reserves in India.
4. What are Indian National Parks? Explain.
5. Discuss Indian Wildlife Sanctuaries
6. What are Sacred Forests? Explain.
7. What are the functions of NBPGR and NBAGR
8. Discuss Project Tiger, India.

Objective Questions. *Select the correct answers*

1. Zoo falls under the conservation type:

(1) *In situ* (2) *In vivo*
(3) *In vitro* (4) *Ex situ*

2. Corbett National Park, India was established in the year:

(1) 1936 (2) 1947
(3) 1968 (4) 2004

3. The concept of Biosphere Reserves under UNESCO's Man And Biosphere (MAB) programme was launched in the year:

(1) 1965 (2) 1970
(3) 1975 (4) 1980

4. The only floating national park of our country is:

(1) Corbett (2) Keibul Lamjoo
(3) Kaziranga (4) Keoladeo Ghana

Answers

1. (4) *2.* (1) *3.* (3) *4.* (2)

23

CHAPTER

India as a Megadiversity Nation and Hot Spots of Biodiversity

Learning Objectives

Introduction • Bio-wealth of India • Mountains as Repository of Biodiversity • Indian Deserts • Indians Wetlands • Indian Mangroves • Indian Coral Reefs • Indian Lakes; Hot Spots of Biodiversity • Hot spots of India-Eastern Himalayan • Western Ghats; Chilika Lake; Status of Wildlife in India.

Introduction

Mittermeir and Werner (1990), using the criteria of species richness, introduced the concept of *Megadiversity Centres*. Following their concept, 12 countries/regions have been classified as *Megadiversity Nations/Regions*. These include Australia, Brazil, China, Columbia, Ecuador, India, Indonesia, Mexico, Malaysia, Madagascar, Peru and Zaire. Brazil occupies the first whereas India is sixth on a list of 12 megadiversity countries. There is undoubtedly great wealth of biodiversity in India as biodiversity is its important strength. The country is recognised to be uniquely rich in genetic, species and ecosystem diversity. Bio-wealth of India is the bedrock of all bio-industrial development in the unusually large rural sector villages numbering about 5,76,000 with 76 per cent of country's population. The great range of geographical regions coupled with variety of ecosystems and climate in India, allow an enormous biodiversity.

India's natural habitats range from the Palaearctic Trans-Himalayas in the north to the Indo-Malayan region in the north east, the Indo-Ethiopian bio-geographic region in the west and the Oriental region in the peninsular land mass. Other than these, there are the coastal and island ecosystems. India represents almost all biogeographical regions of the world. These ecosystems have rendered India 16 major climax forest types and 10 biodiversity rich zoogeographic zones. With the increase in human population coupled with developmental activities, these habitats have been severely fragmented and at many places degraded, causing threat of extinction to many valuable indigenous wild species of plants and animals. What is frightening, is that 33 per cent of these species are facing the threat of extinction. Once lost, we lose them forever. The rich heritage of biodiversity encompasses a wide spectrum of habitats from tropical rainforests to alpine vegetation and from temperate forests to coastal wetlands. With merely 2.4 per cent of the total geographical land area of the world, India contributes globally about 10.8 per cent of plant and 7.31 per cent of animal diversity, a total 8.21 per cent of global biodiversity. India occupies tenth position in the world in the number of plants and mammalian species, fourth in Asia in plant diversity, eleventh in world in terms of endemic species of higher vertebrates and seventh in the world in terms of number of species contributed to agriculture and animal husbandry.

Bio-Wealth of India

India is one of the 12 megadiversity countries of the world. So far, only 70 per cent of the geographic area of the country has been surveyed for biodiversity assessment. Country has 47,000 plant species and 89,451 species of animals (2005-06) with their percentage distribution given in the fig. 23.1. India has 6.7 per cent total number of animal species in the world (fig. 23.2). Among animal species, 2,527 are of protista, 5,070 of mollusca, 68,389 of arthropoda, 210 of amphibia, 390 of mammalia, 456 of reptilia, 119 of protochordata, 2,546 of pisces, 1,232 of aves and 8,329 of other invertebrates. In terms of number of species of arthropod alone, India contributes more than 50 per cent of its biodiversity. There are many species of animals endemic to India, *i.e.,* 36 per cent of species of mammals are not found anywhere else in the world. Similarly, 176 species of endemic birds and 214 species of reptiles are confined to Indian subcontinet, mainly in India. Amphibians have the higher percentage of the endemism-128 species of frogs, toads, salamandars etc. out of 209, which constitute about 61 per cent, are restricted to India (fig. 23.2).

The actual number of species of plants and animals must be much higher since information regarding other flora and fauna is patchy and many biologically rich areas like North East Hills and Western Ghats have not been fully explored. Moreover, most of the microorganisms, many algae and fungi etc. have remained unexplored and are awaiting to be discovered. But one thing is of great significance from conservation point of view of biodiversity that 5,150 plant species and 1,837 animal species need attention.

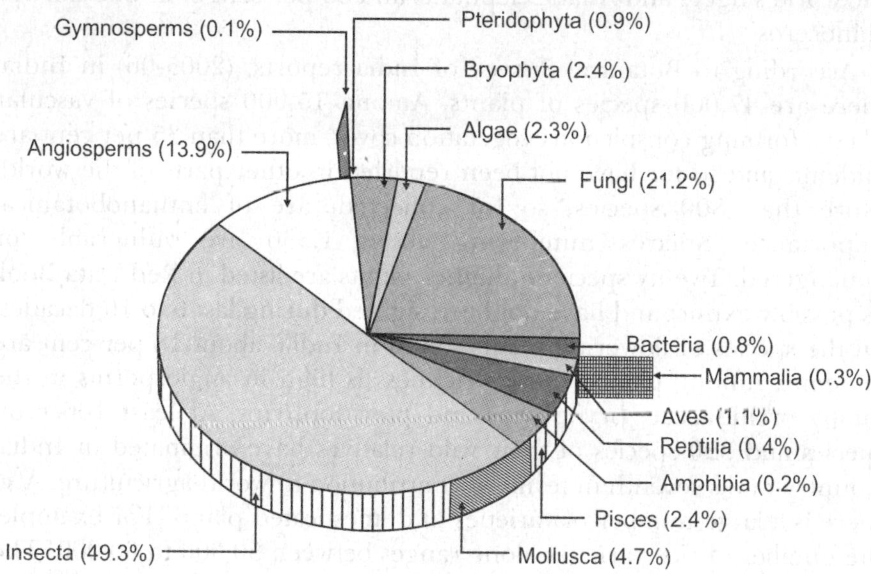

Fig. 23.1. Percentage distribution of biota in India.

Out of 25 hot spots of biodiversity identified in the world, India has two: the Eastern Himalayas and the Western Ghats that contain 5,332 endemic species of higher plants, mammals, reptiles, amphibians and butterflies to name a few. It appears miraculous that a land of a billion human beings and half a billion livestock can also contain 65 per cent of

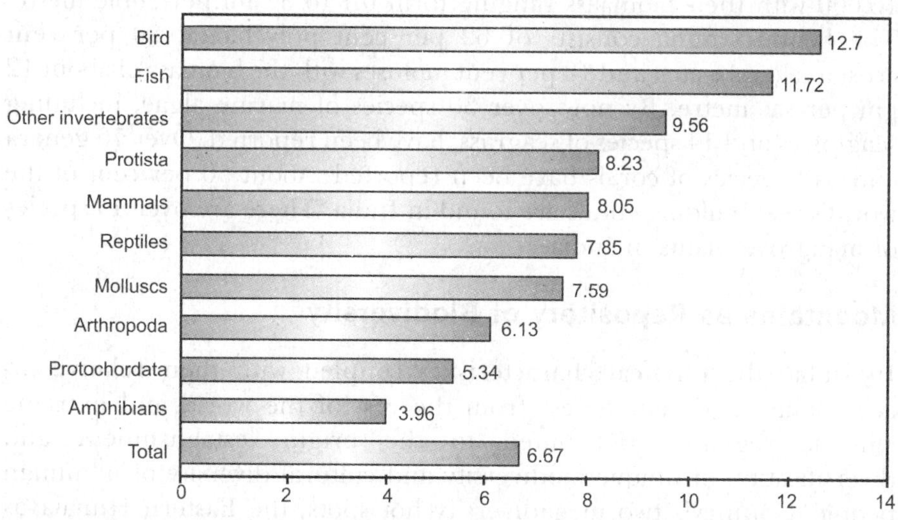

Figure 23.2. Percentage distribution of animal species in India.

the world's tigers and Asiatic elephants and 85 per cent of its one-horned rhinoceros.

According to Botanical Survey of India reports, (2005-06) in India, there are 47,000 species of plants. Among 15,000 species of vascular plants, forming conspicuous vegetation cover, more than 35 per cent are endemic and so for have not been reported in other parts of the world. More than 800 species, so for collected, are of enthanobotanical importance. Species numbering about 1,336 are vulnerable or endangered. Twenty species of higher plants are listed in Red Data Book as possibly extinct and have not been sighted during last 6 to 10 decades. Of the species of flowering plants found in India, about 18 per cent are endemic. Among plants, species richness is high in angiosperms in the family orchidaceae, bryophytes and pteridophytes, At least 166 crop species and 320 species of their wild relatives have originated in India. Country ranks seventh in terms of contribution to world agriculture. Also there is a large number of varieties of domesticated plants. For example, the number of rice varieties alone ranges between 50,000 to 60,000. The crops such as rice, sugarcane, vigna, jute, mango, citrus, banana, several species of millets, spices, medicinal and aromatic plants, first grew in India and then spread throughout the world. India ranks sixth in centers of diversity and origin as far as agro-biodiversity is concerned (Khoshoo, 1990).

India also harbours rich marine biodiversity along 7516.5 km of its coastline with exclusive economic zone of 202 million sq. km, supporting the most productive ecosystems such as mangroves, eustuaries, lagoons and coral reefs. The number of species of zooplanktons recorded is about 16,000 with their biomaass ranging form 0.1 to 37 ml per cubic metre. The benthic fauna consists of 62 per cent polychaeta, 20 per cent crustacea (crabs etc.) and 18 per cent moluscs with the biomass of about 12 gm per sq. metre. By now, over 30 species of marine algae, including seagrasses and 14 species of seagrass, have been reported. Over 76 genera and 342 species of corals have been reported. About 50 per cent of the world's reef building corals are found in India. There are over 45 species of mangrove plants in India.

Mountains as Repository of Biodiversity

In India, the physical characteristics coupled with their relative inaccessibility and remoteness from the rest of the world and extreme climatic regions, contributed to the origin, establishment and diversification of unique biodiversity and cultural diversity of mountain people. Country's two megadiversity hot spots, the Eastern Himalayas and the Western Ghats are the mountainous regions. Also known as water towers, mountains are the treasure houses of endemic species and repositoria of genetic diversity. Mountains also act as critical corridors for many migratory animals. Despite the fact that majority of species of flora and fauna are lost in the valley because these are unable to bear the pressure from the development, the mountains have until now functioned as sanctuaries for a complete rainbow spectrum of biodiversity in our country. The year 2002 was celebrated as International Year of Mountain when India hosted the International Conference of Mountain Children during 15-23 May, 2002 at Uttarakhand.

The North East Himalayas is a treasure house of biodiversity of the country. Several species of oaks, magnolias, larches, laurels, birches, between an altitude of 1500 to 1800 meters and coniferous forest of pine, fir, yew and junipers between an altitude of 1750-3500 meters with their tree trunks covered with mosses and lichens, form the characteristic vegetation. A great variety of ferns surrounding these plants give appearance of a lush green scenery Hoolock gibbon, golden langur, capped langur, macaque, red panda, hog badgers, ferret badgers, crestlers, porcupines etc. are typical animal species of this area. Three kinds of goat antelopes-serrow, goral and takins inhabit various areas of this region. The Western Ghats, as raised land in peninsular India with tropical evergreen rain forests, are treasure house of species diversity and are estimated tc have provided refuge to almost one-third of animal species found in India. The area occupies multiple species forests and provides shelter to elephants, gaur, civet, nilgai, langur and other animals of the hills.

The Indian Deserts

Indian deserts, extending from western part of Delhi to Rann of Kutch and parts of Rajasthan in between, harbour thorny trees, scrubs and plants with small or fleshy leaves-the cacti or succulents and black buck, Indian deserts gerbils, rodents, Asiatic wild ass in Rann of Kutch etc. 'Banni', a saline wasteland that turned sweet and fertile in Rann of Kutch is 25,00 sq. km savanna containing 25 species of grasses including highly productive *Dicanthium*, *Cenchrum* and *Desmostachys* interspread with *Acacia nilotica* (Babool) and *Salvadora* sp. (Pilu). The Forest Department created a large scale plantation of *Prosopis juliflora* and exotic trees, along the boundary to check Rann.

Indian Wetlands

Also known as 'biological supermarket', wetland is the land surface covered or saturated with water for whole or part of the year. Wetlands are shallow ecosystems with often black and rich sediments and abundant nutrients. They have vegetation adopted to thrive in saturated conditions. Wetlands are the most productive ecosystems and essential life supporting systems that provide a wide array of species of plants and animals. They perform several key ecological functions and support rich biodiversity. About 70 per cent of Indian wetlands comprise areas under paddy cultivation. India's most important wetlands are found in two sites – Chilika Lake (Orissa) and Keoladeo Ghana National Park (Rajasthan). Both of them have been declared as *Ramsar sites*.

Traditionally, wetlands are considered wastelands, which bread mosquitoes. They have been continuously drained and developed for agriculture, fish farming or human habitation. It is now only, that their value as providers of ecosystems is recognised. Wetlands prevent flooding by holding excess of water and releasing it slowly. They help in recharging ground water and in purifying water by tapping and holding pollutants in the soil. They are the major breeding, nesting and migration staging areas of many water fowls and shorebirds. They add moisture to the atmosphere. Wetlands can also be used as natural sewage treatment plants that require little technology and maintenance. If they are managed properly, they do not bread mosquitoes because of their predators the fish, insects and birds. Despite their so great importance, they are shrinking rapidly all over the world. Almost half of them have disappeared over the past century and those that remain have been fragmented by 'development'. The worse part of the story is that some of the wetlands have been used as toxic waste dumps and landfills.

India has 2,167 recorded natural wetlands covering an area of 1.5 million hectares or 18.4 per cent of courtney's geographical area. Further, there are 65,245 artificial wetlands spreading over an area of 0.25 million hectares. The reserve tanks, fish ponds, canals, paddy fields etc. could be called man-made wetlands. Although they can not be compared with natural wetlands in terms of biodiversity and other functions, they may still be ecologically as important as natural wetlands. The famous bird

Box 23.1. February 2: World Wetland Day

Every year February 2 is observed as World Wetland Day, which marks the date of the signing of Ramsar Convention on Wetlands in 1971, in the Iranian city of Ramsar. On this day, government agencies, NGOs and citizens undertake action aimed at raising public awareness of the values and benefits of wetlands in general and Ramsar Convention in particular. Each year Ramsar Bureau announces a theme for the year's celebration. The theme for 2003 was "International Year of Fresh water" and for 2004 "From the Mountains to the Sea: Wetlands at Work for Us". The Bureau, on request, sends posters and other materials relevant to the year's theme to help observe the day. The Ramsar Webster www.ramsar.org contains reports from various countries on how the day was observed.

sanctuary of India, Keoladeo Ghana National Park in Rajasthan, is a man-made *marshy wetland*. A few good examples of natural marshes are the 'Terai' located along Himalayan foot hills and adjoining plains, the Rann of Kutch salt marshes of Gujarat and the 'Chushul' and 'Henel' river marshes of Ladaks. Lakes are yet another form of wetlands which harbour submerged, emergent and true plants and plenty of fauna. *Swamps* differ from marshes in possessing woody shrubs and tree which are adopted for life in saline water-logged areas. India has a large areas covered with swampy forests which are known as *mangroves*. Marshes and swamps are much productive than surrounding uplands. *Bogs* are waterlogged areas saturated with ground water or rain water and may be *peatlands* containing *Sphagnum* moss and low shrubs.

India has both *inland wetland* and *coastal wetland* ecosystems. Along coast, there are salt marshes, lagoons, brackish waters, estuaries, mangrove forests and coral reefs. Indian marshes are shallow water areas with abundant vegetation such as reeds, rushes, grasses, sedges etc. Inland wetlands include marshes, swamps flood plains, littoral areas (marginal zones of ponds and lakes) and rivers. The major rivers in India have extensive flood plains. The flat land close to large rivers remains covered with flood waters due to natural floods during monsoon seasons of the year. These areas remain completely innudated even after flood waters recede. The flood plains are more diverse along the lower reaches of rivers. They are the ideal habitats for fish and wildlife. Some of the best forests of the world are located in flood plains. There are also a large number of estuaries, the meeting places between the rivers and the sea.

Taking into consideration, the deterioration of water bodies, a programme on Conservation of Wetlands was initiated by the Ministry of Environment and Forests, Govt. of India in 1987 with the basic objective of assessment of wetland resources, identification of wetlands of national importance, promotion of research and development activities and formulation and implementation of management action plans for identified wetlands. The number of wetlands identified for inclusion in National Wetland Conservation Programme shot up from 27 in 1987 to 71 in 2004-05. There are 25 Indian wetlands which have been identified of international importance under Ramsar sites during CoP 9 held at Uganda during November 2005.

Indian Mangroves

Mangrove plants are those plants which survive in high salinity, tidal streams, strong wind velocity, high temperature and muddy anaerobic soil-a combination of conditions hostile for other plants. These plants are successfully adapted in colonising saline intertidal zone at the interface between the land and sea along the deltas, shallow lagoons, mud flats, bays and backwaters in tropical and subtropical shelter coast lines. The *pneumatophore* roots of these plants, which come out of the soil, are good breeding grounds and prove as nurseries for many fish species like shrimp and sea trout. The plants are commonly *viviparous, i.e.*, seeds germinating on parent plants as such the forest are dense. The branches of trees are the nesting sites for birds like pelicans, snowballs and erget. The plants face *physiological dryness, i.e.*, water is there, but due to its high salinity, plants are unable to absorb it by the roots. Mangroves not only protect coastal communities from the fury of cyclones and coastal storms, but also promote sustainable fishery and prevent the land from sea erosion. In addition, they provide fuel wood and medicines. They also serve the home of a wide range of flora and fauna.

In India, mangroves occur along both the east and the west coasts and represent 7 per cent of the world's mangroves. Three types of mangroves are commonly found: black mangrove with *Avicennia germinans*, the white mangroves with *Laguncularia reacemosa* found at slightly higher tidal elevations and the red mangroves with *Rhizophora mangle* occurring in intertidal zone. The Vedranyam salt swamp in Tamil Nadu and Riparian swamp of Dudhwa National Park in Uttar Pradesh are some other swamps.

In India, by 2005, 38 mangrove areas have been identified, of which Sundarbans in West Bengal are most extensive and prominent (table 23.1). Mangroves are fast disappearing due to coastal development, logging, swamp aquaculture and many ocean disasters. Between 1985 and 2000, India has lost about half of its mangrooves.

TABLE 23.1. State-wise List of Mangrove Areas in India Identified by the Ministry of Environment and Forests (2005-2006)

State/UT	*Mangrove Area*
West Bengal	1. Sunderbans;
Orissa	2. Bhaitarkanika, 3. Mahanadi, 4. Subernrekha, 5. Devi, 6. Dhamra, 7. MGRC, 8. Chilika;
Andhra Pradesh	9. Coringa, 10. East Godavari, 11. Krishna;
Tamil Nadu	12. Pichavram, 13. Muthupet, 14. Ramnad, 15. Pulicat, 16. Kazhuveli;
Andaman & Nicobar	17. North Andamans, 18. Nicobar;
Kerala	19. Vembanad, 20. Kannur;
Karnataka	21. Coondapur, 22. Dakshin Kannada/Honnavar, 23. Mangalore Forests Divisions, 24. Karwar,
Goa	25. Goa;
Maharashtra	26. Achra –Ratanagivri, 27. Devgarh – Vijay Dur, 28. Veldur 29. Kundalika Ravdana, 30. Mumbra- Diva, 31. Vikroli; 32. Shreevardhan, 33.Vaitarna, 34. Vasai Manori, 35. Malvan;
Gujarat	36. Gulf of Kutchh, 37. Gulf of Khambat; and 38. Dumas –Ubhrat.

Indian Coral Reefs

Coral reefs, natural wonders of the ocean, are tropical or sub-tropical shallow water marine ecosystems exhibiting high productivity and biological diversity. They are formed by afterdeath deposition of calcareous exoskeleton of marine coelentrates called coral polyps over thousands of years. On many beaches these beautiful and colourful creations of aesthetic value are on sale. They are among world's oldest, most diverse and most productive ecosystems. The colour of coral comes from alga *Zooxanthellae*. This tiny single-celled alga lives inside the tissue of polyps which is turn get food and oxygen produced as a result photosynthesis from *Zooxanthellae*. The major reef formations in India are located in Gulf of Mannar, Palk Bay, Gulf of Kutch, Lakshadweep and Andaman and Nicobar Islands. In 1987, the National Committee on Mangroves and Coral reefs launched a programme of intensive conservation and management of corals in four areas, *viz.*, Andaman and Nicobar Islands, Lakshadweep Islands, Gulf of Mannar and Gulf of Kutch.

Coral reefs, complex ecosystem, perform many ecological services. During the formation of shells, polyps absorb some carbon dioxide as a part of carbon cycle. These reefs help protect the coastal zone from the impact of sea waves and storms. They act as nurseries for many marine organisms. In terms of their biodiversity and intricacy of relationship, they can be called the rainforests of ocean. They are also valued for providing fish, shellfish, building materials, medicines and employment to people. They bring in valuable foreign exchange through tourism. Coral reefs also give us pleasure of enjoying the world under the sea.

On account of their slow growth coral reefs are very vulnerable to damage and get disrupted easily. They are very sensitive to variations in temperature and salinity. Coral reefs are of course in trouble. Coral bleaching, *i.e.*, expelling of zooxanthellae under stresses, such as increase in temperature of sea water, pollutions, sediment run-off from the land, diseases etc., is the biggest threat resulting them to lose their colour and food and ultimately die. Of the world's reefs, 80 per cent in more than 100 countries are suffering from these stresses.

Indian Lakes

India has several large natural lakes. Wular Lake in Jammu and Kashmir, is the largest natural freshwater lake. The Sambhar is a salt lake in the Great Indian, Desert of Rajasthan in which inflow of salt comes from intermittent streams experiencing flash floods, during occasional intense rain, leaving layer of salt when lake water evaporates.

The Chilika Lake, in Coastal Orissa with an area of 1000 sq. km, is the largest brackish water lagoon in Asia. Rich in fauna, it is home to variety of residential and migratory birds, many fish species and irrawady dolphins. Chilika is one of the largest wintering grounds from October to December for migratory water fowls from all over the world including from Siberia and Australia. It is recognized as a biodiversity hot spots and was designated as a Ramsar site in 1981.

Hot Spots of Biodiversity

The distribution of biodiversity is not uniform across all geographical regions of the earth. An area usually exhibiting species richness coupled with richness of endemic species with its flora and fauna under constant threat of overexploitation, is called biodiversity hot spot. Certain areas of the world, where large number of species are found, are called megadiversity centers. The concept of megadiversity centre was introduced by Mittermeir and Werner (1990), using the criteria of species richness.

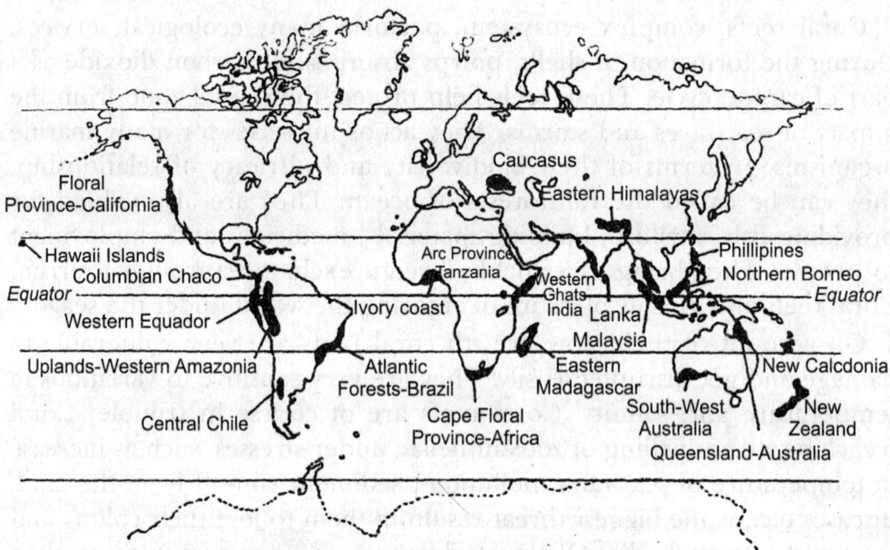

Fig. 23.3. Hot spots of biodiversity.

Norman Myers (1988) introduced the concept of hot spots of biodiversity to designate priority areas for *in situ* conservation and identified only twelve such areas. Later, Myers *et al.* (2000) have brought out an uptodate list of 25 terrestrial hot spots (table 23.2) also shown in the map (fig. 23.3). The list of the hot spots and their vital sings are given in the (table 23.2).

TABLE 23.2 The Vital of Hot Spots of the World

No. Hotspots	*Original extent (sq. km)*	*Vegetation remaining (sq. km)*	*Protected Area (sq. km)*	*Plant Species*	*Endemic Plant Species*	*Terrestrial Vertebrate Species*	*Endemic Vertebrate Species*	*Threatened Species*	*Critically Endangered Species*	*Extinct Species*
1. Caribbean	263500	29840	41000	12000	7000	1581	779	99	32	51
2. California Floristic Province	324000	80000	31443	4426	2125	584	71	12	2	0
3. Mesoameria	1155000	231000	138437	2400	5000	2558	1159	62	12	4
4. Tropical Andrew	1258000	314500	79687	45000	20000	3389	1567	130	20	2
5. Choco-Darien–Western Ecuador	260600	63000	16471	9000	2250	1625	418	32	4	0
6. Atlantic Forest	1477500	121600	33000	20000	8000	1668	563	116	28	1
7. Brazilian Cerrado	1783200	356630	92729	10000	4400	1268	117	22	5	0
8. Central Chile	300000	90000	9167	3429	1605	335	61	8	0	0
9. Caucasus	500000	50000	14050	6300	1600	632	59	10	3	0
10. Mediterranean Basin	23,62,00	1,10,000	42,123	25000	13,000	770	235	42	10	4
11. Madgascar & Indian Ocean Islands	594150	59038	11548	12000	9740	987	771	123	23	46
12. Eastern Arc Mountains & Coast Forests	30000	2000	5800	4000	1500	1019	121	43	7	1
13. Guinean Forest of West Africa	1265000	126500	20324	9000	2250	1320	270	70	7	0
14. Cape Floristic Region	74000	18000	14060	8200	5682	562	53	15	2	2
15. Succulent Karoo in Africa	116000	30000	2352	4849	1940	472	45	12	1	0

16. Mountain of Southwest China	800000	64000	16562	12000	3500	1141	178	34	2	0
17. Indo-Burma	2060000	100000	100000	13500	7000	2185	528	106	20	1
18. Western Ghats & Sri Lanka	182500	12450	12450	4780	2180	1073	355	31	4	0
19. Philippines	300800	21000	25995	7620	5832	1114	555	103	23	0
20. Sundaland	1600000	125000	90000	25000	15000	1800	701	82	13	0
21. Wallacea	347000	52020	20415	10000	1500	1142	529	82	6	0
22. Southwest Australia	309850	33336	33336	5469	4331	456	100	25	1	5
23. New Zealand	270500	59400	52068	2300	1865	217	136	61	5	25
24. New Caledonia	18600	5200	526.7	3332	2551	190	84	9	3	1
25. Polynesia & Micronesia	46000	10024	4913	6557	3334	342	223	88	24	38

The hot spots are the richest and most threatened reservoirs of plant and animal life on the earth. The key criteria for determining hot spots are the number of endemic species they possess and the degree of threat, which is measured in terms of habitat loss. These hot spots together cover only 1.4 per cent of earth's land area yet harbour about 60 per cent of world's biodiversity. Tropical forests appear in 15 hot spots, Mediterranian-type zones in 5 whereas 9 hot spots are mainly or completely made up of islands. As many as 16 hot spots lie in tropical areas. About 20 per cent of human population dwells in the hot spot regions. Hot spots together harbour world's more than 50 per cent of all known terrestrial and endemic species numbering more than 50,000 plant species. The threat to these areas is usually due to natural disasters such as Tsunami. About one-third of terrestrial plant and about one-fourth of vertebrate animal species are found in these hot spots. More than a billion people (1/6 of world's population), living near these areas in poverty and with high rate of human population growth than other areas of the world, pose a serious threat to these hot spots. Any measure of protecting these hot spots needs to be planned, in view the human settlements and tribal pressures.

It is believed that hotspots might have arisen as a result of disturbances during the ice age of the Pleistocene epoch. Worldwide climate fluctuated and cooler conditions occurred in the tropics, causing the rainforests to contract in size. It is through that they got restricted to areas of milder climate mainly in the sheltered lowlands. The species that now form rainforests were, therefore, able to survive in such areas even when they got lost elsewhere. This geographical isolation also led to the evolution of new species in these areas. Thus today we not only find that these refuge areas have high species diversity but show high levels of endemism. These areas contain outstanding evolutionary processes of speciation and extinction and are thus of great scientific interest. The reduction of forests was greater in Africa due to geographical formations than the other parts of the earth, as a result more species were lost and thus this realm today has lower number of species.

Hot spots, usually remarkable and most threatened areas with rich biodiversity, are highly targeted for conservation strategy world over. Many of them have reduced to 10 per cent of their original vegetation and fauna they contained. They are the main areas of focus for biodiversity. The hot spots are quite helpful in solving the extinction crisis in a more manageable manner by conserving the threatened ecosystems and species and preventing loss of biodiversity.

Hot Spots of India

Among 25 hot spots of the world, two regions from India, namely Western Ghats and Eastern Himalayas, are included under hot spots. These extend into neighbouring countries also. Chilika was recognised as a biodiversity hot spot and Ramsar site in 1982. This refers to Convention on Wetlands, an intergovernmental treaty adopted in 1971 in the Iranian city Ramsar. There are more than 1200 Ramsar sites in the world. Both Western Ghats and Eastern Himalayas show a high degree of endemism and are rich in flowering plants, also in reptiles, amphibians, shallow-tailed butterflies and some mammals

(1) Eastern Himalayas. Eastern Himalayan hot spot comes under Indo-Burma hot spot which covers about 2 million sq km tropic Asia, east of Indian subcontinent. Apart from many neighboring countries, Indo-Burma hot spot also covers Andaman and Nicobar Island in Indian ocean. Eastern Himalayas displays an ultra-varied topography that fosters species richness, species diversity and endemism. There are numerous deep and semi-isolated valleys found in this region which are exceptionally rich in endemic species of the plants. Merely in an area of 7,298 sq km of Sikkim, about 2,550 (60 per cent) species are endemic. Being an active center of evolution, it has a rich diversity of flowering plants and is popularly known as *cradle of flowering plants*. Numerous primitive angiosperm families such as Magnoliaceae, Winteraceae, Trochodendraceae, Tetracentraceae etc. and primitive genera of flowering plants such as *Magnolia, Almus, Exbucklandia, Tetracentron, Butea* etc. are found in this region. The forest cover in this region has dwindled to about 1/3 of its original cover. Despite this loss, the region is home to some peculiar plant species. Certain plant species like *Sapria himalayana*, a parasitic angiosperm, was sighted only twice in the last 70 years in this region. During 2005, Botanical Survey of India collected some plant species after 30 years or more. These include:

1. *Agapetes smithiana* Sleum – a rare endangered and a threatened species of Ericaceae (Angiosperm) after 30 years from Sikkim.
2. *Elaeocarpus acuminatus*- a Red Data Book plant from Dibang, Arunachal Pradesh.

Although all Himalayan forests lie well north of the Tropic of Cancer but some of them are the temperate forests as they occur at an altitude between 1,780 to 3,500 meters. Some of the places exceptionally rich in biodiversity in this region are: Namdapha, Tirap, Sirohi, Dzuko, Tura, Senchal and Tale and Buxa vallies.

Eastern Himalayas, the meeting ground of the Indo–Malayan and Indo–Chinese biogeographical realms as well as the Himalayan and

Penisular Indian element, resulted when peninsular plate struck agâinst the Asian landmass, after it broke from Gondawana land.

The tropical vegetation of Eastern Himalayas (which includes the states of Assam, Nagaland, Mizoram, Tripura and Meghalaya- as well as a part of Arunachal Pradesh) typically occurs at elevation upto 900 metre. It embraces evergreen and semi-evergreen rain forests, riparian forests, swamps and grasslands. Evergreen rain forests are found in Assam valley, the foothills and the lower parts of the Naga Hills, Meghalaya, Mizoram and Manipur where the rainfall exceeds 2,300 mm per annum. In the Assam Valley, the giant *Dipterocarpus macrocarpus* and *Shorea assamica* occur singly. Monsoon forests are mainly moist sal (*Shorea robusta*) forests which occur widely in this region.

(2) Western Ghats. Western Ghats extend along a 16,000 km long strip of forest parallel to the western coast of Indian peninsula in Maharashtra, Karnataka, Tamil Nadu and Kerala and has 40 per cent of the total endemic species of plants, 62 per cent of amphibians and 50 per cent of lizards. Forests at low elevation (500 meters above mean sea level) are mostly evergreen and comprise about one fifth of the entire forest expanse. The forest found at an altitude of 500-1500 meters are generally semi-evergreen. Out of 5 locations identified by IUCN as threatened, three area: Agasthymalai Hills-Silent Valley, New Amambalam Reserve basin and Periyar National Park, are found in this region. According to Myers (2000), only 6.8 per cent of the original forests are existing today while the rest have been deforested or degraded. Still worse is the decline of the primary forests as their remaining area seems to be no more than 800 sq km. All but isolated pockets of original forests have been opened up by shifting cultivation causing removal of deciduous species and bamboos among other forms of *degenerate vegetation*. Since a great loss of huge proportion of biodiversity has already taken place, it raises an alarming situation. A declination of forest cover has occurred probably between 1972 and 1985 at a rate paralleling that for India (about 2.4 per cent annually).

The Western Ghat encompasses the montane forests in southwestern parts of India and on the neighboring island of Sri Lanka. The Southern Western Ghats, known as Malabar, is the major genetic estate with an enormous biodiversity of ancient lineage. The western slopes of the mountains recieve heavy annual rainfall, whereas the eastern slopes are driers. The predominant vegetation includes deciduous and tropical rain forests, montane forests and grasslands as well as scrub forests in lower, drier areas. The hot spot abodes a diverse and endemic assemblage of biological diversity. Many animal species such as Asian elephant, Indian tigers and the endangered lion-tailed macaque, inhabit in dense forests of this region. Western Ghat forests contain several tree species of great commercial value such as Indian redwood- *Dalborgia latifolia*, Malabar kino -*Pterocarpus marsupium*, teak-*Tectona grandis* and *Terminalia crenulata*, but they have now been cleared from many areas. In the rainforests, there is enormous number of tree species but at least 60 per cent of the trees of the upper canopy are of species which individually contribute not more than 1 per cent of the total number. Clumps of bamboo occur along streams or in poorly drained hollows thoughout the evergreen and semi evergreen forests of South West India, probably in areas once cleared for shifting agriculture.

(3) Chilika Lake. Chilika Lake, a Ramsar site declared in 1981, in India has also been recognised as one of the hot spots of biodiversity. Chilika is the largest brackish water tidal lagoon that sprawls along the East Coast of India in Mahanadi delta. The lake is pear shaped, with a linear maximum length of 64.3 km and an average mean width of 20km spread over Puri, Khurda and Ganjam districts of Orissa. It is a largest wintering ground for migratory waterfowl in the county. Several rare, vulnerable and endangered species listed in the IUCN Red List of Threatened Animals inhabit the lagoon at least for part of their life cycles. Based on its unique rich biodiversity, Ministry of Environment and Forests, Government of India included Chilika in the list of wetlands selected for intensive conservation and management and designated it as a Wetland of International Importance especially Waterfowl Habitat under the Ramsar Convention, 1981.

Spatial and temporal variations in salinity in combination with nutrient rich shallow water of the lagoon makes it highly productive for biological communities with high species diversity. Chilika supports some of the largest congregations of aquatic birds in the country, particularly, during the winter. These migratory birds come as flocks from as far as remote parts of Russia. Normally the lagoon hosts over 160 species of birds during the peak migratory season of which at least 97 species are international migrants. Total number of fish species vary from 149 to 225. Along with a variety of phytoplankton, 22 species of algae and aquatic plants, the lake region also supports over 350 species of non-aquatic plants including 150 species of vascular plants. Zoological Survey of India, in its survey during 1985-87, recorded over 800 rare, threatened and endangered species of animals including Barakudia limbless skink in and around the lagoon. Irrawady dolphin is an elusive species found in various parts of the Chilika lagoon. Dugong, which is listed as vulnerable species in the IUCN Red List, has been recorded at Chilika in the past but now is a rare or almost extinct species. Much of the lagoon bottom is covered by aquatic weeds, dominated by *Potamogeton pectinatus* in combination with species of *Najas*, *Halophila*, *Pistia*, *Azolla* etc.

The Status of Wildlife in India

All is not well with Indian wildlife. The crush of humanity due to rapid increase in population has affected all habitats of our country. It was in 2005 that tiger finally disappeared from Sariska, a small sanctuary in Rajasthan. Also in 2005 notorious tiger part smuggler Sansar Chand was finally caught after two decades of flouting the system. In late 2004, notorious sandalwood smuggler Verappan was gunned down. The country still has more than half of the world's wild tigers, 65 per cent of Asian elephants, 85 per cent of the greater one-horned rhinos and 100 per cent of the Asian lions. In the past few years the Supreme Court has taken over the role of arbitrator and judge of moral and ethical value systems and even the enforcement of the law of the country. A Supreme Court judge, Arijit Pasayat, in 2002 expounded a strong ethical viewpoint when he declared that "by destroying nature man is committing matricide, having in a way killed mother earth".

Chapter Summary

The concept of Megadiversity centres was given by Mittermeir and Werner (1990) based on species richness and species diversity and listed 12 mega diversity centres/countries in the world. Brazil occupies the first and India occupies sixth rank among Megadiversity countries. Biodiversity is the great wealth and strength of our country. The country has a long range of biogeographic area, wide spectrum of habitats and contributes about 8.22 per cent of global biodiversity. India occupies tenth position in the world in number of plants and mammalian animal species, fourth in Asia in plant diversity, eleventh in the world in terms of endemic species of higher vertebrates and seventh in the world in terms of number of species contributing to agriculture and animal husbandry. The country has 47,000 plants species and 89,451 animal species. About 33 per cent of country's species including 5150 plant species and 1837 animal species need attention from the point of view of their conservation as they are on way to extinction. Due to varied geographical areas and climate, India has mountains as repository of biodiversity as well as a great bio-wealth in its wetlands, mangroves, coastal area, lakes etc. Chilika lake is one of the prominent Ramsar sites.

Norman Myers in 1988 introduced the concept of hot spots and in 2000 gave an uptodate list of 25 hot spots in the world. Hot spots are the geographical regions of the world exhibiting species richness and species diversity with flora and fauna under constant threat due to over-exploitation by humans. These need priority *in situ* conservation. India has two hot spots: Eastern Himalayas and Western Ghats, both endowed with rich diverse flora and fauna with large percentage of endemic species. Some of the plant and animal species found in these areas are found no where else in the world. Many species of plants and animals are waiting for their extinction. Chilika lake, a Ramsar site, declared in 1981, is also a hot spot. It is an estuary, with brackish water, a home of many waterfowls and ground for many migratory birds coming from very far even from Siberia. The flora and fauna of Chilika is quite rich and diverse.

Study Questions

1. Write an account of megadiversity regions of the world and the status of India among them.
2. What are hot spots? Discuss Indian hot spots of biodiversity.
3. Discuss bio-wealth of India.
4. Write in brief about Indian Wetlands.
5. Write in brief about Indian Mangroves
6. Write in brief about Mountains as repository of biodiversity.
7. Write in brief about Chilika Lake

Objective questions: *Select correct answers*

1. According to Myers, 2000, the number of hot spots in the world is:
   (1) 10. (2) 12.
   (3) 24 (4) 25

2. Approximate percentage of species endemic in hot spots of India is:
   (1) 15 (2) 17
   (3) 30 (4) 33

3. Megadiversity regions are assigned as geographical regions of world which have:
   (1) Ecosystem Richness (2) Species Richness
   (3) Richness of genera (4) Community Richness

4. India hosted International Conference of Mountain Children during the year:
   (1) 2000 (2) 2001
   (3) 2002 (4) 2003

5. Commonly known as 'Biological Supermarkets are:
   (1) Mangroves (2) Wetlands
   (3) Mountains (4) Cold deserts

6. World Wethland Day is celebrated on:
   (1) January 2 (2) February 2
   (3) March 2 (4) April 2

Answers

1. (4) *2.* (4) *3.* (2) *4.* (3) *5.* (2) *6.* (2)

24 CHAPTER

International and National Efforts for Conserving Biodiversity

Learning Objectives

Introduction • International Efforts for Biodiversity Conservation • Biodiversity Treaty • Environmental Institutions in Biodiversity Conservation • India's Efforts in Biodiversity Conservation.

Introduction

One of the serious environmental crises around us is the rapid loss of biodiversity, which is posing a threat at local, national and global level. There have been several international treaties and pacts relating to it, the latest and the most comprehensive being the Convention on Biological Diversity (CBD) signed at Earth Summit in Rio de Janeiro in 1992 by 155 nation states that came into force in 1993. This legally binding treaty obliges ratifying countries to protect biodiversity, to move towards sustainable use of biological resources , and to ensure that benefits from such use are shared equitably across local, national, regional and global societies. The other four biodiversity related convention's are the Convention on International Trade in Endangered Species of Wild and Flora (CITES), the Convention on Migratory Species of Wild Fauna Animals (CMS), the Ramsar Convention on Wetlands and the World Heritage Convention.

On the eve of the 2005 Earth Summit, the high level plenary meeting of the 60th session of the UN General Assembly in New York City, 14-16 September, the heads of the secretariats of five global biodiversity related conventions have issued a joint statement calling upon the world leaders *to recognise that to make the Millenium Development Goals a reality in a highly populated planet, biological diversity needs to be used sustainably and its benefits more equitably shared*. The importance of the conservation and sustainable use of biodiversity in achieving Millennium Development Goals has already been recognised by world leaders in their support for achieving a significant reduction in the rate of biodiversity loss by 2010-the so called *2010 target*, the foundation for our well being and continued sustainable existence. We must ensure that the biodiversity will be available for us, and for all future generations.

As a signatory to Convention on Biological Diversity (CBD), India is committed to take appropriate legal and administrative steps to follow its provisions. India ratified this convention in 1994 and during this year itself, preparation of making a Biological Diversity Bill was started. The bill finally entered the Parliament in the year 2000. Soonafter, the National Biodiversity Strategy and Action Plan (NBSAP), founded by Global Environmental Facility (GEF)/ United Nations Development Programme (UNDP) was launched in January 2000 by the Ministry of Environment and Forests. NBSAP is moving along with two guiding principles, that of ecological security of each area and the country as a whole and of livelihood security of the people most dependent on biodiversity. Thus NBSAP process is engaged in covering a whole range of issues relevant to biodiversity: wild plants and animals, microorganisms, crop and livestock diversity, scientific issues, cultural aspects, ethical and economic aspects, intellectual property rights and others. India is also party to other Conventions such as Convention on International Trade in Endangered Species Wild Fauna and Flora (CITES), 1975, Convention on Migratory Species of Wild Animals (CMSWA), 1983, Ramsar Conventions on Wetlands (RCW), 1971 and Convention on Protection of World Cultural and Natural Heritages (CPWCNH), 1972. As regards *in situ* conservation of biodiversity, India has more than has 6,500 national parks, sanctuaries, biosphere reserves etc.

International Efforts for Biodiversity Conservation

Biodiversity conservation is gaining ground precisely because the loss of biodiversity is the present days burning problem. The issue is so vast that it can encompass the interest of many countries. In 1980, three organizations, *viz.*, United Nations Environment Programme(UNEP), World Conservation Union (WCU), previously called IUCN, and World Wide Fund (WWF), prepared a World Conservation Strategy to conserve biodiversity, preserve vital ecosystem processes on which life depends for survival, and develop sustainable uses of organisms and ecosystems. In 1991, four organizations published a new version of the document entitled *Caring for the Earth*. Following the publication of World Conservation Strategy and holding conferences like the 1992's Earth

Summit, several international agreements and global initiatives have focussed on biodiversity conservation.

1. Global Environment Facility (GEF). The global Environment Facility programme was established in 1990 under the auspices of World Bank, United Nations Development Programme (UNDP) and United Nations Environmental Programme (UNEP) on a three year pilot basis. The GEF is expected to commit $400 million towards biodiversity conservation.

2. International Biodiversity Strategy Programme (IBSP). A programme was outlined by World Resource Institute (WRI), World Conservation Union (WCU),United Nations Environmental Progrmme (UNEP) and more than 40 governmental and non-governmental organisations, to stop biodiversity loss and mobilise benefits of biodiversity to human needs sustainably and equitably.

3. Agenda 21. Agenda 21 was developed through a series of inter-governmental preparatory meetings with input from a variety of non-governmental processes including the International Biodiversity Strategy Programme. Agenda 21 is a plan of action on a number of issues including biodiversity.

4. Convention on Biological Diversity (CBD). After international negotiations during Earth Summit, 1992, under the auspices of UNEP, more than 100 nations met at Rio de Janeiro, Brazil to establish a legal framework, which came into force in 1993, establishing international financial support for biodiversity conservation, the identification of international conservation priorities and technology transfer for conservation and sustainable use of biodiversity. The Convention was ratified by 37 countries. India became party to CBD and refitted it in 1994. As on October, 2004, 188 countries were parties to the Convention. The United States signed the convention in 1993, but did not ratified it. The Convention has three main goals: conservation of biodiversity, the sustainable use of components of biodiversity and sharing the benefits from the commercial or other utilisation of genetic resource in a fair and equitable way. Seven meetings of the parties (CoP) to the CBD have been held so far. The eighth meeting of CoP-8 was held in Curitiba, Brazil during March 20-31, 2006. In preparation of a CoP-8, during the year, India has participated actively in the meetings organised under the aegis of the CBD. In preparation of CoP-8, India has sent several submissions on various issues in response to requests by CBD secretariat to the Parties. These include:

1. Indicators for forest biodiversity.
2. Case studies on incentive measures

3. Case studies on sustainable use of inland waters.
4. Review of implementation of the Convention.
5. Definitions on incentive measures.
6. Review of implementations of programme on protected areas
7. Ecological criteria for identification of potential marine area for protection.
8. Development of tool kits for protected areas.
9. Operations of conventions
10. Review of draft guidelines on biodiversity.

Biodiversity Treaty

Under the convention, governments resolved to undertake conservation and sustainable use of biodiversity. They are required to develop national biodiversity strategies and action plans to integrate these into broader national plans for environment and development. Other treaty commitments include:

1. Identifying and monitoring the important components of biodiversity that need to be conserved and used sustainably.
2. Establishing protected areas to conserve biodiversity while promoting environmentally sound development around these areas.
3. Rehabilitating and restoring degraded ecosystems and promoting the recovery of threatened species in collaboration with local residents.
4. Respecting, preserving and maintaining traditional knowledge of the sustainable use of biodiversity with the involvement of indigenous people and local communities.
5. Preventing the introduction of alien species that could threaten ecosystems, habitats or indigenous species and controlling and eradicating alien species already introduced.
6. Controlling the risks passed by organisms modified by biotechnology.

Box. 24.1 International Day of Biological Diversity

Every year, May 22 is observed as an International Day of Biological Diversity. It is supported by the Secretariat for Convention on Biological Diversity. The day provides an opportunity to strengthen people's commitment, participation and action for the conservation of the global biodiversity. Each year the secretariat chooses a theme and provides posters, literature and other related promotional materials. The secretariat's theme for 2004 was 'Biodiversity: Food, Water and Health for All'.

7. Promoting public participation, particularly when it comes to assessing environmental impact of development projects that threaten biodiversity.
8. Educating people and raising awareness about the importance of biodiversity and the need to conserve it.
9. Reporting on how each country is meeting its biodiversity goals.

Despite CBD is a step forward, there are some drawbacks and implementation problems.

1. It excludes the existing gene bank collections and thus 4 million seeds, which are also of high commercial value, are outside the control of CBD.
2. It encourages bilateral agreements, despite many biodiversity issues are regional or global. This has led to some poor countries signing agreements with rich countries or giant corporations giving away their rights over indigenous biodiversity.
3. The implementation of many international agreements of CBD has also been slow and poor because the Convention does not provide for severe penalties for violations, nor does it has an enforcement mechanism.

Following the publication of World Conservation Strategy and Earth Summit, 1992, several international agreements and global initiatives have focused on conservation of biodiversity;

(i) Convention on International Trade in Endangered Species of Wild Fauna and Flora (CITES). CITES has been in force since 1973 with currently 152 countries as signatory to it. Despite many difficulties in enforcement, it has helped to reduce trade in many threatened species including elephant, tiger, chimpanzee, crocodile etc. The country bans hunting, capturing and selling of endangered or threatened species or products derived from them. It lists about 900 species that can not be traded as live specimen or their products. About 29,000 other species, that are potentally treatened are also restricted for international trade. The signatories are bound to pass laws in their country according to the CITES guidelines. Many countries have done so. In many countries, however, enforcement of the laws is lax and even when voilator is caught, the penalties are mild. Further a member country under the convention can exempt itself from protecting any of the listed species. Despite all efforts, trade in many countries, especially in those that have not joined CITES, still goes on. Apart form ban on hunting trading, it is desirable to educate consumers not buy or use the illegal products derived from wildlife.

(ii) Convention on Protection of World Cultural and Natural Heritages. In year 1972, UNESCO's World Heritage Mission held Convention on Protection of World Cultural and Natural Heritages to encourage the identification, protection and preservation of places of cultural and natural heritage around the world considered to be of outstanding value to mankind. A number of sites in the World Heritage list are national parks and reserves that conserve biodiversity. Of the 26 Indian sites in the World Heritage list, five fall under the natural heritage category. These include Kaziranga National Park and Manas Wildlife Sanctuary, both in Assam; the Keoladeo Ghana National Park in Rajasthan; the Sundarban National park in West Bengal; and the Nanda Devi National Park in Uttarakhand.

(iii) The Ramsar Convention In order to create awareness about the importance of wetlands, particularly as a breeding ground of water fowls, International Biological Programme (IBP), International Union of Conservation of Nature and Natural Resource (IUCN) and International Waterfowl and Wetland Research Bureau (IWWRB) organized a series of international conferences in the 1960s. All these paved the way for Convention on Wetlands of International Importance at Ramsar, Iran in Feb. 2, 1971. The conventions was brought into force with effect from December, 1975. In 1990, sixty nations accepted the recommendations of Ramsar Convention. This treaty is aimed at international cooperation for the conservation of wetlands. It also envisages formulation of effective plans and their implementation to protect the dwindling wetlands by the contracting parties. A list of wetlands of international importance (Ramsar sites) has been prepared. Under these conventions till now a total 1200 wetlands sites in 133 countries of international importance, covering an area of 103 million hectares have been designated.

India became a signatory to this convention in 1981. As a signatory, the country has to meet obligations such as: (i) designate wetlands of international importance in the list of so called Ramsar Sites (ii) maintain ecological characters of the listed wetland sites, (iii) organise planning in such a way as to achieve sustainable use of all the wetlands in their territory, and (iv) designate wetlands as natural reserves. Indian wetlands designated as Ramsar Sites are Chilika Lake in Lake Orissa, Keoladeo National park in Rajasthan, Wular Lake in Jammu and Kashmir, Sambhar Lake in Rajasthan and Lotak Lake in Manipur.

Among other treaties that promote biodiversity conservation are Convention on International Trade in Endangered Species of Wild Fauna and Flora (CITES) and Convention on Conservation of Migratory Species of Wild Animals (CCMSWA). United Nation Environment Programme (UNEP) has initiated many regional agreements to protect large marine areas shared by countries. Under the UNESCO's Man And the Biosphere programme, many of the designated reserves are the coastal or marine habitats. In addition World Conservation Union (IUCN) supports 1300 Marine Protected, areas covering about 0.2 per cent of the ocean area.

Promising moves are underway to slow down loss of biodiversity and mobilise its benefits for welfare of mankind. The need to conserve our earth's biological wealth gives nations even more impetus to solve these knotty problems.

Role of Environmental Institutions in Biodiversity Conservation

Many environment institutions are playing an active role in biodiversity conservation. A brief description of these institutions and their efforts in conservation of biodiversity is given below:

1. *United Nations Environmental Programme* (UNEP)

Established by UN is 1972, UNEP has been in the forefront of efforts to protect the world's biological diversity by forging the Convention on Biological Diversity (CBD). By administering the Convention on International Trade in Endangered Species of Wild Fauna and Flora (CITES), UNEP helps to protect over 30,000 of world's endangered species. As a result, the elephant has been brought back from brink of extinction. UNEP's key role led to an increase in number of rhinoceroses in Kenya, Namibia, South Africa, peninsular Malaysia and India. UNEP promoted the preparation of the Global Biodiversity Assessment (1995), a major endeavour mobilising the global scientific community to analyse the state of knowledge and understanding of biodiversity and the nature of our interactions with it.

2. *Worldwide Fund For Nature* (WWF)

Known worldwide by its panda logo, Worldwide Fund for Nature (WWF) is the world's largest privately financed international conservation organisation to protect endangered species and their habitats. WWF works in more than 100 countries around the world to conserve the diversity of life on the earth. It has nearly 1.2 million members in US and another 4 million in the rest of the world. It directs its conservation efforts towards saving endangered species and protecting endangered habitats. From working to save giant panda and bringing back the Asian rhino to establishing and helping to manage parks and reserves world wide, WWF has been a conservation leader for more than 40 years.

3. *International Union for Conservation of Nature and Natural Resources* (IUCN)

IUCN was founded on October 5, 1984 as the International Union for the Protection of Nature (IUPN), following an international conference held in Fontainebleau, France. Later in 1956 the organisation changed its name into International Union for Conservation of Nature and Natural Resources. In 1990 it was shortened to IUCN and now the World Conservation Union (WCU).

The mission of IUCN is to influence, encourage and assist societies throughout the world to conserve the integrity and diversity of nature and to ensure that any use of natural resources is equitable and ecologically sustainable. IUCN builds bridges between governments and NGOs, science and society and local action and global policy. It is truly a world force for environmental governance. A unique Union, IUCN's members are from 140 countries, 77 states, 114 government agencies and 800-plus NGOs. More than 10,000 world recognised scientists and experts from more than 180 countries volunteer their services to IUCN's six global commissions. IUCN is working on more than 500 projects looked after by more than 1,000 staff members in offices around the world. The *Green Web* of partnerships has generated environmental conventions, global standards, scientific knowledge and innovative leadership for last more than 50 years.

IUCN's, objectives in the area of biodiversity and its conservation include:

1. Assessing all new sites nominated by State Parties of Natural World Heritage.
2. Monitoring the state of world's species in the IUCN Red Data List and supporting the Millennium Ecosystem Assessment.
3. Contributing technical assistance to prepare Nation's Biodiversity Strategies and Action Plans.
4. Working with corporate sector on energy, biodiversity and mining and protected areas.
5. Convening the World Parks Congress (2003) and organizing Global Biodiversity Forum held before Convention Conferences.

India's Efforts in Biodiversity Conservation

Biodiversity has direct consumptive value as such attempts have always been in our country to save biodiversity. A scheme on biodiversity conservation was initiated to ensure coordination among various agencies dealing with the issues related to conservation of biodiversity and to review, monitor and evolve adequate policy instruments for the same. Soon after becoming party to CBD and its ratification in 1994,

India became quite active in this area. In January 2000, the Government of India released the National Biodiversity Strategy and Action Plan to consolidate ongoing efforts at conservation and sustainable use of biodiversity and to establish a policy and a programme regime for the purpose. Biodiversity Bill was passed in December 2002 by Indian Parliament. The main intent of this legislation is to protect India's rich biodiversity. It seeks to check biopiracy and to prevent the use of country's biodiversity by foreign individuals and organisations without sharing the benefits with countrymen. One thrust area is the conservation of plants of medicinal value. India has prepared a National Biodiversity Strategy and Action Plan (NBSAP) during 2000-2003. The process was carried out by the Ministry of Environment and Forests, Government of India under the sponsorship of United Nations Development Programme (UNDP) in unique arrangement. Kalpvriksha (KV), a, non-governmental organisation (NGO) based in Delhi and Pune, undertook its technical coordination. 'Kalpvriksha' is an action group working on environmental education, research, campaigns and direct action since 1979. The organization has been working on a number of local, national and international issues. The National Biodiversity Strategy and Action Plan (NBSAP) process involved consultations and planning with thousands of the people across the country, including tribals and other local communities, several governmental organisations, NGOs, academics, scientists, corporate houses, students, the armed forces and other sections of the society. The NBSAP, in order to reverse biodiversity erosion, focuses on three basic goals: conservation of biodiversity, sustainable use of biological resources, and equity in use of biodiversity. Some important measures suggested in NBSAP are:

1. Restoration and regeneration of degraded ecosystems.
2. Recognision of community rights.
3. Development of alternative intellectual rights systems, appropriate for indigenous knowledge.
4. Balancing of local, national and international interests related to biodiversity.
5. Respect for cultural diversity.
6. Preventing deprivation of indigenous people from natural resources.
7. Ecological criteria for identification of potential marine area for protection.
8. Development of tool kits for protected areas.
9. Operations of conventions.
10. Review of draft guidelines on biodiversity.

According to Ministry of Environment and Forests report, during 2005, India has prepared the Third National Report to the Convention on Biological Diversity, following extensive consultations with the members of Consultative Group on Biodiversity issues, other experts, concerned Ministries and other organisations. A national workshop for preparation of the Third National Report was also organised in Gurgaon, Haryana on May 20-21 , 2005 . The report after obtaining necessary approval was submitted to the CBD secretariat. Five meetings of the Consultative Group on Biodiversity issues have been held so far to advise the Government for preparing briefs for international meeting under the CBD, Cartagena Protocol on Biosaftey, Genetic Engineering, Approval Committee, Ramsar Convention, National report to CBD and other related matters.

Group of Like-minded Megadiverse Countries

Seventeen countries, Bolivia, Brazil, China, Colombia, Costa Rica, Democratic Republic of Congo, Ecuador, India, Indonesia, Kenya, Madagascar, Malaysia, Mexico, Peru, Philippines, South Africa and Venezuela, rich in biological diversity and associated traditional knowledge, have formed the Group of Like Minded Megadiverse Countries (LMMC). The LMMC Group holds nearly 70 per cent of global biodiversity and is a duly recognised negotiating block in the UN and other international fora. India was invited to take over presidency of the LMMC on 19 February 2004 during the Ministerial meeting of the Group in Kualalumpur in the margins of CoP-7 meeting of Convention on Biological Diversity for a period of two years. India in its capacity as the Chair of LMMC, had organised an Expert and Ministerial level meeting of the LMMC in New Delhi from 17-21 January 2005 under the patronage of Hon'ble Minister of Environment and Forests and President of LMMC. The report of LMMC held in New Delhi was printed and released by the Minister of State for Environment and Forests, on the occasion of the World Environment Day on 5 June, 2005.

Biological Diversity Act, 2002

The Act is to access the genetic resources and associated knowledge and ensure benefit sharing arrangements apart from developing policies and programmes for long term conservation and protection of biological resources and associated knowledge. In order to facilitate implementation of the Act, National Biodiversity Authority (NBA) has been set up at Chennai. Biological Diversity Rules have been notified, in compliance to the provisions of the Act. Nine states namely State of

Karnataka, Goa, Punjab, Madhya Pradesh, Arunanchal Pradesh, Nagaland, Himachal Pradesh, Kerala and West Bengal have already established State level Biodiversity Boards. Other states are in process of establishing State Boards. Local Level Biodiversity Management Committees are also required to be established to deal with any matter concerning with conservation of biological diversity, its sustainable use and fair and equitable sharing of benefits arising out of the use of biological resources and associated knowledge.

The Cartagena Protocol on Biosafety

The Cartagena Protocol on Biosafety, the first international regulatory framework for safe transfer, handling and use of Living Modified Organisms (LMOs) was negotiated under the aegis of the Convention on Biological Diversity (CBD). The Protocol seeks to protect biological diversity from the potential risks posed by living modified organisms resulting from modern biotechnology. It establishes an Advance Informed Agreement (AIA) procedure for ensuring that countries are provided with the information necessary to make informed decisions before agreeing to the import of such organisms into their territory. It further incorporates produce for import of LMOs with respect to Food Feed and Product (FFP), Risk Assessment and Risk Management Framework and Capacity Building. The Protocol contains reference to a precautionary approach. The Protocol also establishes a Biosafety Cleaning House to facilitate the exchange of information on living modified organisms and to assist the countries in the implementation of the Protocol. The Protocol was adopted on 29 January, 2000 and has been signed by 103 countries (except USA). India signed the Biosafety Protocol on 23 January, 2001 and acceded to Protocol on 17 January, 2003. The Protocol has come into force on 11 September, 2003. So far 131 countries have ratified the Protocol. The first meeting of the conference of the parties serving as the members of parties (CoP-MOP-1) was held at Kualalumpur during 23-27 February, 2004 and it was decided to hold the second and third meeting of conference of parties serving as the meeting of parties (CoP-MOP-2) and (CoP-MOP-3) on an annual basis to expedite the process of addressing resource issues of the Protocol, which is required for its effective implementation. The CoP-MOP-2 was held in Montreal, Canada from 30 May to 3 June, 2005. The meeting adopted important decisions on substantative issues related to Notification, Risk assessment, Handling Transport, Packaging and Identification of LMOs, Compliance, Socio-economic considerations, Public participation and Liability and Redress. The third meeting CoP-MOP-3 was held in March 2006 at Curitiba, Brazil.

The Genetic Engineering Approval Committee (Biosafety Regulatory Framework in India)

The ministry of Environment and Forests, Govt. of India had constituted a Task Force on Recombinant Pharma Sector under the chairmanship of RA Mashelkar, DG, CSIR for streamlining the procedure for recombinant pharma industry. The final report was submitted to the Ministry on 13th September 2005. The recommendations of Task Force were adopted in the inter-ministerial meeting held on 23rd January 2006. The salient features of the recommendations include step-wise protocol for five scenarios: timelines for taking decisions, construction of a standing advisory committee etc. The meetings of Genetic Engineering Approval Committee (GEAC) are being held in every second Wednesday of the month as per guidelines on best practices for environmental regulations adopted by the Ministry. During 'Kharif' 2005, the GEAC approved environment release of 20 Bt Cotton hybrids for commercial cultivation. Bt cotton was approved for the first time for commercial cultivation in the north zone. The GEAC has also approved large scale trials of number of Bt cotton varieties containing cry '1' Ac gene MON531 developed by various companies. The GEAC has also approved Bt cotton hybrids containing new genes for large-scale trials.

GEF-World Bank Capacity Building Project on Biosafety

The Capacity Building Project will enhance the India's national capacity in order to implement the Cartagena Protocol on Biosafety. India already has a place in Biosafety Regulatory Framework in the form of the Rules of Manufacture, Use, Import, Export and Storage of Hazardous Microorganisms/Genetically Engineered Organisms or Cells 1989 notified under the Environment (Protection) Act, 1986. The Project will address the capacity building needs of the country for implementing the national biosafety framework related to the transboundary movement of LMOs in the context of the Cartagena Protocol and coordination of the implementation of the Biosafety Cleaning House (BCH). Specifically, the project will develop national capacities in biosafety required to (i) strengthen the legislative framework and operational mechanism for biosafety management in India; (ii) enhance capacity for risk assessment and monitoring; (iii) establish the Biosafety Database System and Biosafety Clearing House Mechanism (iv) support centres of excellence and a network for research, risk assessment and monitoring; and (v) establish the Project Coordination and Monitoring Unit (PCMU). The development of national capacity in these areas will enhance the national capabilities for implementation of the biosafety issues.

During the year 2005-2006, the progress made in this area includes:

1. The four identified institutions namely National Bureau of Plant Genetic Resources (NBPGR), National Research Center on Plant Biotechnology's (NRCPB), Central Food Technological Research Institute (CFTRI) and G.B. Pant Agricultural University, are being strengthened in terms of equipment and manpower, so as to enable them to enhance their capability of handling of Living Modified Organisms (LMOs)
2. A project specific website has been launched which could be assessed at URL:http:// www.envfor.nic.in/divisions/csurv/biosafety/default.htm. The first issues of Biosafety Newsletter has been published and can be seen in the project website.
3. Training programmes have been conducted to enhance the capacity of the policy makers and technical personals at Central and State Governments/Agencies.
4. A Biosafety Clearing House website (www. Indbch.nic.in) nas been launched for information dissemination in the field of LMOs and also in compliance to the Cartagena Protocol on Biosafety.
5. A training need assessment survey has been carried out to identify the gaps and needs for carrying out for training programmes in biosafety.

India has made its conservation efforts thorough national parks, sanctuaries, biosphere reserves, zoos, deer parks, safari parks, aquaria, NBPGR, New Delhi and NBAGR, Karnal and various projects such as Tiger Project, Elephant Project etc. Snakes and crocodiles are being protected from extinction by the efforts of a dedicated individual, Romulus Whitaker by establishing Madras Snake Park at Chennai in 1972. Many NGOs, local communities and voluntary organisations are now promoting biodiversity conservations and sustainable use of bioresources through indigenous knowledge. In India, a programme is underway to develop a system of community register of local informal innovations related to the genetic resources as well natural resource mangement in general. In 1996, the villagers of Pattuvam in Kerala created history by declaring obsolete ownership over its all genetic materials currently growing within the jurisdiction of the village. Earlier, the villagers prepared a detailed register of all species found in the village. Also the village set up a Forum for Protection of People's Biodiversity, which has been authorised to control all activities related to the biodiversity of the village. All over India, efforts are being made to document indigenous knowledge about local biodiversity. To name a few organisations working in this area are: Foundation for Revitalisation of Local Health Traditions, Banglore (publishes magazine 'Amruth'); the

Green Foundation, Bangalore; the Centre for Madras Knowledge System, Chennai; Beej Bachao Andolan, Tehri Garhwal; and Navadanya, Dehradun. Annadana in Auroville near Pondicherry, is the base of South Asian Network of Soil and Seed Savers established with the help of association Kokopelli of France. The land races and diverse food and medicinal plants are also being conserved successfully by the tribal people and women working individually or with various non-governmental agencies. The women particularly have played a significant role in conservation of agrobiodiversity. Use of GIS (Geographical Information System) and remote sensing have greatly helped in determining the rates, scales and causes of biodiversity loss in India.

Chapter Summary

Rapid loss of biodiversity is one of the most serious global environmental crises. This vast issue was first taken seriously in 1980 when UNEP, WCU (IUCN) and WWF prepared a World Conservation Strategy to conserve biodiversity. Most comprehensive effort in this regard was made by the Convention on Biological Diversity (CBD) signed at Earth Summit, Rio in 1992, posing legal binding on the countries ratifying the convention to protect biodiversity. Several organisations and conventions such as Ramsar Convention, CITES etc. took active interest in promoting sustainable use of biological resources and saving the biodiversity. India, being a party to Convention on Biological Diversity (CBD), released the National Policy and National Biodiversity Strategy and Action Action Plan (NBSAP) in 2000 and passed a Biodiversity Bill in 2002. India is a party to Group of Like Minded Megadiversity Countries (LMMC)-17 countries holding about 70 per cent of global biodiversity. India's National Biodiversity Strategy and Action Plan (NBSAP) sponsored by UNDP was initiated with a unique arrangement with NGO Kalpvriksha. Besides, many NGOs, organisations, institutions and individuals are taking active part in conservation of biodiversity in India.

Study Questions

1. What international efforts have been made to conserve global biodiversity?
2. Discuss India's efforts in conservation of biodiversity.
3. What is National Biodiversity Strategy Action Plan (NBSAP) of India? Discuss the measures suggested by NBSAP.
4. Discuss Agenda 21.
5. Discuss Convention on Biological Diversity (CBD).
6. Discuss Ramsar Convention.
7. Discuss World Wide Fund (WWF)
8. Discuss World Conservation Union (WCU).

Objective Questions. *Select the correct answers*

1. Seventeen countries holding abut 70 per cent of gloval biodiversity have formed an organisation which is named as:

(1) NBSAP (2) IBSP
(3) LMMC (4) CITES

2. India became a party to CBD and ratified it in the year:
   (1) 1990 (2) 1992
   (3) 1994 (4) 1996
3. LUCN is now called as:
   (1) IBCN (2) WCU
   (3) WWF (4) CITES
4. Convention on Internation Trade in Endangered species of Wild Fauna and Flora came in force in the year:
   (1) 1970 (2) 1973
   (3) 1957 (4) 1975
5. International Day of Biological Diversity is celebrated on:
   (1) June 5 (2) February 02
   (3) May 22 (4) August 15
6. Convention on Wetlands of Internation Importance is also called as Ramsar Convention as it took place first in:
   (1) India (2) Iran
   (3) Iraq (4) America
7. The nation not a party of Group of Like Minded Megadiversity Countries is:
   (1) India (2) Canada
   (3) Kenya (4) Mexico
8. India signed cartagena Protocol on Biosafety on:
   (1) January 23, 2001 (2) January 26, 2005
   (3) August 15, 2003 (4) November 10, 2002
9. W W F stands for:
   (1) World Water Forum (2) World Wide Fund for Nature
   (3) World Wildlife Fund (4) World War Fund

Answers

1. (3) 2. (3) 3. (2) 4. (2) 5. (3) 6. (2)
7. (2) 8. (1) 9. (2)

Environmental Pollution

- *Pollution: An Introduction*
- *Air Pollution*
- *Water Pollution*
- *Soil Pollution*
- *Marine Pollution*
- *Noise and Thermal Pollution and Nuclear Hazards*
- *Solid Waste Management*
- *Role of an Individual in Prevention of Pollution*
- *Pollution Case Studies*
- *Disaster Management*

25

CHAPTER

Pollution: An Introduction

LEARNING OBJECTIVES
Definition of Pollution • Sources of Pollution • Cost of Pollution • Kinds of Pollution • Management of Environmental Pollution.

Introduction

EP Odum (1971) defined pollution as *an undesirable change in the physical, chemical and biological characteristics of air, water and soil which affects human life*. However, it is now considered as a change in the physical, chemical or biological features of air, water and soil which directly or indirectly affects the human welfare. It includes harmful effects on human life or on the lives of the desirable species or on industrial processes, living conditions, cultural assets or raw material resources etc.

Pollution may be a result of human activities or natural catastrophies. Examples of natural pollutions are *volcano eruptions, forest fires, flood caused* etc. Incomplete technology lacking close integrated systems is the main cause of man-made pollution.

Today, the problem of pollution has become a major challenge to scientists, environmentalists and humanists as the pollution of various components has gone to such an extent that we are unable to breath fresh air, drink fresh water and eat pure food. If man has to survive, he has to fight and overcome this gigantic problem before it swallows him and his very existence.

SOURCES OF POLLUTION

Pollution may be caused by several sources depending upon the nature of pollutants:

(1) Solid wastes as source of pollution. Solid wastes may be domestic or industrial in nature. Various solid wastes can be categorised as follows.

(i) Industrial wastes, *e.g.* particulate wastes from various industries such as glass fragments, leather pieces, rubber pieces etc.

(ii) Domestic wastes, *e.g.* garbage of kitchen, slaughter houses etc. These may be *combustible* (such as leaves, twigs, papers etc.) or *non-combustible* (such as crockery, plastics, glass etc.) in nature.

(iii) Sewage, *e.g.* human and animal excreta, domestic effluents, detergents etc. (Solid faecal matter is called sludge.)

(iv) Agricultural wastes, *e.g.* plant and animal residues, broken twigs, wood fractions, fruits , pesticides, fertilizers etc.

(2) Liquid wastes as a source of pollution. Industrial effluents and domestic wastes in the form of liquid are the major sources of water and soil pollution. *Industrial pollution* may reach the lethal magnitude. Liquid discharges from chemical factories, refineries, breweries, tanneries etc. contain acids, alkalies, oil and dissolved heavy metals which enter the river water and adversely affect the aquatic life and impair its self-purification system.

Liquid wastes from *domestic sources* may be of inorganic or organic nature. Inorganic liquid wastes include soap water and detergent water from bathing and washing clothes whereas organic wastes include kitchen garbage, faecal water, urine etc.

Run off from *agriculture fields* carry residual fertilizers, pesticides, biocides etc. which enter the water streams and harm the aquatic life.

(3) Gaseous wastes as a source of pollution. The common gaseous pollutants like carbon monoxide, sulphur oxides, nitrogen oxides, hydrogen sulphide etc. are frequently released from various industries and automobile exhausts. Pollution resulting from gaseous wastes is one of the most dangerous and lethal types. Life on earth is dependant upon the air we breathe and if this source of life is contaminated and polluted by lethal gaseous discharges, the very existence of all living organisms is threatened.

(4) Energy wastes as a source of pollution. A significant addition to the sources of pollution are the invisible pollutants, *i.e.*, pollutants without mass or weight and invisible to the eye. Examples of such pollutants are heat and radio-active emissions. Radioactive emissions are most hazardous and their commulative effects are far reaching and damaging to the genetic makeup of living organisms.

(5) Noise as a source of pollution. Unwanted sound or noise above the particular level in the atmosphere is an important pollutant. Indiscriminate and continuous use of radios, traffic horns, public broadcasting systems etc. are common sources of noise pollution.

Aeroplanes and supersonic jets also produce noise of high intensity which may sometimes rupture the ear drum and cause irreparable damage to the brain.

COST OF POLLUTION

Magnitude of pollution effect may be small, temporary and invisible to hazardous, catastrophic and global destructive. In general, the cost of pollution is evaluated on the basis of following lines:

(1) Medical care of health to meet the challenges of pollution-caused diseases such as tuberculosis, typhoid, lung cancer, diarrhoea, jaundice etc.
(2) Loss of resources by unnecessary and wasteful exploitation.
(3) Corrosion of metals and metallic structures.
(4) Involvement of monetary funds and man power for disposal and control of pollutants.
(5) Damage to orchard and agriculture resources.
(6) Damage to historical monuments and textiles.
(7) Damage to biotic forms or biodiversity.

Ultimately man is the target of various environmental pollutants (fig. 25.1).

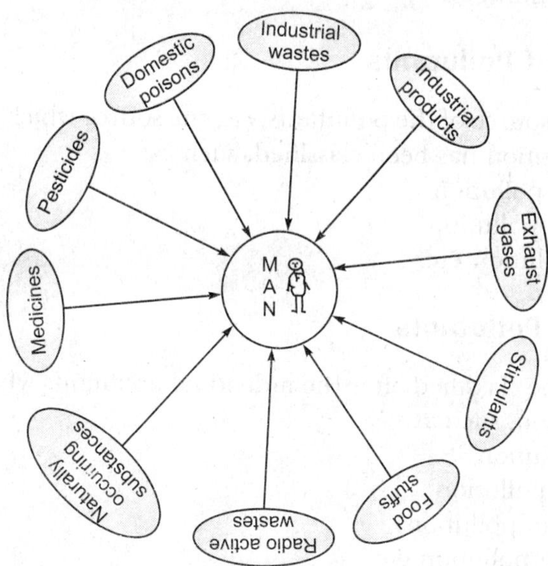

Fig. 25.1. Various environmental pollutants and victim man.

KINDS OF POLLUTION

Pollution has been classified in various ways based on different factors, such as on the basis of :–

(1) Nature of Pollutants

Two categories are recognised on the basis of degradability of pollutants:

(i) **Decomposable** or **Biodegradable**. These are naturally occurring organic compounds which are degraded by biological or microbial agents. Such pollutants include domestic sewage, dead remains of plants and animals, food residues etc.

(ii) **Non-decomposable** or **Non-biodegradable**. These are not decomposed or destroyed over a long period of time. These may be simple wastes such as iron, glass, aluminum, plastic, polythene, poisonous wastes, heavy metals etc.

(2) Components of Environment

Pollution has been classified on the basis of a particular component of the environment being polluted, such as:

(i) Air pollution
(ii) Water pollution
(iii) Soil pollution etc.

(3) Sources of Pollutants

On the basis of source of the pollutants, *i.e.*, the source which produces the pollutants, pollution has been classified, such as:

(i) Fertilizer pollution
(ii) Pesticidal pollution
(iii) Noise pollution etc.

(4) Types of Pollutants

Pollution may be classified after the individual pollutants which is a major cause of pollution, such as:

(i) Lead pollution
(ii) Mercury pollution
(iii) Chromium pollution
(iv) Cadmium pollution etc.

MANAGEMENT OF ENVIRONMENTAL POLLUTION

At present what is required is not so much the ability of human beings to conquer the nature, rather there is a need of a balanced and harmonious collaboration with its forces. The ultimate goal of environmentalists should be to manage the environment in such a way that it can contribute to man's happiness, health and enjoyment and to improve the quality of human life.

The various activities of civilisation invariably interact with the environment predicting various forms of environmental problems. These problems may be local, national or international in nature and have to be managed accordingly. Following are a few possible measures to manage environmental pollution.

(1) Environmental education must be made compulsory in all education systems, from primary to university levels.
(2) Regular information about environmental problems should be imparted through mass communication media.
(3) Specific standards for each kind of pollution should be set and breaking of these standards should be a punishable offence.
(4) Protection of environment must be enforced by mass awareness and by imposing constitution.
(5) Industries, factories, air ports and such other establishments should be far away from the city to minimise pollution.
(6) Private and public enterprises should be given legal and financial incentives to adopt antipollution measures.
(7) Adulteration of food products, drugs and general commodities should be made a punishable offence.
(8) Nuclear testings resulting in the production of radio-active wastes should be restricted to the minimum by international agreements.
(9) Wildlife board and environmental cell should be established in all important cities to popularise antipollution research.
(10) Legal advises, scientific assistance and alternative procedures to reduce the use and release of pollutants should be made known.

The entire world is now involved in combating environmental problems in various ways. National and International efforts, Important Environmental Dates, Indian Centres for Environmental Studies, Centres of Excellence, Important National Awards etc. have been discussed in the first unit of the book.

Chapter Summary

Pollution is an undesirable change in the physical, chemical and biological characteristics of air, water and soil which affects human life. It may be a result of human activities or of natural catastrophies. Examples of natural pollutions are volcano erruptions, forest fires, flood caused etc. Man made pollutions are of great concern today. Sources of such pollution may be solid wastes (industrial, domestic, sewage, agricultural), liquid wastes, gaseous wastes, energy wastes, noise, radioactivity etc. The cost of pollution is the targeted man.

Pollution can be of several types and it can be classified on the basis of *(i)* nature of pollutants (biodegradable and non-biodegradable), *(ii)* component of environment (air, water and soil), *(iii)* source of pollutants (fertilizer, pesticides, noise etc.) or *(iv)* type of pollutants (lead, mercury, chromium, cadmium etc.)

Management of environmental pollution is now essential to keep the balance and harmonious collaboration with nature to save human beings. For this, environmental education, enforcement of legal acts, mass communication, minimization of discharge of pollutants, treatments etc. are must to be obeyed. Pollution is a global problem and it requires everyone's attention.

Study Questions

1. What is pollution? Describe the different types of pollutants and pollution.
2. Define pollution and briefly describe various types of pollutants, pollution and their management.
3. Write a brief account on the various types of pollutants.
4. Write a note on cost of pollution
5. Describe briefly the various kinds of pollution.
6. Define pollution and suggest the possible ways of management of environmental pollution.
7. What is pollution? Define.
8. Name some solid waste pollutants.
9. Name three sources of pollutants.
10. What do you understand by nature of pollutants? Define.

Objective Questions. *Select correct answers*

1. The major pollution causing agents are:

(1) Men (2) Animals

(3) Green house gases (4) UV-rays.

2. An example of natural pollution is:

(1) Forest fire (2) Volcanic eruptions

(3) Natural organic decays (4) All of these

3. An example of non-biodegradable pollutant is:

(1) Sewage (2) Plastic

(3) Litter (4) Vegetable wastes

4. The ultimate sufferer of pollution is:

(1) Plant (2) Terrestrial animal

(3) Aquatic animal (4) Man

Answers

1. (1) *2.* (4) *3.* (2) *4.* (4)

26

CHAPTER

Air Pollution
(Causes, Effects and Control Measures)

LEARNING OBJECTIVES
Introduction • Sources of Air Pollution • Major Air Pollutants • Effects of Air Pollution • Control of Air Pollution.

Introduction

Nearly 2/3 biological species are terrestrial and air is life for them as it makes breathing possible. Air is a mixture of gases that form the atmosphere. A clean air has the following composition:

TABLE 26.1. Natural Components of Air

Gas	*Volume (%)*	*Gas*	*Volume (%)*
Oxygen	20.92	Nitrogen	78.1
Argon	0.9325	Carbon dioxide	0.03
Hydrogen	0.01	Neon	0.0018
Helium	0.0005	Crypton	0.0001

(Source Anonymous, 1987)

Air pollution originated on earth with the beginning of civilisation. The day man learned to use fire, the first pollutants were released into the atmosphere. Now air pollution is an international problem. Although contribution to air pollution by man accounts for only 12-15 per cent of the total air pollutants being released into the atmosphere, the main problem is that the sources of pollution are concentrated in or close to thickly populated cities where they cause harmful to lethal effects on mankind. These deadly poisonous pollutants affect not only to the man but also to the flora and fauna of the locality.

Air pollution may be defined as *that quantity of pollutants which is sufficient to cause injury to human beings and their living creatures*.

Air pollution may be categorised into two types on the basis of origin of pollutants: (I) *First type* of pollutants are released into the atmosphere from a specific source, and (ii) the *second type* of pollutants are resulted from chemical changes that take place in the atmosphere. According to the formation of pollutants, two categories are recognised-

(i) **Primary pollutants**. These enter in the air directly.

(ii) **Secondary pollutants**. These are created in the air from other pollutants under the influence of electromagnetic radiation from sun.

SOURCES OF AIR POLLUTION

Major sources of air pollution are as follows:

(1) Stationary combustion sources. Such air pollutants are produced mainly by burning of fuels and these may be gaseous or particulate in nature. The largely used fuel is coal and petroleum. **Coal** is largely a carbon mixed with some incombustible minerals, sulphur and nitrogen. **Petroleum** consists of mainly hydrocarbons, sulphur and nitrogen. Both coal and petroleum are fossil fuels which on burning produce a mixture of oxides in addition to carbon dioxide and water. Coal also produces mineral ash, some of which are discharged as *fly ash*. Oxidation of sulphur and nitrogen in the fossil fuel produces mainly sulphur dioxide (SO_2), sulphur trioxide (SO_3), nitrites (NO_2), Nitrates (NO_3) etc. Carbon monoxide (CO) is also produced by these fuels. The incomplete combustion of hydrocarbons produces many gaseous and particulate (soots) pollutants. The particulate matters released in the atmosphere include a number of trace elements contained in fly ash such as Sb, As , Be, Cd, Ge, Pb, Hg, Ni, Se, Va, Yttrium etc.

(2) Mobile combustion sources. Such sources include automobiles, locomotives, air crafts etc. Largest source of air pollution in cities are automobiles which cause about 80 per cent air pollution and 75 per cent noise pollution. The major pollutants released through these sources are –

Carbon monoxide	(77.2%, the highest emitted gas)
Nitrogen oxide	(7.7%)
Hydrocarbons	(13.3%)

Combustion of petroleum containing lead products such as tetraethyl lead [$Pb(C_2H_3)$] and tetramethyl lead [$Pb(CH_3)_4$] emits various particulate lead compounds. Several products produced due to incomplete combustion of fossil fuels (coal and petroleum) undergo photochemical reaction with oxides of nitrogen to produce photochemical smog which contains ozone (O_3), peroxyacetyl nitrate (PAN), aldehydes etc. These are often called the **secondary pollutants.**

(3) Industrial processing and other sources. All industries that process organic chemicals using high temperature release various gases causing pollution. The chemical industries are the main source of pollutants such as carbon monoxide and sulphur oxides. These also release other gases and particulate pollutants. Compounds containing chlorine and fluorine, especially CFCs (Chlorofluorocarbons) are widely used as propellants for aerosol cans and as refrigerants. These escape in air and produce pollution.

A kind of particles, ***ice nuclei*** are emitted from steel mills and automobiles. These modify clouds and precipitation.

Fig. 26.1. Various sources of air pollution.

MAJOR AIR POLLUTANTS

Air pollutants may broadly be divided into two types-

(I) Gaseous pollutants such as oxides of sulphur (SO_2, SO_3), oxides of carbon (CO, CO_2), oxides of nitrogen (NO, NO_2), H_2S, NH_3, fluorine, ozone, hydrocarbons, aldelydes etc.

(II) Particulate pollutants such as dust, fly ash, smoke, soot, aerosol droplets, mist, fog, substances in air etc.

(I) Gaseous Pollutants

These are grouped into: *(a)* inorganic gases and *(b)* organic gases.

(A) *Inorganic gases*

The various inorganic gases are –

(1) Oxides of Sulphur

The important oxide of sulphur in the atmosphere is sulphur dioxide.

Sulphur dioxide (SO_2). The composition of fossil fuels used for domestic and industrial purposes has resulted in sulphur dioxide (SO_2) becoming a regular and integral part of the environment. About 80 million tonnes of SO_2 per year is added globally. SO_2 emission in India is also on increase and it was around 14 million tonnes in the year 2000.

SO_2 is a colourless gas with a suffocating and strong pungent odor. Under certain conditions, SO_2 reacts with oxygen (O_2) to form SO_3 (sulphur trioxide), which in turn combines with H_2O (water) to form H_2SO_3 (sulphurous acid) and H_2SO_4 (sulphuric acid). SO_2 would then be called *primary pollutant* and H_2SO_3 and H_2SO_4 as *secondary pollutants*.

(2) Oxides of Carbon

The two common types: CO (carbon monoxide) and CO_2 (carbon dioxide) are found in the atmosphere. Air contains about 0.03 % CO_2.

(a) Carbon monoxide (CO). The major source of the production of CO is the incomplete burning of fossil fuels (coal and petroleum) and wood charcoal. Other common sources are cigarette smoke, motor vehicles, industries, domestic heating appliances, oil refineries etc. According to an estimate, about 6 billion tonnes of CO is produced annually and emitted in the global atmosphere. It is a colorless, odorless and tasteless gas and is an important air pollutant. CO when absorbed into the lungs, combines with haemoglobin in the blood and produces carboxyhaemoglobin which reduces the O_2 carrying capacity of the haemoglobin (blood).

(b) Carbon dioxide (CO_2). It is not a pollutant in itself but human activities are dangerously increasing its amount in the atmosphere. One of the main reasons for its increase in the atmosphere is the widespread use of combustible fuels. At present the concentration of CO_2 in the air is approximately 350 ppm (0.035%) which was about 275 ppm in the mid of 19th century. Presently the concentration of CO_2 in the air is increasing at the rate of about 1.5 ppm annually or 0.7 mg/l/year. A continuous increase of unabsorbed CO_2 in the atmosphere would have a disastrous effect on the flora and fauna, play havoc with the ecological balance and bring about a catastrophic warming effect to the atmosphere resulting in the **green house effect** or **global warming.**

(3) Oxides of Nitrogen (NOx)

The main sources of NOx are thermal power stations, factories, automobiles and air crafts where fossil fuels are used. According to an estimate, each tonne of coal after burning produces 5-10 kg of nitrogen oxides, and as such about 10 million tonnes of nitrogen containing gases like NO and NO_2 are entering into the atmosphere every year. On global level about 60 per cent of nitrogen oxides are produced from natural sources, including biomass burning, fixation by lighting, inflow from the stratosphere, chemical conversion from ammonia in the troposphere and loss from the soil.

NO and NO_2 above requisite level in the atmosphere become pollutants. Both combine with water to form nitrous acid or nitric acid which are highly corrosive and reactive in nature.

$$2NO_2 + H_2O \rightarrow HNO_2 + HNO_3$$

These also give rise to the most dangerous pollutants, *i.e.* photochemical smog and peroxyacetyl nitrate (PAN). PAN (CH_3 -CO-O-O-NO_2) is the most notorious organic oxidant.

(4) Ozone (O_3)

Ozone is a colorless gaseous secondary pollutant. It is produced by chemical reactions between reactive organic gases and oxides of nitrogen, particularly through the interaction of nitrogen oxide, sunlight and hydrocarbons-

$$\text{Nitrogen oxides} + \text{Hydrocarbons} \xrightarrow{\text{sunlight}} \text{PAN} + O_3$$

It is a major component of *photochemical smog*. Ozone is a life savior, if present in stratosphere but a pollutant, if present in troposphere

(5) Other Inorganic Gases

Inorganic gases like hydrogen fluoride (HF), hydrogen sulphide (H_2S), ammonia (NH_3), chlorine (Cl), arsine and phosgene etc. are strong pollutants of the atmosphere. These are emitted from various industries.

(B) *Organic gases*

The important organic gaseous pollutants are-

(a) **Aldehydes.** The thermal decomposition of fats, oil and glycerol releases aldehydes in the form of organic gases. They affect nasal and respiratory tract causing extreme irritation.

(b) **Hydrocarbons.** Various hydrocarbons which pollute the atmosphere are methane, volatile terpenes, ethylene etc. Anaerobic degradation of organic matter produces about one billion metric tonnes of methane annually. Hydrocarbons cause irritation and injury to the mucous membrane.

(II) Particulate Pollutants

These may be in the form of solid particles or liquid droplets including fumes, smoke, fogg, dust, pollen, bacteria, fungi and aerosols. This category includes about 5 per cent of the weight of air pollutants present in the atmosphere. It is estimated that about 8 billion solid particles penetrate into the atmosphere every day.

(i) Smoke and grit. The release of smoke and grit (atmospheric ash or fly ash) from domestic hearths and factory chimneys is a major source of pollution in industrial areas. These are released due to incomplete combustion of fuels. These pollutants cause lung irritation, asthma, bronchitis etc.

(ii) Photochemical smog. The term *smog* was given to the atmospheric condition to describe the unpleasant combinations of *smoke* and *fog*. The toxicity of smog is due to the presence of various chemical compounds such as PAN and ozone mixed in it.

(iii) Biocides and pesticides. These chemicals are commonly applied in the form of sprays and as such an appreciable quantity of these chemicals diffuses into the atmosphere in the form of granules. Since such chemicals are non-biodegradable, these seriously affect the human health when inhaled and absorbed in the body.

(iv) Heavy metals. The gaseous products of fuel combustion emitted from factories and automobiles contain particles of toxic heavy metals harmful to human beings. Lead poisoning or *plumbism* develops in man when he is exposed to an atmosphere containing lead.

(v) Radio-active elements. The radio-active dust that falls on earth (after nuclear tests etc.) or remains suspended in the atmosphere is called *radio active fall out*. It is one of the most dangerous pollutants.

(vi) Liquid particulates. Particulate of liquid nature are released by sprays and liquid aerosols. **Aerosols** are chemicals which are released in the air with force in the form of a mist or vapours. In recent years, serious environmental pollution has been caused by aerosols and emissions from jet planes. The liquid aerosols contain fluoro-carbons which deplete the ozone layer permitting UV radiations to reach the earth's surface.

EFFECTS OF AIR POLLUTION

Air pollution affects all forms of life from man to plants and also to materials related to human welfare. Some of the important effects are as follows:

(1) Effects on human health

Air, which is essential for life, may also have life damaging properties due to the presence of harmful pollutants. Polluted air affects directly to eyes, nose, throat and lungs. Respiratory system sufferings due to toxic air inhalation cause long and deep effects to our body systems internally.

Some important effects on human health are as follows-

(1) SO_2 produces drying of mouth, scratchy throat and smarting eyes. In the form of H_2SO_4, it damages tissues.

(2) Nitrogen oxide at higher concentration impairs the functioning of lungs by causing accumulation of water in the air pores.

(3) Higher level of CO leads to symptoms like laziness, exhaustion of body, headache, disturbances of psychomotor function, decrease in visual perception, serious effects on cardio-vascular systems etc. Prolonged exposure to CO may lead to death of a man.

(4) SO_3, NO_x and CO diffuse into the blood, combine with haemoglobin and impede oxygen transport.

(5) Radio-active elements emitted in the atmosphere cause severe skin diseases and deformities which are inherited in succeeding generations.

(6) Silicon particles, fibres of asbestos and those of cotton cause **silicosis, fibrosis** and **byssinosis**, respectively.

(7) Air borne spores, pollen grains, bacteria, fungi, fur, hairs etc. cause various allergic reactions like bronchial asthma, hay fever and diseases like tuberculosis, dermatitis etc.

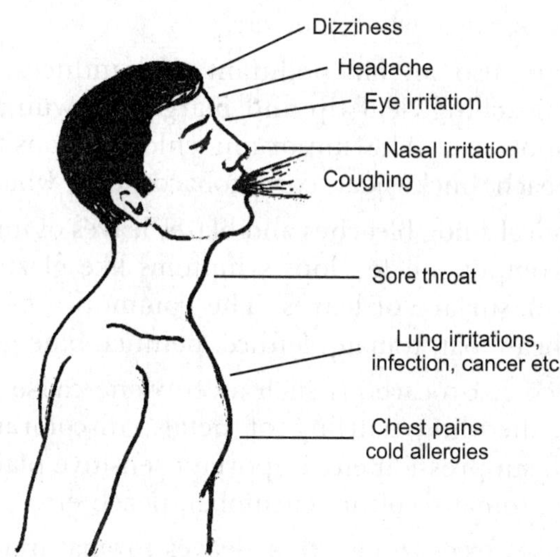

Fig. 26.2. Human body reactions due to inhalation of polluted air.

(II) Effects on plants

Like human beings, plants too are faced with the harmful effects of polluted air. Some important damaging effects of air pollution on plants are as follows-

(1) SO_2 pollution develops symptoms like bleached spots on leaves, chlorosis, early abscission, reduced yield etc. Plants sensitive to SO_2 pollution are barley, pumpkin, alfalfa, cotton, wheat, lettuce, apple, oats, aster, zinnia, birch, elm, pine, grapes etc.

SO_2 has considerable deleterious effects on forest trees. It corrodes membrane surface and adversely affects various metabolic activities including stomatal physiology and photosynthesis.

(2) Oxides of nitrogen reduce the crop yield badly. The common symptoms of such pollution are brown spots on leaves and suppression of growth. The important sensitive plants are azalea, sunflower, mustard, tobacco, pinto beans etc.

(3) Every year economically important plants worth billion of rupees are damaged due to air pollution. In temperate regions, alfalfa has been found susceptible to SO_2, tobacco to O_3 and petunia to photochemical smog. In tropics, mango is severely damaged by polluted air, especially due to CO_2 and CO released from brick fields.

(4) Fluorides reduce the crop yield and damage leaf tissues. Symptoms like leaf tip and margin yellowing (chlorosis), dwarfing, leaf abscission, decreased yield etc. are found in plants due to fluoride pollution. The common fluoride sensitive plants are gladiolus, tulip, apricot, corn, blue berry, grape, blue spruce, white pine etc.

(5) Chlorine is also an air pollutant and induces developing of symptoms like bleaching, leaf tip and margin browning, dropping of leaves, yellow spots etc. The important chlorine sensitive plants are radish, alfalfa, peach, buckwheat, corn, tobacco, oak, white pine etc.

(6) Photochemical smog bleaches and blazes leaves of important foliage crops. Its PAN component develops symptoms like glazing, silvering or bronzing of lower surface of leaves. The common sensitive plants are mustard, pinto bean, oat, tomato, lettuce, petunia, blue grass etc.

(7) Unsaturated hydrocarbons such as ethylene cause premature leaf fall, floral bud shedding, curling of petals, discoloration of sepals, chlorosis, growth suppression etc. Important sensitive plants are orchids, carnation, azalea, tomato, cotton, cucumber, peach etc.

(8) Ozone causes trees to lose their leaves prematurely. It is harmful specially to young plants.

(III) Effects on animals

(1) Air pollution also causes wide spread damage to live stocks. The general effects of air pollution on animals are similar to those on human beings.

(2) Fluorosis, *i.e.* fluoride toxicity is commonly found in domestic animals which ingest various fluoride compounds that fall on leafy part of the plants. Fluorosis results in lameness, loss of weight frequent diarrhoea, and abnormal calcification of bones and teeth.

(IV) Effects on materials

Air pollution damages materials mainly by corrosion of metals, stones, marbles etc. due to acidic compounds in polluted atmosphere. Some important effects are –

(1) The acid rains and the products of photochemical smog cause considerable effect on metals and buildings.

(2) Acidic products of air pollutants cause disintegration of textiles, paper and marbles.

(3) Air pollutants damage historical monuments, *e.g.* the famous marble structure Taj Mahal of Agra is endangered due to SO_2 pollution produced by Mathura Oil Refinery, small industrial units and locomotive exhausts. Books and papers turn yellow due to SO_2 pollution.

(4) Hydrogen sulfide decolorises silver and lead paints.

(5) Ozone oxidises rubber goods and causes damaging effects on fabrics and dyes.

(V) Effects on climate

Air pollution causes changes in atmospheric conditions, *e.g.* climate of big cities is evidently different from rural areas. Some important effects are as follows-

(1) The release of green house gases from various sources are causing rise in global temperature (global warming or green house effect).

(2) An increase of global temperature by 2-3°C may lead to melting of glaciers and polar ice caps, flooding of low lying coastal plains, increase in the flow of rivers, and change in the rainfall pattern and possible submersion of some islands.

(3) Freon gas from aerosol sprays and nitrogen oxide in the atmosphere deplete ozone layer.

(4) Warming of earth surface, depletion of ozone layer, rising of sea level and acid rains are major effects of air pollution at global level.

(VI) Aesthetic insults

Aesthetically valued requirements of the human beings are a clear view of natural and man made objects, breathing fresh air or smelling a floral fragrance. A dust haze or hanging smoke causes poor visibility. Smoke and foul smelling odours emitted by factories, automobiles, drains and garbage dumps make urban life unpleasant.

CONTROL OF AIR POLLUTION

In most of Indian cities, the primary standards of air quality are not maintained. The reason is not only the need of better control technologies but in majority cases, there is a failure of strategies to be adopted. Some important measures to reduce or minimize the pollution of air are as follows:

(1) Use of purified petrol. Use of purified unleaded and good quality of petrol reduces the level of sulphur and lead oxides in automobile exhausts.

(2) Modernisation of industries. The industrial machines and vehicle engines should be made energy efficient to use fuels at the maximum level to reduce the release of pollutants.

(3) Installation of air treatment plants. Large number of treatment plants for air purification must be installed and run throughout the clock.

(4) Use of alternative energy sources. The common alternative sources of energy are wind, water and sun. Automobile industry should manufacture battery operated and solar energy operated vehicles to reduce the use of fossil fuel run vehicles.

(5) Treatment of emissions. The emissions from factories and industries should be treated by various techniques before being released into the atmosphere. Some important techniques are-

(*i*) ***Settling chambers***. These are large chambers in which the large particulates can settle down, *e.g.* gravitational settling chamber (fig. 26.3).

(*ii*) ***Cyclone separator***. This apparatus works on the principle of cyclones, in which the polluted air is introduced into a conical cylinder and made to whirl at a high speed. As a result, the larger particles strike against the wall and drop down to the bottom of the conical. The partly cleaned air exits from the upper region (fig. 26.4).

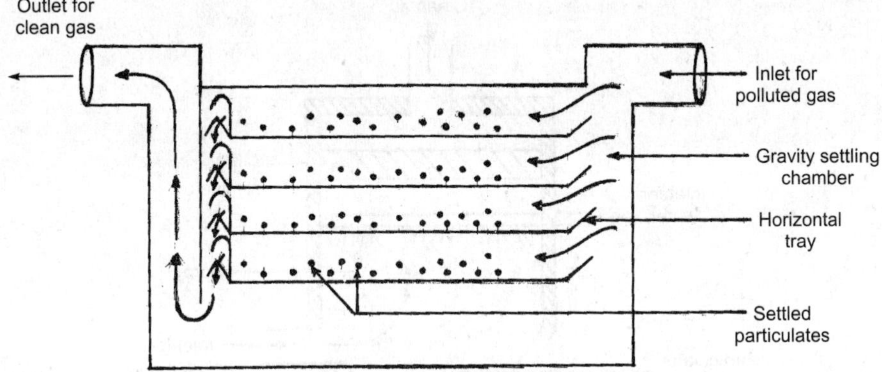

Fig. 26.3. Gravitational settling chamber

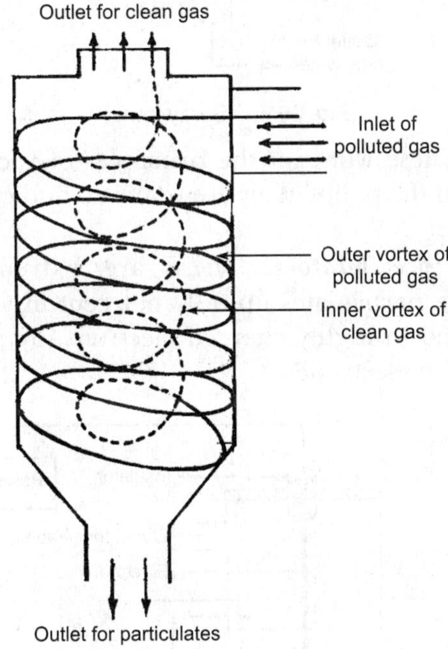

Fig. 26.4. Cyclone separator

(iii) ***Wet and gas scrubbers***. It is an effective and simple method for removing large particulates and effluent gases. The polluted air is introduced through a narrow opening and water sprayed through the upper end. As a result of collision, the particulates are scrubbed out from the air and settle down at the bottom. The poisonous gases dissolve in water to form by products (fig. 26.5).

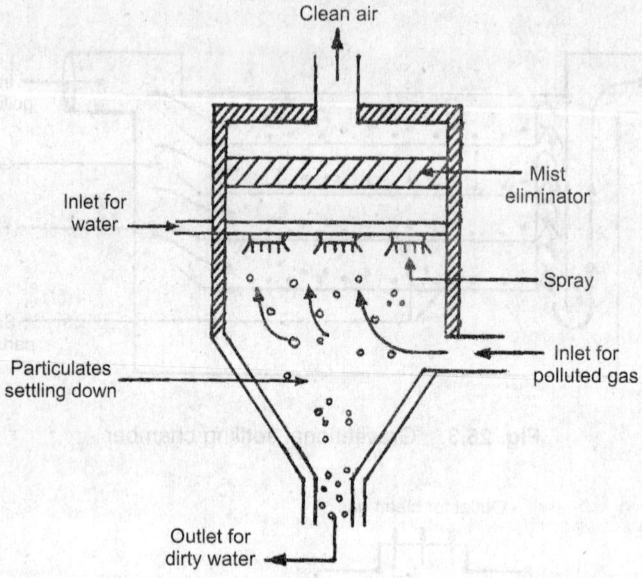

Fig. 26.5. Spray tower

(*iv*) ***Bag filters***. These work on the principles of vacuum cleaners. All particulates of the polluted air are removed and clean air is released from the top.

(*v*) ***Electrostatic precipitators***. These are extremely effective in removing fine particulates upto 90 per cent in which the polluted air is subjected to highly charged electrons, by passing through a pipe fitted with electrodes.

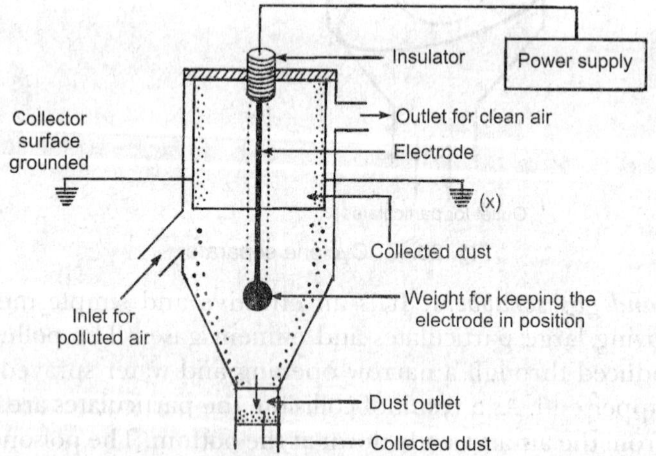

Fig. 26.6. Electrostatic precipitator

The particulates collect in the pipe and are removed at intervals. Electrostatic precipitators are used in power plants (fig. 26.6), carnet and paper mills etc.

(vi) ***Adsorption***. In this process, the pollutants are captured from the polluted air through adsorption by the use of activated carbon. The process is useful for removing organic pollutants.

(vii) ***Incineration and catalytic combustion***. Gaseous pollutants are removed by burning them to CO_2, H_2O and inert gases, or a catalyst may be used to absorb the pollutants.

(6) Plantation of trees. Tree plantation should be made a national priority. It is said *plant a tree and you save a nation*. Green plants absorb CO_2 for photosynthesis and thus reduce CO_2 pollution from the air. Besides, broad leaved ornamental, fruit and forest trees trap large amount of gases and dust on their leaves and twigs. Tree line functions as dust or particulate filters. Some important plants tolerant to various pollutants are given in table 26.2. Techniques of growing *speedy trees* can be adopted for growth.

TABLE 26.2. Plants Tolerant to Various Air Pollutants

Pollutant	*Plants*
Ammonia	Peach
Chlorine	Begonia, egg plant, olive, pepper
Fluoride	Cherry, pear, plum, strawberry
Hydrogen chloride	Pear
Hydrogen sulphide	Apple, cherry, carnation, peach, strawberry
Nitrogen oxide	Orange
Particulate matter	Mango, litchi, banyan and other broad leaved trees.

(after Chundawat, 1989)

(7) Change in life style. Life style change may have considerable effect on the reduction of air pollution. Some important measures can be like– (i) use of energy more efficiently, (ii) use of alternative source of energy, such as solar and wind energy, (iii) industries to install away from cities, (iv) plantation of trees as green belts, (v) ideal traffic planning, (vi) removal of unnecessary speed breakers and check posts, (vii) checking of undesirable burning of vegetation, (viii) car pooling etc.

(8) Enforcement of Air (Prevention and control) Pollution Act, 1981. Air quality standards as recommended by CPCB (Central Pollution Control Board) must be maintained strictly. Some important standards are given in table 26.3.

(9) Environmental education. The most important and fundamental programme for dealing with environmental problems is to impart environmental education. Every citizen from his very childhood should be made aware of the problems of environmental pollution. All schools and colleges should include environmental programmes in their curriculum

TABLE 26.3. Air Quality Standards as Recommended by CPCB

Category	*Permissible concentration* ($\mu g/m^3$) *SPM (Suspended particulate matter)*	SO_2	*CO*	NO_x
Industrial area	500	120	5000	120
Residential area	200	80	2000	80
Sensitive area	100	30	1000	30

and every student should participate actively and be involved so that love to environment and to his country becomes an integral and inseparable part of his character. A fair and healthy atmosphere is a right of the every child of the nation.

Box 26.1 **Air Pollution Tolerant Trees**

Trees having petiolated leaves with thick, fleshy and broad lamina and compact branchings are air and sound pollution tolerant. Some common examples of such trees are-

(1) Sulphur dioxide
Acer pseudoplatanoides
Albizzia lebbeck
Ailanthus excelsa
Alstonia scholaris
Azadirachta indica
Ficus religiosa
Largerstroemia flos–reginae
Mimusops elengi
Polyalthia longifolia
Quercus palustris
Terminalia arjuna

(2) Ozone
Acer platanoides
Quercus rubra

(3) Oxides of nitrogen
Tectona grandis
Fagus orientalis
Quercus rubra
Robinia Pseudoacacia
Sambucus nigra
Ulnus species

(4) Peroxy acetyl nitrate
Acer pseudoplatanoides
A. negunda
Quercus palustris
Q. rubra

(5) Lead
Cassia siamea
Zizyphus mauritiana

(6) Hydrogen fluorides
Ailanthus excelsa
Juniperus virginiana

(7) Dust pollution
Alstonia macrophylla
Cassia siamea
Dalbergia sissoo
Ficus benghalensis
F. infectoria
Mangifera indica
Polyalthia longifolia
Shorea robusta
Syzygium cumini

(8) Noise Pollution
Alstonia scholaris
Azadirachta indica
Butea monosperma
Erythrina variegata
Grevillea robusta
Populus ferolinesis
Pterospermum acerifolium
Syringa vulgaris
Tamarindus indica
Terminalia arjuna
Viburnum lantana

(*Source* : N. Sivasamy and V. Srinivasan 1997)

Chapter Summary

Contaminated air harmful to human beings is called polluted. Sources of air pollution may be stationary combustion sources, mobile combustion sources, industrial processing and other sources etc. Air pollutants may be gaseous or particulate in nature. Among inorganic gaseous pollutants, the common ones are oxides of sulphur (SO_2, SO_3), nitrogen (NO_x) and carbon (CO, CO_2) and ozone. Both oxides of sulphur and nitrogen combine with rain water and cause acid rains. CO_2 is largely responsible for global warming. Organic gases like aldehydes and hydrocarbons are also pollutants. Particulate pollutants include smoke and grit, photochemical smog, heavy metals, radioactive elements etc. Aerosols are liquid particulate.

Air pollution affects human beings, plants and animals and various materials including historical buildings. SO_3, NO_x and CO diffuse into the blood, combine with haemoglobin and impede oxygen transport. Several diseases like silicosis, fibrosis, byssinosis, tuberculosis, bronchial asthma, hay fever, skin diseases etc. may be caused due to air pollutants. Plants are also sensitive to various aeropollutants. These may cause bleached spots on leaves, chlorosis, early abscission (due to SO_2), brown spots on leaves, suppression of growth (due to NO_x), leaf tip and margin yellowing, chlorosis, dwarfing (due to F), glazing, silvering or bronzing of leaf lower surface (due to PAN) etc. Fluoride pollution causes flourosis in animals. Acid rains damage leaf surface and buildings of architectural importance. Air pollution is solely responsible for global warming or green house effects that brings about changes in climate.

Control of air pollution is a difficult problem and requires all round precaution, prevention and treatment. Use of purified petrol, modernisation of industries, use of alternative energy sources are some preventive measures. Treatment of emissions can be tackled through settling chambers, cyclones, wet and gas scrubbers, bag filters, electrostatic precipitators, incineration and catalytic combustion etc. Plantation of trees and change in life style also help in minimizing air pollution. Govt. of India has constituted Central Pollution Control Board (CPCB) and has enforced "Air Pollution (Prevention and Control) Act, 1981. Environmental education also helps in reducing air pollution.

Study Questions

1. What do you understand by air pollution? Describe its various causes.
2. What are various air pollutants? Discuss their effects on vegetation and human beings.
3. Describe pollution and give an account of sources and pollutants of air pollution. Suggest suitable methods to control air pollution.
4. Briefly describe the major air pollutants.
5. What possible measures can be taken to control air pollution? Describe.
6. Describe various effects of air pollution.
7. What are various sources of air pollution? Describe.
8. Describe the effect of air pollution on human health.
9. Briefly describe the effects of air pollution.
10. Describe any three treatment process of polluted emissions.
11. What are gaseous pollutants of air? Describe.

12. Describe in brief the particulate pollutants of air.
13. Explain the following terms:
    (i) Smoke and grit
    (ii) Photochemical smog
    (iii) Electrostatic precipitators
    (iv) Mobile combustion sources of air pollution.
14. Write the effect of following pollutants on human beings:
    (i) SO_2 (ii) CO
    (iii) Photochemical smog (iv) CO_2
    (v) Nitrogen oxides

Objective Questions. *Select the correct answers*

1. A pollutant not released by exhaust of automobiles is:
   (1) SO_2 (2) CO
   (3) Fly ash (4) Hydrocarbon gases
2. Ozone in lower atmosphere is an example of:
   (1) Primary pollutant (2) Secondary pollutant
   (3) Tertiary pollutant (4) Not a pollutant
3. Component not released in the burning of coal is:
   (1) NO_2 (2) SO_2
   (3) Fly ash (4) O_2
4. Largest source of air pollution in cities are:
   (1) Industries (2) Sewage
   (3) Automobiles (4) tanneries
5. The most widely found pollutant in the air is:
   (1) CO_2 (2) CFC
   (3) CO (4) SO_2
6. The major aerosol pollutant present in the jet plane emission is:
   (1) SO_2 (2) CFC
   (3) CCl_4 (4) CO
7. Some pollutants combine with haemoglobin and impede oxygen transport. Such pollutants are:
   (1) SO_3 (2) N_2O
   (3) CO (4) All of these
8. The causes of depletion of ozone layer in the atmosphere are:
   (1) Freon gases (2) Supersonic jet travels
   (3) Nitrogen oxides (4) All of these

Answers

1. (3) *2.* (2) *3.* (4) *4.* (3) *5.* (4) *6.* (2)
7. (4) *8.* (2)

27

CHAPTER

Water Pollution

LEARNING OBJECTIVES

Introduction • Definition • Sources of Water Pollution • Effects of Water Pollution • Control of Water Pollution.

Introduction

To the best of scientific knowledge life can not exist without water, and that is why *water* is popularly called *liquid of life*. Water covers more than 70 per cent of earth's surface, of which 97.3 per cent is in ocean and 2.7 per cent is as fresh water. The fresh water is held up in ice caps and glaciers (72.2 per cent), ground water and soil moisture (22.4 per cent), lakes and swamps (0.35 per cent), atmosphere (0.04 per cent) and stream channels (0.01 per cent). Water is a precious requirement of the living world. A continuous and abundant supply of clean water is essential for the survival and health of all living organisms. A polluted water is a curse to life.

Water pollution is defined as *a change in the quality or composition of water directly or indirectly as a result of man's activities, so that it becomes unsuitable for drinking, domestic, recreational and agricultural purposes*.

In India, most rivers are highly polluted but Damodar and Ganga are the heavily polluted rivers. The sources of pollution of major Indian rivers are various river bank based industries (table 27.1).

TABLE 27.1. The Sources of Pollution of Major Indian Rivers

No.	*River*	*Sources of pollution*
1.	Bhadra	Pulp, steel and paper industries.
2.	Coovm, Adyar	Automobile workshops and domestic sewage
3.	Cauvery	Distilleries, tanneries, paper and rayon mills and sewage

4. Dajora	Synthetic rubber factories
5. Damodar	Fertilizers, thermal power stations, fly ash from steel mills.
6. Godavari	Paper mills
7. Ganga	Jute mills, chemical factories, tanneries, textile mills etc.
8. Gomati	Paper and pulp mills, sewage
9. Hooghly	Paper, jute, textiles, paints, metal, steel and rayon mills and detergents and sewage.
10. Kali	Sugar, paint, silk, yarn, tin, glycerine and rayon industries, distilleries etc.
11. Kula	Tanneries, rayon mills, chemical factories
12. Periyar	Various industries
13. Sone	Paper mills, cement factory
14. Siwan	Sulphur, cement and sugar industries and paper mills.
15. Suwao	Sugar industries
16. Yamuna	DDT factory, power station and sewage.

SOURCES OF WATER POLLUTION

The various sources of water pollution may be simply grouped into-

(i) Uncontrolled dumping of solid degradable and non-degradable wastes,

(ii) Indiscriminate flow of effluents from various industries, and

(iii) Increase in the organic loading and deterioration of the self purification process of water.

The important sources of water pollution are shown in fig. 27.1.

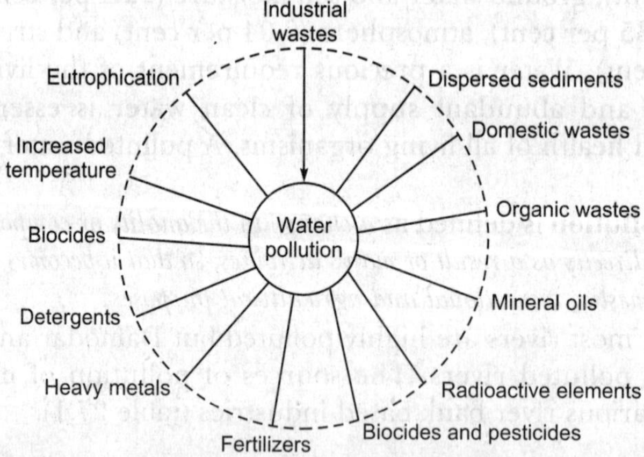

Fig. 27.1. Schematic representation of some important pollutants causing water pollution.

Some common sources of water pollution are as follows:

(1) Domestic wastes

A common and widespread source of water pollution is the discharge of domestic wastes directly into the rivers. Such pollutants may be biodegradable or non-biodegradable in nature.

(i) **Biodegradable**. These can be broken down into simpler organic matters by microbial activities, *e.g.* wastes of organic matter such as vegetables and fruit peelings, waste food, waste paper etc. Such pollutants increase the organic and inorganic contents of the water and damage the aquatic life.

(ii) **Non-biodegradable**. Such pollutants are not decomposed by microbes, *e.g.* plastic goods, polythene bags, glass metal containers etc. These are perhaps the worst kind of pollutants and their number is steadily increasing and posing a serious threat for the future.

(2) *Sewage*

The cloudy fluid consisting of human faecal matter, urine, minerals and organic nutrients in a dissolved state or dispersed in a solid condition is called **sewage**. It is a direct consequence of demophora, *i.e.*, an unabated growth of human population. Sewage disposal is a very serious problem in big cities, and as such it is proving as a major pollutant of inland waters, soil and aquatic ecosystems. The amount of sewage produced in India is about 4.8 m.m^3/day, of which, only 20 per cent is treated and 80 per cent is released untreated. From this untreated amount, if manure is produced, it will be sufficient to fertilize some 0.1 million hectare land per year.

Certain characteristics of sewage water are-

(i) BOD (biochemical oxygen demand) is very high.

(ii) OC (oxygen consumption) for the oxidative decomposition of organic matter is very high.

(iii) Due to high oxygen demand and low rate of photosynthesis, sewage water becomes totally depleted of dissolved oxygen (DO).

(iv) Aerobic organisms are greatly reduced in sewage water due to low content of DO.

(3) I*ndustrial wastes*

While industrial revolution has given advancement, it has also produced an ever-growing problem of water pollution, especially in rivers and streams. Tanneries, distilleries, paper and pulp, sugar, textiles, rubber, petrochemicals, chemical industries etc. conveniently discharge their effluents into the water bodies without any consideration of consequences. Some of these effluents contain toxic chemicals which are detrimental not only to the aquatic life but also to the human beings and terrestrial animals.

Industrial wastes from metal finishing and electroplating plants contain toxic heavy metals and poisons like cyanides; tanneries contain large quantities of chromium, tannin, calcium and chlorides and coke works, liquors and effluents from plastic factories have a high content of phenolic compounds.

About 180 million litres of toxic effluents are discharged every day into the Periyar river by the various industrial units in the greater Cochin area (Kerala). The effluents contain acids and alkalies, fluorides, free ammonia, ammoniacal nitrogen, insecticides, dyes, mercury, chromium, radio nuclides etc. As a result, BOD of water has gone upto 16 as against the normal value of 5.

Fig. 27.2. Direct release of industrial effluents into the river causing water pollution.

(4) *Fertilizers and Detergents*

A fairly large amount of fertilizers added to increase soil fertility is washed off through the irrigation, rainfall and drainage and ultimately reaches the rivers. These pollute the water and make it toxic.

Detergents enter the water in the form of waste water from domestic and industrial washings. Most detergents contain phosphates and stimulate a luxuriant growth of algae developing water blooms. On death of algae, their inorganic and organic components are released back into the water and further increase pollution. Detergents also increase alkalinity and pH of the water making the aquatic habitat inhabitable to both flora and fauna.

(5) *Pesticides and Herbicides* (Biocides)

The use of **pesticides** has become common now –a –days. Popularly used pesticides are DDT, Dieldrin, Eldrin, DDE, BHC, Heptachlor etc. These are non-biodegradable in nature. Of these DDT is most potent and is passed on from one to other organisms without undergoing any chemical change in itself causing extensive damage, so much so, that in recent years, its use has been banned in several countries. It accumulates in the fatty tissues. In India due to its prolonged use, 13-31 ppm DDT can be detected from the fatty tissues of human beings, this value being highest in the world. It happens due to its biomagnification in the Indian ecosystem.

Herbicides are used to kill or eradicate weeds but they are not specific in their action as they destroy both the desirable species as well as the other plants. Their influence is limited not only to the plant community but also to microbes.

(6) *Radioactive wastes*

Radioactive wastes enter into the water bodies in various ways, *e.g.* processing of uranium ore, wastes from radio isotopes using research laboratories, wastes from hospitals using radio isotopes, wastes water released from nuclear power stations or wastes generated during nuclear weapon testing.

Disposal of radioactive wastes is a great problem because their radioactivity is indestructible. They are indiscriminately disposed into water bodies causing fatal damage to the aquatic life. These radioactive wastes accumulate in the cells of aquatic plants and small fishes and from there pass on to animals and man which feed on them. Accumulation and exposure to radioactive isotopes causes ionisation of various body fluids, gene mutation and chromosomal aberrations resulting in distorted morphological and functional changes.

(7) *Thermal pollution*

Pollution arising from a sudden increase in the temperature of water is known as thermal pollution. Several industries utilize water for cooling purposes and release it to the river at a higher temperature. Such releases lethally affect the aquatic biotic communities. Some animals are killed outright by the hot water and as a result entire ecological balance is disturbed. Hot water adversely affect the daily and seasonal behavior, metabolic responses and reproductive rates of aquatic organisms. Increase in temperature of water decreases its oxygen dissolving capacity and its BOD also rises.

(8) *Oil*

Pollution arising from oil spillage (such as from super tankers in sea) is not only dangerous but far reaching in its consequences. The aquatic life and the self purification process of water is so impaired that it may take several years to restore normalcy.

Besides, about 3.5 million tonnes of oil is discharged into the sea every year around the world which is a significant amount to cause pollution

(9) *Eutrophication*

Eutrophication is the enrichment of the water bodies resulting from addition of organic and inorganic nutrients. It leads to the increased growth of algal blooms, which on their death become a medium for bacterial growth and for decomposition process. It inturn leads to oxygen depletion and associated form of water pollution. This results in the death and decay of aquatic organisms and consequently the water becomes murky, turbid and foul smelling and unfit for life activities.

EFFECTS OF WATER POLLUTION

Water pollution affects not only the aquatic ecosystem but human life as well. Some important effects are as follows:

(1) *Domestic sewage*

A number of epidemic diseases such as *cholera, typhoid, dysentery, diarrhoea, infectious hepatitis, jaundice* and *filariasis* are caused by water polluted by domestic sewage.

(2) *Industrial effluents*

Industrial wastes contain a large number of toxic chemicals, heavy metals, non-biodegradable wastes etc. The toxic chemicals are detrimental to aquatic life to terrestrial life, directly or indirectly and disturb the whole ecosystem.

Heavy metal contamination of water causes severe ailments of human beings. Mercury poisoning causes **minamata disease** which was first detected in 1952 in Japanese soldiers consuming mercury contaminated fish of Minamata Bay of Japan. Methyl mercury causes numbness of limbs, lips and tongue of human, deafness, blurring of vision, mental derangement etc.

Heavy metal cadmium causes *cancer* of liver and lungs. It also causes **itai-itai** or **ouch-ouch,** a very painful disease of bones and joints.

In general, the process of biopurification of water is greatly reduced due to after effects of toxic industrial effluents. These kill aerobic forms of life.

(3) *Ground water pollution*

Seepage of industrial and municipal wastes has contaminated even the ground water at many places in our country. Sewage channels and agricultural run off have further aggravated the problem.

Leaching of inorganic fertilizers leads to the accumulation of nitrates in the water. When this water is consumed by man and animals, the nitrates are reduced to toxic form, *i.e.*, nitrites by intestinal bacteria. Accumulation of nitrites in the body causes a serious disease known as **methaeglobinemia** manifested by damage of respiratory and vascular system resulting in suffocation. The disease is also known as **blue baby syndrome**.

Excess fluoride in drinking water causes **teeth deformity** and **skeletal fluorosis** in which bones become stiff, hardened and joints painful. Disease **knock knee syndrome** (out ward bending of knee) is another common disease found due to fluoride pollution.

In West Bengal (India), the presence of excess of arsenic in ground water causes **black foot disease**. Arsenic also causes *diarrhoea, peripheral neuritis, hyperkeratosis* and *cancer* of lungs, skin, mouth, oesophagus, larynx and bladder.

Chronic lead poisoning symptoms include fatigue, weakness, muscular atrophy, paralysis etc. Copper causes hypertension and uremia and zinc causes vomiting and renal damage.

(4) Eutrophication

In nutrient rich water reservoirs, algae grow abundantly and develop **water blooms** or **algal blooms** which, causes loss of species diversity. Many blooming blue green algae secrete toxin in the water and induce oxygen deficiency, as a result aquatic animals die. Alga *Microcystis aeruginosa* secretes microcystin and other toxins, poisonous to aquatic animals including fishes.

(5) Biomagnification

Bio or biological magnification is a phenomenon through which certain pollutants are accumulated in tissues in increasing concentration along the food chain. Such pollutants are non-biodegradable hence continue their accumulation. One such example is DDT, an insecticide use to kill mosquitoes. In an island of USA, the regular use of DDT has been found to reduce the population of fish eating birds. In India, 13-31 ppm of DDT has been found in fat tissues of man, being highest in the world. Likewise, many other persistent pesticides and radionuclides also show biomagnification.

(6) Thermal pollution

Release of hot water from thermal power stations and various industries directly to the water bodies often kills both aquatic plants and animals. It can thus exert a disruptive effect on water ecosystems.

Box 27.1 Water Borne Diseases

Following is a short list of water borne diseases caused due to infectious organisms found in polluted water.

No.	Causal Organisms	Disease	Primary source
	BACTERIAL		
1.	*Campylobacter jejuni*	Gastroenteritis	Human and animal faeces
2.	*Escherichia coli*	Gastroenteritis, Diarrhoea	Human faeces
3.	*Legionella pneumophila*	Legionellosis (acute respiratory illness)	Thermally enriched water
4.	*Mycobacterium tuberculosis*	Tuberculosis	Human respiratory exudates
5.	*Salmonella typhi*	Typhoid fever	Human faeces
6.	*S. paratyphi*	paratyphoid fever	Human faeces
7.	*Shigella dysenterae*	Bacillary dysentry	Human faeces
8.	*Vibrio cholerae*	Cholera	Human faeces
9.	*Yersinia enterocolitica*	Gastroenteritis	Human and animal faeces
	VIRAL		
1.	*Adenoviruses*	Gastroenteritis	Human faeces
2.	*Coxsackie viruses A*	Aseptic meningitis	Human faeces
3.	*Coxsackie viruses B*	Aseptic meningitis	Human faeces
4.	*Echo viruses*	Aseptic meningitis	Human faeces
5.	*Hepatitis A virus*	Infectious hepatitis	Human faeces
6.	*Norwalk* and related viruses	Gastroenteritis	Human faeces
7.	*Polioviruses*	Poliomyelitis	Human faeces
8.	*Reoviruses*	Mild upper respiratory illness	Human and animal faeces
9.	*Rotaviruses*	Gastroenteritis	Human faeces
	PROTOZOANS		
1.	*Balantidium coli*	Dysentry	Human faeces
2.	*Cryptosporidium*	Cryptosporidiopsis	Human and animal faeces
3.	*Entamoeba histolytica*	Amoebic dysentry	Human faeces
4.	*Giardia lamblia*	Giardiasis	Human and animal faeces
	HELMINTHS		
1.	*Dracunculus medinensis*	Dracontiasis	Human and animal faeces
2.	*Echinococcus*	Echinococcosis	Human and animal faeces
3.	*Schistosoma*	Schistosomiasis	Human and animal faeces
	ALGAE (BGA)		
1.	*Anabaena flos-aquae*	Gastroenteritis	Eutrophic Water
2.	*Aphanozomenon flos -aquae*	Gastroenteritis	Eutrophic Water
3.	*Microcystis aeruginosa*	vomitting, gastro-enteritis , paralysis	Eutrophic Water
4.	*Schizothrix calcicola*	Gastroenteritis	Eutrophic Water

(7) *Some other effects of water pollution*

Some other effects of water pollution are as follows:

(i) Pollution of water results in decreased dissolved oxygen content which kills fish and other aquatic organisms.

(ii) The consumption of pesticides and insecticides may cause *cancer, nervous disorders, leukemia* and other serious diseases.

(iii) Solid and liquid wastes from domestic and industrial flush lines choke the water supplies and cause death of fish, cattle and human beings. The toxic materials present in these wastes destroy eggs and the larval stages of fish and other organisms.

(iv) Radioactive elements in the form of wastes cause permanent and far reaching consequences, both to the human and aquatic population.

(v) Polluted waters are turbid, unpleasant in taste, foul smelling and most unsuitable for bathing, drinking and water sports.

CONTROL OF WATER POLLUTION

As such there is no single method to control pollution of water. Various measures can be adopted according to the type and nature of water pollutants. Some of them are as follows:

(1) Effluents should not be discharged directly into water bodies rather only after proper treatments (fig. 27.3). Common treatment process involves three important steps:

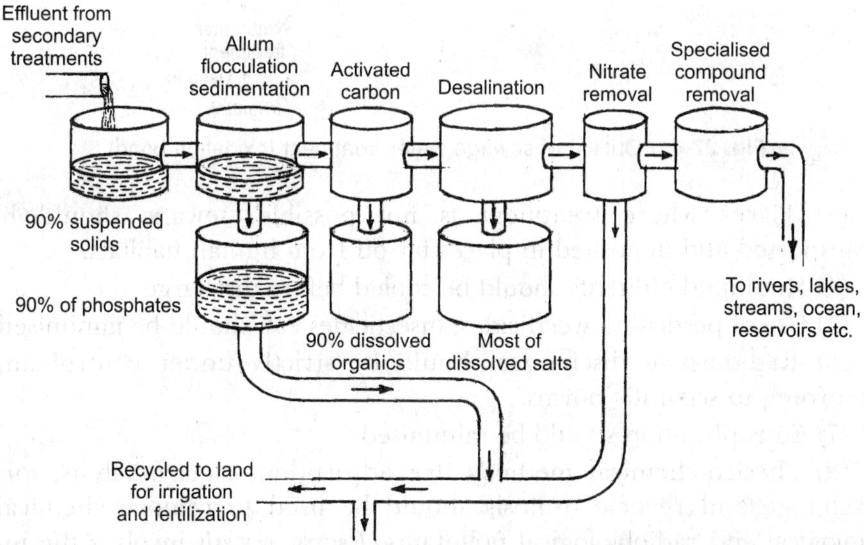

Fig. 27.3. A general scheme of treatment of waste waters in Effluent Treatment Plant (ETP). The process involves physical and chemical treatments.

(i) Removal of suspended large particles.

(ii) Supply of aeration to promote bacterial decomposition followed by, chlorination to eliminate bacteria.

(iii) Chemical treatments to remove nitrates and phosphates.

(2) Proper treatment of sewage must be carried out before disposing it into water bodies. Treatment should be both, physical and biological.

In the **treatment process**, sewage is collected in tanks or **stabilisation ponds** where large particles are allowed to settle down. The sewage water is then chemically treated by coagulants, chelating substances etc. and then led off to the **oxidation ponds** where the blue green alga *Spirulina* is inoculated and allowed to grow. The alga not only purifies the water but itself becomes protein rich and may serve as food for milk yielding cattles. It can be outlined as in fig. 27.4.

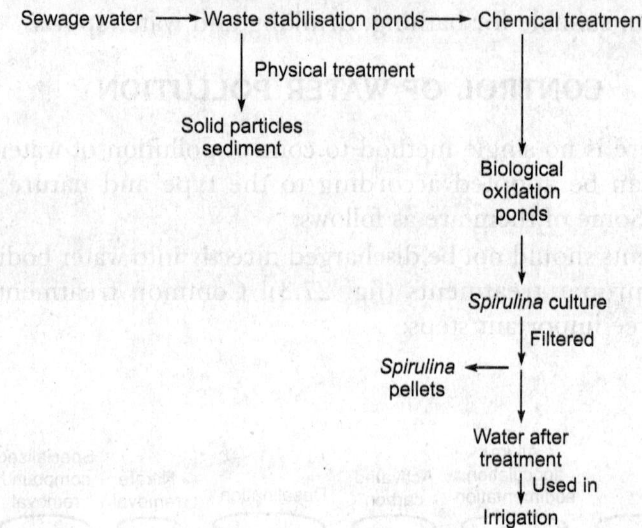

Fig. 27.4. Outline of sewage water treatment (oxidation pond).

(3) Places where treatment is not possible, sewage should be transported and deposited in places far off from human habitats.

(4) Hot liquid effluents should be cooled before discharge.

(5) Use of pesticides, weedicides, insecticides etc. should be minimised.

(6) Radio-active discharges should be strictly under control and according to scientific norms.

(7) Eutrophication should be minimised.

(8) Physico-chemical methods like adsorption, electrodialysis, ion-exchange and reverse osmosis should be used to remove chemical, biological and radiobiological pollutants. *Reverse osmosis* involves the use of chemical porous membrane which attracts solvent molecules but repulses solute molecules. It is successfully used in water pollution control, water renovation, waste purification and water reclamation.

(9) One of the most exciting fields of energy research is the use of **plasmas** for effective control of pollution. *Plasmas* are thin, hot gases produced during thermonuclear reactions which may be used to completely vapourise waste products and converting them to electrified particles of their constituent elements. These particles are then recovered and recycled. Plasmas may well become the major source of pollution control in future.

(10) Algae, water hyacinth and duck weeds grown in ponds can be used to consume nitrates and phosphates present in waste water.

(11) Pollution control through education and law is an effective method to reduce water pollution. In India, **The Water (Prevention and Control of Pollution) Act, 1974** has been enforced.

Chapter Summary

Water pollution is a change in the quality and composition of water, directly or indirectly as a result of man's activities so that it becomes unsuitable for drinking, domestic, recreational and agricultural purposes. Most rivers of India are highly polluted but river Ganga tops the list. Some common sources of water pollution are *(i)* domestic wastes which may be bio-degradable and non-biodegradable, *(ii)* sewage, *(iii)* industrial wastes, *(iv)* fertilizers and detergents, *(v)* pesticides and herbicides, *(vi)* radioactive wastes, *(vii)* thermal pollution, *(viii)* oil, *(ix)* eutrophication etc. Domestic wastes cause a number of epidemic diseases like cholera, typhoid, dysentery, diarrhoea, infectious hepatitis, jaundice, filariasis etc. Industrial wastes contain heavy metals and toxic substances. Mercury pollution causes *minamata disease* and cadmium pollution causes *itai itai* disease. Ground water pollution has been found to cause methaeglobinemia, fluorosis, black foot disease, cancer of skin, mouth and lungs etc. Eutrophication promotes algal growth (water blooms) and makes water impotable. By biomagnification process toxic pollutants increase in concentration along with food chain and ultimately human beings suffer or become victim.

Control of water pollution involves three important steps: *(i)* removal of suspended large particles, *(ii)* supply of aeration to promote bacterial decomposition and *(iii)* chemical treatments. For sewage treatment oxidation ponds are quite useful.

Education and enforcement of laws are also very effective methods to reduce water pollution.

Study Questions

1. What is water pollution? Describe its causes and effects.
2. What do you understand by water pollution? Describe its various effects.
3. What is water pollution? What are its various causes and methods of control.
4. Outline the different sources of water pollution. Discuss the nature of pollutants and suggest the remedial measures.
5. Describe the various sources of water pollutants and the methods to control water pollution.
6. Describe the sources of water pollution.

7. Enumerate the effects of water pollution.
8. Describe various methods to control water pollution.
9. Write a note on ground water pollution.
10. Explain the following terms:
    (i) Biodegradable and non-biodegradable
    (ii) Sewage
    (iii) Eutrophication
    (iv) Oxidation ponds
11. Define the terms:
    (i) Water pollution
    (ii) Biomagnification
    (iii) Fluorosis
    (iv) Thermal pollution
    (v) Algal blooms

Objective Questions. *Select the correct answers*

1. BOD stands for:
(1) Biochemical oxygen demand (2) Biome of deserts
(3) Boron and oxygen depletion (4) None of these

2. A plant preferably grown in oxidation pond is:
(1) *Eichhornia* (2) *Spirogyra*
(3) *Spirulina* (4) Anaerobic bacteria

3. BOD is used for the measurement of:
(1) Atmospheric pollution (2) Sewage pollution
(3) Thermal pollution (4) Nuclear pollution

4. Water pollution is caused by:
(1) CO (2) PAN
(3) Fertilizers (4) Fossil fuels

5. Spraying of DDT to kill insects causes pollution of:
(1) Air (2) Water
(3) Soil (4) All of these

Answers

1. (1) *2.* (3) *3.* (2) *4.* (3) *5.* (4)

Soil Pollution

LEARNING OBJECTIVES

Definition • Sources of Soil Pollution • Effects of Soil Pollution • Control of Soil Pollution.

Introduction

Soil is as important as air and water for the sustenance of life for both, the plants and the animals. It is nature's home for plants to which all forms of life directly or indirectly depend for their nourishment.

Soil is a natural medium of inorganic and organic nutrients and has an inbuilt system of spontaneous recycling of matter. It is affected by changes in the atmospheric conditions as well as water contents and microbial population.

Soil pollution is defined as *an undesirable change in the natural, physical, chemical or biological components of the soil*

Sources of Soil Pollution

The various sources of soil pollution may be categorised into two groups;
(i) Natural sources and *(ii) Artificial sources*

(I) Natural Sources

Natural pollutants are of plant and animal origin. They are-

(1) Plant residues. Normally plants, on death and decay, contribute organic matter to the soil and thereby increase soil fertility. Sometimes residues from crops, fields and orchards carry plant pathogens and pests. These on death and decay cause soil pollution. Naturally occurring or man made forest fires and crops add unwanted organic matter to the soil and destroy soil microflora which create unfavorable conditions leading to pollution. Burning of crops yields residues with CO (8.3 per cent), NO (1.5 per cent), hydrocarbons (5.3 per cent) and particulate matter (8.5 per cent) etc.

The high content of carbon monoxide (CO) is highly toxic to root growth and soil microbes. Nitrogen oxide (NO) combines with soil water and forms nitrous and nitric acids which increase soil acidity (low pH) and make the soil inhabitable to plants and soil micro flora.

(2) Animal wastes. Animal wastes, such as faecal matter, urine, blood, slaughter house waste in the form of liquid or particulate matter, bodies of dead animals etc., are all indiscriminately dumped into the soil. These create unhealthy conditions detrimental to the growth of plants and soil organisms. Excessive organic contents are harmful for healthy growth of roots as they create hypertonic conditions in the soil causing wilting or stunted growth of the plants.

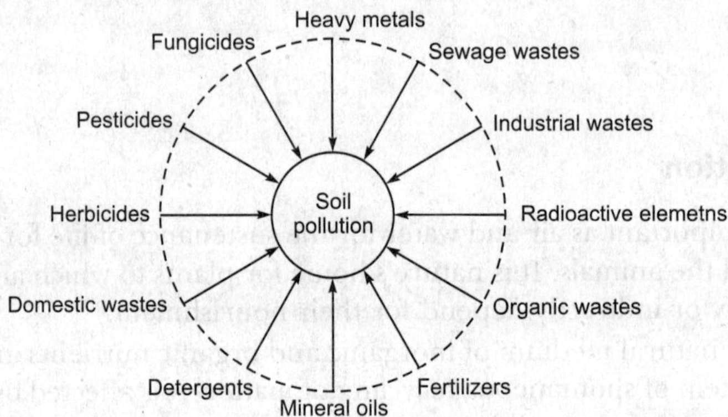

Fig. 28.1. Schematic summary of pollutants causing soil pollution.

(II) *Artificial Sources*

Pollutants produced due to man's activities, *i.e.*, artificial sources may be of the following types-

(1) Industrial wastes. Rapid industrialisation has given birth to the problem of waste disposal. Most Indian industries release their effluents directly into the surrounding fields. These effluents are often enriched with inorganic chemicals that enhance the salt content of the soil. Addition of excessive salts into the soil changes its pH, which, in turn adversely affect the soil organisms. Continuous inflow of such effluents into the soil gradually decreases the fertility of the soil rendering it unfit to support plant life.

At places slow and steady percolation or seepage of toxic heavy metals deep into the soil pollute the ground water. In the industrial areas of Punjab where woolen and bicycle industries are located, a high concentration of nickel, iron, chromium, cyanide and copper has been detected in the ground water at toxic level. Such conditions are common in industrial areas of several other cities of the country. In Kanpur, tannery effluents containing chromium are causing havoc in farmers inhabiting nearby.

(2) Solid wastes. Solid wastes include all substances in a solid or semisolid state disposed off in the soil. These are often garbages from homes and commercial places. Major sources of solid wastes in our country come from domestic wastes, shops, offices, restaurants, hospitals, educational institution and small scale industries. Often solid wastes contain hazardous substances that cause the degradation of soil leading to the level of harmful pollution.

(3) Radioactive wastes. Pollution resulting from radioactive wastes is perhaps the most dangerous and far fetched process. Sources of such pollutants are nuclear explosions, nuclear testings, nuclear reactors etc. Radioactive dust particles produced after an atomic explosion slowly settles down on the soil surface and gradually permeate the soil strata. When absorbed by the plants, these enter the human body via the food chain and cause serious damages. These also disrupt the growth and multiplication of soil microorganisms.

The great problem with radioactive elements is that they can not be destroyed once they have been created and it is only the passage of time which can reduce their harmful properties. Once these are discharged into the ground, these are held physically and chemically in the soil particles.

(4) Detergents. In developing nations, use of detergents is becoming increasingly common. Wastes water from domestic sources is usually rich in detergents. It is allowed to flow unchecked into the ground. Presence of detergents in the soil increases its alkalinity and phosphate contents which affect root growth of the plants and depresses the growth of soil micro-organisms.

(5) Agrochemicals. Several chemicals are used in agriculture to boost the crop productivity to keep pace with the ever increasing population of human beings. These chemicals contribute soil pollution considerably. Some of these chemicals are:

(i) ***Fertilizers***. Increase in crop yield requires the use of inorganic and organic fertilizers. Common fertilizers are urea, NPK, super phosphate, bone meal etc. Accumulation of excessive fertilizers particularly inorganic types disabalances the nutrient contents of the soil. Excess phosphorus and nitrogen affects adversely the plant growth and soil flora and fauna.

(ii) ***Herbicides***. Herbicides are chemicals applied to kill herbs or crop weeds. These are basically of two types *(a)* **selective herbicides** such as 2,4-D, 2,4,5-T and growth regulating substances that kill only

certain types of plants and *(b)* **non-selective herbicides** such as sodium arsenite, sodium chlorate etc. that kill all herbs. Herbicides are poisonous in nature and their indiscriminate use completely wrecks the ecosystem of soil.

(iii) ***Pesticides***. Pesticides are chemicals used to destroy pests which damage or kill crop plants. Commonly used pesticides are DDT (dichlorodiphenyl trichloroethane), BHC, aldrin, dialdrin etc. These often contain toxic heavy metals which increase the toxicity of the soil.

The use of pesticides is increasing steadily and posing health hazards to man and cattles. Excessive use of pesticides have adverse effects on soil fertility, plant growth and soil microbes.

(6) Acid rains. Percolation of air borne sulphuric and nitric acids along with rain water changes pH of the soil and causes adverse conditions for plant growth and soil microbes.

Effects of Soil Pollution

Some important effects of soil pollution are-

(1) Soil pollution decreases soil fertility. Loss of soil fertility makes the soil inhabitable for plants and other organisms.

(2) Soil pollution has adverse effects on soil microbial population that plays important role in soil formation and keeping it fertile.

(3) Indiscriminate use of agro-chemicals destroys the soil flora and fauna and soil properties.

(4) SO_2, SO_3 and oxides of nitrogen of the atmosphere are washed down by rain in the form of H_2SO_4 (sulphuric acid), HNO_2 (nitrous acid) and HNO_3 (nitric acid) which reach the soil and increase the acidity. Increased acidity develops unhealthy conditions for plant growth.

(5) Dumping of radioactive wastes onto the soil is lethal to plants and microbial life. Such pollution has far reaching effects and even human beings are not spared.

Control of Soil Pollution

Some common measures adopted for controlling soil pollution are –

(1) Control of pesticide pollution can be carried out by-

(i) Lowering its use in agriculture.

(ii) Use of degradable insecticides like organic phosphates.

(iii) Use of short-lived chemical pesticides.

(iv) Adopting biological control, *e.g.* use of predators or parasites of pests into practice in place of insecticides.

(*v*) Releasing sterile males into the natural population of insects or pests and

(*vi*) Rotation and diversification of crops etc

(2) Control of fertilizer pollution can be practiced by using long term planning of agriculture. Natural procedures to boost soil fertility, *e.g.* fallowing of fields, growing leguminous crops (to increase soil nitrogen), crop rotation, use of organic fertilizers or farm yard manures or biofertilizers etc. should be encouraged.

(3) Control of pollution due to garbage and other resistant material dumped on land can be adopted by *control tipping* methods or sanitary landfill methods. Another method is *burning of garbage* and utilizing heat to warm residential units and for generation of electricity. Organic wastes can be used for preparing manure and biogas.

(4) Control of acid rains can be achieved by controlling air pollution.

(5) Control of radioactive based soil pollution can be achieved by controlled disposition of radioactive isotopes and minimisation of their use.

(6) Discharge of hot liquids and oil leakage in the soil should be avoided.

(7) Sewage should be biologically treated and converted into manure instead of its uncontrolled release into the soil

Chapter Summary

Soil pollution is an undesirable change in the natural, physical, chemical or biological components of the soil. Soil may be polluted by natural or artificial sources. Natural sources are of plant and animals origins (plant residues and animal wastes). Artificial sources include *(i)* industrial wastes, *(ii)* solid wastes, *(iii)* radioactive wastes, *(iv)* detergents, *(v)* agrochemicals, and *(vi)* acid rains. The common effects of soil pollution are loss of soil fertility, flora and fauna. Soil microbes are also lost and soil becomes inhabitable to plants and other microorganisms. Control of pollution requires measures like control of pesticide pollution, fertilizer pollution, garbage pollution, acid rains and radioactive based soil pollution.

Study Questions

1. Write a brief account of soil pollution.
2. What is soil pollution? Describe their causes, effects and remedies.
3. Describe in brief the various sources of soil pollution and ways to control it.
4. Write short notes on:
   (i) Sources of soil pollution
   (ii) Effects of soil pollution.
   (iii) Control of soil pollution

Objective Questions. *Select correct answers*.

1. Soil pollution is caused by:

(1) Industrial wastes (2) Detergents

(3) Agrochemicals (4) All of these

2. Soil pollution is caused by:

(1) Aerosols (2) Ozone

(3) Acid rains (4) PAN

Answers

1. (4) *2.* (3)

29

CHAPTER

Marine Pollution

Learning Objectives
Introduction • Definition • Sources of Marine Pollution • Industrial Effluents • Oil Spillage • Effects of Marine Pollution • Effects of Industrial Effluents • Effects of Oil Spillage • Control of Marine Pollution • Control of Oil Spillage.

Introduction

Although marine aquasystem covers about four times more area (71% of earth) than land or terrestrial system of earth, yet civilised man by his uncivilised activities has caused marine pollution to the alarming level. *Marine pollution* is a kind of aquatic pollution in which undesirable changes of physical, chemical or biological properties harmful directly or indirectly to human welfare are caused. The term refers to the pollution of sea, ocean and estuaries. The main sources of marine pollution are-

(1) Industrial effluents
(2) Marine ship effluents
(3) Oil spillage
(4) Inflow of fertilizers and pesticides
(5) Nuclear tests and dumping of nuclear wastes
(6) Eutrophication

SOURCES OF MARINE POLLUTION

The important sources of marine pollution are:

(1) Industrial Effluents

The oceans have become the final settling ground for millions of tonnes of waste products from human activities. For example, in late 1960's, West Germany was dumping some 370 tonnes of sulphuric acid, 750 tonnes of iron sulphate, 20 tonnes of chlorinated hydrocarbons and 16,000 tonnes of gypsum waste every day in the ocean.

The table 29.1 shows annual discharge of effluents that occurred between 1970-75.

TABLE 29.1. Industrial and Agricultural Pollutants Discharged Annually into the Ocean of the World between 1970-75.

No.	*Pollutants*	*Approximate discharge in metric tonnes*	*Source*
1.	Hydrocarbons	15,000,000	Industries, automobiles, power plants etc.
2.	Petroleum and hydrocarbons	3,405,000	Industrial wastes oil tankers, off shore wells
3.	Lead	350,000	Automobiles
4.	Mercury	100,000	Industrial operations
5.	Aldrin toxaphene	25,000	Agricultural and public health departments
6.	DDT	25,000	Public health operations
7.	Benzene hexachlorides	50,000	Agricultural and public health operations
8.	Polychlorinated biphenols	25,000	Plastic industries biphenols.

In USA alone, millions of pounds of plastics have been discharged in sea or ocean. It is estimated that in between 1981 to 1994, some 39,676,000 tonnes of wastes have been discharged into the ocean by them.

(2) Marine Ship Effluents

A large number of ships and tankers operate in sea and ocean round the year around the world. They dump or discharge a huge amount of garbage, toxic chemicals, detergents, sewage, plastics etc. directly in ocean water. Such discharges considerably disturb the marine ecosystem.

(3) Oil Spillage

Oil spillage is a worst kind of pollution found in sea and oceans. The marine route is used for transporting huge shipments of oil and petroleum. An spill in oil or petroleum products due to accidents or a deliberate discharge of oil polluted waste cause serious polluted conditions. It has been estimated that about 285 million gallons of oil are spilled every year which is enough to coat a 20 feet wide with half an inch thick oil layer for 8633 miles over the water surface.

The oil spill may be caused by an oil well blow out, a platform accident, a large marine pipe line rupture or a heavy leaking tanker. Some major marine oil spills are given in table 29.2.

TABLE 29.2. Major Marine oil Spills of the World

No.	Year	Source or Tankers name	Location	Approx. spill size in tonnes
1.	1992	Oil well	Fergana valley Uzbekistan	2,72,000
2.	1991	Persian Gulf war	Persian Gulf	8,16,000
3.	1983	Iran-Iraq War	Nowruz Field Persian Gulf	2,72,000
4.	1983	*Castillo de Bellver*	Cape Town S. Africa	2,67,000
5.	1981	Storage tanks	Kuwait	1,06,000
6.	1979	*Aegean Captain*	Tobago	1,66,000
7.	1978	*Amoco Cadiz*	France	2,33,500
8.	1972	*Sea star*	Gulf of Oman	1,15,000
9.	1970	*Othello*	Baltic sea	1,00,000
10.	1968	*Torrey Canyon*	England	1,17,000

The severity of oil pollution came to light when an American supertanker *Torrey Canyon* struck the rocks of South West coast of England and spilt its oil, forming a 40 cm thick blanket over the sea water. It was a superhuman task to clear the wreckage. British Govt. has to spend over 5 lac ponds to deal with the problem. But what government couldn't reclaim was the thousands of aquatic animals which died of choking. The aquatic life and the self purification process of water is so impaired that it may take years to return to its normal condition. Thus, pollution arising from oil spillage is not only dangerous but has its far reaching consequences.

During Persian Gulf War, thousands of barrels of oil was burnt when oil wells were set on fire. Due to the burning, thousands of tonnes of smoke, SO_2, NO_2, CO and hydrocarbons were added towards atmospheric pollution,

(4) Inflow of Fertilizers and Pesticides

Rivers carry agricultural and commercial wastes to the sea which may contain fertilizers, pesticides and toxic heavy metals. These compounds cause adverse effects on the marine life. Such chemicals tend to accumulate in living matters that form part of the food chain, especially when the animals are sedentary, such as benthos and shellfish. They also reduce the shell thickness of the eggs of sea birds and impair the reproductive capacity of some sea mammals. By biomagnification process, these toxic chemicals increase their concentration in sea food chain, step by step, and ultimately cause damage human welfare. Also it may result into loss of species diversity.

(5) Nuclear Tests and Dumping of Nuclear Wastes

Nuclear tests and nuclear waste dumping is another major cause of pollution. Nuclear wastes remain radioactive for a very long period of time and pose serious threat to marine organisms.

(6) Eutrophication

Coastal pollution and sewage discharges from ships enrich the nutrients in the marine water. Wastewater and agricultural run-off into coastal waters contain increasing amounts of nitrogen and phosphorus. Such nutrients and higher ocean temperature, due to global warming, out burst the algal growth and cause agal blooms.

Red sea is named due to abundant growth of a toxic blue green alga, *Trichodesmium erythreum*. It imparts orange, red or brown colour to sea water. Several other algae including dinoflagellates contribute significantly to cause algal blooms or **red tides**. Common toxic algae found in red tides are *Gymnodinium, Gonyaulax, Aphanizomenon* etc.

EFFECTS OF MARINE POLLUTION

The curse of pollution has now began to engulph marine ecosystem in manifold ways. Some important effects are as follows-

(1) Effects of Industrial Effluents

Industrial discharges and dumping of waste products into the sea badly and adversely affect the dissolved oxygen and develop oxygen poor or depleted conditions. Oxygen deficiency causes suffocation to the aquatic life leading to their death. Such situation further augments the pollution level.

Wastes containing toxic substances and pathogens further aggravate adversities to marine life. Marine plants and animals die. Human beings develop various diseases by eating fishes of such area. One such example is **Minamata disease** caused due to mercury pollution.

Dumping of non-biodegradable wastes, such as plastic articles and floating nets cause various problems. Large number of birds, turtles and marine mammals are killed when they ingest such wastes.

(2) Effects of Oil Spillage

Most affected communities of oil spillage are shore based communities including various species of crabs, starfish, bivalves, lobsters, barnacles, sea weeds etc. Mass-scale fish kill occurs during oil spills. The most common victims are various sea birds. The sea birds are affected by direct oiling of their plumage, ingestion of oil or oiled prey or transfer of oil to nesting birds. The oil spillage causes high mortality rates of sea birds. About 50,000 to 250,000 birds are killed every year by spilt oil. It is found that badly oiled shore become largely denuded of animals and sea weeds. Surface oil film prevents the exchange of oxygen with the atmosphere which leads to deoxygenation and poses serious threat to marine life. Shore animals suffer due to toxic hydrocarbons which render failure of respiratory systems. Hydrocarbons and benzpyrene accumulate in food chain and consumption of contaminated fish by man may cause cancer.

(3) Effects of Agricultural Wastes and Pesticides

Toxic agricultural wastes and pesticides reaching sea with the inflow of river water cause disastrous effects on marine life. The killing of the various marine forms develops secondary pollution. Entry and accumulation of toxic substances or pesticides in the food chain has far reaching effects.

Recently, mariculture or marifarming in India has been banned by T.N. High Court because of excess usage of chemicals and antibiotics in the process. These have been found damaging marine ecosystem.

(4) Effects of Eutrophication

Nutrient rich conditions developed due to discharge of sewage and other organic garbage in the sea enrich its water with nitrogen and phosphorus. Such nutrients favour the abundant growth of algae called **algal blooms**, such as red tide. **Red tides** have several toxic algae. Some of these algal species produce toxins that attack the nervous system of fish leading to massive mortality. When these fish are eaten by sea birds, they also die. Consumption of such fishes end up killing marine mammals and affecting fish eating humans.

Algal blooms are dangerous even after they die. They sink to the bottom and reduce the oxygen supply. Such conditions become dangerous to bottom-dwelling organisms such as crabs, oysters and clams.

Algal blooms greatly affect fisheries, aquacultures, tourism and water sports.

(5) Effects of Nuclear Tests and Dumping of Nuclear Wastes

Developed countries, which produce radioactive substances, test them for their potential and finally discharge their nuclear wastes in the oceans considering that silent ocean is the most suitable place for testing and dumping, forget the hazardous consequences of their activities. Nuclear pollution is most hazardous as its effects persist from generation to generation.

(6) Effects on Coral Reefs

Marine pollution affects coral reefs adversely. The increased inflow of sewage and sediments leads to excessive growth of algae and the submerged vegetation which clogs the corals and reduce oxygen supply to the coral animals. The corals become brittle and stunted in such area. A rise in temperature of water around the reefs lead to the bleaching of corals. Marred coral reefs adversely affect the marine tourism.

CONTROL OF MARINE POLLUTION

Control of marine pollution is not an easy task rather it requires awareness, cooperation, skilled training, Govt. support etc. Following control measures may be adopted.

(1) A ban should be imposed against dumping of waste garbages along sea shore.

(2) Input of fertilizers, pesticides and industrial effluents into the river should not be allowed.

(3) Effluents should be discharged only after the proper treatments.

(4) Standard quality of oil tankers and cargo ships should be set and followed to minimise oil spillage accidents.

(5) Ban should be imposed on nuclear testing and on dumping of nuclear wastes into the ocean.

(6) Causes of eutrophication of sea water should be minimised. Sewage and organic waste discharges must be done after proper treatments or in proper systems.

(7) During recent years, several techniques have been developed to minimise loss due to oil spillage

(i) **Physical containment** of oil is carried out mainly by the placement of specially designed floating booms. The oil can then be mechanically collected off the water surface by specialised oil skimming barges, surface pumps, floating absorbants etc.

(ii) **Chemical control** methods have also been worked out. Chemical dispersants help in breaking up surface oil slicks into numerous droplets which are then dispersed into the large volumes of water and carried away.

(iii) **Bioremediation process** is a recent method to reduce loss occurring due to oil spillage. The process involves the degradation of organic contaminants as a result of biochemical activities of micro-organisms. Such bacteria convert complex organic substances into substances like fatty acid and carbon

dioxide. Some new strains of *Pseudomonas* have been developed by genetic engineering techniques which break hydrocarbons into simpler compounds rapidly and reduce such pollution. Such bacteria have been popularly called **superbugs**. The super bugs are capable of decomposing hydrocarbons at a faster rate than the natural strains.

(iv) Some cheap **absorbants** like saw dust, peat moss, pine bark etc. are sometimes used to absorb and prevent further spreading of the spilled oil. At times, synthetic absorbants like polyethylene and polyurethanes are also used. A large amount of oil can be removed from the sea surface by collecting out these absorbants after they have absorbed sufficient oil.

Chapter Summary

Marine pollution is a kind of pollution of marine water due to undesirable physical, chemical and biological changes harmful to human welfare. The main sources of such pollution are *(i)* industrial effluents, *(ii)* marine ship effluents, *(iii)* oil spillage, *(iv)* inflow of fertilizers and pesticides, *(v)* nuclear tests and dumping of nuclear wastes and *(vi)* eutrphication. The common industrial and agricultural pollutants are hydrocarbons, petroleum, lead, mercury, aldrin, toxaphine, DDT, benzene hexachloride, polychlorinated biophenolds etc. Oil spill is a worst kind of pollution which may be caused by an oil well blow out, a platform accident, a large marine pipe line leakage or oil tanker accidents. Several accidents of marine oil spills have occurred world over. Inflow of fertilizers, pesticides and toxic heavy metals from agricultural and commercial wastes through rivers causes adverse effects on marine life, especially of coastal regions. Nuclear tests and dumping of nuclear wastes in oceans by developed countries is another cause of marine pollution. Sewage discharges from ships and coastal cities enrich marine water with nutrients and promotes the formation of algal blooms. The Red sea is named after a blooming blue green alga, *Trichodesmium erythreum*.

Industrial wastes are hazardous to marine life not only by depleting the oxygen in water but also due to powering of toxic chemicals and metals. The effects of oil spillage is not restricted only to marine life but also to various sea birds. Some 50,000 to 250,000 birds are killed every year by spilt oil. Nuclear pollution is most hazardous as its effects persist from generation to generation. Pollution affected Coral reefs adversely affect the marine tourism.

Control of marine pollution is not an easy task as it requires awareness, cooperation, skilled training, Govt. support etc.

Inflow of agrochemicals and industrial effluents must be minimised and these should be discharged after proper treatment. Ban should be imposed on nuclear testings and on dumping of nuclear wastes. Norms should be set to reduce oil spillage accidents. Various available techniques such as chemical control, physical containment techniques, bioremediation process and use of cheap absorbants etc. should be adopted to minimise loss of the aquatic life.

Study Questions

1. What are the main sources of marine pollution? Describe their effects and suggest methods of control.
2. 'Oil spillage is a major cause of marine pollution'. Explain the methods to minimise its effects.
3. Write a brief account of-
   (I) Sources of marine pollution
   (II) Effects of marine pollution
   (III) Control of marine pollution
4. Define or explain the following terms in relation to marine pollution:
   (i) Marine pollution
   (ii) Oil spillage
   (iii) Industrial pollutants
   (iv) Marine algal blooms
   (v) Minamata disease
   (vi) Superbugs

Objective Questions. *Select correct answers*

1. Sources of marine pollution are:

(1) Industrial effluents (2) Oil spillage
(3) Nuclear tests (4) All of these

2. Superbugs are:

(1) Synthetic bugs (2) Bacteria
(3) Radionuclides (4) Industries

3. Torrey Canon is famous for:

(1) Gulf war (2) Industries
(3) Electric potential (4) Oil spillage

4. Red sea is named after:

(1) Red alga (2) Blue green alga
(3) Oil spillage (4) Brown alga

5. Minamata disease is an example of:

(1) Air pollution (2) Water pollution
(3) Marine pollution (4) All of these

Answers

1. (4) *2.* (2) *3.* (4) *4.* (2) *5.* (3)

Noise and Thermal Pollution and Nuclear Hazards

LEARNING OBJECTIVES
NOISE POLLUTION-Introduction • Sources of Sound Pollution • Effects of Noise Pollution • Control of Noise Pollution. THERMAL POLLUTION- Sources of Extra Heat in Water • Effect of Warm Water on Aquatic Life • Control of Thermal Pollution • NUCLEAR HAZARDS- Sources of Radiation Exposure • Biological Effects of Radiation • Protective Measures.

NOISE POLLUTION

Sound is a physical disturbance in a medium such as gas, liquid or solid which can be detected by human ear. **Noise** is a wrong sound at wrong place at wrong time. Noise is thus an unwanted and unpleasant sound. **Noise pollution** may be defined as *unwanted sound released into the atmosphere without any concern to the adverse effects it may have* or *the unwanted noise, released into the environment and affecting our lives adversely*. Though sound is an integral part of our lives but its high intensity has a great physical and physiological effects which can be extremely damaging.

It has been estimated that if sound of high intensity, as emitted by supersonic jetplanes, automobile horns, industrial machines, radios and loud speakers, continues for a prolonged period, it may cause temporary or permanent damage to the hearing with several other side effects. As such, in today's highly mechanised world, noise pollution has assumed alarming and frightening proportions. The condition is more severe in third world countries like India, where noise pollution is not accounted under the Environment Protection Act. World Health Organisation (WHO) has recommended the following noise levels for different localities of the city (table 30.1).

TABLE 30.1. Permissible Noise Levels in Different Localities of the City

No.	Locality	Noise level (dB) permitted	
		Day	Night
1.	Hospitals and Educational institutions, churches etc. (*i.e.,* silence zones)	45	35
2.	Residential zone	55	45
3.	Commercial shopping zone	65	55
4.	Industrial zone	75	65

Measurement of sound. The best known measurement of sound has been devised on log scale with a reference to 10 and is called *decibel* (*deci*=10 and *bell* after Graham Bell). The decibel (dB) measures a sound intensity by a ratio of loudness with reference to the softest audible sound. In log scale log 1=0, log 10=1, log 100=2, log 1000=3 etc.

From the faintest audible zero level, dB value exceeds to 30 in calm and quiet places such as libraries, operation theatres and radio recording rooms, whereas in quiet rooms and light traffic, the dB of sound is about 50. A dB of about 90 is created by moving scooters, trucks and buses. Noise produced by running machines in factories has a dB value exceeding 100 dB. It has been observed that sound beyond 85 dB falls under pollution level as it harms our hearing ability, beyond 100 dB, it becomes uncomfortable, and beyond 120 dB, it is painful.

The loudness of sound as generated by different sources is elucidated in table 30.2.

TABLE 30.2. Sound Levels of Different Sources

No.	Sources	Sound level (dB)	Sound level (as understood by man)
1.	Self whisper	25	Very Quiet
2.	Ordinary conversation	60	moderately sound
3.	Food blender	85	very loud
4.	Motor cycle	90	very loud
5.	Jet plane	105	uncomfortable loud and hearing damage
6.	Thunder clap	120	uncomfortable loud and hearing damage
7.	Hooting of trains	120	uncomfortable loud and hearing damage
8.	Jet plane taking off	150	Painful, ear drum rupture
9.	Rocket engine	180	Painful, ear drum rupture

Sources of Sound Pollution

The list of indoor and outdoor sources of high noise generation is infact endless. The common sources in our houses are gadgets like mixtures, vacuum cleaners, washing machines, coolers, air conditioners, radio and music systems, TV etc. The outdoor sources are factories, vehicles, aeroplanes, trains etc. The noise produced by machines in factories is a major occupational hazards. Loud speakers and crackers are used in many festivals and festive occasions like Diwali, marriages, religious and political ceremonies etc.

Effects of Noise Pollution

Noise exposure is an additional stress to man which affects according to its intensity, frequency and periodicity. Some illustrations are as follows-

(1) Effect of high intensity sound (80-100 dB). The sound of such intensity is emitted by industrial machines, motor cycles, high speed music systems etc. It can cause emotional and behavioral changes by producing nervous tension and cardio-vascular problems like heart disease and blood pressure due to constriction of peripheral blood vessels. It also affects the central nervous system (CNS) and aggravate peptic ulcers, allergies and eczema. Regular exposure to machine sound causes complete sensorineural hearing loss, rise in serum cholesterol and body plasma concentration in factory workers. A broadband noise of 90 dB adversely affects the efficiency of work but high frequencies impair functional performances.

(2) Effect of explosive sounds. The sounds above 110 dB is referred to as explosive sound which may be generated by crackers, trains, motor cycles etc. Such sounds cause acoustic trauma leading to severe concussion of internal ear, vomiting, vestibular symptoms and profound deafness.

(3) Effect of loud and sudden noise. A *sonic boom* produces startle effect and damages brain. It can also cause physical damages to property such as window breaks.

(4) Effect of intermittent noise. Such sound is very disturbing and may be a cause of psychiatric illness as found in areas having regular sounds of aeroplanes.

(5) Effect of low noise. Common noise produced by crowds, highways, radios, TVs etc. interferes with conversation and causes emotional and behavioral stress. It is believed that such noise is responsible for the increased consumption of alcohols, drugs, tranquilizers and sleeping pills. Emotional breakdown and insomnia may arise due to too much noise.

(6) Effect of absolute silence. Such is a nature of man that absolute silence is also undesirable. If in a perfect *anechoic room* (sound level 0-5 dB), one stays alone for a few minutes, he will be overcome by an uncomfortable sensation as pounding of heart will be felt and ringing noises start in the ears. One can become uncomfortable too. Mild sound (about 10 dB) has been found necessary for human existence. He feels pleasant in sounds produced by nature such as bird songs.

Control of Noise Pollution

The efforts to control noise pollution are usually aimed at lowering the sound intensity of the noise source. There can be three ways to eliminate or reduce noise pollution –

(*i*) eliminate the noise at the source,

(*ii*) modify the path of sound transmission and

(*iii*) provide the receiver with some forms of protection

The simple ways for reducing noise pollution are as follows:

(1) To reduce noise produced by industries. Noise of industry can be reduced by replacing old machineries by newer and more efficient ones. These machines have modern methods of rivetting and chipping in place of nuts and bolts. The noisy generators should be located far away from the place of work. Factory workers must wear ear-muffs (for sound above 90 dB) or ear plugs (for sound below 90 dB). Along with these adaptations, trees like ashok, banyan, neem, kadamb etc. should be grown around factory to minimise noise and dust.

(2) To reduce community noise. Use of loud speakers during marriages or other festive occasions should be banned or else permitted with certain restrictions like sound level and period of operations. Measures should be taken to prohibit the manufacture, sale and use of crackers of high sound intensities. Public should be made aware of their damages.

(3) To reduce traffic noise. Old and sound producing vehicles should be denied to operate on road. Hooting and blowing of horns needless should be restricted legally. In USA, vehicles having more than 82 decibel horns are not allowed. All along highways 50 feet wide plantation strips should be developed. In these strips, bushy shrubs and evergreen and dense trees should be grown. Such strips not only absorb noise but also deflect the sound upwards. Studies reveal that trees like *Casuarina* attenuate noise upto 10 dB.

(4) To reduce aeroplane and jet noise. Aerodromes should be located far away from the residential areas. Heavy and thick green belts around aerodromes not only function as noise absorbers but also serve as air purifiers. These can be needfully developed.

(5) Planning of cities and housing system. Cities should be developed in a way that they can escape from noise of industries and highways. Newly constructed buildings should have sound proofing mechanisms. There should be sufficient number of green belts in between residential areas.

(6) Legal control of noise pollution. According to the opinion of developed countries like USA, Switzerland, Germany etc. Indians are noisy people, every activity of theirs (social, religious, family) is performed in noisy way at the cost of health and peace of many others. It is perhaps because they are not aware of the assault of noise on the environment. However, recently Central and State Pollution Control Boards have been given power to frame certain rules and regulations to control noise pollution. But nothing appears to come out as yet. Certain rules and regulations are–

(i) Silence zones should be created near hospitals and educational institutions.

(ii) Use of sound amplifiers should be strictly restricted.

(iii) All vehicles must have effective silencers and uncontrolled blowing of horns should be declared illegal. Double sirens in and around residential area should be prohibited.

(iv) Mid-night air craft flights should be minimised.

(v) Restriction on factory noises should be handled legally. Government should encourage use of new soundless machines.

(vi) Development authorities should find legal provisions to establish sufficient green belts in and around cities.

THERMAL POLLUTION
(Thermal Water Pollution)

Water pollution due to heat is popularly referred to as thermal pollution. Hot water from different sources when released in the water bodies affects the aquatic life adversely. Both animals and plants are damaged due to hot water.

Sources of Extra Heat in Water

The industries, nuclear power plants and thermal power stations use large quantities of water for cooling. As a result water gets heated by absorbing energy. For them, it is easier to let off this hot water directly into the nearby water bodies. In thermal power stations, upto 70 per cent of heat is lost as waste which is absorbed by the cooling water. The water becomes hot and is then released in water reservoir. Such hot water often causes a rise in temperature upto 10-15°C and becomes fatal to aquatic life. Similarly an average sized nuclear power plant which produces 3000 megawatts of electricity, produces waste heat at the rate of about 20×10^9 BTU (British Thermal Units) per hour. This heat when let off in nearby water bodies has a profound effect on aquatic ecosystems. Sewage and industrial waste effluents when discharged in the water also increase temperature by 4-6°C.

Effect of Warm Water on Aquatic Life

Some important effects of hot water on aquatic plants and animals are as follows-

(1) Small animals, fishes, zooplanktons and phytoplanktons and small plants are killed outright by hot water.

(2) Mild hot water may also affect the aquatic organisms, which may be growth promoting or growth retarding. A bacterial disease *Salamon* is thought to aggravate by a slight rise in water temperature.

(3) Cellular proteins coagulate and semipermeable membrane loses its properties by hot water.

(4) Hot water adversely affects the enzyme system and metabolic activities.

(5) The quantity of dissolved oxygen in water is reduced as oxygen bubbles out. It is often called *gas bubble diseases* of water. This results in suffocation of animals and increase of BOD value.

(6) The susceptibility of organisms to toxic material present in water increases.

(7) The normal algal population may disappear and less desirable population may come up. Bloom forming blue green algae are more tolerant to heat. Thus the heat affects species diversity as well.

(8) Thermal pollution may affect the entire ecological balance, *e.g.* heat sensitive eggs may not hatch as those of trousts. Similarly eggs of certain fishes may hatch early. Several insects begin hatching with slight rise of temperature in water. The motile aquatic forms migrate away to the other water bodies etc.

Control of Thermal Pollution

Water should be allowed to cool down before it is discharged. It can be passed through fountains for rapid cooling. Large tanks or reservoirs should be constructed to retain the water for a little longer time. Time is the best precaution to cool down the water before being discharged outside. It should be preferably discharged in running water systems.

NUCLEAR HAZARDS
(Radiation pollution)

Radiation is the emission of rays and particles from a source, and the source of ionising radiations is the group of radioactive elements. The radiations are of two types:

(*i*) Corpuscular radiations (particulate in nature, *e.g.* α and β radiations).

(ii) Electromagnetic radiations (waves of shorter wavelengths, *e.g.* X-rays, UV rays, *r*-rays, infra-red rays etc.)

High energy radiations which have ionising property, are emitted by radioactive elements and are called ionizing radiations.

Sources of Radiation Exposure

The sources of radiation, to which man is exposed, can be-

(I) natural or (II) man-made.

(I) Natural sources include *(i)* cosmic rays, *(ii)* environmental and *(iii)* living organisms. Radionuclides of radium, thorium, uranium and isotopes of potassium (K^{40}) and carbon (C^{14}) are very common in soil, rock, air and water. Uranium and thorium ores are found in Kerala.

Man is also exposed to internal radiation due to presence of small amounts of uranium, thorium and isotopes of K, C and strantium (Sr) in the body. Internal radiation values vary from 25 to 75 m rads/yr.

(II) Man-made sources include *(i)* X-rays machines, *(ii)* radioactive fallouts (nuclear tests), *(iii)* nuclear reactor wastes *(iv)* industrial, medical and research uses of radioactive materials and *(v)* miscellaneous (TV, tube light, radium watches etc.).

Of these, nuclear power sources are important and of concern as their radionuclides are released in water as well as in the atmosphere.

Sources of Radioactive Pollution

Three main sources of radioactive pollution are as follows-

(i) Radioactive fall out from atomic weapon testings.

(ii) Radioactive fall out from small atomic weapons used for peaceful purposes.

(iii) Atomic waste material.

(i) Radioactive fallout from the atomic weapons. After atomic explosion (after testing of the atomic weapons), the radioactive dusts containing radioactive isotopes fall on the earth. These materials mix and interact with the natural particulate matter and the matter released by the man made industries etc. in the atmosphere and then continue to fall out gradually over broad geographic areas for a very long time. They produce ionising radiation and pollute the air. The extent of pollution depends upon type and size of bomb and the amount of environmental material which gets mixed up in the explosion.

A number of nuclear explosions have already been made during recent past in different parts of the world. In an explosion, roughly about 50% of the energy goes to the blast, 33% as heat and rest 17% or so to radioactivity. The radioactive dust that falls to the earth after atomic explosion is called **radioactive fall out**. These fall outs comprise various types of radioactive elements with different half life time. *Half life means the time taken for half the atoms in a given sample to disintegrate*.

Some important radionuclides with their target tissue and half lives is given in table 30.3.

TABLE 30.3. Some Important Nuclides and their Half Lives

Radionuclide	*Half-life*	*Target tissue*
Calcium-45	165 days	Bones
Carbon-14	5760 years	Whole body
Caesium-137	27 years	Soft tissue, genital organs
Iodine-129	17 Million years	Thyroid
Iodine –131	8 days	Thyroid, spleen, lymph
Plutonium-239	24,400 years	Bones, liver, spleen
Radium-226	1620 years	Bones
Strontium-90	28 years	Bones
Tritium (3H)	12.3 years	Whole body

Two most dangerous of radionuclides of the nuclear fall out are strontium–90 and cesium-137. Both of them contaminate environment and their effect persists for several years.

(ii) Radioactive fallout from small atomic weapons. Although most of the nuclear weapons are used for peaceful purposes, *e.g.* excavation of harbors and canals, in surface mining etc., these also pollute the environment.

About 80% fall out from such experiments is deposited on the ground, about 5% passes into lower atmosphere whose debris settle down in a few weeks and the remaining 15% goes to the upper surface and spread widely with air, come down with rains and enter the food chain.

(iii) Atomic waste material. In nuclear power plants, the radioactive isotopes are used as fuel. The radioactive wastes from nuclear plants may be in the form of gas, liquid or solid. These power plants are designed in a way that there is no leakage of radioactive wastes in any form. However, man-made procedures are not permanently leak proof, as has been evidenced by **Three Mile Island** nuclear power plant leakage in USA in 1979 and melt down of **Chernobyl** nuclear power plant in USSR in 1986. Both had hazardous effects.

Liquid effluents with dissolved and suspended materials enter water bodies and contaminate them. These enter the food chain and harm the organisms at all trophic levels including man. Radionuclides with long half lives continue to harm for equally long time.

Waste disposals containing radionuclides is a very tedious problem. There is no suitable and cheap procedure of disposing them, particularly for spent nuclear fuel, gaseous effluents and low level wastes. And also there is no reliable method for storing them, particularly when hundreds of tonnnes are being produced every year. At any time, radioactivity is likely to escape from the waste in water bodies, concrete cases and salt formations in high mountains and pollute the biosphere.

In modern procedures, the burnt up nuclear fuel (spent fuel) is sent to the reprocessing plants and then to the burial ground or to some other form of containers. The reprocessing plants and their disposing systems are often located some distance away from the nuclear power plant. By such process, there is always a danger of accidents resulting in the contamination of the environment during transportation of spent fuel.

Biological Effects of Radiation

The effect of radioactive pollutants depend upon-

(i) half-life time
(ii) energy releasing capacity
(iii) rate of diffusion and
(iv) rate of deposition of contaminants

Various atmospheric and climatic conditions also determine the pollution effects.

Biological effects of ionising radiations may be-

(i) Short range effects or
(ii) Long range effects

The **short range** effects are acute and expressed within few days or weeks after the exposure to radiation. The effects may be-

(a) physical crippling or
(b) immediate death

The **long range** effects take longer time to express. Such delayed effects of radiations are now centers of world's interest. These include-

(i) Genetic changes
(ii) Point mutations and chromosomal aberrations
(iii) Increased incidence of tumours and cancers
(iv) Shortening of life span
(v) Loss of vitality
(vi) Anaemia
(vii) Haemorrhages etc.

The most sensitive tissues for acute doses of radiation are intestines, spleen, lymph, nodes and bone marrow. Thyroid glands of children are badly damaged due to such exposures.

Two main radioactive elements in the nuclear fall out are Iodine –131 and Strontium–90 (I-131 and Sr-90).

I-131 easily enters the food chain at any level and gets concentrated in the terminal component through the process of *biological amplifications*. Unfortunately, man is most often the last link of many food chains. Once I-131 becomes accumulated in the human body, it can **damage–**
(i) White blood cells etc. *(ii)* Bone marrow etc. *(iii)* Spleen etc. *(iv)* Lymph nodes etc. and cause- *(a)* Lung tumours *(b)* Skin cancer *(c)* Sterility *(d)* Defective eye sight etc.

Strontium can replace calcium in plants and animals, and **SR-90** is a bone seeker. It causes-

(i) Bone cancer, and
(ii) Tissue degeneration.

Sr-90 reaches dairy products through vegetation and use of it by cattle and then to man by consumption of contaminated food, meat, milk and dairy products.

Transmutations of radioactive isotopes incorporated in the structural component of the body of the organisms cause complete chaos in the system as the very basis of internal structure of the body may crumble down.

Organism infested with radioactive substances is a moving source of radioactivity and health hazards for others who come in contact with him.

The biological effects of ionising radiation may also be divided into two main groups, the somatic and genetic effects, as follows

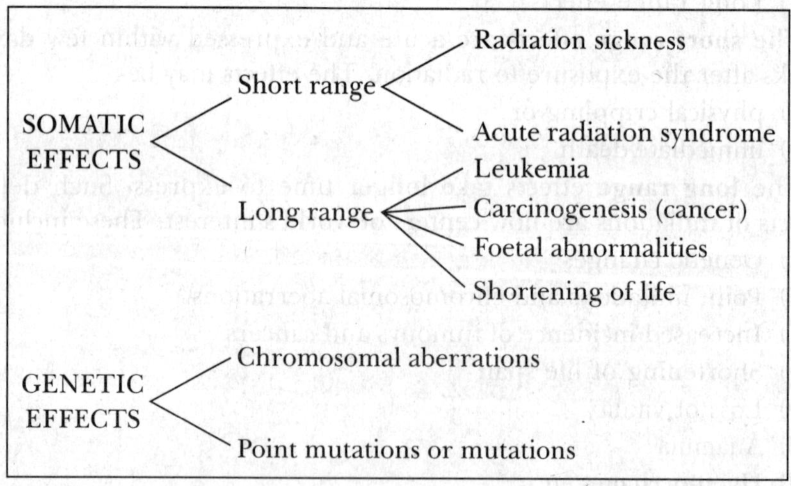

Protective Measures

Following are a few important measures to protect from damage done by radio-active substances-

(1) While handling UV lamps, dark glass spectacles or goggles must be worn. Ultra violet rays can not penetrate dark glasses.

(2) Visible light neutralises UV damage considerably. Thus, exposure to sun light to any individual exposed to UV could be a good remedial measure to treat the exposed individual.

(3) Nuclear fall out hazards must be minimised by adopting certain precautions, *e.g.*

(i) **High level wastes** are long-lived and have high radioactivity per unit volume. These must be contained somewhere as follows-

(a) In underground tanks without treatments.

(b) Liquids should be converted into inert solids like ceramics and then buried in deep underground.

(c) Stored in deep salt mines.

(ii) **Low level wastes** which have a very low radioactivity must be dispersed into the environment in such a way that these can not enter the food chain. The high level components of these wastes should be separated and disposed like high level wastes.

(4) The radioactive wastes should be stored in places where these gradually decay to their final stable products.

(5) These should be used in minimum and subjected to natural disintegration.

Chapter Summary

Noise pollution is referred to as the unwanted noise released into the environment that affects our lives adversely. Sound above 85 dB is considered as noise pollution. Sources of such intense sounds are many including machineries, automobiles, aeroplanes, supersonic jets etc. The intense sound (80-100 dB) can cause emotional and behavioral changes by producing nervous tension and cardiovascular problems. Explosive sounds (above 110dB) cause severe concussion of internal ear, vomiting, profound deafness etc. Control of noise pollution is possible by reducing *(i)* noise of industries, *(ii)* noise of community, *(iii)* traffic noise, *(iv)* aeroplane and jet noise etc. Legal remedies should also be enforced.

Water pollution due to heat is called **thermal pollution.** Sources of extra heat in water are industries, nuclear power plants and thermal power stations. Hot water affects the aquatic life adversely. Aquatic organisms may be directly killed or may die due to depletion of oxygen.

Radiation is the emission of rays (electromagnetic radiations) or particles (carpuscular radiations). Natural sources of radiations include cosmic rays, environment and living organisms. Radionuclides of radium, thorium, uranium, C^{14} and K^{40} are commonly found in soil, rocks and air. Man-made sources of radiation include *(i)* x-ray machines, *(ii)* radioactive fall outs, *(iii)* nuclear reactor wastes, *(iv)* uses of radioactive materials in testing nuclear weapons, research and treatments etc. Biological effects of radiation are of short and long range and on somatic and genetic cells. It is most hazardous type of pollution having its effects transmitted from generation to generation. These may cause mutations, chromosomal aberrations, cancer etc. Protective measures are only methods to control nuclear pollution of which risk free disposal is most effective.

Study Questions

1. Give a brief account of noise pollution.
2. What are various natural and man made sources of radiation pollution? Describe few methods of protection against them.
3. What is pollution? Describe sources and pollutants of radioactive pollution. Suggest methods of control.
4. Describe briefly the thermal pollution.
5. Write short notes on:
   (i) Radioactive wastes
   (ii) Effects of noise pollution
   (ii) Control of noise pollution
   (iv) Control of thermal pollution.

Objective Questions. *Select the correct answers*

1. Ear muffs or cotton plugs are used to reduce pollution of:
   (1) Air (2) Nuclear
   (3) Noise (4) Thermal

2. Increased heart beat , constriction of blood vessels and dilation of pupil are symptoms of:
   (1) Air pollution (2) Noise pollution
   (3) Nuclear pollution (4) All of these

3. Aquatic life is damaged by:
   (1) Sound pollution (2) Air pollution
   (3) Thermal pollution (4) All of these

4. Exposure of human body to radiation may cause:
   (1) Polio (2) Hemophilia
   (3) Gastroenteritis (4) Leukemia

5. A bone seeker nuclear fall out is:
   (1) Sr^{90} (2) $C^{14}O2$
   (3) P^{35} (4) Sr^{85}

Answers

1. (3) *2.* (2) *3.* (3) *4.* (4) *5.* (1)

31 CHAPTER

Solid Waste Management

LEARNING OBJECTIVES
Introduction • Classification of Solid Wastes • Types and Sources of Solid Wastes • Causes of Solid Wastes • Effects of Solid Wastes • Solid Waste Management • Recycling of Solid Wastes.

Introduction

The term **solid waste** refers to all discarded and thrown away solid and semi-solid wastes arising from human and animal activities. Its quantity depends upon season and standard of people living in an area. Advanced societies believe in *use and throw culture*, that further increases solid wastes.

Classification of Solid Wastes

Solid wastes may be classified into the following categories:

(1) Garbage. It includes putrescible (easily decomposable) organic wastes like animals, fruits or vegetable residues, meat, bones, spoiled food etc.

(2) Rubbish. These do not decompose rapidly and can be of two types:

(i) ***Combustible*** like paper, cardboard, textiles, wood items, leather, plastic containers etc.

(ii) ***Non-combustible*** such as crockeries, metals, metallic cans, empty glass bottles etc.

(3) Industrial wastes. These arise from industrial activities, such as fly ash, chemicals, sludge, paints, toxic substances etc.

(4) Agricultural wastes. These are crop residues and include materials like jute, cotton, rubber, tea, coffee, coconut, straws, sugarcane wastes, cattle-shed wastes etc.

(5) Pathological wastes. These include slaughter house wastes such as meat, hairs, bone chippings, hides, skin and excretions.

(6) **Hazardous wastes**. These are solid wastes which affect adversely to living beings, such as radioactive wastes, toxic chemicals, flammable wastes, explosives and hazardous wastes from hospitals and research institutions.

(7) **Demolition and construction wastes**. These are building material wastes like bricks, stones, concrete, plasters, plumbing and sanitary parts etc.

(8) **Miscellaneous wastes**. Solid wastes from uncertain sources such as abounded vehicles, street sweepings, dead stray animals etc. are included in this category.

Types and Sources of Solid Wastes

Three general categories of solid wastes are-

(i) Municipal wastes,
(ii) Industrial wastes and
(iii) Hazardous wastes.

(i) Municipal wastes. Municipal wastes arise from domestic activities, restaurants, institutions, public places etc. and typically include rubbish, garbage, ashes, building materials, treatment wastes, dead animals etc. Sources of such wastes are residential, commercial and open areas. Average composition of such wastes is given in table 31.1.

TABLE 31.1. Average Composition of Municipal Refuse

Solid waste	*Approximate composition (%)*
1. Paper, wood, cardboard	53-55
2. Garbage	22-25
3. Glass, crockery, ceramics etc.	8-10
4. Metals	5-8
5. Plastic, rubber, discarded textiles, vehicles junks etc.	5-7

(ii) Industrial wastes. These arise from industrial activities and typically include ashes, rubbish, building material wastes, toxic wastes, metal containers, plastic containers, paints etc.

(iii) Hazardous wastes. These are substantially dangerous and hazardous to plants, animals and human beings. Substances which have characteristic ignitablility or corrosivity or reactivity or toxicity are called hazardous. Radioactive substances, chemicals, biological wastes, flammable wastes and explosives are common hazardous wastes.

Causes of Solid Wastes

Quantitative increase of solid waste is quite rapid due to:

(1) Over population. Population increase is a great problem nowadays and is causing a rapid growth of pollution of all kinds including solid wastes. Advanced countries have adopted a culture of **use and throw** which is further aggravating the problem. The equation is very simple that an increase in population increases pollution and these can be combined to form one word, *i.e.*, **popollution**.

(2) Urbanisation. Solid waste is primarily an urban problem where people develop the habit of using a variety of items and discarding them afterwards. These comprise almost all kinds of solid wastes.

(3) Affluence. In an affluent society, the per capita consumption is very high and there is a general tendency to declare items as obsolete. Such discarded items increase solid wastes regularly. Removal of such wastes develops a serious problem.

(4) Technology. Technologies today have changed the culture of using things. It is distinctly apparent in package industry for most economic goods. There is a shift in technology from the returnable packaging to non-returnable packaging. For example, the returnable glass containers or bottles are being replaced by non-returnable cans, plastic containers, plastic bottles etc. Since packaging materials like those made from plastic are non-biodegradable, they are largely responsible for causing solid waste pollution.

Effects of Solid Waste Pollution

Solid wastes can pollute air, water and soil; leave several environmental impacts; and cause health hazards. Improper handling and transportation of solid wastes further aggravate these effects. Some of the effects are as follows:

(I) Health Hazards

(1) During handling and transfer of hospital and clinic wastes, disease transmission may take place. Several diseases are possible through their pilferage.

(2) Vectors like rats and insects invade refuse dumps and can spread various diseases.

(3) Diseases like bacillary dysentery, diarrhoea and amoebic dysentery may result in humans from food and water contamination through flies.

(4) Rats dwelling with infectious solid wastes may spread diseases like plague (black death), salmonellosis, trichinosis, endemic typhus etc.

(5) Water supply, if gets contaminated with pathogens present in solid wastes, may result in cholera, jaundice, hepatitis, gastroenteric diseases etc.

(6) Choking of drains and gully pits by the solid wastes results in water logging which facilitates breeding of mosquitoes and results in the spread of diseases like malaria and plague.

(II) Environmental Impacts

(1) Organic solid wastes emit obnoxious odour on their decomposition and makes the environment polluted.

(2) Burning of waste products like plastic and rubber pollutes the atmosphere with noxious fumes.

(3) Scavengers and stray animals invade the roadside garbage and litter the waste over large area causing much aesthetic damage to the atmosphere.

(4) Leachates from such refuge dumps percolates into the soil and contaminates underground water.

Solid Waste Management

Solid waste management (SWM) may be defined as the application of techniques that will ensure the orderly execution of the functions of collection, processing and disposal of solid wastes. Its main objective is to reduce or minimise the harmful effects and find ways for their better utilisation.

The three basic functional elements of SWM are:

(i) Collection

(ii) Disposal

(iii) Reutilisation or Recycling

(I) *Collection of Solid Wastes*

Collection is the first fundamental function of solid waste management. It is to be made from various places such as residential, commercial, institutional and industrial areas. There are three basic methods of collection:

(i) Community storage points. The refuse is first stored in small storage bins and at some allotted time, the stored refuse is made available to collection agency. The agency collects it in vehicles for proper disposal. This procedure is mainly followed for municipal refuse.

(ii) Kerbside collection. In this process, the refuse is brought in containers and placed on footway from where it is picked up by collection agencies.

(iii) Block collection. In this system, the refuse container is emptied into collection vehicles at the point of collection.

In advanced countries and in some of Indian metropolitan-cities, biodegradable and non-biodegradable refuses are collected and are disposed separately.

(II) *Disposal of solid wastes*

Solid waste disposal is a big problem and a long process, as it has to be carried out with following objectives:

(i) The process should be economically viable.

(ii) It should not create any health hazard.

(iii) It should not cause adverse environmental effects.

(iv) It should not be aesthetically unpleasant.

(v) It should involve opportunities for recycling of materials or waste utilisation.

At present, the common methods adopted in solid waste disposal are as follows-

(1) Manual component separation or **salvage**. Reusable articles are manually separated. Such items are cardboards, paper, newspaper, glass, metals, wood, aluminum cans etc. These are manually sorted out or salvaged either for recycling or for resale.

(2) Compaction. After separation, volume of wastes is mechanically reduced by compactors. The compressed waste material is filled either in large containers or solid bales are formed. Such process facilitates transportation.

(3) Incineration. The combustible wastes are subjected to incineration, *i.e.*, burning at a very high temperature (above 670°C). Plastics, cardboards, rubber, cartons, wood, scrap, floor sweepings etc. are incinerated for volume reduction.

Since incineration results in air pollution by producing fly-ash, corrosive fumes of SO_2, hydrogen chloride and organic acids etc., adequate procedures should be adopted to avoid contamination of air.

Incineration of solid wastes is carried out in properly constructed hearth of furnaces. Since it requires relatively a high initial cost, the method is generally used when suitable dumping land areas are not available and disposal in sea is not possible. The final products of incineration are ashes and clinkers. The ashes are dispersed by dumping in low lying areas while clinkers can be used as aggregate for low grade concrete or as road material.

Incineration process has following advantages:

(a) No odour or dust nuisance is produced.

(b) The method is sanitary, as all vectors and pathogens are destroyed.

(c) Some revenue can be generated by raising steam power and selling of clinkers.

(4) Open dumping. Solid wastes are directly dumped in low lying areas of the earth. Since this method is cheaper, it is extensively adopted in India. Such dumpings have certain disadvantages like-

(*a*) Public health hazards are caused because such dumping promotes breeding of flies, mosquitoes, rats and other pests.

(*b*) Open dumping requires larger areas, thus causes wastage and shortage of lands.

(*c*) These dumps directly and on burning cause air pollution.

(5) Sanitary land filling. It involves the disposal of municipal wastes on or in the upper layers of degraded areas in need of restoration. In the process, solid wastes are compacted and spread in thin layers. The final layer is covered by one meter thick earth layer to prevent burrowing by rodents. It is a biological method of waste treatment which results in decomposition products like CO_2, CH_4, NH_3, H_2S and H_2O which can be harnessed as renewable sources of energy.

This method has certain advantages like-

(*a*) It does not cause environmental damages.

(*b*) It does not cause health hazards.

(*c*) It prevents breeding of pests and disease vectors.

It has certain disadvantages like leaches can cause-

(*i*) Pollution of ground water and

(*ii*) Pollution of surface water

(6) Pyrolysis. It is a kind of destructive distillation in which the solid wastes are heated in pyrolysis reactor at 650-1000°C in oxygen depleted environment. By this process, the chemical constituents and chemical energy of some organic wastes are recovered. The organic constituents split up into gaseous liquid and gaseous fractions (CO, CO_2, CH_4, tar, charred carbon etc.)

(7) Composting. Bacterial decomposition of the organic components of the municipal wastes results in the formation of humus or compost and the process is known as composting. This method is best suited to Indian conditions, especially for small and medium size towns. It solves three problems simultaneously, *(i)* disposal of solid wastes, *(ii)* disposal of night soil and *(iii)* production of valuable manure for crops. Composting may also be called **biodegradation**.

(III) *Recycling of solid wastes*

The recovery of solid waste components for possible use as raw materials is called recycling salvaging. It has two important steps: *(i)* separating materials from refuse, and *(ii)* reprocessing them for reuse. Separation can be done either at the point of generation or at a control place with processing facility. In India, several poor children do the process of segregation round the year. They segregate paper, metal pieces, plastic etc. in separate containers.

Recycling is an integrated part of solid waste management. By proper utilization of solid wastes, a developing country like India can earn several advantages, *e.g.* Waste utilisation –

(*i*) Contributes, directly or indirectly, to economic development.
(*ii*) Generates opportunities for employment.
(*iii*) Helps in conservation of natural resources.
(*iv*) Helps to make many useful products of basic necessities.
(*v*) Reduces environmental hazards and air and water pollution.

Examples of Recycling and Waste Utilisation

Some important examples are as follows:

(1) Construction material from wastes. Construction material can be produced or harnessed from wastes such as silt from water works, red mud from aluminum industry, agricultural wastes, fly ash etc. Recently, fly ash from thermal power plants has find uses in multiple ways, *e.g.* it is used in/ or as (i) manufacture of portland pozzolana cement, (ii) manufacture of oil well cement, (iii) making sintered light weight aggregates, (iv) clay binding bricks, (v) precast fly ash concrete building units, (vi) filter in mines, (vii) in bituminous concretes, (viii) as plasticizers, (ix) as water reducing and sulphate resisting concrete, (x) amendment and stabilization of soil, (xi) in insulating bricks etc.

(2) Utilising agricultural wastes. Agricultural wastes can be used in different ways. In general, they are used for the manufacture of paper and black boards. Sugarcane bagasse is used as chief sources of cellulose. Attempts are being made to produce or recover protein from such wastes etc. Such wastes are also utilized for the production of energy.

(3) Medicines from agricultural wastes. Filyal is readily available from agricultural wastes such as from corn cobs and oak hulls. Furfural is produced commercially by the reaction of corn cobs with sulphuric acid. It is the basic material used for the synthesis of nitrofurans, a germicide used for treating cattle diseases.

(4) Utilising slaughter house wastes. Wastes from slaughter houses can also be utilised, *e.g.* blood is used in pharmaceutical industries and hides and skin are used for leather production.

(5) Recovery of heavy metal ions. Presence of toxic heavy metal ions in industrial wastes is of major concern. These can be recovered by chemical treatments or by bio-extractive technology.

(6) Liquid fuels from agricultural wastes. Techniques are being developed for the production of ethanol by fermentation of agricultural wastes for use as a liquid fuel. This will reduce the use of petroleum. In Brazil, gasohol (alcohol + gas) is frequently used in automobiles.

(7) Energy from wastes. Industrial and urban wastes can be utilized as an enormous source of potential energy. A national programme in India has been launched since June, 1995 for proper treatment of these wastes and the resultant recovery of energy from these sources. An incineration plant has been installed in Delhi. It has a capacity of using 300 tonnes of solid municipal wastes per day. A pyrolysis plant is proposed to set up in Bombay which will be used to extract liquid fuels. Several projects for power generation from bagasse (produced in sugar mills from sugarcane) are under execution in the country.

(8) Conversion of agricultural wastes into cheap and efficient fuel. Indian and Dutch scientists have developed a technology of converting agricultural wastes like rice husks and groundnut shells into briquettes to be used as an efficient economical and non-polluting fuel. The process is known as screw-press technology. These briquettes are potential environmental friendly fuel resources.

(9) Rubber from old tyres. Used tyre casings are reused in the manufacture of synthetic rubber. Sulphur mounting bacteria *Seuephotobees* consume sulphate from used tyre wastes and leave the polymer back bone of carbons (rubber) intact during recycling.

(10) Proteins from cellulose wastes. Scientists of Louisiana State University of USA have developed techniques for making protein from cellulose wastes with the help of cellulomonas bacteria. The bacterial enzymes disintegrate the cellulose chains into digestible form.

(11) Utilising aquatic weeds. Common aquatic weed water hyacinth (*Eichhornia crassipes*) can be utilised by conversion into fertilizers, biogas production and in the manufacture of papers.

(12) Oil from plastic wastes. German scientists have shown possibility of converting plastic wastes into oil. Plastic bags, yogurt cartons, computer casing etc. are used to produce synthetic oil which is used to produce high quality oil based products.

(13) Plastic for heat and electricity generation. European Association of Plastic Manufacturers have found that greater the proportion of plastics in waste, the greater is the heat produced. This heat can be used for generating electricity.

(14) Recycling of plastics. Plastic is easily recycled to make new packs, soft boxes, greases and adhesives etc.

(15) Recycling of metals. Various metal pieces collected from the wastes are reintroduced into the production cycle for the production of the same products.

(16) Cooking gas and electricity from organic wastes. Government of India is promoting and subsidising to install biogas plants for generating cooking gas and generating electricity. Cattle dung is frequently used for biogas plants.

Chapter Summary

All discarded and thrown away solid and semi-solid wastes arising from human and animal activities are called solid wastes. These include garbage, rubbish (combustible and non-combustible), industrial wastes, agricultural wastes, pathological wastes, hazardous wastes, demolition and construction wastes etc. On the basis of types and sources three general categories can be recognized *(i)* municipal wastes, *(ii)* industrial wastes and *(iii)* hazardous wastes. The main causes of solid wastes are *(i)* over-population, *(ii)* urbanization, *(iii)* affluence and *(iv)* technology. Solid waste pollution causes health hazards, and environmental disturbances. Vectors inhabiting solid wastes spread various human diseases like bacillary dysentery, diarrhoea, amoebic dysentery, plague, salmonellosis, trichinosis, endemic typhus, cholera, jaundice, hepatitis, gastroenteritis etc.

Solid waste management is a difficult task. Its main objective is to reduce the harmful effects and find ways for their utilisation. The three basic functional elements of SWM are *(i)* collection *(ii)* disposal and *(iii)* reutilisation or recycling. The three basic methods of collection are *(a)* community storage points, *(b)* kerbside collection and *(c)* block collection. In collection process, biodegradable and non-biodegradable refuse must be separated and disposed separately. At present, the common method adopted for disposal of solid wastes are *(i)* salvage or manual component separation, *(ii)* compaction, *(iii)* incineration, *(iv)* open damping, *(v)* sanitary land filling, *(vi)* pyrolysis, *(vii)* composting and *(viii)* recycling. Some important examples of recycling and waste utilization are *(1)* construction material from wastes, *(2)* utilising agriculture wastes, *(3)* medicines from agriculture wastes, *(4)* utilizing slaughter house wastes, *(5)* recovery of heavy metal ions, *(6)* liquid fuels from agriculture wastes, *(7)* energy from wastes, *(8)* conversion of agricultural wastes into cheap and efficient fuel, *(9)* rubber from old tyres, *(10)* proteins from cellulose wastes, *(11)* oil from plastics, *(12)* recycling of plastics, *(13)* cooking gas and electricity from organic matter etc.

Study Questions

1. What are solid wastes? Suggest various methods to reduce their hazardous effects.
2. What are various sources, causes and effects of solid waste pollution? Describe.
3. Write a brief account of solid waste management.

Objective Questions. *Select correct answers*

1. Solid wastes include:
   (1) Garbage (2) Rubbish
   (3) Hazardous wastes (4) All of these

2. Causes of solid wastes are:
   (1) Urbanisation (2) Over population
   (3) Use and throw culture (4) All of these

3. Solid waste management includes:
   (1) Collection (2) Disposal
   (3) Recycling (4) All of these

4. Solid wastes can be made to use for:
   (1) Packing (2) Transportation
   (3) Recycling (4) Municipal earnings

Answers

1. (4) *2.* (4) *3.* (4) *4.* (3)

32

CHAPTER

Role of an Individual in the Prevention of Pollution

LEARNING OBJECTIVES

Introduction • Direct Role of an Individual • Preventive Measures for Air Pollution • Preventive Measures for Water Pollution • Preventive Measures for Solid Waste Pollution • Indirect Role of an Individual • Icons of Environment India • Anna Hazare • Rajendra Singh • Sundar Lal Bahuguna.

Introduction

Pollution is a man-made disease that can be cured only by man and remedies adopted by them. Man's quest for advancement has entered in a competition of industrialisation and information technologies. Incomplete and inefficient technology coupled with increased population are the main causes of man-made pollution.

Today, the problem of pollution has become a major challenge to scientists, environmentalists and humanists. The pollution of various components has gone to such an extent that we are unable to breathe fresh air, drink pure water and get healthy food. If man has to survive, he has to fight and overcome this gigantic problem before it overcomes him and wipes out his very existence.

Scientists all over the world are awfully concerned over the unmindful over-exploitation of our heritage, the beautiful nature, which must be nurtured for the future generation. Making this their central theme, over 1,500 representatives of 114 countries met in Stockholm from 5 to 14 June, 1972 and adopted the motto **Only One Earth** for the entire humanity. This *International Conference on Human Environment* declared 5th June as the **World Environment Day**. The Indian representative of the conference, our then, Prime minister Smt. Indira Gandhi stated, that

"*Modern man must re-establish an unbroken link with nature and with life. He must again learn to invoke the energy for growing things and to recognize, as did the ancients in India, centuries ago, that one can take from the earth and atmosphere only so much as one can put back into them.* Our Father of Nation, Mahatma Gandhi stated that *Earth provides enough to satisfy one's need, but not any man's greed.*

Administrative director of UNEP, M. Kolba (1992) suggested that **care and share** is the simple and best method to control pollution. This is being followed in advanced countries but in developing countries, people have began to understand what is *care* but they don not follow to *share*. The main causes of pollution in developing countries are-

(i) Increasing population
(ii) Illiteracy and poverty
(iii) Rapid industrialisation
(iv) Failure in pollution management.

Individually man can participate in solving the problem of pollution mainly in two ways:

(I) Direct role or direct participation
(II) Indirect role or indirect participation

Direct Role of an Individual

Direct role of an individual involves those functions which are done by himself. Since nature can not solve the problem of pollution on its own basis, it is the duty of man to find means to solve this problem.

Since the cause of pollution is mainly human beings, only they can minimise it by preventing its increase and spread. One can remember the old saying that **prevention is better than cure**. A man should regulate and reduce the causes of pollution at every step. Some simple principles to reduce the pollution are as follows-

(1) *Preventive measures for air pollution*

1. Establish or transfer factories or industries away from residential areas and follow treatment methods to reduce pollution.
2. Reduce the use of autovehicles, by moving in pool based common groups and not driving the vehicle at least one day in a week.
3. Keep vehicle filters clean and use only quality fuels.
4. Make vehicle's engine off while taking or standing at any crossing.
5. Use CNG (Compressed Natural Gas) as fuel for auto engines instead of liquid petroleum.
6. Use self or low noise horns and avoid blowing them unnecessarily.
7. Use efficient silencer in vehicles.
8. During celebrations, loud speaker should be avoided or use them below sound pollution level (below 85dB).

9. Use LPG for cooking instead of coal.

(2) *Preventive measures for water pollution*

1. Sewage disposal is the main cause of water pollution in big cities. One should take care for proper disposal of sewage.
2. Sewage should be diverted towards oxidation ponds or treatment plants.
3. Sewage should be released to the river only after treatment.
4. Industrial effluents should be released into the water bodies only after the proper treatment.

(3) *Preventive measures for solid waste pollution*

1. Divide the domestic wastes into biodegradable and non-biodegradable components before discharging.
2. Collect regularly and systematically the biodegradable part of wastes and dispose it every day in closed plastic bags.
3. Non-biodegradable wastes should be collected and disposed for recycling processes.
4. One should never allow scattering of domestic wastes.

Indirect Role of an Individual

Man can participate in various activities for solving the problem of pollution, *e.g.*

1. One should support expansion of environmental education by participating or supporting the system.
2. One should make efforts to develop awareness of environmental pollution. It can be done either by-
   - *(i)* Publishing information on pollution
   - *(ii)* Organising various exhibitions and competitions or
   - *(iii)* Sponsoring environment related advertisements.
3. One should make Non Governmental Organisations (NGO) or encourage and help in developing NGO for social services against environmental pollution. Green friends, Ecofriends, Green Brigade, Eco Task, Socleen etc. are various NGOs serving the society and developing awareness against environmental pollution in public.
4. One should celebrate environmental dates as festivals of society and develop them as non-caste and non-religious occasions for all.
5. One should organise special programmes and rallies on *Tree Plantation day* (7 July) and *World Environment Day* (5 June) and raise the slogans in the society, *e.g.*

Each one Tree one
Donate one tree
Keep your city clean and green
Nature for future
Nature nurture etc.

6. One should encourage to follow Environmental Laws by making them popular and help people to respect.

SOME ICONS OF ENVIRONMENT INDIA

Every one has right to live peacefully in healthy environment. For this, individual as well as community involvement is extremely effective. This fact can be substantiated by the contribution of following three great environmentalists of the country.

Anna Hazare

A retired military man of village Ralegan Siddhi of district Ahmednagar, Maharashtra decided to uplift his village. In 1976, Anna Hazare started integrated development programs based on watershed management, soil conservation, afforestation, promotion of use of biogas and solar cooking devices. The popularity of using alternate energy strategy led his village declared as an **Urja Gram** by the government. All credit goes to dedicated and well determined Anna Hazare who successfully formulauted sustainable development of the village and its people.

Seeking support and cooperation of his villagers was a difficult task. He had a modest beginning with the renovation of the village temple and developed it as a centre where people could assemble to discuss how to bring about the welfare of the village. With the support of villagers, Hazare began watershed development to enhance irrigation facilities. A well managed systems of cooperatives formed around the wells dealt with the distribution of water on the basis of possession of a ration card and against the payment of water tariff. Irrigation benefitted to all villagers. Watershed management was followed by massive tree plantation which provided enough food for cattles and developed a healthy environment. A pasture land of about 500 acres has also been developed. Soon the prosperity of village became evident and agriculture produce became surplus. Surplus grains are stored in **Grain Bank** from where villagers can borrow at the time of need.

Voluntary services of villagers, dedication of Hazare and well planned cooperative system developed Ralegan Siddhi as a symbol of community development and cleanliness. This village plays a role model for other villages which have developed under **Adarsh Gram Yojna** established by the government of Maharashtra.

Rajendra Singh

General secretary of Tarun Bharat Sangh, Rajendra Singh helped Bhaonta Kolyala village of district Alwar of Rajasthan to become an example of sustainable development and prosperity. Some 60 years back the village had dense forest and rich wildlife. Deforestation and lack of care to retain rain water turned the village suffering with starvation and poverty. The dried village wells and water scarcity compelled most of villagers to migrate at other places.

Luckily villagers came to know about Gopalpura village, some 20 km. away, where water in wells was available round the year with the help of an NGO, **Tarun Bharat Sangh** under the guidance of Rajendra Singh. Villagers of Bhaonta Kolyala approached him to help while he was on **Pani Yatra** (march for water).

Rajendra Singh promised to help them officially, provided the villagers unite and cooperate him. Under his direction, the villagers united and took the task of regenerating the *johads* (an earthen check dam) so that they could collect every drop of rain that fell there. In 1988, the villagers began to repair johads. When the mansoon came, the results were overwhelming. Now the village has over 238 johads and a 7m. tall concrete dam in the upper catchment area of the Arvari river. The whole picture of the village so changed that wells are filled with water, agriculture is now well irrigated and prospering and milk production has increased by 10 times. Every rupee invested in reviving the johads has almost tripled the annual income of the people.

Bhaonta Kolyala village now stands for message that if the people work with co-operation to regenerate the environment, there are unexpected blessings. This also means that a village with healthy environment is a village with economic prosperity. The individual who spearheaded the integrated effort towards sustainability in this village was **Rajendra Singh**. The village has been awarded the **Down to Earth-Joseph C Award** in 2002 and Dr. Singh has been honoured with Magsaysay Award.

Sundar Lal Bahuguna

In the early 1970s, the **Gangotri Gram Swarajya Sangh** in Uttarkashi and **Dasholi Gram Swarajya Sangh** in Gopeshwar agitated against contractor system of extraction of forest resources. The movement ended with the formation of UP Forest Development Corporation. However, felling of trees by the corporation created some serious problems. By now, people learned that the main products of the forest are **soil, water** and **oxygen** and not the timber as understood by the foresters.

The first effective movement against felling of trees started in December 1970, from a village, Advanis of Tehri Garhwal district of

Uttarakhand. Under the leadership of Sundar Lal Bahuguna and Chandi Prasad Bhat, the women of village tied the sacred thread around the trees and hugged (*hindi* chipko) them against the axemen coming to cut the trees. They faced police firing in February 1978 and courted arrest for this cause. This movement became famous as **chipko movement**. Soon the movement spread to different villages of Tehri Garhwal with wider sense of ecology.

Bahuguna marched more than 3000 km from Kashmir to Siliguri (W.B.) to develop awareness and started movement in all parts of Himalayas. This march became famous as **Kashmir Kohima chipko foot march**. During his march, environmentalists from different countries like France, Germany, Switzerland, Sweden etc. visited him to offer moral support.

The Chipko movement became popular in other parts of the country like Arawali region, Thane, Chhatisgarh, Santhal, 24 Parganas, Western Ghats etc. It developed an ideology of planting 5Fs **Food, Fodder, Fuel, Fibre** and **Fertilizer** trees and got involved in protecting the environment and bringing peace, prosperity and happiness to mankind.

In 1982, Bahuguna presented the plans of Chipko Movement at UNEP (United Nations Environment Programme) held at London, UK. On 5th June (1972) while celebrating World Environment Day in a exhibition at Stockholm (Sweden), following sentence was displayed about Chipko Movement.

> *"A powerful environmental movement has grown up on the slopes of mountains of Himalayas. Villagers have created an effective non-violent way to stop the devastation by forest industries. When the axeman comes, the people form a ring (circle) around the trees and embrace the trees. This has given the movement* its name *Chipko Andolan* -the tree hugging movement".

S L Bahuguna earned several national awards for his contribution including Sanjai Gandhi Award for Environment, Padam Shri, Rashtriya Akta Puraskar, Padam Vibhushan (2009) etc.

Chapter Summary

Pollution is a man made problem created as a result of irresponsible ways of over exploitation of nature. The problem has become a major challenge to scientists, environmentalists and humanists as we are now unable to breathe fresh air, drink pure water and get healthy food. If man has to survive he has to overcome this gigantic problem before it overcomes him and wipes out his very existence.

Though global efforts are being made, the responsibility of an individual is equally significant. The individuals of developed countries behave with care and precautions but those of developing and underdeveloped countries are lagging behind. They require to learn as to how to reduce and control pollution. It requires not only facilities and training but also mass awareness. For mass awareness, several types of programmes are organized round the year in various countries.

Since the cause of pollution is only human beings, only they can minimise it by preventing its increase and spread. An individual should learn that prevention is better than cure.

There can be several preventive measures to control air pollution, individually, *e.g.* *(i)* establish industries away from residential areas, *(ii)* maintain automobiles fuel efficient, *(iii)* minimise and cooperative use of the vehicles, *(iv)* use CNG based autoengines, *(v)* stop use of loud speakers and other noise producing systems, *(vi)* use LPG for cooking medium etc.

An individual can adopt several preventive measures to reduce water pollution, *e.g.* *(i)* adopt proper system of sewage disposal, *(ii)* divert sewage towards oxidation ponds or treatment plants, *(iii)* dispose properly the domestic effluents etc. and to reduce solid waste pollution, *e.g.* *(i)* divide domestic wastes into biodegradable and non-biodegradable components for separate disposal, *(ii)* use closed plastic bags for waste disposal, *(iii)* collect recyclable articles separately for recycling process etc.

Individuals can contribute several indirect roles to minimise pollution. For developing mass awareness, one can participate in publishing information on pollution, organising various competitions and exhibitions sponsoring environment related advertisements etc. Individual can further participate in developing various eco-friendly organisations such as Eco-friends, Green Friends, Green Bridade, Eco-Task, Soclean etc. and in organising rallies on various environmental dates.

Several Indians have shown exemplary contribution towards developing pollution free environment and bringing social prosperity. Rajendra Singh, Anna Hazare, S.L. Bahuguna etc. are such icons.

Study Questions

1. Describe the role of an individual to control pollution.
2. Briefly describe the role of individual in preventing:
   (i) Air pollution
   (ii) Water pollution
   (iii) Solid waste pollution
3. Briefly describe the contribution of following environmentalists:
   (i) Anna Hazare
   (ii) Rajendra Singh
   (iii) Sunder Lal Bahuguna
4. Explain the following terms:
   (i) NGO
   (ii) Urja gram
   (iii) Chipko movement

Objective Questions. *Select correct answers*

1. June 5, is celebrated as:
   (1) World Food Day (2) World Science Day
   (3) World Environment Day (4) Tree Plantation Day

2. A famous environmentalist is:
   (1) Jahangir Baba (2) Anna Hazare
   (3) Arjun Singh (4) All of these
3. Tree plantation Day is celebrated on:
   (1) June 5 (2) February 28
   (3) July 7 (4) November 14
4. Pani Yatra was organised by:
   (1) S.L. Bahuguna (2) Rajendra Singh
   (3) Medha Patekar (4) Sandeep Pandey

Answers

1. (3) *2.* (2) *3.* (3) *4.* (2)

33

CHAPTER

Pollution Case Studies

LEARNING OBJECTIVES

Bhopal Gas Tragedy • The Ganga Pollution • The Yamuna Pollution • Minamata Tragedy • Taj Mahal issue • London Smog Disaster (Classical smog) • Photochemical Smog • Aerosols • Pollution Related Diseases.

BHOPAL GAS TRAGEDY

It is an example of industrial pollution accident wherein a large scale leakage of toxic gas took place in the midnight of 2/3rd December 1984. The toxic gas was of a pesticide chemical – **methyl isocyanate** (MIC) produced in the Indian factory of Union Carbide. Due to failure of net scrubber outlet, the gas leaked for 40 minutes and some 30 tonnes of it was released in the atmosphere. The gas, instead of diffusing to higher atmosphere, started spreading in the lower region due to winter and low temperature. Spread of MIC gas over thickly populated area became the cause of catastrophe. Every thing happened in just two hours.

It caused the death of over 8000 people (only 2500 according to government records) due to suffocation, cardiac failure and pulmonary disorders. Some 200 children were born dead, 400 died after birth, 47 per cent pregnant women suffered abortion, more than 10,000 people became permanently and some 1,50,000 partially disabled and hundreds of people lost their eyesight. It is believed that some 5 lakh people became sufferers. Gas also got dissolved in different open water bodies and continued to cause the damage further. Even today several limping, blind and coughing people can be observed in old Bhopal region.

Government of India sued the case against the American company and demanded compensation for the loss about US$ 47,000 million and to pay Rs 750.00 per month per disabled person. So far no payment has been made to the sufferers. In July 2004, Supreme Court of India issued an order to Reserve Bank of India to immediately pay Rs 1503 crore to the sufferers from the amount government received from Union Carbide Limited, U.S.A. The help has been given to 5.72 lakh sufferers and compensation to 15,274 killed person.

As per TV channel IBN-7 (2-4 July, 2007), Government of India is effortful to recommend to withdraw the case against American company, DOW Chemicals, who purchased UCL Ltd. USA in 1999, to avoid payments for losses towards this tragedy, *i.e.*, a tragedy over tragedy.

THE GANGA POLLUTION

River Ganga is so important and closely associated with Indian culture and civilisation that it is often designated as 'Maa-the mother' and is worshipped in India and now the National River of India (2008).

The Ganga is the ninth largest river in the world and second largest in India with a length of 2,525 km. from Gangotri (place of origin) to Ganga Sagar (place of meeting the ocean). A number of Himalayan rivers including Mandakini, Alakanda, Yamuna, Ghagra, Gandak, Koshi etc. join it during its course. About 692 cities/towns are located along its bank of which 27 cities have population more than one lac. More than 600 km part of the river, particularly between Kanpur (UP) and Patna (Bihar) is highly polluted. Everyday it receives about 1300 million litres sewage and 250 million litres chemical effluents. The main causes of pollution of Ganges water are the mixing of the industrial effluents (20%) and domestic and municipal effluents (80%) into it.

Following are main causes of the Ganga Pollution:

(1) Sediments load of the Ganga river and other associated rivers.
(2) Sewage disposal of villages, towns and cities.
(3) Discharge of industrial effluents.
(4) Release of dead bodies of animals and human beings.
(5) Cremation along the bank of river and disposal of cremation material and burnt and half burnt dead bodies of human beings.
(6) Surface run off of toxic chemicals, agro-chemicals, pesticides etc.

In 1985, the Ministry of Environment and Forests started the Ganga Action Plan (GAP) for preventing irreversible damage and restoring the water quality of Ganga. GAP has the following **objectives:**

(i) All round environmental improvements.
(ii) Installation of sewage treatment units and their proper operation and maintenance.
(iii) The basic facilities of sewage treatment to be coupled with production of energy and manure and provision of pisciculture, aquaculture and irrigation-treated water.
(iv) Economic benefits to the local population.

GAP helped in estimating the magnitude of the Ganga pollution. In an estimate, it was calculated that about 10,90,000 kg toxic effluents and 13,00,000 kg domestic and municipal effluents are discharged every year from the cities of UP only. Storey of Kanpur is frightening. From the city about 200 million litres of sewage and 5.8 million litters of toxic effluents (from 151 tanneries) are disposed everyday. The analysis of Kanpur Ganges water revealed that it has *(i)* 63.7-173 mg./L total solid wastes, *(ii)* 9.33 to 17.33 mg/L chlorine, *(iii)* 3.05-5.5 mg/L dissolved oxygen (DO), *(vi)* 2.86-30.33 mg/L biochemical oxygen demand (BOD), and *(v)* 0.66 – 89.14 mg/L chemical oxygen demand (COD) . The quality standard of water is C and E type, *i.e.*, water is not fit for any purpose.

The first step of GAP started in 1985 and second in 1995 to complete the plan by the end of 1999. Despite, Rs 416.33 crore has been spent to clean the water, the results achieved remain insignificant. Various other projects have also been subsidised by the Government of Netherlands and Japan to clean the water of the Holy Ganges.

THE YAMUNA POLLUTION

River Yamuna after being orginated for Yamunotri (Uttarakhand) covers a passage of 1376 km. before joining the river Ganga at Allahabad. The river has been divided into five segments from the point of view of pollution level.

(i) **Himalayan segment** from Yamunotri to Tajewala, 172 km, non-polluted zone.

(ii) **Upper segment** from Tajewala to Wazirabad, 224 km polluted by agrochemicals used in Haryana.

(iii) **Delhi segment** from Wazirabad barrage to Okhala barrage, 22 km, highly polluted zone due to Delhi effluent discharges.

(iv) **Eutrophicated segment** from Okhala barrage to joining place with river Chambal, 490 km., highly polluted with microbes and with highest BOD, Mathura and Agra covered in the segment.

(v) **Mixed segment** from Chambal sangam to Allahabad 468 km, pollution level decreases.

The pollution problem in Delhi segment is acute and highly dangerous. About 1900 million litre (m.l.) sewage is discharged every day in this segment of 22 km. The total treatment capacity of Delhi plants is about 1270 m.l./ day, *i.e.*, about 630 m.l./day of it is still discharged directly to the river Yamuna without any treatment. Moreover, the treated water is also not pure. It remains partially untreated. Further 2800 m.l. effluents/ day are discharged into the river by eleven main *nullahs* (out of total 19) which result in an increase of 200 BOD and 160 tonnes suspended solids. Najafgarh and Shahadra *nullahs* are most problematic.

The pollution indicator coliform test revealed the presence of 24,000,000 *Escherichia coli* (bacteria) per 100 ml (maximum permitted 7500 coli/100 ml) in the down stream of Okhala region which is indicative of the magnitude of water pollution. Conditions become more severe during summer.

In 1993, in the second step of Ganga Action plan (GAP) of Government of India, Yamuna Action Plan (YAP) was also taken up. It included 127 working projects of which 48 are in UP, 76 in Haryana and 3 in Delhi. After 13 years of works on these projects, the pollution problem still remains serious. Pollution is further nurtured due to barrages built in Delhi, Mathura and Agra for civic water supply and number of canals made for irrigation. All this has made river Yamuna as *sewage vahini*.

MINAMATA TRAGEDY

Minamata is a small coastal town of Kyushu in Japan. In 1950s, town people noticed a strange behaviour of their cats. These cats were twitching, stumbling and jerking. They called them, *the dancing cats*. Studies revealed that such cats were suffering from brain damage due to methyl mercury poisoning.

Methyl mercury may cause fatal poisoning by developing crippling disease, now popularly called **Minamata disease**. The disease occurred in epidemic form in Kyusyu (Japan) during 1953-1961 and in Niigata (Japan) in 1965.

The Chisso Chemical Plant of Minamata Bay (Kyushu) use to release mercury in the bay which was converted into soluble methyl mercury by sedimentary bacteria like *Clostridium cochearum* and *Pseudomonas* species. Such form of mercury was absorbed by fishes and concentrated within their bodies to the dangerous levels. These fishes, containing about 50 ppm of mercury, when consumed by people lead to an epidemic of nervous problems including numbness, tingling sensations, headaches, blurred vision, loss of muscle control and slurred speech. About 50 people died and more than 3500 became disabled. The children suffered from mental retardation, cerebral palsy and convulsions.

People of Japan, Sweden and Canada are still suffering from Minamata disease because these countries widely use mercury compounds as algicides and fungicides in agriculture, pulp and paper industries and chloro-alkali plants.

Minamata tragedy is an example of *marine water pollution, mercury pollution* and *bio-magnification*.

TAJ MAHAL ISSUE

Taj Mahal, being one of the seven wonders and the most beautiful monuments of the world (2007), attracts international tourists throughout the year. It was built along the bank of river Yamuna by Mughal Emperor Shahajahan in memory of his beloved wife Mumtaz more than 350 years back in Agra. It is now included in the list of World Heritage due to its archaeological importance.

In 1972, Government of India established Mathura Oil Refinery in Mathura. This step was most resented by various environmentalists of the country and protested that its air pollutants were harmful to Taj Mahal.

During the oil refining process, sulphur dioxide (SO_2) is released in a large quantity along with the smoke. Air containing SO_2 blow all round and reacts with water during rains. SO_2 combines with water to form sulphurous acid (H_2SO_3) and sulphuric acid (H_2SO_4) and produces acid rains. Environmentalists protested that acid rain causes damage to the marbles of Taj Mahal.

Government constituted a committee in 1974 to find out the fact and authorised Italian company Technico to evaluate the changes occurring in the air due to Mathura Oil Refinery of Indian Oil Corporation. According to the report of the committee and determinations carried out by Italian company, it was concluded that in Agra the quantity of SO_2 would increase upto 1-3 microgram which is negligible and harmless to the Taj Mahal. As a precautionary measure, the nearby factories have been now removed to sustain healthy air around this tourist centre. Also government adopted a liberal attitude of giving LPG connections to replace burning of coal and wood fuel in the surrounding area.

LONDON SMOG DISASTER

On Thursday, 4 December 1952, a high temperature air mass covered Southern England and due to temperature difference a white fog settled over London. It blackened due to rise of particulate matter and SO_2 level arising from coal-fired heating and power production systems. A coal induced smog was formed by interaction of sulphur dioxide, smoke and water to produce sulphuric acid mist. Such condition lasted for five days. On Friday morning visibility was zero.

The smog was intensely irritating to the human beings, particularly to their respiratory systems. People soon developed red eyes, burning throats and nagging cough. Elderly and sensitive people were first victims. More than 4,000 people died before the fog disappeared (December 9, 1952).

The term *smog* is a synchronym of *smoke* and *fog*. It can be coal induced or photochemical smog. The coal induced smog is popularly referred to as **London smog**. Since its nature is chemically reducing with high levels of SO_2, it is also called **reducing smog** and because it was truly a major air pollution disaster, it has also been referred to as **classical smog.** From this incidence, people learned that air pollution can no longer be ignored. The episode was enough to pass the Clean Air Act, 1956 in England.

However, London has experienced many air pollution disasters after December 1952, *e.g.* in-

January, 1956 – 1000 people died
December, 1957 – 700 died
January, 1959 – 200 died
December, 1962 – 700 died and
January, 1963 – 700 died.

All these casualities reveal the magnitude of loss due to air pollution in London.

PHOTOCHEMICAL SMOG

During 1950, the city of Los Angeles, California, became afflicted with serious air pollution due to large volume of traffic on streets. A *photochemical smog* was formed by the interaction of hydrocarbons and oxidants (like NOx, CO, O_3) in the presence of sunlight. It caused eye irritation, visibility reduction and damage to crops and rubber crackings. It was an *oxidising smog*, also called **photochemical smog** or **Los Angeles smog** (unlike reducing London smog).

Formation of Photochemical Smog

Hydrocarbons and oxides of nitrogen react in the presence of sunlight to produce photochemical smog which contains ozone (O_3), hydrogen peroxides, organic peroxides, organic hydroperoxides, peroxyacetyl nitrate (PAN) etc. Various reactions taking place in the formation of photochemical smog are as follows:

(1) $NO \rightarrow NO_2 \xrightarrow[UV\ light]{Sunlight} NO + O$

(2) $O + O_2 \rightarrow \underset{(ozone)}{O_3}$

(3) $NO + O_3 \rightarrow NO_2 + O_2$

(4) $\underset{(Hydrocarbon)}{RCH_3} + O_3 \rightarrow \underset{(Hydrocarbon\ free\ radical)}{RCH_2}$

(5) $RCH_2 + O_2 \rightarrow \underset{(Oxygenated\ free\ radical)}{RCH_2O_2}$

(6) $RCH_2O_2 + NO \rightarrow RCH_2O + NO_2$

(7) $RCH_2O + O_2 \rightarrow \underset{(Aldelyde)}{RCHO}$

(8) $RCHO + O_2 \rightarrow \underset{(Hydroperoxy\ radical)}{HO_2}$

(9) $HO_2 + NO \rightarrow \underset{(Hydroxyl\ radical)}{HO} + NO_2$

(10) $HO + RCH_3 \rightarrow RCH_2 + H_2O$

These photochemical generated radicals are reactive intermediates and undergo a series of reactions in the presence of ultravoilet light to form still more radicals which combine with number of pollutants in the air producing NO_2, O_3, H_2O_2 and a number of toxic chemicals like aldehydes, formaldehydes, peroxy acetyl nitrate (PAN:$C_2H_3O_5N$) etc.

Characteristics of Photochemical Smog

The toxicity of photochemical smog is due to the presence of various toxic compounds. Following are some distinctive features of such smog.

(1) PAN (peroxy acetyl nitrate) is the most toxic chemical found in the smog. It causes irritation to eyes and lungs and can cause even cancer. In plants, it causes necrosis or chlorotic spots on the upper surface of leaves and blocks Hill reaction (photosynthesis) and checks growth and productivity.

(2) SO_2 and H_2S are readily oxidised to sulphates in a smoggy atmosphere. Sulphates (SO_4) cause acid rains, corrosion, reduced visibility and general health hazards.

(3) Nitric acid (HNO_3) formed in the smoggy atmosphere reacts with ammonia to form ammonium nitrate. Both compounds cause extensive damage to plants, human health and are extremely corrosive in nature. Nitric acid further contributes to the acid rain.

(4) Nitrogen oxides are also associated with diseases like cancer, cardiac disorders and diabetes mellitus.

(5) Various oxidants and free radicals present in the smog enter the human body with inhaled air and cause severe damage to the respiratory system.

(6) Ozone (O_3) present in the smog causes yellow spot formation on leaves, reduction in plant growth and retardation in the rate of photosynthesis, pollen germination and in the yield of plants. In human beings, it causes nose and throat irritation, eye watering, throat dryness, headache, inflammation in lungs with breathing difficulties etc.

Photochemical smog is an excellent example of *synergism* in which total effect of interactions between two or more substances exceed the sum of the effects of individual substance. Smog is characteristics of big cities like Tokyo, Los Angeles, Chicago, Mumbai, Kolkata, Surat, Ahmedabad, Delhi, Chennai, Kanpur etc.

In India, all big cities have serious smog problems during winter due to automobiles, industrial and domestic exhausts that pose a potent health hazard. As such the control of photochemical pollutants, especially the primary precursors (hydrocarbon and nitrogen oxides) is of prime importance to safeguard the human health and economic plants.

AEROSOLS

Aerosols are certain chemicals released into the air with force in the form of a mist or vapour. An important source of aerosols in the upper atmosphere is the jet aeroplane emissions. Such aerosols contain fluorocarbon propellants which deplete ozone layer in the stratosphere. Ozone acts as preventive layer against the entry of UV rays towards the earth. Depletion of ozone layer is also affected by oxides of nitrogen and sulphur released from supersonic aeroplanes. Some chlorofluorocarbons such as CCl_3F and CCl_2F_2 are commonly used as refrigerant and propellant. Aerosol pollutes atmosphere largely as a result of human activities. CCl_4 (carbon tetrachloride) has been shown to destroy stratospheric ozone. It may originate from natural sources as well as well as from human activities.

Important sources of aerosol production are-

(i) Thermal power plants
(ii) Automobile exhausts
(iii) Supersonic aeroplanes
(iv) Mining and other industries

Accumulation of aerosols in the atmosphere causes various problems, such as–

(1) Aerosols enter and accumulate within the respiratory system and cause diseases like fibrosis, silicosis, asbestosis, T.B. etc.
(2) Condensation of aerosols accelerates precipitation.
(3) Pathogenic spores and allergenic agents are disseminated widely by aerosols.

The dry or wet precipitations considerably clear the aerosols from the atmosphere

POLLUTION-RELATED DISEASES

Now several pollution borne disease are well-known. In developing countries, main diseases are infectious and are parasitic in nature but in developed counties, diseases occur due to industrial, technological, radioactive and chemical pollutions.

Some diseases caused by air and water pollution are as follows:

TABLE 33.1. Diseases caused by Air and Water Pollution

Name of disease	*Particulars*	*Symptoms*
(1) Asbestosis	Found in asbestos workers, talc miners, rubber workers etc. (air pollution)	Inflammation, pigmentation, fibrosis in lungs
(2) Allergy diseases *e.g.* hay fever,	Caused by allergenic pollens, weeds, house dust,	Various allergic reactions such as cough, cold, asthma, reddish

bronchitis, eczema, migraine, hives	animal hairs, certain drugs and dyes found in air and water pollution	blisters, fever, watering of eyes.
(3) **Beryliosis**	Found in workers of beryllium mines	Various reactions and respiration problems.
(4) **Black lung cancer**	Found in coal mine workers	Pulmonary fibrosis, may lead to respiratory failure
(5) **Bronchitis and Asthma**	Found in people exposed to air pollution for a long term	Various respiratory problems and allergy
(6) **Byssinosis**	Found in textile mill workers	Various respiratory problems and fibrosis
(7) **Cholera**	Water borne disease caused due to bacteria *vibrio cholerae*	Vomiting, high fever diarrhoea, leading to death
(8) **Diarrhoea**	Water borne disease caused due to infectious bacteria like *Escherichia coli*.	Acute intestinal disorder
(9) **Gout**	Found in people inhaling manganese dust and due to dietary disorders	Painful swelling in joints especially in toes, knees and fingers.
(10) **Hepatitis-Jaundice**	Water borne diseases caused by virus	Hepatitis B is fatal and cause death. Jaundice damage livers. Both are hepatic (liver) diseases.
(11) **Minamata disease**	Caused due to mercury poisoning	Sensory disturbances, muscular rigidity, involuntary movements, can lead to death.
(12) **Silicosis**	Found in silicon quarries and coal miners.	Form nodules in lungs.
(13) **Typhoid**	Water born disease caused due to bacteria *Salmonella typhi* and other species.	Paralysis of gastro-intestinal tract. Boils and ulcers are formed in stomach and bacteria enter in blood stream.

Some Diseases of the Modern Age

Now a days, some diseases are frequently found in human beings which may occur due to pollution as one cause. Some examples are –

(i) **Cancer**- Smoking is the mot common cause of lung cancer, eating preserved food might lead to oral and stomach cancer, some industrial pollutants cause cancer, mammography may induce breast cancer and UV radiation can cause skin cancer.

(ii) **Leukemia**- Sometimes, it has been found linked to benzene which is ingredient of gasoline. It is a serious disease in which too many white cells are produced which cause acute weakness and some times death.

(iii) **Viral diseases**- Several viral diseases borne due to various kinds of pollution, *e.g.* HIV/ AIDS, Polio, Dengue fever, haemorrhagic fever etc. are becoming curse to the common man.

Chapter Summary

Bhopal Gas Tragedy occurred in the midnight of 2/3 December 1984, due to leakage of poisonous gas MIC (methyl isocyanate). It killed more than 8000 peoples and disabled more than 1.5 lakh people. The gas was released from the factory of Union Carbide.

The most polluted river of the country is river Ganga. The main causes are release of industrial effluents from tanneries, chemical factories etc. and sewage from various cities. Ministry of Environment and Forests in 1985 started Ganga Action Plan (GAP) to clean Ganga by giving grants of more than 400 crore rupees. The Netherland and Japan have also extended their help. River Yamuna is another polluted river, with its precarious condition in Delhi segment. Govt. of India has started Yamuna Action Plan (YAP).

Minamata tragedy is famous due to marine water pollution by methyl mercury which caused epidemics of nervous problems, numbness, blurred vision etc. in human beings. Now disease is known as Minamata disease, named after the place Minamata of Japan.

One of the most beautiful buildings of the world is The Taj Mahal of India. Its beauty is endangered due to SO_2 released from Mathura Oil Refinery.

London smog is a reducing smog famous for hazardous environmental pollution occurred in December, 1952 in London. More than 4000 people died. Photochemical smog or Los Angeles Smog is oxidising smog rich in PAN which is a toxic chemical. It affects plants, animals, human beings etc. badly. Aerosols are certain chemicals released into the air with force in the form of a mist or vapour. It is one of the major causes of ozone layer depletion in stratosphere.

Pollution is hazardous that directly or indirectly causes several diseases. A few of them are – allergic diseases, black lung cancer, bronchitis, asthma, cholera, diarrhoea, typhoid, jaundice, cancer etc.

Study Questions

1. Describe briefly any three study cases, one each related with air pollution, river pollution and marine water pollution.
2. Write short notes on:
   - (i) Bhopal Gas Tragedy
   - (ii) London Smog Disaster
   - (iii) Los Angeles Smog
   - (iv) Ganga Pollution
   - (v) Minamata Tragedy
   - (vi) Photochemical Smog

Objective Questions. *Select correct answers-*

1. MIC is related with:

(1) Taj Mahal (2) London Smog
(3) Aerosols (4) Bhopal Gas Tragedy

2. Minamata disease is caused due to pollution by:

(1) Lead (2) Mercury
(3) PAN (4) SO_2

Answers

1. (4) *2.* (2)

34

CHAPTER

Disaster Management

LEARNING OBJECTIVES
Introduction • Disaster Prone Regions of India • Major Disasters • Govt.'s Efforts towards Disaster Management • Floods, Earthquakes • Causes of Earthquakes • Cyclones • Landslides • Orissa Supercyclone –1999 • Tsunami Disaster –2004.

Introduction

Disaster may be defined as "*An occurrence arising with little or no warning, which causes or threatens serious disruption of life, and perhaps death or injury to large number of people and requires therefore a mobilization of effort in excess of that normally provided by the statutory emergency services*." Disaster is thus an event which afflicts a community, the consequence of which are beyond the immediate financial, material or emotional resources of the community.

The disasters can be of two types-

(i) **Natural disasters**, *e.g.* earthquakes, volcanic eruptions, hurricanes, tornados, avalanches, floods, droughts etc.

(ii) **Man-made disasters**, *e.g.* Chernobyl disaster, Bhopal tragedy, dam leakage or collapse etc.

The World Disaster Report (2004) has shown that disasters have increased in their frequency and intensity. The face of disasters keeps on changing. This happens in both, the developed and the developing countries. The report reveals that from 1994 to 1998 average disasters per year were 428 whereas from 1999 to 2003, their number has increased to 707 per year. The less developed countries face maximum disasters and the maximum damage. The **hydro meteorological disasters** such as avalanches, land slides, droughts, extreme temperatures, floods, cyclones, forest fires are becoming more common and frequent. Such disasters have multiplied *seven times* in last four decades (1962-2000). The **geographical disasters,** which include earthquakes, Tsumani etc. have multiplied *five times* in the same period. Among the natural disasters, floods are most common.

Asia suffers most from natural disasters, killing the highest number of people. The number of people killed from 1994 to 2003 in Asia is about 4,80,000 which is distinctly higher than death of 65,526 in Europe. During this period 23,42,91,3000 people were affected in Asia. The worst flood affected countries are China, India and Bangladesh. In July, 2003, the flood in China affected nearly 150 million people and in India during August and September in the same year affected 7.6 million people.

During 1994 to 2003, the disasters reported worldwide were 3055, of which floods were maximum (1160) followed by windstorms (802) and drought and earthquakes (257). Drought is the deadliest disaster accounting for 48 per cent of all deaths from natural disasters. The following table 34.1 reveals the number of people killed in various continents due to disasters during 1994-2003

TABLE 34.1. Total Number of People Reported Killed by Continent and by Year, 1994–2003

	1994	*1995*	*1996*	*1997*	*1998*	*1999*	*2000*	*2001*	*2002*	*2003*	*Total*
Africa	3144	2962	3484	4064	7006	2688	5756	4462	8272	5810	47648
America	2874	2628	3541	3069	21865	33989	1820	3460	2285	2026	76557
Asia	13328	75590	69706	71033	82373	75890	11608	29255	13358	37860	480001
Europe	2340	3366	1204	1166	1434	19448	1627	2196	1699	31046	65526
Oceania	103	24	111	388	2227	116	205	9	91	64	3338
HHD	2462	8145	2062	2450	2893	3250	2039	1849	1895	31619	58664
MMD	15874	18815	16465	18122	42244	72023	13198	33498	14754	41275	286268
LHD	3453	57610	58519	59148	69768	56858	5779	4035	9056	3912	329138

Source: World Disaster Report 2004; HHD: High Human Development, MMD, Medium Human Development, LHD: Low Human Development

Disaster Prone Regions of India

India is a disaster prone country and its region wise details are as follows-

(1) **Northern region**. It is confronted by avalanches, land slides, floods, droughts and earthquakes (Area falls under seismic zone III to IV).

(2) **Eastern region**. It is affected by severe floods (due to Brahmputra and Gangetic rivers), droughts, heavy wind, heat waves, hail storms, cyclones and earthquakes.

(3) **North Eastern region**. It is affected by floods, land slides, wind damages and earthquakes (area falls under seismic zone IV to V)

(4) **Western region**. It is confronted with severe droughts, wind erosions, floods, cyclones and earthquakes (Gujarat, Bhuj).

(5) **Southern region**. It is vulnerable to cyclones, sea erosions, tsunami, landslides etc.

(6) **Islands of Andaman, Nicobar and Lakshadweep**. The area is prone to sea erosion, tsunami, sea level rise etc.

As such some 600 districts of the country are prone to disasters. Of these, 80 districts are coastal and 100 districts are hill located.

Impact of Disasters

The direct impacts of disasters are death, injury to people and animals and destruction of properties. Life line supports, *e.g.* destruction of communication, power supply, drainage etc. are also damaged. Commercial, industrial and agricultural activities come to a stand still. Poors are worst affected by losing their residences, employments and life amenities. They are psychologically terrorized and become helpless.

Major Disasters in India

India has long been affected by various disasters. However, major disasters of last two decades are-

(1) October 1991 – Uttarkashi earthquake
(2) September 1993 – Latur earthquake
(3) March 1999 – Chamauli earthquake
(4) October 1999 – Orissa super cyclone
(5) January 2001 – Bhuj (kutch) earthquake
(6) December 2004 – Tsunami
(7) August 2005 – Bombay Gujarat flood
(8) October 2005 – Srinagar Muzafarabad earthquake

Government Efforts Towards Disaster Management

The central government. of India plays a facilitating role in the disaster management to the states. Government assists states with all required supports like defence services, air dropping, searching, air lifting, rescuing, transport of relief goods, health personnel and medical support, telecommunication restorations, availability of transport services like trains and ferries. Government has also constituted a crisis management committee headed by a Cabinet Secretary. Secretaries of various central ministries like Defence, Home, Health, Telecommunications, Shipping, Road and Transport, Agriculture, Animal Husbandry, Environment and Forest are also member of the committee. Under Home Ministry, a Central Relief Commission has been constituted which coordinates with all ministers. There is yet another division under Home Ministry, *i.e.*, National Disaster Management Division which deals with all kinds of disaster managements other than drought. The nodal ministry for drought is the Ministry of Agriculture.

Government of India constituted a *High Powered Committee on Disaster Management* in 1999, under the leadership of Mr. J.C. Pant (former Secretary). He discussed the problem with experts of all branches such as NGO's, State governments, district collectors, academicians, policy makers and international specialists. The High Power Committee has constituted five sub-committees dealing separately-

(i) Water and climate related disasters,
(ii) Geologically related disasters,
(iii) Chemical, Industrial and Nuclear related disasters,
(iv) Accident related disasters and
(v) Biological related disasters.

Following these guidelines, state governments have been requested to prepare at state level and district level disaster management plans. Besides, a *National Institute of Disaster Management* has been set to provide various trainings related to the management. A *National Disaster Management Authority* has been constituted and a *National Disaster Management Act* has been enacted. Meteorological department has been strengthened with most advanced scientific systems and a *Tsunami Warning Centre* will be set up in next two years (2008). A separate battalion of paramilitary forces has been created and deployed regionally for quick relief services. The Govt. has also introduced Disaster Management in the academic syllabus, right from the school level.

In addition, a *Calamity Relief Fund* (CRF) has been constituted in which Union Government and State Governments contribute in the ratio of 3:1. The Eleventh Finance Commission has recommended nearly Rs. 11,000 crores for the period of five years and twelveth Finance Commission has recommended a Rs. 23,000 crore for next five years. So far CRF (of eleventh finance commission) has been made eligible for disasters related with cyclone, drought, flood, earthquake, fire and hailstorm but in twelveth commission, CRF will also deal with landslides, avalanches, cloud bursts and pest attacks.

However, Govt. has so far not formulated to implement insurance systems in disaster prone zones. In future, insurance of life, health and properties must be done to really help poor and downtrodden people. Government. help never exceeds more than 10 per cent of individual damages.

FLOODS

In flood, a long part of the earth is submerged under the water for several days. India is the worst flood affected country in the world after Bangladesh. About 40 million hectares or nearly $1/8^{th}$ part of the country is flood prone, of which about 8 million hectares of land is affected annually and about 3.5-10 million hectares of cropped area faces regular damage.

Flood is very common in UP, Bihar, West Bengal, Assam and Orissa. Due to intense monsoon, floods in other non-flood prone areas are also becoming common as in Andhra Pradesh, Karnataka, Rajasthan, Gujarat and Maharashtra. Flood causing common rivers of India are Ganga, Yamuna, Ghagra, Kosi, Gandak, Gomti, Brahmaputra, Mahanadi, Godavari, Krishna etc.

Causes of Floods

The main causes of flood can be grouped into two types-

(I) *Natural causes*

The important natural causes of floods are:

(1) Prolonged rainfall
(2) High intensity rainfall
(3) Meandering passage of rivers
(4) Sudden change in the gradient of river water floor or break in slope of longitudinal profiles
(5) Landslides and volcanic eruptions
(6) Blockage of river passage, and
(7) High tides, storm surge or tsunami in coastal areas.

(II) *Man-made causes*

The important man made causes are:

(1) Construction work
(2) Urbanisation
(3) Alteration of river passage
(4) Construction of dams, bridges and storage bodies
(5) Deforestation
(6) Farming and
(7) Changes in uses of lands.

Human activities in the coastal low lands and the river valleys can intensify the hazard of flooding. Such activities include:

(i) drained and reclaimed wet land for agriculture
(ii) removal of sand from beach
(iii) ground water extraction and mining of gas and oil
(iv) destroying natural vegetation in muddy coastlines
(v) destroying or damage of coral reefs and
(vi) deforestation and soil erosion.

Flood Damages

In India, a gradual increase in frequency, intensity and dispersion of flood has been found. As compared to losses due to flood during 1950 to 1965, a three-time increase was found in between 1971 to 1975 and five times in between 1976 to 1978. India suffers an annual average loss of 8000 to 10,000 million rupees and death of human beings in between 3-4 thousands. Maximum damage due to flood is found in Bihar (23.9 per cent), followed by UP (23.8 per cent), Andhra Pradesh (15.4 per cent) and West Bengal (7.0 per cent) of the total annual loss. Various kinds of flood damages include death of human beings and animals, destruction of urban constructions, electricity supply, water supply, roads, crop fields etc. Flood may often be followed by epidemic diseases.

Management of Floods

Following are some control methods to prevent flood:

(1) Delay in reaching of the surface-run-off water to the river. Afforestation and watershed management help prevent rapid surface run off water. Forests prevent flood in several ways, *e.g.*

(i) Forests hinder rapid flow of surface run off water.

(ii) Forests increase infiltration of rain water and thereby decrease the quantity of surface run off water.

(iii) Forests prevent soil erosion, thus prevent silting of the river and other water bodies.

(iv) The reduction of silting and sedimentation in the river increases water holding capacity of rivers and more water is adjusted and regulated.

(2) Management of rapid discharge of river water. Rapid flow of water can be facilitated by reducing meandering of the river. Sharp and sinuous curves of rivers should be amended to straighten the river passage.

(3) Reduction of water volume of the river. Volume of water in the river can be regulated by constructing flood control storage reservoirs, such as dams. Besides controlling the flood, dams have multiple uses, *e.g.* irrigation and power generation or hydro-electric generation. Such measures have been successfully taken up by Damodar Valley Corporation (DVC) by constructing four dams on the river before it reaches the flood prone areas. Several other dams have been constructed on the various rivers of the country.

(4) Development of flood diversion system. Systems are developed to divert excess water during flood into artificially constructed channels flowing to different directions. Such diversion reduce volume of water, flood crest and flood magnitude. This practice has been successfully carried out by Ghaggar flood diversion project of Rajasthan in which 340 cubic metre per second water is diverted to different low lying areas, around sand stoops and various water bodies.

(5) **Minimisation of flood effects**. By constructing levees and embankment using stones, rock pieces, cement, concrete etc. it is possible to prevent outflow of water during rains. Such structures have been made in and around various Indian cities on different rivers, *e.g.* Delhi, Lucknow, Allahabad, Gorakhpur, Azamgarh, Patna, Jalpaiguri etc. Several kilometer long embankments have been constructed along several rivers such as Kosi, Brahamputra, Bangmati, Hoogli etc.

(6) **Development of forecasting system and management**. Though disaster management is a state problem, Government of India facilitates and supports them by providing various helps from different ministries. Government has constituted a Central Flood Control Board in 1954 under which State Flood Control Boards have been formed. Flood forecasting and warning system in India started functioning since 1959. The Board keeps on monitoring flood position of different rivers and informs or warns to the people though radio, TV, news paper etc. so that proper precautions may be taken up. Union Govt. provides supports to the state governments like defence services, air dropping, searching, rescuing, transport of relief goods, health personnel and medical support etc. These supports come from ministries like Defence, Home, Health, Telecommunications, Shipping, Road and Transport, Agriculture, Animal Husbandry, Environment, Forest etc. However, Home Ministry is the nodal ministry for disaster management.

EARTHQUAKE

Strahler (1976) stated that earth is a major demonstration of the power of the *tectonic forces* caused by endogenic thermal conditions of the interior of the earth. As such, an earthquake is a motion of ground surface, ranging from a faint tremor to a wild motion capable of shaking buildings apart and causing gaping fissures to open in the ground.

An earthquake is a sudden motion or trembling of the ground produced by abrupt displacement of rock masses. Most of them result from the movement of one rock mass passing another in response to *tectonic forces*. Rock is elastic and resists against pulling apart. When the stress exceeds the strength of the rock, the rock breaks along a pre-existing or new fracture place called a *fault*.

Magnitude of Earthquake

The magnitude or intensity of energy released in the earthquake is measured by *Richter scale* devised by Charles F Richter (1935). The number indicating magnitude (M) on Richter scale ranges between 0 and 9 but infact the scale has no upper limit because it is a logarithmic scale. The effects of Richter magnitude and corresponding damages, in general, are as follows (table 34.2).

TABLE 34.2. Richter Magnitude and Characteristic Effects (Damages)

	Magnitude	*Description*
(1)	3.5	Earthquakes noticed only by sensitive people.
(2)	4.2	Vibration felt at rest especially in upper floors of the building
(3)	4.3-4.7	Felt by people, Rocking of loose objects
(4)	4.8 (rather strong)	Most sleeping people awakened, bells ring
(5)	4.9-5.4 (strong)	All suspended objects swing; vehicles overturns; falling of loose objects
(6)	5.5-6.1 (very strong)	Walls crack, plaster falls, general alarm
(7)	6.2 (destructive)	Poorly constructed buildings damaged, chimneys fall, causalities
(8)	6.9 (ruinous)	Some houses collapse, pipes break open, ground begin to crack
(9)	7-7.3 (disastrous)	Ground cracks badly, rail bents, landslides, many buildings destroyed, high casualities
(10)	7.4-8.1 (very disastrous)	Bridges destroyed, railways, pipes, cables go out action, great landslides and floods
(11)	Above 8.1 (catastrophic)	Total destruction, ground rises and falls in waves, objects thrown into air

Real damages caused by earthquakes begin at magnitude 5 and continue to increase to nearly total destruction (more than 8 magnitude). The frequency of earth quake occurrence is quite high, *e.g.* earthquakes of 2-3 magnitude occur about 800,000 per year and these are not felt by human beings. However, earthquake of magnitudes over 8 occurs once in every 5 to 10 years. The 1934 Bihar earthquake (India) measuring 8.4 and Good Friday earthquake of Alaska, USA of 8.4–8.6 on Richter scale are among the greatest earthquakes of the world ever recorded.

Some major earthquake hazards of India are as follows-

TABLE 34.3. Major Earthquake Hazards of India

Time	*Place*	*Effects (figs. are approximate)*
October 11,1737	Calcutta	The worst earthquakes of India; some 3,00,000 death of human beings.
May 30, 1885	Kashmir	3000 people killed.
April 4, 1905	Kangra	20,000 people killed
January 5, 1934 (8.4 R.s.)	Bihar-Nepal	10,700 people killed

August 15,1950 (6.7 R.s.)	Assam	1500 people killed, floods in rivers
August 21, 1988	Darbhanga (Bihar)	850 people killed
October 20,1991 (6.6 R.s.)	Uttar Kashi	Over 2500 killed
September 30, 1993 (6.3 R.s)	Latur (Maharashtra)	11,000 killed
January 26, 2001 (8.1 R.s.)	Bhuj (Gujarat)	Upto 1,00,000 people killed
October 8, 2005 (7.4 R.s.)	Pak-India	More than 40,000 people killed

Types of Earthquakes

Natural earthquakes caused by endogenic forces are –

(1) Volcanic – caused due to volcanic eruptions.

(2) Tectonic –caused due to stress and strain along Earth's plates or dislodging of rocks during faulting.

(3) Isostatic – caused by isostatic imbalance due to sudden geological activity at a regional scale.

(4) Plutonic – earthquakes originating deep inside the earth between 250 to 720 km deep.

Causes of Earthquakes

Earthquakes are caused due to disequilibrium in any part of the crust of the earth. The disequilibrium may occur due to –

(i) Volcanic eruptions

(ii) Faulting and folding

(iii) Upwarping and down warping

(iv) Hydrostatic pressure of man-made water bodies (like dams, reservoirs, lakes etc.)

(v) Plate movements

The **Plate Tectonic Theory** has been accepted as the most plausible explanation of earthquakes. According to the theory, the crust of the earth is composed of solid and moving plates. The earth's crust comprises six major plates (*i.e., Eurasian plate, American plate, African plate, Indian plate, Pacific plate* and *Antarctic plate*) and twenty minor plates. These plates are constantly moving in relation to each other due to thermal convective currents originating deep within the earth. Thus, all tectonic events occur along the margins of plates.

The rate of plate movement is around a millimeter per year but strain builds up for decades or centuries until it reaches a critical level, and then everything happens at once. The strongest seismic waves are generated at the initial breakpoint. The breakpoint is called *focus* and the point on the surface of earth directly above the focus in referred to as *epicentre*. The seismic waves move away from the source of the earthquake (focus or also called hypocentre) in the form of:

(*i*) Primary or pressure waves (P waves)

(*ii*) Secondary, Shear or Transverse waves (S waves)

(*iii*) Long or Surface waves (L waves)

P waves are fastest and travel down into the earth, S waves travel down in wavy motion into the earth and from side to side of the rocks and L waves cause up and down or side to side motion of the earth surface. L waves are always last to arrive. Most of the damages caused by an earthquake are done by L waves which travel along the surface of the ground. Scientists measure these waves through an apparatus called *seismometer* to get definite information about the origin of earthquake (*i.e.*, focus), magnitude and destructive power of earthquake and probable cause of earthquake.

Hazardous Effects of Earthquakes

Hazardous effects of earthquakes are not evaluated in terms of their intensities on Richter scales but on the basis of quantum of damages done to human lives and property. The disastrous effects include deformation of ground surfaces, damages and destruction of buildings, rails, roads, bridges, dams, factories, destruction of towns and cities, loss of human and animal lives, violent fires, landslides, floods etc. Some important impacts are as follows:

(1) Slope instability and failures and land slides. Earthquakes in hilly areas cause slope instability and slope failure and ultimately cause land slides and debris falls which damage settlements and transport systems on the lower segments of hills. These effects, due to loose lithologies, become more common in hills during monsoon.

(2) Damage to human structures. Earthquakes cause heavy loss of human properties by damaging buildings, roads, rails, factories, dams, bridges etc. Weaker structures are damaged more. The earthquake that occurred in Bihar on 15 January 1934 records one of the major seismic events of the world. Its magnitude was 8.4 on Richter scale and the epicentre was close to Darbhanga. The damages caused by this earthquake include 10,700 human deaths, landslides and slumping in an area of 250 km length and 60 km width, ruptures in the ground surface, faults etc. Some impacts have been listed in the table (34.2).

(3) Damages to the town and cities. Earthquakes have worst effects on human population and buildings in the town and city areas. Due to tremors, large buildings collapse and men and women get buried under the large debris of collapsed structures, ground water pipes are bent and damaged and water supply is disrupted, telephone and electricity poles are uprooted causing total disruption of telecommunication and electric supply, road blocks disrupt transport system and helpline supply etc. On October 11, 1737, Calcutta city was badly damaged due to severe earthquake damaging thousands of buildings and killing about 3,00,000 people.

(4) Loss of human lives and properties. The destructive magnitude of earthquake is largely evaluated on the basis of human casualities in terms of deaths. Some of the death records are given in table (34.3).

Earthquakes causing human death above one lac people occurred in 1556 (in Shen-Shu, China, 8,30,000 deaths), in 1737 (in Calcutta, 3,00,000 deaths), 1908 (in Messina Italy, 1,60,000 deaths), 1923 (in Tokyo, Japan, 1,63,000 deaths), 1932 (in Sagami Bay, Japan, 2,50,000 deaths), 1976 (in T'ang Shan, China, 7,50,000 deaths).

A great damage of human property worth 5.6 billion US dollars occurred in China (in T'ang Shan) by July 1976. In 1980, property worth 10 billion US dollars was damaged in Italy by the earthquake of November 1980.

(5) Tsunamis. The seismic waves (tsunami) caused by earthquakes travelling through sea water generates high waves and cause great loss of life and property. The undersea earthquake that occurred under the Indian Ocean on December 26, 2004, produced tsunamis devastating the shores of Indonesia, Sri Lanka, India, Thailand and other countries. Tsunami waves upto 15 metres in height travelled at a speed of 800 km/hr. (More details on tsunami given in case studies).

The Kutch (Gujarat) earthquake of June 16, 1819 generated strong tsunamis which submerged a huge coastal area. During this happening, a land area measuring 100 km in length was raised upward (because of tectonic movement) which provided shelter to the stranded and marooned people. Now people call this raised land as **Allah's bund.**

Prevention, Control and Mitigation of Earthquake

So far neither forecasting of earthquake has become possible nor the exact place, time and magnitude can be predicted. However, records show that where an earthquake has occurred, it will occur again. Seismic history of Gujarat reveals that severe earthquake occurs every 30 years, *e.g.* Bhavnagar earthquake (1872), Kutch earthquake (1903), Dwarka earthquake (1940), Bharoch earthquake (1970) and Bhuj earthquake (2001).

The total mitigation programme to reduce the impact of a severe earthquake can be splitted in three phases, as follows-

(I) Prevention Phase Before Earthquake Disaster

Following are a few important actions to be taken or considered:

(1) Preparation of earthquake catalogues with details of epicentres and geological tectonic maps.
(2) Installation of seismological observatories for moderating seismic activity.
(3) Identification of seismic zones with seismic risks.
(4) Development of antiseismic codes of design and construction of various structures.
(5) Training of architects and engineers in earthquake engineering principles and use of antiseismic codes.
(6) Promulgation of laws and by laws for adopting earthquake resistant features in new constructions.
(7) Development of methods for existing structures to strengthen against seismic movements.
(8) Earthquake insurance of various structures and human lives to reduce the economic impact, in general.

(II) Emergency Phase Workings After Disaster

Following are valuable actions that can be taken after the occurrence of earthquake:

(1) Restoration of information and communication lines.
(2) Evacuation of the people.
(3) Maintenance of law and order and support of local people.
(4) Medical care of injured.
(5) Recovery of the dead bodies and their disposal.
(6) Construction of temporary shelters.
(7) Restoration of water and food supply lines.
(8) Restoration of transportation system.
(9) Preventing people from entering into buildings liable to collapse.
(10) Quick assessment of damage.
(11) Management of required sources for reconstruction and rehabilitation and development of self reliance among affected people.
(12) Collection of scientific data to monitor the after shocks.

(III) Consolidation and Reconstruction Phase

This phase involves the following actions:

(1) Detailed survey of damages done to the man made structures to prepare proposals regarding their repairs, restoration, strengthening or demolition.

(2) Selection of proper site for new-settlements.
(3) Adoption of strategy for new constructions.
(4) Execution of reconstruction programmes.
(5) Review of seismic codes and norms of constructions.

CYCLONES

Cyclones are raging spirals of wind and rain that are born near the Equator over the sea, creating an area of low atmospheric pressure with a series of closed isobars around its centre (M.M. Shukla, Science Reporter, 2001). Wind blows from all sides towards the low pressure centre resulting in fast spiral motion and the whirling air to rise up. The high speed wind moves in *anticlockwise direction* in northern hemisphere and in *clockwise direction* in southern hemisphere. Each cyclone comprises a relatively calm area in the centre called *eye* surrounded with spiral cloud bands.

The severity of cyclones is measured on the *Beaufort scale* (named after Francis Beaufort), based on easily observable indicators such as tree movement and damages to structure. Cyclones are categorised on the basis of severity into *(i)* normal *(ii)* severe and *(iii)* very severe. The word *supercyclone* has been coined by IMD (India Meteorological Department) assessing severity of Orissa cyclone of 1999.

In other parts of the world, cyclones are termed as *hurricanes* and *typhoons*.

Hurricanes are tropical cyclones with surface wind speed reaching *Force- 12* of Beaufort scale in excess of 64 knots (about 120 km/hr), that occurs around the Caribbean sea and Gulf of Mexico.

Typhoons (a Chinese word for great wind) are also tropical cyclones with winds much above *Force –12* of the Beaufort scale. These occur in China sea and in the Pacific oceans.

On a global scale, 80 to 100 cyclones are formed each year, of which 3-4 severe cyclones are formed in the Bay of Bengal between April to June or September to December. According to World Meteorological Organisations (WMO), India suffers only *six per cent* of total cyclones whereas China and Japan face *30 per cent* and America *23 per cent* but magnitude of damages in India is too high.

Genesis of Cyclones

Shukla (2001) stated that the cyclones are the product of collective phenomena, in which the lighted heated air mass from the ground level rushes towards higher altitude, the vertical velocity being as high as 20 metres per second. A cyclone begins to form when warm moist air rises from the surface of the seas and is funneled upwards in a natural updraft.

As soon as, this moist and hot air rises, it cools and condenses into rains. The upward move of air generates *low pressure area* above the hot sea water area which remains stationary for 3-4 days and drawn energy from sea surface.

As the pressure in the centre falls, the wind speed increases to 40 km per hour and *cloud bands* start spiraling around the centre, causing squalls. The warmed air rises upto 12,000 m or more inducing the wind from the surrounding areas to rush inwards creating spirally moving storms. The cyclone then moves landwards to area of low pressure. The outcome is high speed winds accompanied by extremely heavy rainfall, often as high as 30 cm per day. Cyclones also cause storm surges upto 7.5 metres high causing floods. Such surges are caused by a combination of factors, mainly winds pushing water ahead of them and the sucking effect of the cyclone's action raising the sea level.

Cyclone Retrospect and Damages

Cyclone comprises fierce spiralling high speed winds, sometimes more than 100 km/hr, accompanied with heavy rainfall often exceeding 25 cm/ day and may be as high as 100 cm/ day. The highest wind speed recorded in the Indian subcontinent was that of Orissa cyclone of October 29,1999 with wind speed ranging from 200 to 300 km/hr. which blew men and animals, scooters, bicycles, roof tops etc. in the air.

An average area covered by various cyclones ranges from 40 to 800 km wide and average life span of a cyclone may be few hours to 9 days.

Like other natural disasters, cyclones too cause large scale killings and damages to man made structures. The direct effect of cyclones on

Box 34.1 Categories of Cyclones (Hurricanes)

Cyclones are categorised on the basis of their intensity which reveals the potential property damage and expected flood level. The intensity is measured by Saffir–Simpson Scale which gives rating in five categories:

Category 1. Winds 119-153 km/h, No real damage to property, Flood in some coastal parts, Storm surge 1.2 to 1.5 above normal.

Category 2. Winds 154-177 km/h, Some damage to property, Some higher flood effects, Storm surge 1.8-2.44 m above normal.

Category 3. Winds 178-209 km/h, Some structural damage and large trees uprooted, Flood destroying small structures of coastal region, Storm surge 2.7-3.7 m above normal.

Category 4. Winds 210-249 km/h, Extensive damage to residential structures, Shrubs and trees blown away, Flood affecting coastal regions upto 10 km inland area, Storm surge 3.96-5.5 m above normal.

Category 5. Winds more than 249 km/ h, Devastating to property and vegetation, Flood damage high affecting 8-16 km inland areas of the shoreline, Storm surge greater than 5.5 m above normal.

economy is reflected by damages to infra structure, crops and productive assets of local people. Huge financial burden is imposed on relief and rescue operations. The worse suffers are poor who not only loose their property and employment but also face increased cost of goods and services.

Rainfall associated with cyclones can be intense and may continue over a long period. Rain seeping into homes and buildings and attacking foundations may cause severe damage to them. Insufficient drainage may lead to local flash floods leading to precarious conditions.

Mitigation Measures

Following are a few important points to follow to reduce the damage:

(1) **Prediction and Early warning system**. One of the short term mitigation measures against cyclone disaster is the efficient warning system. Govt. of India has established a separate Indian Meteorological Department (IMD) to forecast not only where and when precisely the cyclone would strike the land but also the expected speed of the winds and the magnitude of the storm surges. Cyclones can be predicted very reliably much in advance to enable evacuation of vulnerable populations.

In India, a 24 km wide coastal belt has been identified as a cyclone hazard zone. In *eastern coast*, it stretched from Chingleput and Tanjavur districts of Tamil Nadu, East Godavari, Krishna, Nellore, Prakasm and Srikakulam districts of Andhra Pradesh, Balasora, Cuttack, Ganjam, Jagatsinghpur and Puri districts of Orissa, to Midnapur and 24 Parganas in West Bengal and in Western coast it includes districts of Junagarh, Amreli and Bhavnagar of Gujarat.

The IMD has deployed 12 cyclone detection radars on east and west coasts of the country. These use INSAT satellite to track the cyclones. Using satellite pictures and the cyclone detection radars, it is now possible to provide adequate advance warning about cyclones.

(2) **Maintaining telecommunication**. Cyclone disrupts communication and other services badly and cuts off the affected area from the rest of the world. Andhra Pradesh has a system of HAM radio operators. Ham is a common term for amateur radio communicators. They communicate with each other using voice, computers, morse code, pocket radios and satellite links. During the super cyclone disaster in Orissa, where telecommunication department was ineffective, there was only one satellite phone in whole of Orissa state set up by Ham radio operators on midnight of October 1. It's role was found very effective in maintaining telecommunication in cyclone effected areas for rescues of people.

(3) **Community education and awareness**. Public education and community involvement at gross root levels could play vital role in

evolving a predisaster mitigation programme and post disaster management operations. People should be educated as to how to face such problems. They should be trained about stocking of food grains, drinking water medicines etc. and to keep trade implements safe. They should be made to become dedicated volunteers crucial for management and maintenance of shelters, evacuation of community kitchens and camp's management etc.

(4) **Structural improvement and afforestation**. In hazard prone areas, there is a need of integrated efforts to construct hazard resistant structures. It has been found that round and polygonal houses are most suitable for the reduction of wind pressure. The whole structure should be so tightly in built to move as a unit during cyclone attack.

The settlements should be at a safe distance from the sea coast and should be located behind natural windbreaks such as sand dunes or clumps or multiple rows of trees to reduce the impact of winds. Forest belt comprising tall and deep rooted trees serves as protective walls. Forests provide not only protection but also fuel and fodder to the local people. Mangrove forests are most suited to cyclone prone areas.

As compared to earthquakes, tackling of cyclone disaster is relatively easy provided accurate prediction, proper planning and adequate preparedness are taken care of. Awareness should be developed to follow disaster management experts, technocrats and policy makers. A preventive policy and sustainable programmes can effectively reduce the magnitude of damage.

LANDSLIDES

The term landslides is used to cover a wide variety of landforms and process involving the down slope slipping of soil and rock or rock pieces under the influence of gravity. They are caused due to the disturbance of the equilibrium between stress and strength of material on slope.

Landslide is a major environmental hazard, found especially in the mountain region. Depending upon the nature of dominant material in the movement of surface matter, they are known as *rock falls*, *landslides* (rock and soil) or *avalanches* (snow and ice). The more rapidly moving landslides cause great hazard to life as compared to slower ones.

Himalayas are a dynamic mountain range prone to regular landslides particularly during heavy rains. In *Western Himalayas*, (August 18, 1998), more than 350 people were killed including 60 pilgrims who perished at the tiny hamlet of Malpa while on way to Kailash-Mansarovar. At Malpa, the entire camp comprising about half a dozen fiberglass huts sheltering pilgrims was wiped out and buried under tonnes of rocks and muds after days of torrential rains.

In Northeast part of Himalayas, recurring landslides are one of the biggest stumbling blocks hindering the smooth flow of traffic and supply lines. Sonepur on National Highway (NH 44), which links Shillong with Agartala, is the most landslide prone area. In 1988, a massive landslide dumped large boulders, trees and mud upto a height of 50 feet on highway at Sonepur blocking almost half a kilometer of the road. The road remained closed for about two months till the 50,000 cubic metres of debris could be cleared. It has been observed that in Himalayas (Indian part) an average of 300 landslides occur every year.

Causes of Landslides

All landslides involve failure of earth material under shear stress. The causes which contribute to increase shear stress may be many but a few important ones are as follows:

(1) Many hazardous landslides are induced by earthquakes.

(2) Prolonged rainfall or incessant heavy rain increases in water content of slope material and results into landslides.

(3) Many slopes are weakened by human activities such as timber harvesting and other causing deforestation.

(4) Man-made structures like making roads, heavy loading of weak surface due to construction, garden irrigation, sewage effluent systems and constructions of water holding structures like lakes and dams etc. are also important factors inducing landslides.

Effects of Landslides

Landslides can be as disastrous as earthquakes causing various types of damages. These causes:

(1) Killing of human beings,

(2) Destruction of man-made structures,

(3) Blockage of vital roads,

(4) Cutting of supply lines,

(5) Blockage of river passage causing flood,

(6) Destruction of vast useful area,

(7) Denudation of land etc.

Forecasting and Warnings

A perfect forecasting of landslides is not possible. However, it is focused on identifying location, magnitude and timing of specific events so that emergency actions may be achieved efficiently. Prediction provides the framework for planning management, engineering and investment decision making. A forecast makes it possible to issue generalised warnings of debris and mud flows following storms and heavy rainfall. In general, many landslides are preceded by a period of soil or surface creep before slope failure occurs, often giving rise to surface cracking. Monitoring of such conditions provide warning.

Disaster Mitigation

Landslide mitigation options aim not only at reducing risks but at the same time also controlling artificially induced aggravation of landslide risks. Knowledge and awareness must combine to cope with the problems.

Now means are available for assessing the risks of landslides of a particular area. Once the information is available such areas can be avoided for habitation, road building or pilgrim resorts to prevent causalities. In some stitches of landslide prone national highways, particularly in north-east India, slope stabilisation has been tried out with some success.

Today, Sonepur land slide is no longer a cause for worry. A multipronged, well executed strategy has helped stabilise the hill slopes. The measure taken was provision to *channelise rain water down the slope* into the river to prevent ground percolation of the water. In mud areas, special bitumen based binders were used to *increase the cohesion*. It was followed with a *massive plantation programme* at the landslide site to bind the soil more effectively. Better technologies and better monitoring systems are required to mitigate the damage.

In landslide prone areas, people should be prevented to inhabit and already inhabited areas should be shifted to safer place. Also people should be made to learn to identify early signs such as rock falls or cracks or subsidence in the ground which usually precede landslides.

Uncontrolled construction of roads, large scale denudation of land and uncontrolled terracing for cultivation make the area vulnerable to landslides. These must be managed properly.

Case Studies:

ORISSA SUPER CYCLONE–1999

October 29, 1999, the most tragic day of Orissa history, a super cyclone hit the western part of Orissa. It lasted for 36 hours with highest wind speeds ranging from 200 to 300 km per hour, and devastated 12 districts which resembled a non-descript war-torn country with a glory landscape of rotting dead bodies and animal carcasses. The cropland was rendered saline and covered with beach sands. Electricity, water supplies and telecommunication were completely snapped. Rail tracks were twisted or broken and roads were washed out and untraceable. The highest wind speed recorded in the history of India was this one. It was so strong that it blew men and animals, along with scooters and bicycles in the air often dashing them against wall and killing them.

By this super cyclone, life and properties of 18,000 villages were damaged, 20 million people were affected, 10,000 people were killed, 15 million became homeless, 0.5 million cattle lost or killed, 90 million trees were uprooted and some 1.8 million hectare of land was damaged.

The super cyclone was predicted well in advance. The IMD (India Meteorological Department) officials in Calcutta could trace the depression as early as October 25 and alerted the Orissa state administration of the impending disaster. But these early warnings were ignored leading to heavy loss of human lives. Had the authorities taken timely action to move people to safer areas, a large scale loss of life would have been averted. Not only this, the delay in providing timely relief to the affected people led to further hardships and miseries for the desperate victims.

TSUNAMI DISASTER 2004

Tsunami is a Japanese word meaning harbour waves (*Tsu*=harbour, *nami*=waves). A rare mega-thrust earthquake took place in the Indian ocean off the western coast of northern Sumatra (Indonesia) at the magnitude of 8.9 Richter scale. The hypocentre of this undersea earthquake, which occurred on December 26, 2004, was at 3.316° N and 95.855°E, some 160 km of Sumatra, at a depth of 30 km below mean sea level. This is at the extreme western end of the *Ring of Fire*, an earthquake belt that accounts for 81 per cent World's largest earthquakes. In the quake, an estimated 1200 km fault line slipped 20 m along the subduction zone where the India plate dives under the Burma Plate. As a result sea bed of Burma plate, have risen to several metres vertically up over the Indian plate creating the shock waves in the Indian Ocean. These Tsunami waves travelled at a speed of 800 km/hr with height being upto 10 m in Indian coasts.

These Tsunami waves devastated the shores of Indonesia, Sri Lanka, India, Thailand and other countries with waves upto 15 meters in height hitting coasts as far as East Africa (4500 km or more) or more west of the epicentre. The quake was also felt as far away as India, Malaysia, Myanmar, Thailand, Singapore, Bangladesh, Maldives, Sri Lanka etc. This disaster was the deadliest natural disaster in modern history.

This Tsunami disaster caused damages in 12 countries, and killed more than 1.66 lac people (41,000 in Indonesia, 22,000 in Sri Lanka, 1657 in Thailand, 3,000 in Andaman Nicobars (India) and 8,009 in Tamil Nadu, 593 in Pondichery, 171 in Kerala (India). These waves took 35 minutes to reach from Indonesia to Andamans and 2.5 to 3.0 hrs to South Indian coasts. In India, maximum damage was done in Nagapattinam (T.N).

It was noted that maximum damage occurred along the denuded coasts of Indian sea and places which have 3-4 km belt of mangrove forests and *Casuarina* like trees along sea coast were relatively least affected. Thus, ecological balance becomes essential to combat such tragedies.

Government of India received foreign relief fund worth Rs. 2939.85 crore, of which only Rs. 249.88 crore have been allotted to various states by June 2006.

Chapter Summary

Disasters are unpredicted causes that seriously disrupt to life. These kill or injure a large number of people and animals. Disasters can be broadly of two types *(i) Natural disasters, e.g.* earthquakes, volcanic eruptions, hurricane, tornado, avalanche, flood, droughts etc. and *(ii) man made disasters, e.g.* Chernobyl tragedy, Bhopal gas tragedy, dam leakage etc. Several disasters occur every year. The World Disaster Report (2004) revealed that from 1994 to 1998 average disasters per year were 428 whereas from 1999 to 2003, the number has increased to 707 disasters per year.

Asia suffers most from natural disasters killing the highest number of people as compared to the rest of the world. During 1994 to 2003, the disasters reported were 3055 of which floods are maximum (1160), followed by windstorms (802), droughts and earthquakes (257).

India is a disaster prone country in which basically six zones have been recognized (the Northern region, Eastern region, North eastern region, Western region, Southern region and Islands of Andaman Nicobar and Lakshadweep).

Government of India assists states with all required supports like defence services, air dropping, searching, air lifting, rescuing, transport of relief goods, health personals, medical support, telecommunications, restorations, availability of transport services etc. It has constituted a *Crisis Management Committee* comprising secretaries of various ministries. Home ministry looks after such crisis problem and it has developed *National Disaster Management Division* (NDMD) to deal with all kinds of disaster management. To facilitate the management, government of India has further constituted a *High Power Committee on Disaster Management*. Besides, a *National Disaster Management Authority* has been constituted and a *National Disaster Management Act* has been enacted. A *Calamity Relief Fund* has been constituted to be contributed by Union and State Govts. in the ratio of 3:1. Some Rs. 23,000 crores have been recommended by twelveth Finance commission for calamity Relief fund.

Flood is the submerged long part of the earth under the water for several days. India is the worst **flood** affected country in world after Bangladesh. About $1/8^{th}$ part of the country is flood prone. Flood is common in UP, Bihar, West Bengal, Assam and Orissa. Causes of flood can be *natural* (*e.g.* prolonged rainfall, land slides, volcanic eruptions, tsunami etc.) or man made (*e.g.* deforestation, dam accidents, urbanization etc.). Flood causes various kinds of damages that include death of human beings and animals and destruction of urban constructions, electric supply, water supply, roads, crop fields etc. Flood may often follow the epidemic diseases. Some methods to prevent flood are *(i)* delay in reaching of surface run off water to the rivers, *(ii)* minimisation of flood effects, *(iii)* development of forecasting system and management etc.

An **Earthquake** is a sudden motion or trembling of the ground produced by abrupt displacement of rock masses. Its magnitude is measured on Richter scale. Real damages caused by earthquakes begin at magnitude 5 and total destruction occurs at 8 or so. Causes of earthquakes may be volcanic eruption, faulting and folding, upwarping and down warping, hydrostatic pressure, plate movements etc. The Plate Tectonic Theory has been accepted as the most plausible explanation of earthquakes which explains that earth is composed of solid and moving plates. There are six major plates and twenty minor plates. The rate of plate movement is around a millimeter per year but strain builds up for decades or centuries and then everything happens at once. Earthquakes cause several hazardous effects, *e.g.* *(i)* slope instability and failures and land slides, *(ii)* damage to human structures, *(iii)* damages to the town and cities, *(iv)* loss of human lives and properties, *(v)* Tsunamis etc. Earthquake management comprises prevention, control and mitigation. *Prevention phase* includes, the preparation of earthquake catalogues, installation of seismic observatories, identification of seismic zones, training of architects and engineers, promulgation of laws and by laws, earthquake insurance etc. *Control and Mitigation* phase comprise emergency phase working after disaster and consolidation and reconstruction phase.

Cyclones are raging spirals of wind and rain that are born near the Equator over the seas, creating an area of low atmospheric pressure with a series of closed isobars around its centre. The high speed wind moves in *antilock wise* direction in northern hemisphere and in clockwise direction in southern hemisphere. In other parts of the world, cyclones are termed as hurricanes and typhoons. Severity of cyclones is measured on the Beaufort scale. A cyclone begins to form when warm moist air rises from the surface of the sea and is funneled upwards in a natural updraft. Cyclones cause damage to wildlife, construction and killing or injuring human beings and animals. Mitigation measures to reduce the damage includes- *(i)* prediction and early warning systems, *(ii)* maintaining telecommunications, *(iii)* community education and awareness, *(iv)* structural improvement and afforestation etc.

Landslides term is used to cover a wide variety of landforms and process involving the downslide slipping of soil and rock or rock pieces under the influence of gravity. These are common in mountain regions. Himalayas are prone to regular landslides. Causes of landslides may be the earthquakes, prolonged heavy rainfalls, deforestation, construction of water holding structures etc. These cause various types of damages, *e.g.* killing of human beings, destruction of man-made structures, blockage of roads, rails and supply lines, flood, denudation of land etc. Landslide mitigation includes controlling artificially induced aggravation of landslide risks, slope stabilisation, channelising rain water, afforestation, checking uncontrolled construction of roads and land terracing etc.

Orissa Supercyclone 1999 and Tsunami 2004 are good examples of case studies.

Study Questions

1. What are disasters? Explain their causes and ways to manage them.
2. Mention the various disasters of India, describe the Government's efforts towards disaster management.
3. What is flood? Describe its causes, damages and management.
4. Earthquake is a major kind of disaster causing maximum damages. Describe its hazardous effects and methods to mitigate earthquakes.
5. What is a cyclone? How is it formed? Describe the various mitigation measures.
6. What are the various causes of landslides? Describe the various disaster mitigations.

7. Briefly describe:
   - *(i)* Orissa SUPERCYCLONE –1999
   - *(ii)* Tsunami disaster –2004
   - *(iii)* Disaster prone regions of India
   - *(iv)* Govt. of India and disaster management
   - *(v)* Causes of floods
   - *(vi)* Flood management
   - *(vii)* Causes of earthquakes
   - *(viii)* Mitigation of earthquakes
8. Write short notes on:
   - *(i)* Formation of cyclones
   - *(ii)* Cyclone mitigation measures
   - *(iii)* Causes of landslides
   - *(iv)* Landslide mitigation

Objective Questions. *Select correct answers*

1. An example of natural disaster is:
   (1) Bhopal gas tragedy (2) Chernobyl disaster
   (3) Dam collapse (4) Earthquakes

2. A man made disaster is:
   (1) Cyclone (2) Earthquake
   (3) Chernobyl disaster (4) Flood

3. Orissa super cyclone occurred in the year:
   (1) 2004 (2) 1999
   (3) 1945 (4) 1560

4. National Disaster Management Division works under the ministry of:
   (1) Environment & Forests (2) Finance
   (3) Home (3) Family welfare

5. Tsunami disaster of 2004 was the result of:
   (1) Super cyclone (2) Earthquake
   (3) Landslides (4) None of these

Answers

1. (4) *2.* (3) *3.* (2) *4.* (3) *5.* (2)

Social Issues and the Environment

- *From Unsustainable to Sustainable Development*
- *Urban Problems Related to Energy*
- *Water Conservation, Rain Water Harvesting, Watershed Management*
- *Resettlement and Rehabilitation of People, Its Problems and Concerns, Case Studies.*
- *Environmental Ethics: Issues and Possible Solutions*
- *Climate Changes, Global Warming, Acid Rains, Ozone Layer Depletion, Nuclear Accidents and Holocaust, Case Studies.*
- *Wastelands Reclamation*
- *Consumerism and Waste Products*
- *Environment Protection Acts*
- *Forest Conservation and Wildlife Protection Acts*
- *Issues Involved in Enforcement of Environmental Legislation and Public Awareness*

From Unsustainable to Sustainable Development

Learning Objectives

Introduction • Concept of Sustainable Development • Concern Over Unsustainable to Sustainable Development • Consequences of Unsustainable Development • The Key Aspects for Sustainable Development • Path of Sustainable Development • Need for Sustainable Development for Improving Quality of Life • A Strategy for Sustainable Development • Guidelines to Campaign for Sustainable Society • Status of Sustainable Development in India.

Introduction

The concept of sustainable development became well known through the new famous Brundtland Report entitled *Our Common Future* published in 1987, also known as the World Commission on Environment and Development. The report recognised that *natural resources are not inexhaustible and development should be aimed to meet the needs of the present generation without compromising the ability of future generations to meet their own needs*. It also highlighted that the resources should be enhanced by gradually changing the ways in which technologies are developed and used. The developing countries must be allowed to meet their basic needs of employment, food, energy, water and sanitation. It is to be done in a sustainable level of population. Economic growth should be revived and the developing nations should be allowed a growth of equal quality to that of developed nations. The resultant shift in the developmental paradigm led to a paradigm shift in the environmental science. Humans were seen as a part of the ecosystem functioning and impacting upon ecosystems to a greater or lesser extent. In this view of ecosystem/landscape level functions, traditional societies, living close to nature and natural resources, were viewed as better integrated into the system than the industrialised societies. One of the first efforts to recognise the paradigm shift in the ecology in this direction was made by PS Ramakrishnan as early as in 1970s while working on interdisciplinary case study on shifting agriculture ('Jhum')–centred sustainable management of natural resources in North Eastern Hill States of India.

There is a greater realisation today than ever before about the need for working in the interphase area of ecological and social processes, if we are to effectively manage our natural resources. This is particularly important for effectively managing the biodiversity rich *hot spots* that are largely located in the developing tropical countries where the traditional societies act as the custodians of this biodiversity. These societies used the biodiversity to meet their livelihood concerns in a sustainable manner, until the industrialised humans interfered with their life patterns by exploiting them indiscriminately. There is an increasing concern amogst a small section of the scientific community on understanding and validating the *traditional ecological knowledge* which formed the anchor for managing biodiversity by the local people while conceiving it and enhancing it through human selection.

Concern for sustainable management of natural resources has assumed greater significance in the context of global change – a large variety of anthorpogenic changes involving climate, biological invasion, land use and land cover including site desertification and biodiversity depletion, that are happening today at a rapid rate. Global change has put a big question mark on our ability to sustainably manage our biosphere, with concerns for an improved quality of life for all humans on this planet. Globalisation of economics, that has been initiated in a unpopular political world, has also raised a number of issues about sustainable development of the biosphere, with equity considerations. It has also raised the likelihood of rapid erosion in human cultural diversity. There is a fear that homogenisation of societies may lead to loss of flexibility and our ability to chose pathways for development according to a value system based on the cultural heritage. Some feel that globalisation may create a whole range of opportunities and benefits for many, thus improving the environmental quality in some situation, but the impacts may not be uniform. It is in this context that dangers involved in economic dominance, by a few over a vast majority who are still struggling to have a reasonably good quality of life, is becoming an issue for intense debate. For a changed perspective to emerge, we need to go through a series of reconciliations for sustainable development of our biosphere – we need to reconcile between ecology, economics and ethics and connect the broken links between them, lest we fall prey to the kind of anthropocentred thinking that has been the back bone of traditional ecological paradigm which dominated past millennium. Ecology and development in their true sense imply designing strategies for natural resource management, by linking up ecology with social processes.

Concept of Sustainable Development

The term sustainable development was introduced in the year 1987 in the World Commission on Environment and Development (the Brundtland Commission) in its report Our Common Future. According to this report sustainable development is the *development that meets the needs of the present without compromising the ability of future generations to meet their own needs*. This concept of sustaining the earth, raised public awareness and created a responsibility for better stewardship towards the environment. But this begs more questions than it answers. What is development itself? What are needs? Whose needs, present or future? What is to be sustained? What about the context and condition of the natural world of which humans are a tiny part?

There are other definitions of sustainable development as well. The IUCN/UNEP/WWF *caring for the earth* gave the definition (1991) as *Improving the quality of human life while living within the carrying capacity of the supporting mechanisms*. Agenda 21, adopted during the United Nations Conference on Environment and Development (UNCED) held in Rio de Janeiro, Brazil in 1992, is a blueprint for achieving the goal of sustainable development. Principal I of Rio Declaration declared that human beings are at the central of concerns for sustainable development. They are entitled to a healthy and productive life in harmony with nature. According to Goldwin Freeman's (1994) definition *Biospherically compatible and socially equitable improvement in the quality of life*.

These definitions appear merely slogans, yet sustainable development is really useful. We should **treat the earth as if we intended to stay.** One problem faced by environmental managers is that its fundamental concepts are still debated and the goal of sustainable development is not fully formed.

Concern Over Unsustainable to Sustainable Development

The concern between unsustainable and sustainable development is not new. Malthus's speculation on earth's capacity to sustain an overgrowing population is widely known. Since early part of 20th century, the economists have been keeping in their mind the issues such as – How to optionally exploit a non-renewable resource, and how to make the polluter pay? In late 1960s and early 1970s, economists focused on the present opposition of economic growth at the cost of environmental degradation. The reason which caused the people to think so is obvious, *i.e.,* there exists a relationship between environmental degradation and economic activity. And the focus of debate changed to achieve economic growth and development in more environment-friendly way. This shift resulted after realisation that environmental protection and well being of human race are the two sides of a coin and man's future can not be secured with the indiscriminate exploitation of nature and natural resources. But the development can not be stopped and we have to follow a middle path, *i.e.*, sustainable development.

Consequences of Unsustainable Development

Consequences of unsustainable development are many. Climate changes in our planet, that had been occurring as slow roller coaster, have now speeded up by human interference. There are six main processes: population increase, demanding food security, deterioration of land quality and accumulation of wastes; pollution in the oceans, rivers, ground water etc.; depletion of biodiversity; growing energy needs; and the changes we have brought about in the chemistry of the atmosphere, that are responsible for unsustainable development. Since the industrial revolution, we have been using sky as a waste unit by enhancing the natural greenhouse effect with carbon dioxide, CFC (Chloro-Fluoro Carbon), methane oxides of nitrogen and related molecules into the atmosphere. Rapid growth of mega cities in poor countries, requires technologies for transport and utility which must be dealt at low costs. These require low-tech solutions, which can be easily adopted, applied and distributed widely. Until now, development has been human oriented, that too mainly for a few rich nations. They have touched the greatest heights of scientific and technological development, but at what cost. The people who suffer from damage may be different from those who enjoy the benefits of economic growth. There is an urgent need to assess the costs to human welfare caused due to the environmental

Box 35.1 Brundtland

Dr Gro Harlem Brundtland of Norway, the first former environment minister to head a government and a member of the UN Foundation, won Blue Planet Prize 2004 for 'putting forward globally the innovative concept of sustainable development', notably through her chairing of the World Commission on Environment and Development in 1980s. She is also cited for her later work as Director General of the World Health Organisation and recalls how 'as a young doctor, I became aware of public health significance of a safe and secure environment'. The term sustainable development, introduced in 1987 by World Commission on Environment and Development (Brundtland Commission) in its report 'Our Common Future', arouse public awareness and created a responsibility for better stewardship towards the environment. According to the Brundtland report 'sustainable development is the development that meets the needs of present without compromising on the ability of future generations to meet their own needs'.

damages, particularly for the poor. The air we breath, the water we drink and the food we eat have all been badly polluted. Of what use this development if developers themselves find it difficult to sustain their lives on this planet? Our natural resources are just dwindling due to overexploitation. If growth and development continue in the same way, very soon we will be facing the 'dooms day' as suggested by Meadows *et al.* (1972) in their world renouned academic report 'The Limits to Growth'. This is unsustainable development which will lead to collapse of the interrelated systems of this earth.

The Key Aspects for Sustainable Development

The key aspects for sustainable development may be inter-generational equity or intra-generational equity.

(1) *Inter-generational* Equity

Inter-generational equity emphasises us to minimise any adverse impact on resources and environment for future generation, *i.e.,* it suggests us to handover a safe, healthy and resourceful environment to our future generations. This can be made possible only when we stop overexploitation of resources, reduce waste discharge and emissions and maintain ecological balance.

(2) *Intra-generational* Equity

Intra-generational equity emphasises us to minimise the wealth gaps within and between the nations during the developmental process. The Human Development Report of United Nations (2001) emphasises that the benefits of technology should seek to achieve the goals of intra-generations equity. The technology should address to the problems of developing countries, producing drought tolerant varieties for unpredictable climates, vaccine for infectious diseases, clean fuel for domestic and individual use etc. This type of technological development will support the economic growth of the poor nations and will help in narrowing the wealth gap and lead to sustainability.

Sustainability of a system largely depends upon its carrying capacity, *i.e.,* if the carrying capacity of a system crossed due to overexploitation or some other causes, environmental degradation is ought to occur till it reaches to the point of the no return. The carrying capacity of the system has two basic components: the supporting capacity, *i.e.,* the capacity to regenerate and assimilative capacity, *i.e.,* the capacity to tolerate various stresses, which play a significant role in deciding the resource utilisation patterns in order to attain sustainability. Consumption should not exceed regeneration and change should not be allowed to occur beyond tolerance capacity of a system. It is only by treading on the road to sustainable consumption that we can hope to save our oceans, rivers, forests, mineral resources, wildlife and above all, the dignity of majority of humankinds. If consumption has to be sustainable, it should not only regenerate natural capital but also meet the needs of the poor. Thus sustainable consumption movement has to be finally a people's movement. People will force the change in consumption patterns in various ways, be they consumers, householders, workers or voters. The sustainable consumption will require changes of mindsets in the government, in industry, among citizens, every where. The struggle is long and arduous and the strong and affluent will be hard to convince and change their mindsets. Taking into mind UN Academy of Science headed a conference on 'Sustainable Consumption' in Tokyo in 2000 which was attended by 80 organised inter-academy panels of academics around the world.

Path of Sustainable Development

Main stream of sustainable development urges the maintenance of ecological integrity; integration of environmental care and development; adoption of an internationalist stance; satisfaction of atleast basic human needs for all; utilitarian consumption, inter-generational, inter-group and inter-species equity; the application of science and technology and environmental knowledge to world development; the acceptance of some economic growth, *i.e.*, within limits; and the adoption of a long term view. There is a great need to restructure for sustainability and reshape our economy to reflect environmental cost and values involving every one. It requires industries to use and manufacture mere efficient and innovative goods, utilisation of fewer resources and offering better values. Sustainability involves governments to set high environmental standards in its energy consumption and purchasing policies. Most important is that all individuals must recognise that the products they buy and use do not tax the environment in any way.

Need for Sustainable Development for Improving Quality of Life

All of us consume goods and services, and as consumers we hold the responsibility of what we consume and in what quality because our consumption habits and life styles influence others as well as our environment. Our need and requirements will reflect the manner of development that occurs in our society. And if we gear our needs and requirements towards sustainability, then only can sustainable development be possible. Our today's consumption pattern will inevitably have its implications tomorrow. As a consumer, one should be conscious of products which are eco-friendly and resource conservation oriented. While assessing eco-friendly products, one should take care of the following points:

(1) Is the product recyclable?
(2) How much pollution would the product cause during its production, consumption and disposal?
(3) What impacts the products have on your health?
(4) Is the product resource conservation oriented?

A Strategy for Sustainable Development

An ethic based sustainable living and care for each other and the earth is the foundation for sustainable living. All life on earth, with soil, water and air, constitute a great independent system-the biosphere and disturbing one component can affect the whole system. There is no other rational option except to use the resources of the earth sustainably and prudently. Some of the principles of sustainable development are: respect and care for community of life, improve the quality of human life, conserve the earth's vitality and diversity, minimise the depletion of non-renewable resources, keep with the earth's carrying capacity, change the personal attitudes and practices, enable the community to care for their own environment, provide a national framework for integrating development and conservation and care a global alliance (fig. 35.1)

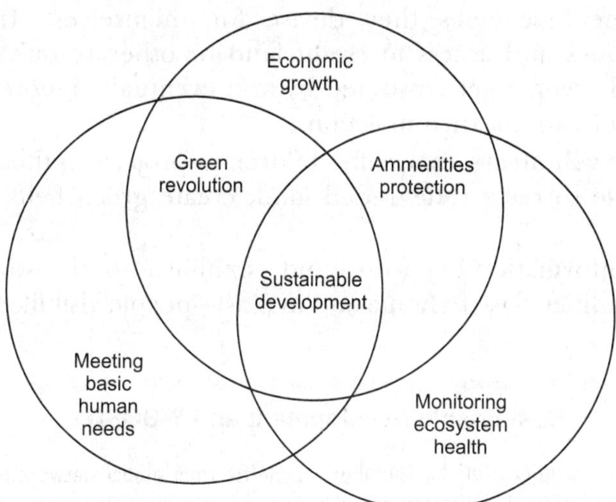

Fig. 35.1. A model of strategy for sustainable development.

Guidelines to Campaign for Sustainable Society

Everyone is a participant in the quest for a sustainable society and, therefore, the campaign should encourage a two-way flow of information, enabling people to contribute as well as to receive ideas and information. Methods used will inevitably vary with country, cultural traditions, religion and stage of development but the following guidelines and methods cover the spectrum.

(1) Involve every one and encourage their ideas, if needed using local language.
(2) Use all available media (print, radio, TV, film, video tapes, theatre, street theatre, *Nookad Natak*, dance, song, traditional story telling, puppet show etc). Face to face and audiovisual means of communication should be used in areas of low literacy. Traditional methods can work well. Postal campaigns and environment literacy programmes can give useful backing.
(3) Relate national and global issues to local situations, using familiar examples and experiences.
(4) Get people to interact and discuss their vision for their areas. Explain that how future may be threatened by current global and local trends and the solutions.
(5) Give people summaries and synthesis of the facts in appropriate form. Encourage development of synthesis for teachers, labour unions, business groups, government officials and politicians. Include case studies of what has had not been worked in the past.
(6) Make sure that people have access to clear and comprehensible information. Help them with advice and practical support to implement schemes they devise for themselves. Training in techniques and access to credit land or other resources may be needed. People get frustrated by and eventually ignore proposals that they can not turn in action.
(7) Involve volunteers, especially children in projects in their areas, for example to restore degraded land, create green belts and plant trees.
(8) Use information centres and exhibits, both within local communities close to home and in places people visit like museums,

Box 35.2 Sustainable Development and WBCSD

World Business Council for Sustainable Development, a coalition of 170 international companies of more than 35 countries and 20 major industrial sectors united by shared commitment, is taking an active interest in sustainable development via the three pillars namely economic growth, ecological balance and social progress. WBCSD also derives benefits from global network of 50 national and regional business councils and partner organizations involving more than 1000 business leaders throughout the world. The activities of this Council reflect its belief that the pursuit of sustainable development is good for business and business is good for sustainable development.

zoos, botanic gardens and national parks. These are especially effective because people choose to go to them and expect to learn.

Status of Sustainable Development in India

India is a unique country with tremdendous natural diversity associated with all kinds of climates and rich flora and fauna. The human societies in this country have evolved within magnificent environment. As such there occurs a great cultural diversity and reverence to nature, inherent in our cultural ethos. The very first verse in *Ishopanishad*, which was Mahatma Gandhi's favourite, says *Everything in the universe belongs to the Lord. Enjoy what is left for you as your share and do not covet what is not your's*. It means leading a simple life. Gandhiji drew inspiration from this perspective and tried to set an example for others by leading a simple life (Nadkarni, 2006).

Kautilya's famous treatise *Arthashashtra* describes what may be considered as the world's first conservation strategy for natural resource. During his regime those who conserved forests, water and other natural resources were given relaxation in taxes. Contemporary Mauryan kings also have similar provisons for conservation practices.

We basically concentrate on sharing with others. We start with giving, *i.e., danam*. *Bogah*, consumption for self, comes later. Even after giving, the remainder has to be consumed with discipline, *i.e., maryada*. Thus, there is a discipline for the household, *i.e., grihasth*, for community and even for the ruler. In *Bhagwatmahpurana*, it is said that one is entitled to only that amount of wealth which is enough for one's upkeep, the rest has to be shared, and if one does not do so, the same can be taken away.

Yavad bhriyate jatharam tavat savatvam hi dehinam
Adhikam yobhimanyet sa steno dandmarhati

Those who possess and give it back to the community and nature, part of our existence, are respected. In our tradition, people share food with animals and birds. Even snakes are given importance because they have role to play in balancing the nature. It is perhaps to underline this that the divine was born as tortoise (*Kachhapavtar*), the boar (*Varahavtor*) and even a horse (*Hayagriva*). A person who had no worldly asset was respected more than a king. We can see this in or times in our scientific community Dr JC Bose, Prof PC Ray, Prof Satyen Bose, Dr CV Raman, Dr. Homi Bhabha, the past President of our country APJ Abdual Kalam, to name a few as the true *Karamyogis*-non-possessive and entirely devoted to their mission. A beautiful verse for *Vayupurana* illustrates the law of consumption.

Na jatu kamah kamanam upbhogena shamyeti
Havish krishnavartmeva bhutya evabhivardhate

From this we understand how we run after unsustainable consumption because desires never come to an end or satisfied by fulfilling them. In other words, it in itself does not give satisfaction because by enjoying the desires, the desires grow more and more as fire grows more by the offerings of *ghee* and other objects into it. Traditionally, consumption is severely conditioned by the simplest outlook of the life and that is why the consumption is qualified by 'sustainable'. Even now, in villages, people do not prefer to cut and destroy young trees.

The country has to follow old traditions and has still to go a long way in implementing sustainable consumption oriented development. As a developing country, with a large population that has already crossed the one billion mark and is still expanding, India needs to lay emphasis on framing a well-planned strategy for the developmental activities which may increase economic growth. The task of sustainable development in such an eco-cultural diverse and thickly populated nation appears to be complex. In 1985, Ministry of Environmental and Forests has formulated guidelines for various developmental activities keeping in view the sustainability principles.

India faces many challenges in context of natural resources. The demands from sheer number of people affect land use dynamics at a local level, leading to resource degradation. In this context we need to look at the overall consumption pattern of the different sections of the society and link it up with market level. With much of our natural resources now being largely confined to upland regions of the country, the pressure on these resources is immense. Rapid industrialisation is already leading to pollution problems and impacting upon air, water and soil quality. We have to face the scenarios emerging from 'global change' and 'globalisation' concerns and the problem of coping up with uncertainties arising therefrom.

Sustainable Agriculture

India, as a developing country of tropics, has so far seriously pursued only the high energy input pathway for agriculture, in order to attain self sufficiency in food for its increasing population pressure. We have achieved a lot in this direction. But it has left many negative impacts. What is needed is that the modern agriculture has to co-exist with other additional pathways of agricultural development designed to meet with specific socio-ecological situations, based on a whole variety of traditional ecological knowledge-based agroforestry model. Landscape level diversity with a whole variety of land use systems-natural and human managed should co-exist. This alone will help in conserving both natural and human managed biodiversity, an issue which has become critical for human survival. Such a landscape management approach alone will provide the necessary stability and resilience to ecosystem to cope up with uncertainties in the context of global change and globalisation of economics.

Sustainable Forestry

India has come up with a number of policy guidelines not only to strengthen the environmental value of forests but also to increase forest cover in the country to a more reasonable level of one-third area, more effectively with community participation and ensure benefit accruing to the local community in terms of fodder, fuel wood and other non timber forest products, apart from production of timber for the country's increasing needs. Conservation and management of forests represent two sides of the same coin. What is applicable to forest management is also important for conservation. Apart from local initiatives, regional and global efforts are called for, in order to conserve the remaining hot spots of biodiversity through initiatives based on environmental holism. Ours is the last generation with the options to reverse the trends in environmental degradation and transform the world into a healthy and sustainable state.

India contains many elements of hope for sustainable development. Several individuals, communities, NGOs and government agencies have been leading a quite revolution in all sectors of development. Millions of hectares of land have been regreened and made productive by rural communities on their own or with the aid of innovatively designed Joint Forest Management, *Van Panchayat* or other schemes; a series of energy and construction alternatives have been developed and tried out by NGOs; decentralised water harvesting has eradicated drought in hundreds of villages in the driest parts of India and roof top rain water harvesting is slowly becoming a norm in many cities; under-funded government agencies, such as those on non-conventional energy success, have shown remarkable results with meagre resources; and thousands of farmers and farmer groups are showing that it is possible to produce enough food and make a decent living by practicing organic biodiverse agriculture. Many mass movements have forced a rethinking on infrastructure, energy, industrial development, forestry, fisheries and other development sectors by blocking one destructive project after the other. The obnoxious plastic carry bags, by now choking every urban water way and even messing up remote beaches and mountains, is being banned in one place after the other. Increasing consumer awareness, although slow is creating the demand for sustainable forms of products. Health care is being decentralised, as is watershed development, and even education in some states. On the biodiversity front Ministry of Environment and Forests has moved ahead with progressive legislation and with uniquely participatory strategy and action plan. Such development usually started as sporadic and individual initiatives, and at some point reached a critical mass. When this happens, it affects overall policy directions, and that is the ultimate hope. Thus people have to be made for increasingly aware that they can hold the development accountable to their promises. But for more basic rethinking on sustainable development, serious policy level changes and action on the ground, will be needed. Only then can we make this first decade of twenty first century a memorable one, for taking a road to sustainability.

International Cooperation and Sustainable Development Division is the nodal point within the Ministry of Environment and Forests, Government of India to coordinate all international environmental cooperation and sustainable development issues. Ministry of External Affairs, Government of India is the nodal agency of Commission on Sustainable Development (CSD) set up in 1993 in Agenda 21 of UN Conference on Environment and Development held in Brazil in 1992.

Chapter Summary

The concept of sustainable development became well-known through the new famous Brundtland Report entitled 'Our Common Future' published in 1987, also known as World Commission on Environment and Development. The report recognised judicious management of natural resources in the course of development. Although various definitions have been given to sustainable development, but all of them centre around the principle of economic growth and development in more environment-friendly way. Consequences of unsustainable development are many but climate changes in our planet is occurring as slow roller coaster. There are many reasons for unsustainable development but population increase is most responsible. Sustainable development can be achieved only by sustainable consumption. Path of sustainable development can only be decided by human beings as to how to adopt their life styles which reduce the consumption of resources and check pollution. Sustainable development for achieving quality life can only be possible if we gear our needs and requirements. Campaign for sustainable development should be targeted to common people by involving every one and making them aware of the consequences of unsustainable development. Status of sustainable development in India is much higher as reverence to nature is inherent in our cultural ethos. 'Caring and sharing' is the basic concept hidden in all religions and all Indian societies. Sustainable development can be achieved only through changing our consumption patterns which match with the carrying capacity of our planet.

Study Questions

1. What is sustainable development? Explain.
2. What are the possible ways to achieve sustainable development? Describe.

3. Discuss the status of sustainable development in India.
4. What are the consequences of unsustainable consumption? Explain.
5. Briefly describe:
   (i) Key aspects of sustainable development.
   (ii) Need for sustainable development.
   (iii) Brundtland Report.
   (iv) Gudelines of sustainable development.

Objective Questions. *Select correct answers*

1. The title of the Brundtland report is:
   (1) 'Caring for Earth'
   (2) 'Caring and sharing'
   (3) 'Our Common Future'
   (4) 'Environment-friendly Resource Consumption'
2. The blue print for achieving the goal of sustainable development was prepared in:
   (1) Stockholm Conference 1972
   (2) Rio Summit, 1992
   (3) Nairobi Conference, 1982
   (4) World Summit on Sustainable Development.
3. Brundtland Report is concerned with:
   (1) Water conservation (2) Air pollution
   (3) Sustainable development (4) Environmental protection
4. Jhum is:
   (1) A folk dance (2) A type of cultivation
   (3) A desert plant (4) An endangered animal

Answers

1. (3) *2.* (2) *3.* (3) *4.* (2)

36

CHAPTER

Urban Problems Related to Energy

LEARNING OBJECTIVES
Introduction • Energy Consumption Patterns • Energy Production Patterns • India's Efforts in Fulfilling of Energy Needs • Integrated Energy Management • Sustainable Energy Consumption • Mismanagement in Power Distribution.

Introduction

Energy is the basic need to national economy. Until recently, a huge majority of human population lived in rural areas. Their economic activities centred around agriculture, cattle rearing, fishing or cottage industries. With the dawn of industrial era, cities have become the main centres of economic growth, trade, education, innovations and employment. The rapid development of cities attracted the movement of rural folks to cities in search of employment and other needs which could not be fulfilled in the villages. Now about 30 per cent of human population in India is living in urban areas which have well established industries and trading avenues. Rapid rate of industrialisation and economic development has caused the cities to expand tremendously. Unprecedented rate of population increase in these urban areas coupled with migration of rural masses is making cities difficult to accommodate them. Expansion of cities is taking place both vertical to provide residential facilities for the people and horizontal to give space for housing, industrial estates and commercial complexes. Thus, cities are engulfing nearby sub–rural or rural areas even agricultural and forest lands, the *urban sprawl.*

Energy Consumption Patterns

In developing countries the urban population has the largest share of energy consumption. Energy is needed for household purposes, comfort, transport, telecommunication, industries, defence establishments, development projects and many other purposes. Energy plays a vital role in socio-economic development of urban communities. Today none of us can thrive of a comfortable life without energy. As compared to rural areas, the energy requirement in urban areas is much higher. Urban people have higher standard of living. Their life style demands more energy inputs in every sphere of life. City people demand energy for:

(1) Residential and commercial lighting apart from municipal road lighting.
(2) Public and private transportation such as local train, metro, truck, bus, taxi, car, scooter etc. for movement.
(3) Operation of electric gadgets and appliances to sustain their modern life styles.
(4) Air conditioners for comfort.
(5) Running of heavy industries needing higher energy demands, small scale industries, corporate bodies, companies and various trades.
(6) Disposal of waste and operation of sewage treatment plants.
(7) Operation of devices to control air and water pollution.
(8) Proper supply of municipal water, etc.

Today, in the modern world, the trend towards consumerism to make the life more comfortable, efficient and convenient, is the main cause of increased energy demand. In cities, constant supply of electricity has become indispensable for high rising multistoried buildings, apartments, multiplexes and shopping malls. Many of these establishments have their own high capacity power generating sets for power back up. These generators cause air and noise pollution and produce hazardous chemicals which are harmful for human health.

Urban people demand low cost energy for their comfortable living and operation of wide range of electric appliances readily available to them. The home appliances, such as refrigerator, washing machine, geyser, cooler, electric iron, television, CD player, radio, water purifier, computer, electric stove, microwave oven etc. are proving as a measure of quality life. People with higher standards of living require the comfort and luxury of air conditioners in addition to a whole lot of electric gadgets and appliances. This causes an increased demand of energy as the consumption of energy increases with the increased operation of each and every appliance people use. The quality of life of city people depends on the use of machinery and gadget to perform their activities quickly, efficiently and comfortably. Urban people are able to perform more tasks with the help of these appliances and machines. The kind of transportattion used by people has become symbol of status and reflects their quality of life they lead. Although in modern developed cities, economically advanced people prefer using air services and car but middle class people use metro, local trains, public transport buses, taxies etc. As the demand for energy in the urban areas is continuously rising, there is a serious concern about the supply of energy. But the increase in the installed generation capacity has not matched the demand rate. Under the circumstances, a major thrust of energy policy must be emphasised in increasing commercial energy supply.

Energy Production Patterns

About 70 per cent of the Indian population living in rural areas consume only 10 per cent of the total commercial energy as compared to urban areas and industries which consume 90 per cent of the total commercial energy. About 75 per cent of villages are electrified, out of which only 15 per cent use electricity for their household purposes. The rapid rate of consumption of commercial energy, not matching the rate of its production, has compelled us to think of producing more energy. In order to fulfill this increased demand of commercial energy, many attempts are being made at government and private under taking levels. Additional micro/mini hydro-projects are being developed for decentralised power generation. A considerable progress has been made in trapping other sources of energy especially non-conventional or alternate energy sources. Attempts are also being made in optimising energy utilisation. Adopting 'hard path', *i.e.*, generating energy from fossil fuels, such as coal and petroleum, has its own limitations. Adopting 'soft path', *i.e.*, use of alternate energy sources, has been given priority in response to the pressure on foreign exchange reserves due to the increased consumption of oil. In recent years Organisation of Petroleum Exporting Countries (OPEC) is continuously causing an abrupt rise in the rate of the oil. The soft path relies mainly on renewable energy and prefers to explore sunlight, wind, ocean, biomass etc. to trap their energy. The renewable energy programmes not only develop new sources of energy but also provide opportunity in the area of energy conservation, employment generation and human health.

India's Efforts in Fulfilling of Energy Needs

According to the Government report released on '*Rajiv Gandhi Akshay Urja Diwas*' on August 20, 2006, India had an installed capacity of 8,600 MW of renewable power. The country occupies second position in the renewable power in the world. Globally, India occupies second position in biogas plants, fourth in wind power, fifth in small hydro, seventh in solar photovoltaic and ninth in solar water heating. The country has more than 1500 Vestas Type Wind Electric Generators of various capacities at different location in the states of Rajasthan, Gujarat, Maharashtra,

Madhya Pradesh, Orissa, Tamil Nadu, Kerala and Karnataka. Besides, many indigenous developments in the area of non-conventional use such as thermal systems for cooking, heating, drying and process heat/stream; solar cells, battery operated vehicles; biogas systems for electricity generation; biomass gasification systems for electrical, thermal and mechanical applications; and hydrogen powered motorcycles and engines, have been developed.

Indian Renewable Energy Development Agency (IREDA) was established in 1987 under the Ministry of Non-Conventional Energy Sources, Government of India with the following objectives:

(1) promote renewable sources of energy.
(2) extend financial support for the generation and conservation of energy.
(3) provide financial support to manufacture non-conventional energy source systems and devices.
(4) facilitate leasing of non-conventional energy source equipments.

We are fortunate to fulfill our growing energy needs, particularly in urban areas, as we can generate 70,000 MWs through wind power, 1,50,000 MWs through hydropower and five million MWs through solar energy. It is expected that 50,000 MWs will be generated through non-conventional energy sources by the year 2030. In many cities of India, conversion of municipal waste into energy is also planned. This will also improve environmental quality and help in easy disposal of the solid organic wastes, a major problem in big cities.

Integrated Energy Management

It is difficult to predict energy supplies and demand as technical, economical, social and political assumptions are constantly changing. Also there are seasonal, annual and regional variations in energy utilisation. Keeping in view these facts, it becomes necessary to have integrated energy management which recognises that no single source can possibly provide all the energy required by the source nation. It also emphasises locality sited energy generation patterns and exchanges among states based on their consumption patterns. But the basic objective of the energy management should be sustainable energy development and consumption. The sustainable energy development should be aimed at providing reliable source of energy causing no harm to our local, regional and global environment and should consider that the future generations inherit the quality environment with a fair share of earth's resources.

Mismanagement in Power Distribution

In India, an important factor for the gap between the demand and supply of commercial energy, hampering growing energy needs of urban areas, is the widespread mismanagement, inefficiency and corruption in the power generation and distribution systems. Today the gap between supply and demand for electricity has widened from 5.9 per cent to over 9 per cent. The peak power shortage during summer, has made it 12 per cent to 15 per cent. A number of our power plants are very old and poorly equipped. Monitoring systems are inefficient and lack modern devices. The root cause of the relative lack of power modernisation is the antiquated method of State Electricity Boards which are monoliths of socialist planning. These boards are highly inefficient and support themselves by charging minimum usage fees and other associated tariffs that effectively prohibit private investments in and expansion of the power sector. There are many defects in distribution systems and check over transmission losses and thefts. If all these things are managed properly, India's GDP growth could be 2 or 3 per cent higher.

Sustainable Energy Consumption

For sustainable consumption, energy should be used efficiently, economically and intelligently. People living in urban areas, the largest energy consumers, must adopt modifications in their life styles to reduce their energy consumption. A few energy conservation tips to follow are:

(1) Switch off lights, fans or other appliances when not required.
(2) Switch off TV, washing machine, AC, microwave oven, computer etc. to save standby power.
(3) Use task lighting which focuses light where it is needed. For example, a reading lamp lights only reading material rather than the whole room.
(4) Replace the bulbs and tube lights with Compact Fluorescent Lamps (CFLs) to save energy consumption. A 15 watt CFL produces the same amount of light as a 60 Watt bulb.
(5) Reduce usage of electrical gadgets or appliances.
(6) Use natural light during the day time.
(7) Use solar cookers and solar heaters.
(8) Use power factor correction capicitators for water pumps.
(9) When dust builds up on refrigerator's condenser coil, the motor works harder and consumes more electricity. Clean the coils regularly to make sure that air circulates freely.
(10) Battery chargers, such as those for cell phones, laptops and digital camera, draw power whenever they are plugged in. Pull the plug when not required and save energy.

Chapter Summary

Energy is a basic need to national economy. Rapid rate of urbanisation has created much pressure on natural resources. Energy plays a vital role in socio-economic development of urban communities. Urban people have higher standard of living and use a vast variety of electric appliances and as such they have higher demand of commercial energy. Apart from domestic use, electricity is also required to run a large number of industries, trades, transportation, for street lighting and water supply. Energy is also needed to fulfill various needs for comforts of the urban people. About only 30 per cent of population living in cities consumes 90 per cent of total energy budget of India. To cope with increased demand of urban population, India has expanded its power generation programme. A considerable progress has been made in trapping alternate energy sources such as sunlight, wind, ocean etc. An integrated energy programme has been launched to fulfill needs of the people based on seasonal, regional and national level. The important factor in widening the gap between demand and supply of commercial power is widespread mismanagement in State Electricity Boards. Minimising use of electric appliances, developing alternate energy sources and taking care of saving electricity can help conserve energy.

Study Questions

1. What are the various energy production and consumption patterns?
2. Discuss India's efforts in fulfilling energy needs of urban areas.
3. Discuss mismanagement in Power Distribution.
4. Discuss integrated Energy Management.
5. Discuss sustainable Energy Consumption.

Objective Question: *Select the correct answers*

1. Indian Renewable Energy Development Agency (IREDA) was established in the year:

(1) 1983 (2) 1985

(3) 1987 (4) 1989

2. Largest share of energy is consumed by:

(1) Agriculture (2) Rural power supply

(3) Urban power supply (4) Industries

Answers

1. (3), *2.* (3)

37

CHAPTER

Water Conservation, Rainwater Harvesting, Watershed Management

LEARNING OBJECTIVES

Introduction • WATER CONSERVATION • Water Management Practices • World Water Vision • RAIN WATER HARVESTING • Traditional Rainwater Harvesting • Modern Techniques of Rainwater Harvesting • WATERSHED MANAGEMENT • Importance of Watersheds • Objectives of Watershed Management • Strategies for Watershed Management • Integrated Watershed Development Programmes in India.

Introduction

The world, they say, is moving towards water wars. A once abundant resource is increasingly being wasted, misused and polluted. Adequate and regular supply of water is a serious problem, whether one lives in the city, town or a village of India or any other part of the world. India faces an impending water crisis most visible in rural areas of some states and many of its cities. Just think of a rural women, who has to walk several kilometers each day to fetch the water and a housewife in Delhi who has to wake up at 4.00 am to turn on the municipal tap. Water reforms are now on the top of the governance agenda. The search is apace for innovative and sustainable solutions. Fortunately one does not need to look far to find such solutions. Many communities and enterprenures have implemented innovative, local-level solutions to overcome water supply problems over the past couple of decades, as reliance on microwater delivery systems instead of one megasystem. Attempts are being made at different levels to conserve the water before it is on sale.

Rainwater harvesting, by employing practices to store rainwater, both in rural areas and cities and recharge ground water is one of the recent methods to conserve the water. It involves collecting, storing and conserving local surface runoff of rainwater for future use. The need for rain water harvesting was felt with the problem of water shortage in arid and semi arid parts of the country, on being low cost method and for health benefits. The rainwater can be trapped on the roofs of buildings or in small ponds and can be used as and when required.

Watershed areas are the flood plains which are reserved for water storage, aquifer recharge, wildlife habitat and aqua-agriculture. They are used for management of rainwater and resultant runoff. They are characterised by water flow. The Himalayas is one of the most critical watersheds in the world. Watershed management ensures us beneficial developmental activities such as domestic water supply, irrigation, hydro-power generation etc.

WATER CONSERVATION

Among various resource conservation problems related to man's life support system, most important is water resource conservation. Water is elixir of life and is pre-condition for plants, animals and human life. It is an essential input for all economic activities. The status of water resource, in turn, is related to the status of the soil and vegetation system, man and animals around and the pattern of resource utilisation for economic development. Since we can not increase the total volume of earth's finite water, we should conserve our supply and keep our sources clean. Water conservation is the careful use and protection of water resources, both quantitatively and qualitatively. Even simple actions in our daily life, such as turning off the tape while brushing the teeth, can save enough water. Some common approaches of water conservation are:

(1) Reducing domestic water wastage.
(2) Reusing domestic waste water for watering kitchen garden or lawn.
(3) Reducing agricultural water wastage by adopting efficient irrigation practices such as drip or sprinkling irrigation.
(4) Reducing water wastage in industry by recycling the used water.
(5) Harvesting rainwater by adopting practices to store rainwater or recharge groundwater.
(6) Afforestation and watershed protection to improve water economy.

Box 37.1 Book Review: Tal-Taliya Ke Taranhar

The book entitled *Tal-Taliya Ke Taranhar* (1993), written by Anupama Misra who is presently affiliated with *Gandhi Shanti Pratisthan*, is an effort to make people aware of the importance of ponds in water conservation. It throws light on rehabilitation of old ponds and development of new ponds. More than 1.25 lac copies of this book have been sold by now. Inspired with the theme of this book, people of Purulia (WB), Lapodia (between Jaipur and Ajmer) and Malerkotla (extreme West Indo–Pak border) have been successful in rehabilitation of ponds and conservation of water.

Water Management Practices

Water conservation needs effective water management practices. Efficient water management should fulfil the following objectives:

(1) Provision of adequate amount of potable water supply to both rural and urban populations.
(2) Provision of optimal water supply of standard quality to agriculture in time, minimising wastage of water during irrigation, providing proper drainage and preventing water logging and salination.
(3) Provision of adequate water supply of standard quality to industries.
(4) Full exploitation of potential resources for power generation.
(5) Prevention of wastage and pollution.
(6) Measures of check depletion of resources.
(7) Purification of water and control of its pollution.

Some important water management practices to provide sustainable supply of potable water of quality are:

(1) Construction of dams and reservoirs to ensure constant supply of water round the year and controlling the floods and generating hydro-power in addition.
(2) Interlinking of rivers for equitable distribution of water.
(3) Extention of canal systems for proper distribution of water.
(4) Restoring water sources such as lakes, ponds, wells etc.
(5) Regular dredging of wells.
(6) Desalinisation of water bodies, groundwater and sea water, making it suitable for drinking, cooking and other purposes.

In India, Bhabha Atomic Research Centre (BARC) has developed a number of desalinisation technologies based on Multi-Stage Flash (MSF) evaporation, Reverse Osmosis (RO) and Low Temperature Evaporation (LTE) to provide potable water in rural areas and on ships, and water for industrial uses. The on-line domestic water purifier technology developed by BARC met with good success.

World Water Vision

The water available to us on earth is in a finite quantity that has not changed over millennia. This has to be just a justaposed against increasing demands from growing population. Increased population, coupled with the process of urbanisation and 'development', are also expected to vastly increase demand for fresh water. This situation of a finite supply of water and growing demand leads to the projections of water scarcity which could be severe. A whole series of institutions and networks have sprung up to deal with this and related matters. World Water Commission, World Water Council, Global Water Partnership and many other organisations are making several exercises to build up national, regional and global 'World Water Vision' for the year 2025, with the goal that all people have access to safe and sufficient water to meet their needs, including food, in ways that maintain the integrity of freshwater ecosystems. World Water Vision exercise's ultimate purpose is to generate global awareness of the crisis that women and men face and of the possible solution for reddressing from it. This awareness will lead to the development of new policies and legislative and institutional frameworks. The world's freshwater resources will be managed in an integrated manner at all levels, from individual to the international, to serve the interests of human kind and planet earth effectively, efficiently and equitably.

By 2025, the three primary **objectives** of integrated water resource management to be achieved are:

(1) Empowering women, men and communities to decide on the level of access to safe water and hygienic living conditions that they wish and to organise to obtain it.
(2) Producing more food, creating more sustainable livelihoods for women and men per unit of water applied (more crops and jobs per drop), and ensuring access for all to food required for healthy and productive lives.
(3) Managing water use to conserve the quantity and quality of freshwater and terrestrial ecosystems that provide services to humans and all living beings.

The five key actions to achieve the World Water Vision objectives are to:

(1) Involve all stakeholders in integrated water resource management.
(2) Move toward full cost pricing of water services for all human uses.
(3) Increase the public funding for research and innovation in the public interest.

Box 37.2 Fifty Crore Rupees Spent a Month on Potable Water in Chennai

According to South India Packaged Drinking Water Manufacture's Association source, over and the above water supplied by Metro Water, people of Chennai spend a monumental Rs. 50 crore to buy 3.7 billion litres of water each month.

- Five million 250 ml packets at Re. 1 each.
- Seventy five lakh on one litre bottles at Rs. 10-12 each.
- One lakh twelve litre cans at Rs. 18-30 each.
- Twenty five thousands 20-25 litres bubble top.
- Two thousands water tankers carrying 12,000 liters each and making 5 trips a day (120 million litres) at Rs. 700-900 for a tanker load.
- Fifty crore rupees spent a month on potable water.

(4) Recognise the need for cooperation to impose integrated water resource management in international basins and to-

(5) Massively increase the investments in water.

Within India, a consciousness of the importance of water resource developed around 1945 with the establishment of Central Water Commission. In 1997, National Commission on Integrated Water Resource Development Plan was launched which completed its work and submitted the report to the Government of India in 1999.

RAINWATER HARVESTING

Rainwater harvesting is a technique to capture, collect and store rainwater in tanks or divert it to the ground to recharge the ground water, to be used as and when required. This is done by special water harvesting devices depending on the area chosen. In rural areas it is done through a check dam or *johad*, dug well, percolation pit etc. In urban areas rainwater may be diverted to a well or ground near hand pipe or collected in tanks of varying sizes. In *pukka* houses it is collected on the rooftop or terrace and stored in specially made tanks. Surface run off rain water collected in tanks is prevented from pollution. Rainwater harvesting areas and the use of collected water in these areas may be of the following types:

(1) Rain water collected from rooftop, courtyards or similar compacted or treated surfaces is stored in tanks or reservoirs and is used for domestic purposes or for groundwater recharge.

(2) Rain water collected in larger areas in and around rural areas, through check dams, is used for irrigation, live stock and groundwater recharge.

(3) Macro catchment rainwater harvesting, also called harvesting from external catchment, is the case where runoff rainwater from hill slope catchment is diverted to the cropping area located at the foot hill or flat terrain.

(4) Micro catchment rainwater harvesting involves collecting surface runoff rainwater from small catchment areas and storing it in the root zone of an infiltration basin on which are planted trees, shrubs or annual crops.

Rainwater conservation strategy is adopted because of the fact that the average precipitation over Indian land mass is around 1200 mm annually. In most of the places rains are concentrated during June to September during monsoon period. It is an astonishing fact that Cherrapunji, in North East Himalayas, which once considered as the place with highest rainfall of about 11,000 mm annually, now undergoes scarce rains (Box 4.1). The places with high precipitation rate form ideal spots for rainwater harvesting. But before adopting a rainwater harvesting device, it is necessary to carefully study soil characteristics, topography, rainfall pattern and climatic conditions of the area selected.

The rainwater harvesting is **aimed** at:

(1) Reducing run off of water
(2) Avoiding flooding of roads and streets
(3) Meeting the increasing demands of water at cheaper cost
(4) Raising the water table by recharging ground water
(5) Getting pollution free and hygienic water for health benefits
(6) Reducing ground water contamination
(7) Supplementing ground water supplies during lean season
(8) Solving water problem in arid and semi-arid regions
(9) Reducing soil erosion
(10) Improving soil moisture.

Traditional Rainwater Harvesting

In India, rainwater harvesting is an old practice. Kautilya, the ancient Indian economist, during his period used to give relaxation in taxes to those who conserved the harvested rainwater. Indian way of water harvesting can be witnessed even today in the form of *pucca talabs*, *bawaris*, *johads*, *hauz* etc. constructed by many emperors in their territory in

Box 37.3 Navdany's Water Liberation Declaration

Participants at a NGO Navdany's international workshop on "Globalisation and People's Rights to Natural Resources" adopted the Water Liberation Declaration on sixteenth December, 2001. The declaration given below has been signed by over a 100 individuals and organisations across the world.

"Water is life. It is a gift of nature. The access to water is a natural and fundamental right. It is not to be treated as a commodity and traded for profit. People shall have the right to freedom from thirst, and shall have adequate access to safe water for all of their living needs."

Experiences all over the world reveal quite convincingly that water which is "life" is being privatized and brought under corporate control. This will deprive people of water, lifeline for survival. All water resources should be owned, controlled, managed and utilised by the local communities in their natural setting.

We the people from all over the world will not allow our waters to be made a commodity for profit.

We will work together to liberate water from corporate/private agencies, control and return it to people for common good.

We demand "the governments all over the world should take immediate action to declare that they accept waters in their territories a public goods and enact regulatory structure to protect them".

villages, towns, cities and capitals, which were used to collect rainwater and ensured adequate water supply in dry periods.

In India, ground water boom is turning to bust. Half of the traditional hand dug wells and millions of tube wells have dried up across Western India. Also in other parts of the country, the level of ground water has gone quite low. According to an estimate of Central Ground Water Board, if we continue to exploit our ground water indiscriminately then in the next 20 years, 15 states of the country may face acute shortage of the ground water. Government of India has recognised rainwater harvesting as a thrust area. Until early nineteenth century, the rainwater was stored in shallow mud walled reservoirs, the tankers that captured monsoon rains. Some individual success stories of rainwater harvesting with remarkable results have encouraged the modern methods (Box 37.4).

Box 37.4 Community Based Rainwater Harvesting

A Doon-based NGO, Rural Litigation and Entitlement Kendra (RLEK) in Garhwal, in 1996 undertook project 'Swajal' in Satengal. Col. (Retd.) Kuldeep Singh Maan, incharge of RLEK along with his team members worked hard to ensure that the harvesting technique gave the villagers safe drinking rainwater. In turn public had to contribute to only 10 per cent of the cost of the project in the form of voluntary labour or cash. The rainwater poured from the rooftops of the houses that came to the project area was collected in tin sheet troughs and then directed to a ferro-cement tank through PVC pipes. The project involved very small amount of expenditure but gave adequate water both in quality and quantity.

To rescue from hydrological anarchy, people should be promoted for rainwater harvesting for sustainable increased growth in agriculture and meeting water needs of the area facing water scarcity. In urban areas rainwater harvesting should be made mandatory so that the water stored could be used for the purpose (Box 37.5).

Box 37.5 Rainwater Harvesting in Dwarka

A century old tradition of harvesting and conserving rainwater helps residents of Dwarka in Gujarat India to keep the water crisis at bay. Almost every house in this sacred city has an underground tank, which is used to collect rainwater every year. The drainpipes from rooftops and terraces are connected to the tank, which is sealed from all sides except a small opening is left at the top which allows periodical cleaning of the tank with bags of lime, used as an disinfectant. The tanks are large enough (upto 25m^2) to collect ample amount of water even if rainfall is good. Since the water is used strictly for drinking and cooking, it could last upto two years. Such a rainwater harvesting and conservation measure in this region is making a difference with other severely affected parts of Saurashtra in Gujarat.

Modern Techniques of Rainwater Harvesting

In arid and semi arid regions, artificial ground water recharging is done by constructing shallow percolation tanks. Check dams made of some suitable native material, such as loose rocks, rocks, stones, slabs, sacks, wirenets, plants, poles etc. are constructed for harvesting run off rainwater from large catchment areas. Rajendra Singh of Rajasthan, popularly known as '*Water man*' has been doing commendable job for harvesting rainwater by initiating people to construct *johads*, a type of checkdam. For this work he has been honored with the prestigious Magsaysay Award.

For storing water underground, ground water flow can be intercepted by building ground water dams which are advantageous over surface dams. Ground water dams undergo minimum evaporation loss and have lesser chances of getting contaminated.

The roof top harvesting is low cost and effective method for urban rainwater harvesting. The rainwater can be collected in roof tops of houses and buildings and can be diverted to some surface or underground tank or a pit through a delivery system. This water can be later used to recharge underground aquifers by diverting the stored water to some abandoned dug well or the site of a hand pump (fig. 37.1). Technique is low cost and requires little expense for maintenance. Recently Delhi Govt. has announced to establish two awards (Chief Minister's 'Rainwater Harvesting Awards') from 2007 for outstanding contribution in rainwater harvesting.

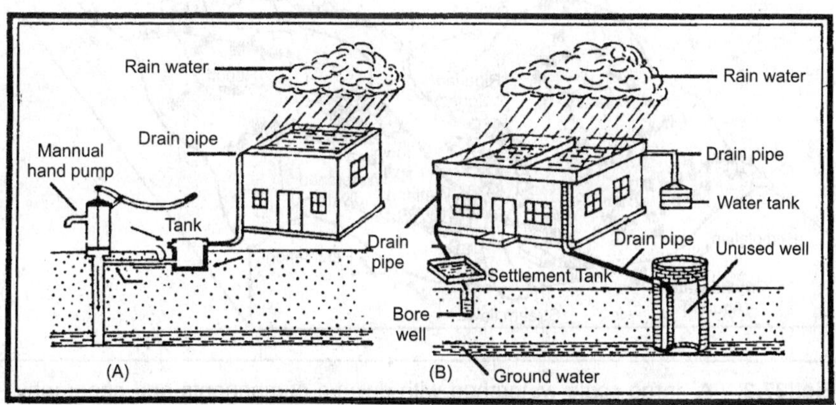

Fig. 37.1. Rainwater harvesting (A,B two methods).

WATERSHED MANAGEMENT

Watershed may be defined as the land area from which surface water drains under gravity to a single outlet such as a stream, a river or a lake.

The rainwater flows from the ridge into *nallas* and finally collects in the ponds. This entire area with a common drainage is called watershed. Thus watershed is a geographic unit that collects, stores and releases water. Source of this water is rain, snow or coastal fog. It gets stored in ponds, lakes, subsurface soil and geological formations. Rivers and streams as well as ground water flow, each of them then releases this water. Watersheds are based on the geography of the area. A watershed is the land area where water flows across on its way to creek, stream, wetland, pond, lake or river. Watersheds are also called drainage basins or catchments because they 'drain' and 'catch' rain and somewhat water that falls onto land. They are defined, on the basis of location, such as the area contributing to a farm pond or to lake contributing to the river at its mouth. The boundaries of the watersheds are determined by the elevation of the surrounding land, with a highest elevation points marking the boundary (fig. 37.2).

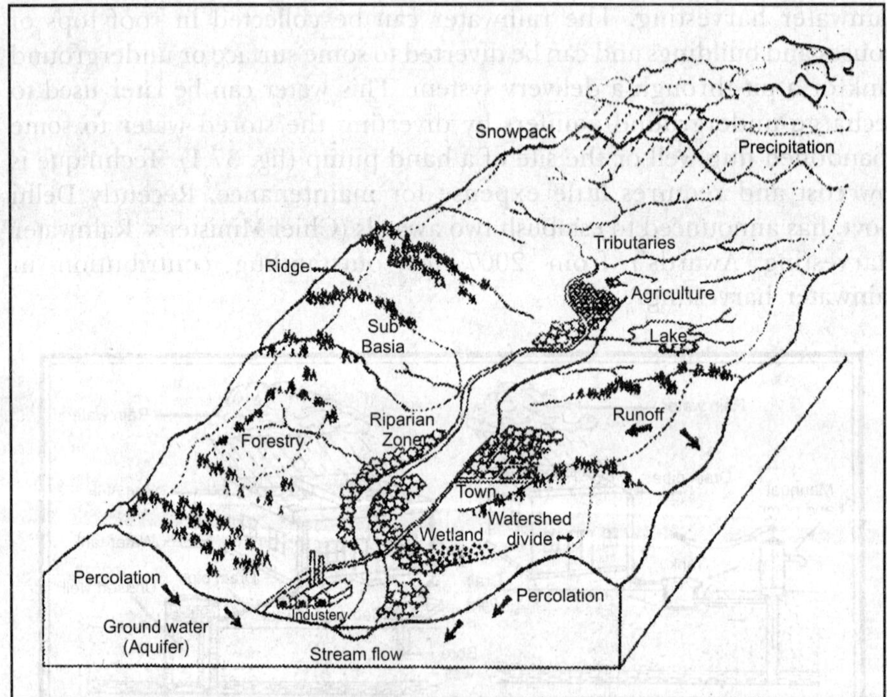

Fig. 37.2. A large-scale watershed with diverse components and geography.

Based on size, shape, land use pattern and location, watersheds can be categorised into different groups. Based on size, it may be mini (1-100 ha) to macro (more than 50,000 ha). Based on shape it may be fan-shaped (near circular) or fen-shaped (elongated). Based on location it may be a hill or flat watershed. It may be a highland, a tribal settlement or a settled cultivation area.

In India, there are mainly three watersheds, *viz.*, Himalayan range with its Karakoram branch in the north, Satpura and Vindhyan ranges in Central India, and Sahyadri or Western Ghats on the west coast.

The hydrological conditions in watersheds are such that the water becomes concentrated within a particular location such as in a river or a reservoir by which the watershed is drained. The land area drained by a river is known as a river basin. The watershed comprises complex interaction of water, soil, land form, vegetation and land activities. People and animals are integral part of a watershed and have mutual impact on each other. We may live any where, it would certainly be a watershed.

Importance of Watersheds

Watersheds play a crucial role in natural functioning of the earth. Hydrologically watersheds integrate the surface water run off of an entire drainage basin. Economically, they play a critical role as a source of water, food, recreational amenities and many more activities. Watershed's, water can be used for irrigation, power generation, transportation etc. These also influence sedimentation and erosion, vegetation growth, floods and droughts. Ecologically, watersheds constitute a critical link between land and sea. They provide habitat within wetlands, rivers and lakes for 40 per cent of world's fish species, some of which migrate between marine and fresh water systems.

Objectives of Watershed Management

Watershed management is aimed at rational utilization of land and water resources for optimum and sustained production, causing minimum damage to natural resources. Its management is an attempt to prevent land degradation and as a holistic process for getting maximum production out of land. The **objectives** of the watershed management are:

(1) Rehabilitation of watershed through proper land use, adopting conservation methods for minimising soil erosion and moisture retention to ensure high productivity of land.
(2) Managing watersheds for beneficial sustainable development activities such as domestic water supply, irrigation, hydropower generation, transportation and recreation.
(3) Minimising the risks of floods, droughts and landslides.
(4) Developing rural areas in the regions with clear plans, in order to improve the economy of the region.
(5) Restoring soil fertility which is vital for agriculture.
(6) Retaining habitats for aquatic flora and fauna.

Strategies for Watershed Management

Watershed management is a project based and ridge to valley approach for *in situ* conservation of land and water according to the suitability or people's benefits as well as sustainability. In includes focus on village common land, institutionalised community participation, emphasis on sustainable rural livelihood support systems and capacity building as a vital component.

(1) *Rain Water Harvesting Approach*

One of the ways of watershed management is rainwater harvesting by allowing the rainwater to move down the slopes slowly, ensuring optimum infiltration and percolation. Some of the procedures one could adopt are reducing the impact of rain on the soil, checking speed of run off of water at various intervals, taking up all operations on the contour and diverting the excess of water to prevent the pressure on the processes that start from the highest point of watershed and end right in the lower parts.

(2) *Watershed Protection Approach*

Watershed protection approach of management is a strategy for effectively protecting and restoring aquatic ecosystems and protecting human health. By this strategy many water quality and ecosystem problems have been solved by watershed level rather than based on an individual water body or discharger level. Important features of this strategy are targeting priority problems, promoting a high level of stake holder involvement, integrated solutions that make use of the expertise, authority or multiple agencies and measuring success through monitoring and data gathering.

Steps in Watershed Management

In general, watershed management includes proper management of water in hilltops, hillslopes, foothills, river basins and plains with appropriate practices in a planned way. Various steps in watershed management are:

(1) Preparation of base maps for carrying out surveys.
(2) Reconnaissance survey of the watershed for over all development.
(3) Assessing rainfall characteristics and classification of lands for different uses based on capability classification for agriculture, forestry, pasture, horticulture etc.
(4) Preparation of inventory of existing land uses and farm sizes.

(5) Appraisal of agricultural production patterns and potentials, present and potential markets and possible group action managements.
(6) Carrying out topographic and hydrologic surveys for engineering works.
(7) Geo-hydrological survey to delineate areas suitable for ground water development.
(8) Formulation of an integrated time-based plan for land and moisture conservation, ground water recharge, development of productive afforestation, agricultural production, grasslands and horticulture.
(9) Assigning of priorities for implementation of the project.
(10) Assessing social costs and benefits.

Watershed management involves effective methods of retaining water by contour bunding/graded bunding; check dams and gully control structures including land levelling/land smoothening; bench terracing; farm ponds; percolation ponds; water ways; and diversion drains. The process starts from the topmost rocky areas where the run off water needs to be drained out by forming diversion drains in order to protect steep slopes below, which are suited only for growing fodder.

Success of Watershed Management

A successful watershed management programme conserves and uses rainfall and fertility of the soil more effectively. It is a developmental alternative which would provide greater stability to rain-fed agriculture, maintain productivity and assure a dependable harvest. In India, there have been many successes in watershed management programmes such as in Alwar (Rajasthan), Ralegaon Siddhi (Maharashtra), Karnool (Andhra Pradesh), Devas (Madhya Pradesh) and Bundelkhand (Uttar Pradesh). The regeneration of the Aravari and other small rivers with *johads* (earthen checkdams) in Alwar district (Rajasthan) brought about a change to the lives of the people. While some of the initiatives were started by Government and NGOs for better management of water resources, others are attempts made by the community/rural population to improve their natural resource base and get more from their land.

Community Based Watershed Management

Beyond the importance of people's participation, the general principles of effective community based watershed management are yet to be defined. There are observable and measurable characteristics of successful and unsuccessful water management which need to be better documented. In addition to the particular physical, social and institutional characteristics of the project setting, project 'success' is also dependant on the extent to which the project objectives and implementation strategies are collectively developed and clearly defined and the extent of participatory processes imparting a sense of project ownership both at the field level and within the project implementing organisation. The emerging view of watershed management within an ecological and sustainable development perspectives is different from conventional methods of development but these differences are not always articulated or fully recognised in project implementation. An integral part of the capacity building task is to develop a new awareness and new or restored set of community values and principles for environmental management of watershed resources. Development assisted projects provide useful opportunities for researchers to test various hypotheses about community based watershed management such as the role of different government and community structures or the alternative types of benefit sharing arrangements, *e.g.* on the effectiveness of watershed interventions. What is needed is a stronger commitment to observing, measuring and comparing actual responses to the variety of approaches currently being used in community based watershed management. Watershed management is an evolving concept in India. Basically it is associated with the proper storage and reharvesting of rain run-off water by appropriate management of natural and man-made watersheds. People's participation is gradually being recognised by government agencies as an essential component of all developmental efforts including watershed management. The implementation of participatory watershed management practices has enabled communities to overcome problems and gain more control over their natural resources and livelihoods. Community participation can ensure effective utilisation of available resources and voluntary sharing by the user group and their time, energy and money on the programme and adopt the recommended measures and practices on a sustained basis. For water resource, participatory process initiates the optimum utilisation of watersheds. In view of the large problem, area to be treated and the present level of allocation of funds, the programme has to be taken up on a continuing basis for project period of five years. Presently 947 blocks of 155 districts in 13 states are covered under the programme. The expenditure of the project is shared equally by the Central and State Governments.

Desert Development Programme (DDP) in India

Keeping in view increased population of human and livestock, resulting great stress on natural resources of arid and semi-arid areas, the programme was launched both in hot desert areas of Rajasthan, Gujarat and Haryana and later extended to Andhra Pradesh and Karnataka and the cold desert areas of Jammu and Kashmir and Himachal Pradesh. Currently 2202 Watershed Projects, covering an approximate 11 lakh hectares, have been targeted for development over 4-5 year period and 2200 new watershed projects are proposed to be allocated to solve the problem of continuous depletion of vegetation cover, increase in soil erosion and fall in ground water table in the areas selected.

Watershed Development in Shifting Cultivation Areas

The scheme of Watershed Development Project in Shifting Cultivation Area (WDPSA) was launched in seven North Eastern States during the Eighth Plan from 1994-95 for overall development of urban areas on watershed basis. During the Ninth Plan, 1.57 lakh hectare has been treated. The new guidelines of the scheme on the basis of new watershed to common approach, have been effective from November 2000. During the Tenth Plan, an area of 0.6 m ha has been treated up to the end of 2004-05.

Watershed Development Under Employment Assurance

As per guidelines of employment assurance, 50 per cent of fund allocated under DPAP and DDP were supposed to be spent on watershed development. Various states are following the norms by spending a fare share towards watershed development.

Training for Watershed Development

Group and village level institutions promote people's participation in watershed development for major schemes of DPAP and DDP involving them in decision making and monitoring of the projects through formal or informal training programmes. For successful development of these Area Development Projects, Government has continued Training for Watershed Development. A Committee for the massive training programme for different levels of functionaries in the implementation of the projects has been formed.

Integrated Watershed Development Programme (IWDP)

In India, the scheme has been under implementation since 1989-90 and has come up in July 1992. The scheme is being implemented on watershed basis from April 1, 1995 under guidelines of Watershed Development to generate employment in rural areas, besides enhancing people's participation in watershed development programmes at all stages. This leads to equitable sharing of benefits and sustainable development. The major activities taken under this scheme are soil and moisture conservation; planting and sowing economically important plants; encouraging natural regeneration; promoting agroforestry and horticulture; wood substitution and fuel wood; conservation measures; measures needed to disseminate technology, training extension and creation of greater degree of awareness among participants; and encouraging peoples participation.

Drought Prone Areas Programme (DPAP)

It is one of the areas development programmes launched by the Government of India in 1973-74 to tackle special problems faced by those fragile areas which are constantly affected by drought. The programme has been implemented in 1994 on watershed basis. The planning, execution and maintenance of Watershed Project is entrusted to local people's organizations, specially instituted for the purpose.

Chapter Summary

The world is moving towards water wars. A once abundant resource, water is being depleted. Adequate water supply has become a problem. The search is a pace for innovative and sustainable solutions. **Water conservation** is the careful use and protection of water resources, both qualitatively and quantitatively. Conservation of this precious resource can be done by its sustainable use and checking its pollution. Adopting water economy in our daily life, in agriculture, industry, rainwater harvesting and watershed management are some effective measures for conservation of water. Water management requires provision of potable water supply to all by full exploitation of potential water resources. World Water Vision is underway of making several exercises with the goal to provide safe and sufficient water to all by 2025.

Rainwater harvesting is a technique to capture, collect and store rainwater in tanks or divert it to recharge groundwater. In rural areas it can be done by check dams whereas in urban areas rainwater can be collected at rooftops of houses and reserve rainwater can be used as and when required. The practice is common in rain fed areas of our country. India has a long history of rainwater harvesting through wells, tanks, 'johads', 'bawaries' etc. most of them got made by many emperores during their periods. Rooftop harvesting in urban houses is being made mandatory. Apart from being low cost and effective, rainwater harvesting provides us pollution free water supply.

Watershed is a geographic unit that collects, stores and releases water which comes into it in the form of rain, snow or coastal fog. The water gets stored in ponds, lakes, sub-surface soil and geological formations. Himalayas is the chief watershed in India. Watersheds play a crucial role in natural functioning of the earth. They can be used for irrigation, power generation, transportation etc. Watershed management involves rational use of land and water resources for optimum and sustainable production.

Study Questions

1. What is water conservation? Discuss the various methods of water conservation.
2. What are various practices of water management in India? Describe.
3. Discuss advantages and methods of rainwater harvesting.
4. What are watersheds? What are their functions? Explain.
5. Write a brief account of objectives and strategies of watershed management.
6. Describe briefly:
   (i) World Water Vision
   (ii) Traditional Rainwater Harvesting
   (iii) Importance of Watersheds
   (iv) Integrated Watershed Development Programme, India.

Objective Questions. *Select the correct answers*

1. Most economic method of water conservation is:
   (1) Construction of Dams (2) Interlinking of Rivers
   (3) Rainwater Harvesting (4) Watershed Management

2. World Water Vision has its goal to provide safe and sufficient water to all by the year:
   (1) 2010 (2) 2015
   (3) 2020 (4) 2025

3. The person popularly known as Water Man in India is:
   (1) Medha Patekar (2) Rajendra Singh
   (3) Sundar Lal Bahuguna (4) Baba Amte

4. Johad is a:
   (1) Dug well (2) Check dam
   (3) Percolation pit (4) Storage tank

5. In India, cold deserts occur at:
   (1) Tamil Nadu (2) Jammu and Kashmir
   (3) Rajasthan (4) Gujarat

Answers

1. (3) *2.* (4) *3.* (2) *4.* (2) *5.* (2)

38

CHAPTER

Resettlement and Rehabilitation of People: Problems, Concerns and Case Studies

LEARNING OBJECTIVES

Introduction • Reasons for Displacement of People • Issues Related to Displacement • Resettlement and Rehabilitation • Displacement due to Dams • Development Project and Resettlement and Rehabilitation of People • Resettlement and Rehabilitation Policy: Objectives and Guidelines • Case Studies

Introduction

Development projects raise the quality and living standard of the people of the country. Many of the development projects have resulted displacement of millions of people worldover. Resettlement and rehabilitation is a process of preparing and helping displaced people to cope with their new environment with physical, sensory and emotional development. However, people replaced due to development projects could not receive that much sympathetic attention and international aid as those who were faced to flee from disaster or conflict, even though the consequences they faced might be much grave as compared with faced by those displaced by other forces. In the past, the dominant view was that large scale development projects accelerated progress towards the brighter and better future, and if people were displaced along the way that was deemed a necessary evil. Large scale displacements, however, give rise to certain socio-economic and environmental problems. In the Article 25 (1) of United Nation's Universal Declaration on Human Rights, it is clearly stated that housing is a basic human right. The World Bank proved as the first multi-layerd lending agency that adopted a policy for Resettlement and Rehabilitation (R&R). The policy is contained in its 'Involuntary Resettlement' document released in June 1990. World Bank's Operational Directive 'Environmental Assessment' adopted in October 1991 has directed categoric need for environmental assessment of various development projects and solving rehabilitation problem.

Reasons for Displacement of People

Millions of people are compelled to leave their homes and move to other places, cities or foreign countries for various reasons:

(1) Conflict such as wars between countries, civil wars, ethnic conflicts and religious persecution.

(2) Natural disasters such as cyclone, hurricane, tsunami, flood, earthquake, volcano and drought.

(3) Environmental disasters such as industrial mishappenings, nuclear accidents, pollution of water bodies, toxic contamination of sites and oil spills.

(4) Development projects in various sectors such as construction of dams and reservoirs, irrigation, urban infrastructures, transportation, energy distribution, agricultural expansion, forest conservation, industrial development etc.

(5) To satisfy economic needs, *i.e.*, migration in search of a livelihood or for improving the quality of life.

(6) Forcible relocation by the governments under population redistribution programmes as happened in South Africa under apartheid, in China during cultural revolution, and in Russia in the Stalinist era.

World Bank estimates that since 1990, every year about 10 million people have been displaced worldwide by development projects for variety of reasons. In India, during last 50 years around 25 million people have been displaced due to development projects. Dams have physically displaced over 40 million people all over the world. The report of World Commission on Dams, 2000, concluded that 'impoverishment and disempowerment have been the rule rather than the exception with respect to resettled people around the world'. The report provides specific recommendations to national governments, NGOs, affected people's organisations, private sectors, funding agencies and development banks on all social aspects of large dam projects, including displacement and resettlement.

When people leave an area, they leave behind many environmental problems such as abandoned agricultural land and demolished habitats. Migration of such population also causes several environmental problems in the new areas into which they move. There are large refugee camps which lack basic amenities at many places in the world. Now these camps have become permanent settlements. In many other cases, the refugees upset the environmental balance of the area by making unsustainable demands on the natural resources. Unless more countries become globalised and offer better opportunities to refuges, the environmental degradation will continue.

Issues Related to Displacement, Resettlement and Rehabilitation

The majority of displacements and their follow resettlement and rehabilitation programmes, involving large population, have taken place in the population dense developing nations because of massive development projects undertaken in these countries, financed by foreign and international agencies. The following are the major issues faced by displaced population.

(1) One of the major issues is the loss of identity and the loss of intimate link between the people and the environment they lived. The age old traditional knowledge inherited in indigenous people towards the respect of nature and natural resources is also lost.

(2) Their families are splitted, during which women are the worst affected. There is also loss of kinship systems, social and cultural functions and folk songs and folk dances.

(3) The tribals, who are already economically poor, are usually the most affected among the displaced. Displacement further increases their poverty due to loss of home, land, jobs and lack of proper assess to their common property assets by rehabilitators.

(4) In being ignorant with the market policies and trends, displaced people, in modern economic set up, get alienated in cash compensation whatever they get.

(5) The land acquisition laws, which are not site specific, often ignore the community ownership of the property, which is an inbuilt system among tribals. And thus, tribal lose their communitarian basis of economic and cultural existence, feeling like fish out of water.

(6) The major problem of rehabilitation, in rural sector, is land degradation and deficiency of water which has accelerated rapid marginalisation and fragmentation of agricultural land of rural communities.

(7) The compensation of lost land is often delayed or not paid to displaced people. Moreover, agents called *bichauleyas* and corrupt officials deprive the poor migrants of their full compensation.

(8) Generally the new land, that is offered in lieu of their own land, is less fertile and unfit for agriculture.

(9) In the areas where such people are rehabilitated, the ethenic and caste differences make it difficult for the refugees to live peacefully with the communities already present in the area.

Displacement due to Dams

In India, most of the displacements have been due to land acquisition by the government for various multipurpose river valley projects such as construction of dams for hydro power generation, irrigation and water supply. Although these dams have been able to increase agricultural production, enhance power generation and solve the cause of adequate water supply, but they have left behind a major social problem, the resettlement and rehabilitation of millions of displaced people, whose eventual influx into urban areas, have made them almost as refugee. During last 50 years, more than 20 million people are estimated to have been affected directly or indirectly by big dams constructed in river valleys. Some of them like Sardar Sarovar, Gujarat; Tehri Dam Project, Uttarakhand; and Narmada Sagar Project, Madhya Pradesh etc. have caused a great problem of resettlement and rehabilitation of displaced people ousted from these areas. For acquiring land to be used for development project, Government of India has the Land Acquisition Act, 1894 which empowers to serve notice to the people to vacate their land.

Box 38.1 **Sardar Sarovar Project**

The much debated Sardar Sarovar Project, which plans to build 30 big, 130 medium and 3000 minor dams on Narmada river and its tributaries, is estimated to submerge an area which is going to affect 573 villages and about 33014 project affected families/ oustees including over three lakh people. Community rights of tribals are also breached. As per Action Plan 1993, about 14,124 project affected families were given resettlement in Gujarat and 18,890 in Madhya Pradesh. Those allocated Gujarat, 2830 project affected families have been allotted agricultural land at the rate of 2 h. each and among them 2731 have also been allotted house plots. Besides 733 project affected families have been allotted house plots in Madhya Pradesh. Furthermore, 2341 project affected families in Gujarat and 733 in Mahya Pradesh have been extended with rehabilitation grant and other financial assistance as provided in the R & R Policy. For resettlement of 14124 project affected families in Gujarat, there is requirement of 82 rehabilitation sites and for 18,890 projects affected families in Madhya Pradesh, there is need of 98 resettlement sites. There is an urgent need of civic amenities like school, 'panchayat bhawan', community hall, internal roads, drinking water facilities, pasture land, religious place, community meeting place, hospitals, cremation grounds etc. for these settlements. Rehabilitants have also to be supplied the electric power and telecommunication facilities. However, there are problems of landless oustees and those natives who were cultivating forest lands.

The Section 16 of the Act, has provision for cash compensation. Sometimes additionally the project may provide employment to the oustees.

According to Sundar Lal Bahuguna, one of the strong opponents of Tehri Dam Project and well known environmental activist, "the big dams are being constructed everywhere in tribal or hilly areas. The people of these regions are being uprooted to provide power to big cities, industries and irrigation for comparatively more prosperous areas. This is unethical. The locals are affected by sudden rise in prices of essential commodities. Their scarce resources of water, fuel etc. are exploited, leaving nothing for them".

Although, these multipurpose river valley projects, such as big dams have provisions for providing suitable rehabilitation to oustees, but there is no suitable provision to rehabilitate those people who live in the villages of the marginal areas of mountain slopes which become very fragile and might slide down in near future after the reservoirs are filled with water. Moreover, the displacement of local residents with their typical mountain living lifestyle and culture and resettlement to other places, having totally different social and environmental set up, causes many socio-economic problems among oustees-

(1) Rehabilitation of oustees in low lying areas and in the plains deprive them of forest produce which they used to get for the sustenance of their life. They are also deprived of drinking water which they used to get from natural springs and streams.
(2) Their productive systems are dismantled and their agricultural land, productive assets and sources of income are also lost.
(3) The hill environment that had conditioned their life style and provided these people some unique qualities such as hard work, honesty, bravery etc., bestowed in the topographic characteristics of the hills, in most of them, is likely to be lost due to competition of resources after they are settled in new areas.

Box 38.2 Tehri Dam

The 260.5 metre high Tehri Dam on the Bhagirathi river in the Garhwal (Himalayas) of Uttarakhand, is being built to bring prosperity by generating 2400 MW peaking power which according to builders of the dam, will help in establishing 140 industrial cities such as Modinagar in Uttar Pradesh (near Delhi), providing 300 cusec of water to meet the water supply needs of Delhi and irrigating 2.7 lakh hectares of land in Western Uttar Pradesh. The project was signalled after three decades of long campaign against the project by, the noted activist and strong opponent of Tehri Dam, Sundar Lal Bahuguna, the propagator of Chipko Movement. The Tehri Dam would affect about 10,000 residents of the Tehri town and cause the displacement to loom over large population. The resettlement and rehabilitation of oustees would become a serious problem.

(4) The oustees lose the grandeur of ecosystem, community life, kinship system, their own devices of recreation and enjoyment such as folk songs and folk dances.
(5) Their cultural identity, traditional authority and society network of interdependency, also diminish.
(6) They are often placed to such localities where their own productive skills prove of less application.

Development Projects and Resettlement and Rehabilitation of People

The development projects are planned to bring benefits to the people of the country. However, many of these projects such as dams, reservoirs, hydro power projects, large industries, thermal and nuclear power stations, refineries etc., that displace people involuntarily, result certain economic, social and environmental problems. Earlier, displacement of people in lieu of development projects was deemed to be a necessary evil but now these projects have to bear the cost of rehabilitation of millions of people who lived in these areas. The World Bank's' Operational Directive 'Environmental Assessment' recognises three categories of development projects, each kind needing specific kind of environmental impact assessment caused by the project.

Projects categorised under 'A' type, are those in which the impacts are likely to be comprehensive, broad, sector-wise or precedence setting. The impacts may be sensitive, irreversible and diverse. Among this category are included Dams and Reservoirs; Forestry Production; Large-scale industrial plants and industrial estates; Mineral resource Development including oil and gas; Port and Harbour Development; Reclamation and New Land Development, Resettlement of Projects; River Basin Development, Thermal and Hydro power Development; and Manufacture,

Box 38.3 Former Hunter-gatherers Became Fishermen

Most people displaced by development projects, especially dams, lose their lands, livelihood and shelter and sometimes do not get adequate or no compensation at all. The case happened with people of about 22 villages, mostly, 'Adivasi' families living in forests, during the development of a dam on the River Tawa, a tributary of the river Narmada in Madhya Pradesh. This contingent, instead of moving to other villages acompained for and secured fishing rights in the reservoir created by the dam. The 'Adivasis', who used to live in the forests, learnt fishing operations and formed the Tawa Matsya Sangh (TMS), a federation of the primary fisherman's cooperative societies with 1200 members, which in 1996 was given the five year lease and charge of overall management of the reservoir with fishes and fishing operations by the government. The case study throws light on as to how the forest hunter gatherers became good fishermen.

Transportation and Use of Pesticides or other Hazardous and/or Toxic materials.

Projects categorised under 'B' type are those in which the projects may have adverse environmental impacts that are less significant than category A impacts. Few, if any, of these environmental impacts are irreversible. The impacts are not as sensitive, numerous, major or diverse as category A impacts. Remedial measure can be more easily designed. Although, they do not require full environmental assessment but do require some environmental analysis. Among this category are included Small-scale Agro Industries; Electrical Transmission; Aquaculture and Mariculture; Small-scale Irrigation and Drainage; Renewable Energy; Tourism; Rural Water Supply and Sanitation; Watershed Management and Upgrading; Rehabilitation; and Maintenance of Small-scale Projects.

Projects categorised under 'C' type, are those in which an environmental analysis is normally not required as the projects are likely to have no adverse environmental impacts. The impacts, if any, are negligible, insignificant or minimal. Among this category are included Education; Family Planning; Health; Nutrition; Institutional Development; Technical Assistance; and Most Human Resources projects.

In India, Environmental Impact Assessment Scheme for Development Project was enforced in 1994 by the Ministry of Environment and Forests. During 2005-06 the Ministry accorded clearance to 946 projects out of 963 submitted for appraisal.

Resettlement and Rehabilitation Policy: Objectives and Guidelines

The objectives of involuntary resettlement and rehabilitation are aimed at providing benefits to the people displaced due to development projects, particularly for those from A category projects. Some of the objectives and guidelines are as follows:

(1) 'Land for land' proving better policy than cash settlements for oustees.

(2) Provision of an appropriate share in the cake of the development to displaced.

(3) Creation of new settlements for oustees with their own environment.

(4) Giving oustees assurance for employment.

(5) Provision of appropriate compensation along with compensation for their losses at full replacement cost.

(6) Providing oustees physical and social rehabilitation in infrastructure including community services and facilities.

(7) The persons to be displaced be given replacement cost prior to their move.

(8) Provision of assistance with the move to the displaced person and support during transition period in the resettlement site.

(9) The displaced people be assisted in their efforts to improve their former living standards, income earning capacity, and production level or at least to restore them.

(10) Particular attention be given to the poorest groups to be resettled.

(11) Efforts to be made to avoid or minimise involuntary displacement by exploring all viable alternative project designs. For example, in planning for construction of dams, realignment of roads or reductions in dam height may significantly reduce resettlement needs.

(12) Community participation be encouraged in both planning and implementation of the project.

(13) While dealing with tribals, the following five principles, the 'Tribal Panchsheel' (designated by Pt. Jawahar Lal Nehru) be kept in mind and special care be taken about that they be

(i) Allowed to develop along the lines of their own genius and as such no legal binding should be imposed on them forcefully.

(ii) Allowed to administer in their own way and should not be overadministered by imposing various tempting schemes for them.

(iii) Encouraged to live in their own traditional way of life, their culture and art.

Box 38.4 **Medha Patkar**

Miss Medha Patkar is a rare combination of a fine intellect and compassionate heart. Patkar is a never say-die activist who has been relentlessly campaigning against big dams in general and the Narmada Project in particular. She holds P.G. degree in Master of Social Work from Tata Institute of Social Sciences, Mumbai and worked for "Rural and Urban Development and Community Projects". As a speaker, she has a capability to turn out a well-reasoned argument in support of her stand. Medha Patkar has been relentlessly opposing all government plans to construct Narmada Dam. She is the leader of Narmada Bachao Andolan (NBA) and the strongest opponent of Narmada Project. Her opposition for the project is because of the damage caused to flora and fauna and the habitats of the tribals, besides submergence of large tracts of agricultural land in Madhya Pradesh, Maharashtra and Gujarat. She has gained a lot of public support because of her efforts for a genuine cause. Medha Patkar has been recipient of various international awards including the alternate Nobel prize, the Right Livelihood Award, and the Goldman Prize. Some people call her 'Eco-terrorist 'but she is doing a lot for the cause of the environment.

(*iv*) resettled in neighborhood of their own environment.

(*v*) allowed to preserve their own rights of forest land.

The encouraging fact is that displacement caused by development for economic uplift are now causing greater attention than they did in the past. In 1998, a team of international law scholars presented to UN a set of Guiding Principles on internal displacement. These were the first guidelines developed within the context of human rights and humanitarian laws to address the issue of internal displacement. In 2002, UN set up an 'Internally Displaced Person Unit' in their office for the coordination of humanitarian affairs.

Recently, Narmada Control Authority gave its nod for raising the height of the Sardar Sarovar Dam on the river Narmada from 110.64 metres to 121.29 metres, prompting Narmada Bachao Andolan (NBA) leader Medha Patkar to begin an indefinite hunger strike. The NBA's contention was that even persons displaced by previous acts had not been rehabilitated.

Chapter Summary

Development projects raise the quality and the living standard of the people of country but many of the them, in achieving so, have resulted displacement of millions of people and created resettlement and rehabilitation problems world over. Large-scale displacements, however, give rise to certain socio-economic and environmental problems. Reasons for displacement of people include natural disasters, conflicts and wars between countries and development projects etc., but the most important is displacement due to development projects. World Bank estimates that since 1990, every year, about 10 million people are displaced worldwide by development projects especially dams. Displacements leave behind them many problems. Problems due to displacement by dams are more serious and require proper resettlement and rehabilitation policies. Resettlement and rehabilitation of oustees deprive them of the environment that they used to get for the sustenance of their life. The oustees lose their identity and traditions. They also lose their productive systems and source of income. World Bank's Operational Directive 'Environmental Assessment' has categorised development projects in three A,B and C types, each kind needing specific environmental impact assessment. In India, Environmental Impact Assessment' scheme for development projects was recognized in 1994 by the Ministry of Environment and Forests. During 2005-06, Ministry accorded clearance to 946 projects. Objectives of involuntary resettlement and rehabilitation should be aimed at providing benefits to the people displaced due to development projects by providing them 'land for land', suitable compensations and helping them to cope with their new environment with physical, sensory and emotional development.

Study Questions

1. What are the various reasons for displacement of people? Explain.
2. Discuss the various issues related to displacement, resettlement and rehabilitation involving large populations.

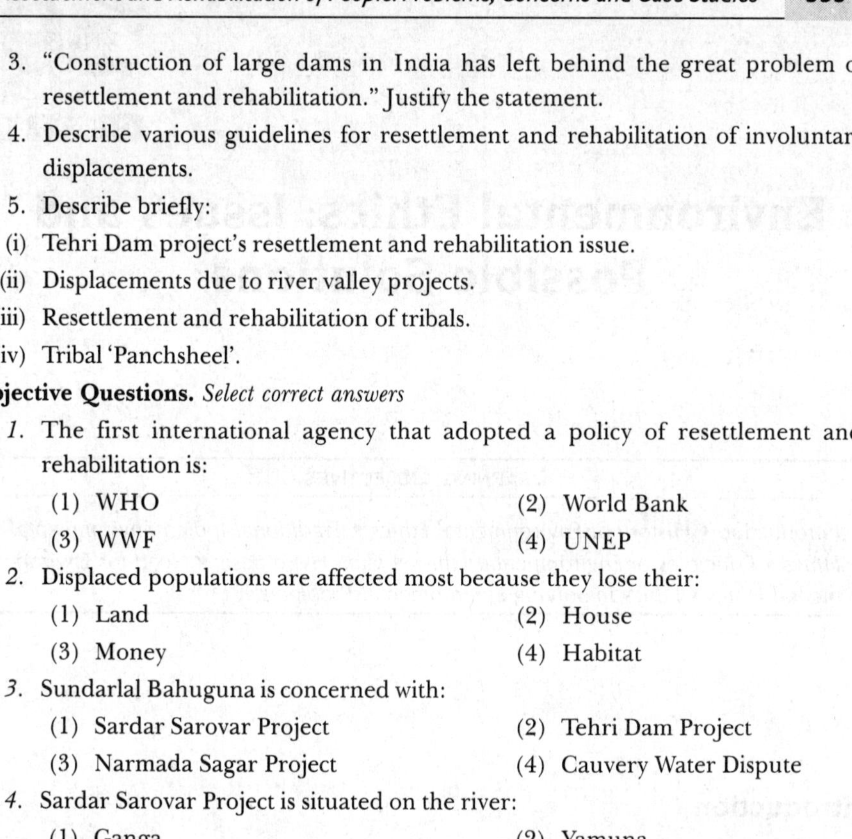

3. "Construction of large dams in India has left behind the great problem of resettlement and rehabilitation." Justify the statement.
4. Describe various guidelines for resettlement and rehabilitation of involuntary displacements.
5. Describe briefly:
(i) Tehri Dam project's resettlement and rehabilitation issue.
(ii) Displacements due to river valley projects.
(iii) Resettlement and rehabilitation of tribals.
(iv) Tribal 'Panchsheel'.

Objective Questions. *Select correct answers*

1. The first international agency that adopted a policy of resettlement and rehabilitation is:
(1) WHO (2) World Bank
(3) WWF (4) UNEP

2. Displaced populations are affected most because they lose their:
(1) Land (2) House
(3) Money (4) Habitat

3. Sundarlal Bahuguna is concerned with:
(1) Sardar Sarovar Project (2) Tehri Dam Project
(3) Narmada Sagar Project (4) Cauvery Water Dispute

4. Sardar Sarovar Project is situated on the river:
(1) Ganga (2) Yamuna
(3) Cauvery (4) Narmada

Answers

1. (2) *2.* (4) *3.* (2) *4.* (4)

39

CHAPTER

Environmental Ethics: Issues and Possible Solutions

LEARNING OBJECTIVES
Introduction • History of Environmental Ethics • Traditional Indian Environmental Ethics • Principles of Environmental Ethics • Gaia Hypothesis • Need for Environmental Ethics • Ethics in Solving Environmental Problems

Introduction

Ethics is a branch of philosophy that deals with the morals, *i.e.*, the distinction between the right and the wrong. The term Ethic, derived from the Greck Word '*Ethos*', literally means *the character of a person as described by his or her actions*. Thus ethics can be defined as a system of cultural values motivating people's behavior at all levels, *i.e.*, individual, institutional, social, regional and international. Traditionally all societies have had strong cultural connections with nature. All religions have much to say about the relationship between mankind and the earth. The cultural ethos of people still remains for many traditional societies as part of their lifeline. Unless environmentalism is firmly rooted in culture and ethics of the societies, it would immediately bring impact upon the livelihood concerns of a very large proportion of already marginalised and socially discarded societies of the developing world, the degradation of the environment will be the end result.

The formulation of a new field 'environmental ethics' began since 1970s with the cooperation of responsible groups in the business world who were concerned about the environment and liked to see environment-friendly attitude among corporations. An example is the Coalition for Environmentally Responsible Economic (CERE) formed in 1989 to promote responsible corporate environmental conduct. The

CERE is a network of over 80 organizations including environmental groups, public interest and community groups and investors. The CERE encourages companies to endorse the principles formulated by it. By 2002, 70 US companies including large multinational corporations, mid-sized companies and small firms had endorsed CERE principles.

History of Environmental Ethics

Every human society has set of beliefs about nature and natural resources. The concept of environmentalism that brings ancient concerns with modern knowledge of environment, seems to be a recent subject but its roots are embedded deep within human history. Virtually all religions have much to say about environmental ethics as to what is right or wrong concerning issues, principles and guidelines relating human interactions with their environment. But no religion, perhaps, lays as much emphasis on environmental ethics as **Hinduism**. The *Vedas*, *Puranas*, *Upanishads*, *Gita*, *Mahabharat*, *Ramayana* and many more ancient scripts contain the earliest message of environmental ethics. For centuries, while praying to Goddess *Durga* Hindus had said *So long the earth has mountains, forests and trees, human race will survive*. Sustainable development, burning topic of current century, has been well advocated in *Atharveda* sayings:

> "What of thee I dig out, let that quickly grow over
> Let me not hit thy vitals, or thy heart."
>
> "Supreme Lord, let there be peace in the sky and atmosphere, peace in the plant world and in the forests; let the cosmic powers be peaceful; let *Brahman* be peaceful, let there be fulfilling peace everywhere."

A stage has come when we need to seriously reconsider our vision of the environment. Are we satisfied with having patches of protected bio-diversity in the form of nature reserves, places as islands in a vast ocean of monoculture? Or are we looking for more heterogeneity in our landscapes, as nature would like to have it for us, so that biodiversity is not merely restricted to nature reserves? The later approach will provide resilience to the environment by strengthening the internal buffering mechanism against uncertainties in the environment. This brings us to *cosmic tree* which is rooted in *Brahman*, the *ultimate* and has been described by the ancient seers of India in the *Upanishad*. *Tat twam asi* (that thou art) in a little sense means that individual is the part of the creation, *i.e.*, humans are well integrated into ecosystem functions giving back what they have taken from the environment-the concept of '*sarvabhutaday*' in **Buddhism**. In *Lotus Sutra* Buddha is presented metaphorically as a "rain cloud covering, permeating, fertilising, and enriching" all living beings, to free them from their misery to attain the joy of peace, joy of present world and joy of '*Nirvana*. "Whoever plants a tree and diligently looks after it until it matures and bears fruits, is rewarded" said **Prophet Mohamed**.

The concept of non-violence is integrated in all religious teachings. The eastern philosophical thoughts have taken it to greater heights through the concept of *Ahimsa*, abjouring violence through both, word and action. From a purely ethical point of view of biodiversity conservation, there are many who would argue that do other organisms too have as much right to live, as human race has. Such a rationalship within the framework of environment conservation and management alone will give meaning and value to human existence, and take us forward through the new millennium.

Traditional Indian Environmental Ethics

Despite, there are many religions in India, people believe in their own religion and are motivated to do the things as they are taught by the religion. **Hinduism** believes that all beings are a manifestation of one essential being called *Brahman*. "Suffering of one life from is the suffering of other life forms, to harm other beings is to harm oneself" formed the basis of 'Chipko Movement', a successful environmental movement to save trees by Bishnoi women. **Jainism**, pioneer in global leadership in environmental ethics supports *Ahimsa*, non-injury to all living beings and believe in sustenance of life on minimum resource use. **Buddhism**, assuming leading role in global conservation movement, realises the interconnectedness and interdependence of all conditional things and hypothesises that happiness is found through restrain of desire and boundless loving kindness for all living beings. According to **Sikh Scripture,** *Guru Granth Sahib*, human beings are composed of five elements, *viz., earth* teaching us – patience, *air* – mobility, *fire* – warmth and courage, *sky* – equality and broad mindedness and *water* – purity and cleanliness. The central concepts of **Islamic Taught** by the *Quran-Tawheed* (unity), *khalifa* (trusteeship) and *akrakh* (accountability) also serve as pillars of the Islamic environmental ethic.

Principles of Environmental Ethics

Leopard (1949) defined ethic as "a limitation on freedom of action". According to philosophic tradition, the answer to as to who should follow the ethics is, only human beings are worthy of ethical consideration. If so, this leads to a concept of 'anthropocentrism, which requires concerted efforts of individuals, corporations and other groups to consider how human actions may directly affect the natural environment and indirectly other human beings. An ethic which is mainly 'life centred' was known as 'biocentrism' (Goodfaster, 1978). Considering man as a member of biotic community, later on led to the hypothesis of 'ecocentrism' in twentieth century. Traditional western philosophers have a non-anthropocentric view and believe in that environment is more systematically integrated with man and ignoring may have untoward environmental consequences.

Gaia Hypothesis

Gaia, a term used for *Greek Goddess –Mother Earth*, hypothesis suggests that life manipulates the environment for the betterment of life. According to Gaia hypothesis, the planet earth works as a larger organism capable of self maintenance. The Gaia hypothesis is actually a series of several hypotheses in which the *first* suggests that the life greatly affects planetary environment, the *second* asserts that the life affects the environment for the good of life and the *third* one is that the life deliberately and consciously controls the global environment. It appears that the systems of positive and negative feedback, that operate in the atmosphere and oceans, are sufficient to explain the mechanisms by which life affects the environment. The hypothesis, as a whole, has some merit as it relates to the present and future, indicating that the future status of life may depend on actions which are being taken now and those which would be taken in the coming years. The people living today are taking conscious decisions concerning the future of the earth.

The Need for Environmental Ethics

The factors which have created the need for environment ethics are:

(1) New Effects on Nature. Currently modern technological civilisation has been affecting the nature to a great extent. This requires analysis of ethical consequences of human actions on the environment.

(2) New Knowledge About Nature. Until a few decade ago, only small segment of society realised that human activities could be changing or altering the global environment. But now the modern science demonstrates how humans have changed and causing change in the environment in ways not previously understood. For example, it has now been proved that burning of fossil fuels and deforestation result increase in carbon dioxide concentration in the atmosphere, and that this may bring about irreversible changes in the global climate. Thus modern knowledge, coupled with better understanding of the environment, is raising new ethical issues.

(3) Expanding Moral Concerns. An important question in environmental ethic arises whether only human have right to survive on this planet. Does it not form a duty of mankind to think so about other organisms and nature as a whole? Is it not our moral obligation to leave the environment in good condition for our descendents? These and some more questions lead to a need of environmental ethics.

According to Darwin (1904), all ethics evolved so far rest upon a single premise-the individual is a member of community of independent parts. It includes soil, water, plants and animals or collectively the land. Influenced by Darwin, Leopold (1949) proposed the concept of land ethic, "a thing is right when it tends to preserve the integrity, stability and beauty of the biotic community. It is wrong when it is otherwise". Man, *Homo sapiens*, as a part of nature or simply a member of land community, is a moral species too and is capable of ethical deliberations and can make choices. Though ecology now acknowledges the normalancy of change and disturbance in nature, the Leopard Land Ethic, appropriately emphasizes the role of humans from conqueror of the land to the protector of the environment, the steward of nature. In 1970s, people started thinking about environmental ethic for the rights of the future generation over earth's resources. From 1980s, the area has grown further by educating people through voluminous literature on environmental ethics. In 1989, G-7 Countries called upon a conference on environmental ethics at Brussels. The conference emphasised the need for individuals and nations towards a responsible stewardship for the environment. The conference formulated a code of environmental practice based on simple environmental ethic of stewardship of the living and non-living systems on the earth. The code stands for environmental monitoring, sustainable development, full accounting of costs of resources, a recognition of independent values and individual and corporate stewardship for the environment.

Ethics in Solving Environmental Problems

Environmental ethics can provide guidelines for putting beliefs into action and help to decide what to do when faced with the crucial situation. It motivates us:

(1) To love and favour the earth, since it has blessed us with life and governs our survival.

(2) Keep each day devoted to earth and celebrate turning of its seasons.

(3) Not to hold ourselves above other living things and have no right to drive them extinction

(4) To limit our offspring's because increased population will over-burden the earth.

(5) To be grateful to plants and animals which nourish us by giving food.

(6) Not to waste our resources to buy destructive weapons.

(7) Not to run after gains at the cost of nature, rather we should strive to restore its damage.

(8) Not to conceal from others the effects we have caused to our action on earth.

(9) Not to snatch from future generations their right to live in a clean and safe planet by impoverishing and polluting it.

(10) To consume the material goods in moderate amounts so that all may share the earth's precious treasure of resources.

Chapter Summary

Ethic deals with the morals, *i.e.*, the distinction between the right and the wrong. Traditionally all societies have had their cultural connections with the nature. The concept of environmental ethics appeared only few years ago to remind people of their duty to respect the environment. Every religion and human society has set of beliefs about the nature and natural resources. India is a unique country with a great cultural diversity and reverence to nature, inherent in its cultural ethos. The roots of the environmental values are deep in our ancient *vedic* literature and *upanishads*. All religion in India are of one and the same underlying principles of the environmental ethic, *i.e.*, respect for nature, care for the environment, sustainable use of resources and path of '*Ahimsa*'. Environmentalism tells us as to how human actions may affect natural environment. Gaia hypothesis relates present and future, indicating that future status of life may depend on actions which we are taking now. The factors which created need for environmental ethics are new effects on nature, our new knowledge about the nature and expanding moral concerns. Leopord's (1949) land ethics emphasizes role of humans from conqueror of the land to protector of environment by stewardship. Ethic helps to solve many environmental problems and teaches us what to do when faced with crucial situation.

Study Questions

1. Throw light on the history of environmental ethics in India.
2. What is the role of ethics in solving environmental problems? Explain.
3. Discuss Gaia Hypothesis.
4. Discuss Buddhism and Environmental Ethics.

Objective Question: *Select the correct answer*

1. '*Ahimsa*' was greatly supported by-

(1) Hinduism (2) Buddhism

(3) Jainism (4) Sikhim

Answer

1. (3)

40

CHAPTER

Climate Changes, Global Warming, Acid Rains, Ozone Layer Depletion, Nuclear Accidents and Holocaust

LEARNING OBJECTIVES

CLIMATE CHANGES • GLOBAL WARMING • Greenhouse Effect • Causes • Greenhouse Gases • Effects • Remedial Measures; ACID RAINS • Formation, Effects, Control Measures • OZONE LAYER DEPLETION • Ozone Layer • Formation • Depletion • Ozone Depleting Substances • Causes • Effects • Protection and Maintenance • Indian Efforts; NUCLEAR ACCIDENTS and HOLOCAUST • Case Studies.

CLIMATE CHANGES

One of the most important global environmental problems is the changing climate of world. It is alarming to note that the average temperature of the earth's surface has risen by 0.6°C since the late 1800's and is expected to increase by another 1.4 to 5.8°C by the year 2100. Even if the minimum predicted increase takes place, it will be larger than any century long trend in the last 10,000 years. During the twentieth century, the sea level has risen on an average by 10 to 20 cm and it is expected that it will further rise by 9 to 88 cm by the year 2100. The Third Assessment Report (TAR) of the Inter-governmental Panel on Climate Change published in 2001 stated that there is new and stronger evidence that the most of the warming over the last 50 years is solely due to human activities. Climate changes are intricately connected with phenomena like **Global Warming, Acid Rains, Ozone layer depletion, Nuclear Accidents and holocaust** etc.

The temperature increase is now becoming inevitable, posing threats to climate changes. Recognising the precarious situation, most countries have joined an International Treaty, the United Nations Framework Convention on Climate Change (UNFCCC) to consider what can be done to reduce global warming. India is a party to UNFCCC, which was adopted in May 1992 and it came into force on March 21, 1994. The objective of the UNFCCC convention is to achieve stabilisation of concentration of greenhouse gases into the atmosphere at a level that would prevent dangerous anthropogenic (human activities) interference with the climate system.

The convention enjoins upon the different countries to protect the climate system on the basis of equity and in accordance with their *common but differentiated responsibilities* and *respective capabilities*. For the furtherance of the cause Parties to the Convention in 1997 adopted the **Kyoto Protocol** in recognition of necessity for strengthening developed country commitments. The Kyoto Protocol commits the developed countries, including economies in transition, to reduce emissions of greenhouse gases by an average of 5.2 per cent below 1990 levels during 2008-2012. The Kyoto Protocol came into force on February 16, 2005. India is a party to this Protocol.

In Eleventh Conference on Parties of UNFCCC held at Montreal (28 November–9 December) 2005, a landmark achievement, was the adoption of various decisions, now popularly called **Marrakech Accords**. India Government has started funding the projects relating to climate changes and organising awareness programmes, seminars, workshops, meetings etc. in this regard.

Under the Clean Development Mechanism (CDM) of the Kyoto Protocol, a developed country with quantified emission limitation and reduction commitments, would take up the greenhouse gas reduction project in a developing country. The CDM aims to assist developing countries in achieving sustainable development by promoting environmental friendly investment from industrialised countries.

Till March 10, 2006, the National CDM Authority has approved 252 projects in various fields including Biomass based cogeneration, Energy efficiency, Municipal solid waste, Renewable energy such as Wind, Small Hydro projects etc. Govt. of India is considering various aspects such as agriculture, water, forestry infrastructure, sea-level rise, extreme climate events, researches, education, public awareness etc. to involve under different projects on climate changes.

The details of various aspects of Global Warming, Acid Rains, Ozone Layer Depletion, Nuclear Accidents and Holocaust are accounted in the text.

GLOBAL WARMING (Greenhouse Effect)

The phenomenon greenhouse effect was first suggested in 1986 by a Swedish Chemist Archeniuss who predicted that the concentration of gases reflecting heat would increase significantly. He presumed that the doubling of these gases would increase the earth's temperature by several degrees. He was right in his prediction except for his timings as he expected the ill-effects to emerge after a few centuries and not so soon. It is feared that when all reserved fossil fuels are burnt in another 300–400 years, the air temperature will rise by a few degrees resulting in catastrophic effects such as melting of snow on the poles and mountain tops causing rise in sea level and unprecedented flooding of land area. The rise in temperature will not be steady, but may mean a 1°C rise by 2026 and 3°C by 2100. Accordingly, global sea level may rise by 20 cm in 2030 and 65 cm in 2100. This would bring a significant effect on the regional variations. The increase in global warming is almost equal to the increase in concentration of greenhouse gases of the atmosphere.

Global Warming or **Green House Effect**

The gaseous mantle, around the earth, permits a considerable portion of solar radiation to enter right upto the surface of earth. Some of it is absorbed by the earth surface, while remaining radiates back. The gaseous envelope allows sunlight (short wave) radiation to reach the earth, but when light radiates in the form of long wave (infra red) radiation, it is absorbed by green house gases and causes warming up. This heat is transferred to layers above, as warm layer rises and in turn passes on to higher and higher layers.

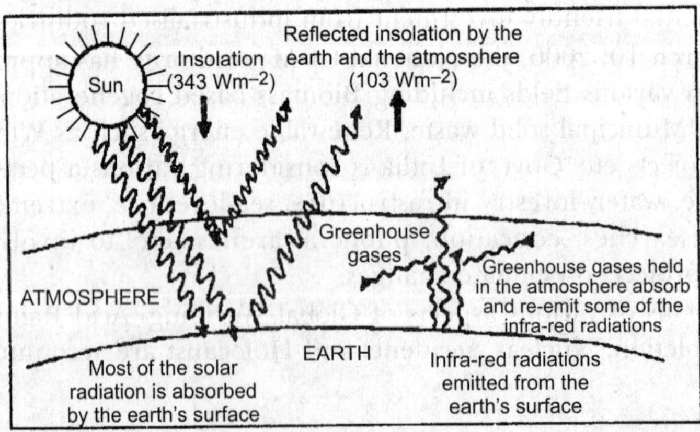

Fig. 40.1. Global warming by greenhouse gases – a schematic representation

The atmosphere radiates part of this energy back to the earth. This downward flux of radiation, called **greenhouse flux**, keeps the earth warm. The phenomenon is called **greenhouse effect**. The mean annual temperature of the earth is about 15°C and if there would have been no greenhouse gases in the atmosphere, the earth's mean temperature would have dropped sharply down to about minus 20°C (–20°C). The rise in temperature of earth's atmosphere, caused due to greenhouse effect, depends on the amount of CO_2 present in the atmosphere. Normal CO_2 concentration (0.03 per cent) keeps the temperature of earth's surface nearly constant due to energy balance of sun rays, which strike on the earth, heat it and then radiate back into the space. This phenomenon is called **energy budget**. The CO_2 layer functions as a glass pane of greenhouse which allows the sunlight to filter through it, but prevents heat from being radiated back into the outer space. As long as, greenhouse gases keep the earth warm and cosy, it is welcome but their increase in earth's gaseous mantle is changing the earth's climate alarmingly, not suitable to human beings.

The global temperature had risen by only 4°C over the last 18,000 years, but because of rapid industrialization, it has increased by about 0.5–1°C since the last century. The rate of global warming during the past quarter century was greater than any previous period since 1880. The six, out of seven warmest years of last 130 years, have occurred in the past seven years. The scientists are concerned about the warming trend and if this trend continues, global warming of approximately 1.5°C by 2015-2050 and 3°C by 2050 – 2100 has been predicted.

In general, global warming refers to *gradual rise in atmospheric and ground surface air temperature and consequent change in global radiation balance caused mainly by anthropogenic progress leading to climate change at different levels*.

Causes of Global Warming

The causes of global warming are the gases that are responsible for increasing global temperature. Such gases are CO_2, CFCs, CH_4 and N_2O. The approximate contribution of percentage of greenhouse gases to global warming is 61 per cent by CO_2 (carbon dioxide), 15 per cent by CH_4 (methane), 11 per cent by CFCs (Chlorofluorocarbons), 4 per cent by N_2O (nitrous oxide) and 9 per cent by others (T Siddiqui, 1990). Ozone in troposphere also acts as a greenhouse gas.

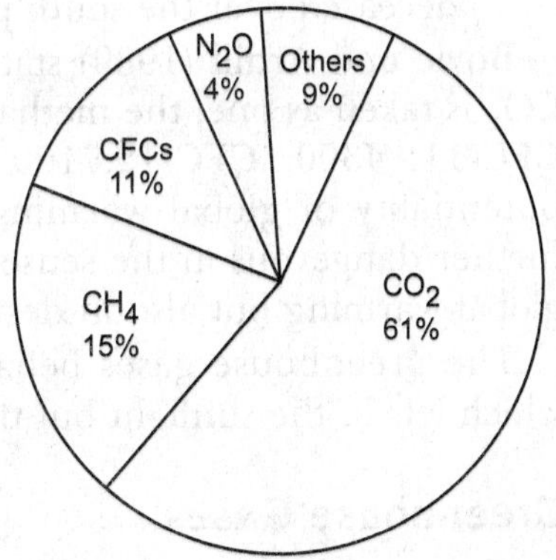

Fig. 40.2. Contribution of percentage of greenhouse gases to global warming (Siddiqui, 1990).

The Citizen's Fifth Report of Centre for Science and Environment (1999), New Delhi revealed that the emission of CO_2 has been held as the most dangerous and most significant greenhouse gas emission causing global warming. In the year 1995, about 22,700.2 million tonnes of CO_2 was emitted in the world. Shares of world CO_2 emission is given in the table 40.1

TABLE 40.1. Share of World Carbon Emission, 1995, in Percentage of Total Carbon Emission of World, (22,700.2 Million Tonnes).

Name of the Country	*% emission*	*Name of the country*	*% emission*
United States	23	Germany	4
China	13	India	4
Russia Federation	7	Brazil	1
Japan	5	Indonesia	1

Five worst CO_2 emitter countries are United States (5648.6 M.T.), China (3192.5 M.T.), Russian Federation (7 per cent of the world total), Japan (5 per cent of world total) and Germany (4 per cent of the world total).

In the report of the Intergovernmental Panel on Climate Change (IPCC), 1996, the trend of concentration of other greenhouse gases in the atmosphere was published. It has revealed the following facts –

(1) The methane release registered a phenomenal increase from the preindustrial level of 700 ppmv (ppmv= parts per million by volume) to 1724 ppmv in 1992.

(2) Atmospheric N_2O increased from the preindustrial level of 275 ppmv to 311 ppmv in 1992.

(3) The concentration of tropospheric ozone has doubled from preindustrial level to 1993 level in northern hemisphere but it has deceased over the south pole.

Boyle and Ardill (1989) stated that if **global warming potentiality** of CO_2 is taken as one, the methane has 11 times, nitrous oxide, 270 times, CFC-11, 4300, CFC-12,7100 and HCFC-22 has 1600 times higher potentiality of global warming in one hundred year times. CFCs are further dangerous in the sense that these are not only involved in direct global warming but also in destroying ozone layer in the stratosphere.

The greenhouse gases behave somewhat like glass in a greenhouse which lets in the sunlight but traps outgoing thermal radiation.

Greenhouse Gases

The global warming causing common greenhouse gases are carbon dioxide (CO_2), methane (CH_4), Chlorofluorocarbons (CFCs), nitrous oxide (N_2O) and ozone (O_3).

(1) Carbon dioxide. The increasing combustion of fossil fuels (coal, petroleum, mineral oil, natural gases) in various industrial and transport systems has resulted a steady increase of CO_2 gas in the atmosphere. The increase has been nearly 25 per cent since 1850. In India, contributions from coal, petroleum and natural gas and transport emissions are about 85,355 and 15 million tonnes respectively of carbon during 1988-89 which nearly causes 2.8 per cent global warming.

CO_2 has been significantly used in photosynthesis by green terrestrial and aquatic plants but even then its concentration in air has risen from about 280 ppm since the early 19th century to 316 ppm in 1960 and 359 ppm in 1994. During 1960s, 70s and 80s, the rate of increase had been 9, 12 and 16 ppm, respectively. With this estimation CO_2 concentration in atmosphere could be 600 ppm by the year 2050. At 400 ppm CO_2 level, global temperature would rise to about 1°C (Dhaliwal and others, 1996).

(2) Methane. Methane naturally emanates from garbage dumps, but it is also a primary component of natural gas. Rice fields and swamps contribute about 25–30 per cent global methane production. India's contribution is nearly 6 per cent. In Indian paddy fields, CH_4 emission is about 2 per cent of global paddy field emissions (*i.e.*, 3.9×10^{12}g CH_4 per year out of 110×10^{12} g per year) and about 9 per cent of global cattle (garbage and faeced matter) contribution (*i.e.*, 7×10^{12}g/ year out of 80×10^{12}g/yr). Methane concentration in the air has been considerably increased from 750 ppb (parts per billion) in 1750 A.D. to 1750 ppb in 2000 A.D. Global warming potentiality of methane is eleven times higher than that of CO_2.

(3) Chlorofluorocarbons (CFCs). Modern world is using CFCs quite frequently. These are used in foams (about 26%), aerosols (25%), solvents (19%), air conditioners (12%), refrigerants (8%) and others (10%). India produces CFCs around 5000 tonnes/year against the global production of 7,00,000 tonnes/year (*i.e.*, less than 1%). Global warming potentiality of CFCs is very high (1600 to 15000 times) as compared to that of CO_2. The concentration of CFCs in the atmosphere is less than 1 ppm, though it is increasing at the rate of 5% per year. World over, efforts are being made to reduce the CFC production, but more than 15 million tonnes have already been released in the atmosphere by 1985.

(4) Nitrous oxides. It is a quite stable gas (average life 150 years) with dual damaging effects. It is not only a greenhouse gas but also a ozone destroyer. The common sources of N_2O production are natural soil, inland waters, oceans, fertilizers and burning of forests, grasslands and other biomass. The global warming potentiality of N_2O is 270 times higher than that of CO_2. The concentration of N_2O in the atmosphere has increased from about 270 ppb in pre-industrial time (1750 A.D) to about

316 ppb in year 2000 AD. Ozone, CCl_4, CH_3CCL_3 and HCFC are other greenhouse gases.

TABLE 4.2. General Sources of Greenhouse Gases

Green house gas	*Sources of discharge*	*Percentage to global warming*
Carbon dioxide (CO_2)	Burning of fossil fuels, deforestation, fire, Respiration	61
Methane (CH_4)	Fresh water bodies, enteric fermentation, rice fields, biomass burning, anaerobic decomposition of organic wastes	15
Chlorofluorocarbons (CFCs)	Leaking air conditioners, refrigeration units and evaporation of industrial solvents, aerosol sprays and propellants	11
Nitrous oxide (N_2O)	Biomass burning, nylon production, agriculture, burning nitrogen rich fuels, nitrogen rich fertilizers, fossil fuel consumption	4

Effects of Global Warning

Scientists are predicting various effects of global warming particularly on climatic changes and human diseases. Climatic changes may have far reaching repurcusions on global nature composition such as changes in cultivation and agriculture systems, biodiversity and in the management practices. Taking into consideration of different predictions, the following generalisation may be made.

(I) Global Warming and Climate Changes

Global warming will have larger effects on climate which may be summarised as follows:

(1) It is predicted that by 2030, CO_2 concentration will reach about 600 ppm (0.06%), and average global temperate will rise by 1.5-4.5°C. Temperature rise will be more evident in the high latitudes in the Northern Hemisphere, and in India, the temperature will go up by 1°C in summer and 3.4°C in winter.

(2) Global warming will affect rainfall pattern. India will witness an increase in rainfall by 5 per cent in summer and upto 15 per cent in winter. However, lower mid-latitudes of the country may have rainfall very limited in summer and 5-10 per cent decrease in winter. In general, the rainfall will be enhanced over northern

region and southern peninsula with an expected decrease in central part of India, which is basically an agriculture zone.

(3) The level of sea water would rise by 20-140 cm in various parts of the world due to global warming. In the last century, sea levels have already risen by about 30 cm. The implied rate of rise is about 4-6 cm/ annum, which is 2-6 times faster than over last 100 years. A rise in sea level will hit the lower coastal region of the country. It has been estimated that the annual mean of sea level rise in India by 2030 AD would be 20 cm along the coast of West Bengal, 10 cm along Andhra Pradesh and about 5 cm along Kerala.

(4) There will be a marked increase in potential evapotranspiration, particularly in water stressed zone of the country.

(5) Ice and snow caps on mountains will melt exposing dark surfaces which will cause a further rise in temperature and reduce the snow fall periods.

Box 40.1 The Big Meltdown

An article *'The big Meltdown'* by R. Vinayak (India Today, Nov. 8, 2006) discusses as to why we should worry about Himalayan glaciers. In 40 years, key Himalayan glaciers have shrunken by 21 per cent. As the rate quickens, hydropower will be badly hit, food crops will face catastrophic water shortage and weather conditions could be irrevocably altered. According to A.V. Kulkarni, a renouned glaciologist and his team of researchers from Indian Space Research Organisations (ISRO) Space Application Centre (SAC), "our research shows that the glaciers are receding at an alarming rate. The Himalayas, home of about 9000 glaciers – one of the largest concentrations of frozen fresh water outside the Polar regions – are most understudied. Considered as the thermometer of global warming, glacial melt is difficult to track". But recent studies of 466 Himalayan glaciers by ISRO Space Application Centre using satellite images and ground investigations show how rapidly they are receding. Field investigations revealed that in just five years (1998 to 2003), the Chhota Shigri glacier has reduced by 800 metres. Close to 127 small glaciers have shrunken by 38 per cent and larger ones by 12 per cent. The total glaciated area has reduced but the number of glaciers has gone up. They are fragmenting into smaller ones.

SAC studies for the first time conclusively demonstrated that the Himalayan glaciers are also retreating due to climatic variations. According to R.K. Pachauri, Chairman of The Energy and Resources Institute (TERI) and Director General of the Intergovernmental Panel on Climate Change (IPCC) "Any evidence that glaciers are melting is a warning bell. We are seeing the phenomenon across the globe-the Arctic's, the Andes, the Alps and now the Himalayas. It is likely to severely alter India's fresh water balance and adverse impact on food and energy production". IPCC's own evaluation predicts that average world temperature would rise by 1.4°C to 5.8°C by the end of this century. Such a rise would upset the environmental balance with catastrophic consequences. While many island countries like Maldives and Bangladesh would lose much of their land to the rising ocean. Dramatic altering of climate patterns could turn fertile areas into drought prone zones and increase calamities such as floods, hurricanes and typhoons. According to SI Hussain, for India, melting glaciers can create a havoc and need an urgent action to reduce harmful emissions across the country.

(6) Irregular rains and melting of snow on mountain caps will increase the flood incidences world over.

(7) Glaciers, considered as the thermometer of global warming, will undergo a big Meltdown which would be difficult to track. The glacier melt will adversely alter fresh water balance and dramatic alteration of climatic patterns (see Box 40.1).

(II) Global Warming and Human Diseases

World Health Organisation (WHO) has expressed concern over global warming which could cause spread of Malaria which already threatened about 2100 million people of the world. Rise in temperature and rainfall changes will affect the distribution of mosquitoes which carry diseases like malaria, encephalitis, filariasis (elephantiasis), chikunguinia etc. Malaria is not found in the highlands of Ethiopia, Kenya, Indonesia, part of China and South Africa etc. and as such these areas are threatened for such diseases due to global warming.

Global warming will certainly bring about a change in the distribution of communicable diseases, and vector borne diseases could invade new areas and threaten the health of millions of people.

(III) Global Warming and Agriculture and Forestry

Climatic changes due to global warming will have significant effects on agriculture, forestry and livestock and there may be two direct effects, *first* by increasing CO_2 in the atmosphere and *second* by changed climate on crops, livestock and water supplies.

(i) **Effect of increased CO_2 in the atmosphere**. An increase of CO_2 in the atmosphere promotes the rate of photosynthesis and reduces the rate of transpiration. Higher rate of photosynthesis enhances the plant growth and it has been estimated that wheat and barley yield will increase as high as 40 per cent. C_3 plants like wheat, barley, groundnut, cotton, potatoes, mungbean, pigeonpea, chickpea, banana etc. will be more benefitted.

It has been observed that the CO_2 concentration above 400 ppm begins to induce the stomatal closure which reduces the rate of transpiration and gaseous exchange between atmosphere and the plants. The reduced rate of transpiration, coupled with increase in temperature due to global warming, will cause adverse effect on plant growth and various metabolic activities.

(ii) **Effect of changed climate**. Seino (1990) and Dhaliwal and coworkers (1996) predicted following effects of global warming on agroecosystems:

1. Evaporation will increase.
2. Temperature of soil and water will rise.
3. Snowing period will be reduced.
4. Agricultural area will increase.
5. Incidence of pests and diseases will become higher.
6. Growth of wild plants will increase.
7. Quick deterioration of soil fertility will take place.
8. Microbial activities will be enhanced.
9. Decomposition rate of organic matter and fertilizers will increase.
10. Irregular rainfall, flood and water problems will increase.
11. Probabilities of occurrence of cyclones and hurricanes due to global warming will increase adversities for agriculture.

In general, climate changes, due to global warming, will affect the crop yield (depending upon moisture and temperature), cropping pattern, insect pests and diseases and management practices which may also affect the socio-economic systems. The net effect of CO_2 and climate change on agriculture in low rainfall areas and coastal low lands need to be determined precisely. The influence of interaction of rising CO_2 and temperature with toxic air pollutants may have far reaching effects on agriculture and forestry.

Remedial Measures to Reduce Global Warming

Remedial measures to reduce global warming means to reduce production of greenhouse gases by limiting the combustion of fossil fuels, deforestation and CFC using systems. Global efforts have been made on these lines and following are some important measures.

(1) The Toronto World Conference held at Toronto, Canada in June, 1988, called for 20 per cent voluntary reduction in the emission of CO_2 by the year 2005, specially by highly industrialised countries like USA, UK, Germany, France, Canada etc. By this reduction, greenhouse effect may be minimised or atleast stabilised. However, some of these countries did not agree to the Toronto Resolutions stating that such measures will adversely affect to the progress of their countries, even developing countries will face serious problems of energy crisis.

Developed countries did not sign the Toronto resolutions in the same way as they did not sign the Montreal Protocol of September, 1987 which called for the reduction in the production and consumption of CFCs to save the ozone layer.

Representatives from 105 countries signed a treaty and resolved to stop warming of the earth (Bankok, May, 2007) and warned that only eight years remain to save the world from its disastrous effects. G-8 countries met at Germany (July, 2007) and promised to reduce the production of greenhouse gases, one half from the present level, by 2050, although USA diplomatically stated that cooperation from India and China is essential in this regard.

(2) To avoid problems which arose against Toronto Resolutions, scientists must discover and develop alternative sources of energy and power and improved better technologies. For example, scientists should evaluate the potential of alcohol as the major source of fuel to be used in the transport sector. Ethanol in combination with natural gas called *gasohol* has been successfully used in Brazil as motor fuel. Ethanol production can be increased by growing energy plants.

(3) Technologies should be advanced and made more efficient to derive maximum energy from fossil fuels to reduce CO_2 and N_2O emissions.

(4) Solar energy may be developed as alternative to the conventional fossil fuel energy, atleast in those tropical countries where sunlight is abundantly available during most period of the year. Biogas may be considered as another alternative source of energy for the domestic sector. (Details of non-conventional and conventional energy given in unit 2 of the book.)

(5) Large-scale afforestation and reforestation should be taken up, as forests are big *natural sink* of carbon dioxide. It is estimated that the forest cover of 120 million hectares can fix about 780 million tonnes of atmospheric carbon dioxide each year. It would reduce release of CO_2 by combustion of fossil fuels and burning of firewood by 47 per cent of the total annual release.

Should we want not to be cursed by future generations, greenery of the landscape is the only immediate solution of ever-increasing greenhouse effect caused by unending increase of atmospheric carbon dioxide.

ACID RAINS

The term **acid rain** was coined by Robert Angus (1872) to refer mixing of rain in the air with pollutant gases released from burning of fossil fuels in power stations, factories and automobiles and bringing about the dilute sulphuric acid or nitric acid or both together in the atmosphere. Acid rain simply means fall out of acids caused by sulphur dioxide and nitrogen oxides with rainfall.

Formation of Acid Rains

The main source of acid rains are oxides of sulphur and nitrogen which are emitted from industrial establishments and different types of autovehicles. In the process, oxides of sulphur and nitrogen released from fossil fuel burning, stay long in the atmosphere and are oxidized into sulphuric acid (H_2SO_4) and nitric acid (HNO_3). The process of acid formation can be summarized as follows-

$$SO_2 + H_2O \rightarrow H_2SO_3 + O \rightarrow H_2SO_4$$
$$NO + O_2 \rightarrow NO_2 + H_2O \rightarrow HNO_3 + H_2O$$

The acids may remain in the atmosphere for a long time in clouds and fogs. On an average, H_2SO_4 constitutes about 60-70 per cent and HNO_3 about 30-40 per cent in acid rain (fig 40.2).

Acid rains are not confined to the source areas of the emissions of oxides of sulphur and nitrogen, rather they cover much larger area far away from source of emission of pollutants because these pollutants being in gaseous form are carried away and spread over larger areas through wind and clouds.

Water of rains is not neutral because of presence of CO_2 in the atmosphere, which dissolves and makes it slightly acidic. When sulphuric acids and nitric acids are dissolved in rainwater, it becomes more acidic and pH of water falls. The pH of rainwater in certain localities of USA, such as West Virginia, sometimes goes down up to 1.5 and in Europe as low as 2.4. The water becomes more injurious when the pH falls below 4.0.

Effects of Acid Rains

Acid rains spread in large areas and cause damages in various ways. For example, acid formed due to pollutants released from several mills in Germany and UK, have caused widespread acid rains in Scadinavian countries like Norway and Sweden with the result that most of Scandinavian lakes have lost their biological communities and such lakes are now biologically known as **dead lakes**.

In Anglo American and West European countries, acid rains are often called **lake killers** because these rains have been identified as main factors of the *deaths of lakes* which means the death and destruction of all aquatic life forms in the lake or water bodies. In Ontario region of Canada, out of 2,50,000 lakes, some 50,000 have been adversely affected by increase of the acidity in water, and of these, 140 lakes have been declared as *dead lakes*.

Acid rains also affect human community adversely. According to Hamilton, an American doctor, about 7500 – 12,000 people die of acidic sulphates coming out of combustion of fossil fuels all over the world every year.

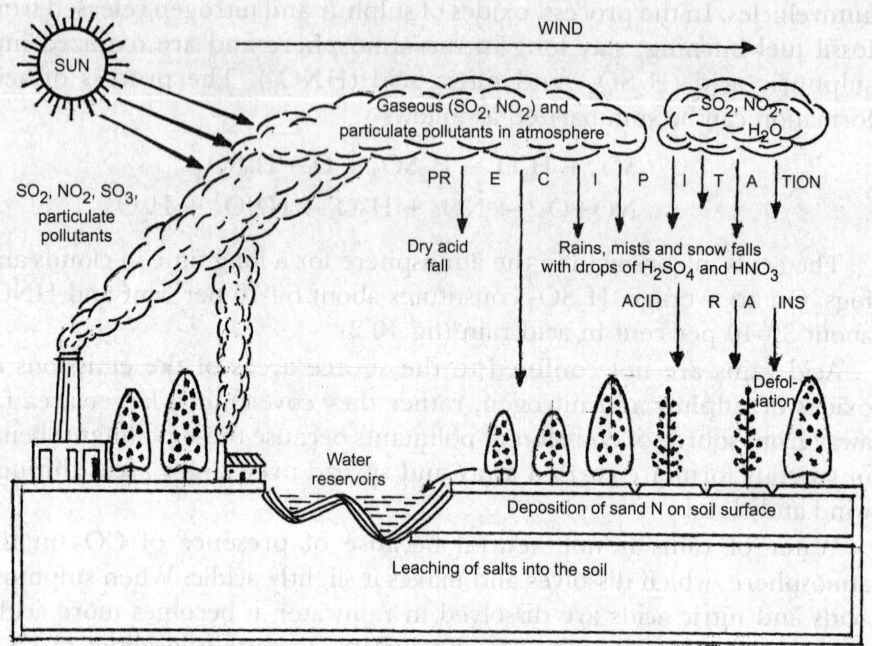

Fig. 40.3. Acids rains-schematic view.

Acid rains significantly affect the soil fertility, as increased acidity destroys mineral elements and other nutrients of the soil. There are reports that forests of Canada, USA, Germany and many countries of middle Europe have been largely damaged due to acid rains. In plants, acid rains retard the rate of photosynthesis, growth of roots and nutrient absorption from the soil. Soil microflora is largely damaged in acid rain affected soils.

Acid rains, particularly *dry depositions*, damage historical buildings and monuments due to corrosion. In Europe and America, several valuable sculptures and antiques of precious marbles and other valuable stones are now being kept in museums to save them from the danger of acid rains. Taj Mahal of Agra is being spoilt by the sulphuric acid fumes formed due to pollutants released from Mathura Oil Refinery.

The problem of acid rains in India is no longer alarming at present as studied by BARC (Bhabha Atomic Research Centre) and WMO (World Meteorological Organisation). However, recent studies in India have revealed that industrial areas of large cities like Delhi, Nagpur, Calcutta, Bombay, Surat, Ahmedabad etc. witness acid rains. It is mainly due to excessive burning of fossil fuels in thermal power stations.

Control Measures

The only remedial measure to reduce acid rains is to reduce pollutant discharges in the atmosphere. Acid rain is not a local problem, rather its dimensions cover international areas. Some possible measures suggested to reduce acid rains are as follows:

(1) Low pollutant producing fuels should be used.
(2) During combustion of fossil fuels, methods should be adopted to reduce the formation of pollutants.
(3) Oil desulphurisation by using hydro-sulphurisation techniques, should be adopted.
(4) Vehicular pollution should be minimised.
(5) Screening of pollutants from exhausts and fuel gases should be practiced.
(6) Instead of fossil fuels, use of alternative energy sources should be adopted.
(7) Liming of lakes, streams and rivers by using lime stones and dolomites should be carried out after regular intervals in acid rain prone areas.

OZONE LAYER DEPLETION

Ozone Layer

C.F. Schonbein discovered ozone and stated that it is a three atom isotope of oxygen or merely a triatomic form of oxygen, *i.e.*, O_3. It is a bluish-coloured gas with a pungent smell which occurs naturally in scant amount between 10 and 50 km above sea level and is most strongly concentrated at an altitude of 20-25 km which is called **stratosphere** zone of atmosphere or **ozonosphere**.

Ozone is considered as a protective shield and earth's umbrella because it prevents ultra violet solar radiation from reaching the earth surface. In absence of this ozone layer, the temperature of lower atmosphere might have risen to such an extent that the *biological furnace* of the biosphere might have turned into *blast furnace* with no life on the earth. Thus, the presence of ozone layer in the stratosphere is of vital significance to the biome of the biosphere.

Formation of Ozone Layer

Ozone gas is unstable and the creation and destruction of ozone gas is a gradual and contiguous natural process. The oxygen molecules in the atmosphere between 80 km to 1000 km are broken down by ultra violet solar radiation or by an electric discharge during thunderstorm. The atomic oxygen can then either recombine with an oxygen molecule (O_2) to form ozone once again, or combine with other substances. It can be explained as follows;

$$\text{High energy radiation from Sun} \rightsquigarrow O_2 \text{ (Oxygen molecule)} \longrightarrow O + O \text{ (The oxygen atoms)}$$

$$O \text{ (Oxygen atom)} + O_2 \text{ (Oxygen molecule)} \underset{\text{UV rays}}{\rightleftharpoons} O_3 \text{ (Ozone)}$$

The formation of ozone (O_3), due to collision of 3 atoms (O_2+O) through the process of photochemical reaction triggered by the sunlight, is more active in the atmospheric zone of 30 to 60 km height from sea level and the maximum *ozone density* (mass of ozone per unit volume) is found between the height of 20 km to 25 km (fig. 40.4).

Most of the stratospheric ozone is formed in the atmosphere over the tropical areas from where some ozone is transported by the atmospheric circulation to the polar areas up in the atmosphere. The quantity of ozone is measured in DU (Dobson Unit). For his work on ozone, KR Ramanathan of Kerala, is nicknamed **Mr Ozone**

Depletion of ozone layer

The British Antarctic Survey Team led by Joseph Forman provided first hand evidence of ozone depletion over the Antarctica in 1985 and discovered **ozone hole**. He reported 40 per cent loss in the spring time ozone layer in the atmosphere lying over Antarctica. (Box 40.2) Soon, a multinational team was sent on expedition to study the ozone layer loss in Antarctica.

The team revealed that-

(i) The average natural concentration of ozone dropped by 50 per cent between August 15 (1987) to October 7 (1987) and

(ii) In some patches, the ozone concentration dropped by 100 per cent which resulted in the formation of ozone less patches, now called **ozone holes**.

The startling revelations of ozone hole shocked the world community and soon a **Montreal Protocol**, on substances that deplete the ozone layer, was signed in September, 1987 by 35 countries. This was the first International Agreement to limit the production and consumption of ozone depleting chemicals under the UNEP (United Nations Environment Programme).

Ozone Depleting Substances

In 1974, scientists discovered that chlorofluoro-carbons (CFC), trade name *Freon*, used in refrigerator, air conditioning, aerosol and as cleaning agents, are continuously destroying the ozone layer. Two commonly used CFCs are CFC-11 and CFC-12. They have typically only 2 to 5% of the ozone destroying potential of typical CFCs.

Another substance that attacks stratospheric ozone is halon, a class of fluorocarbons with bromine atom. Nitrogen oxides emitted from supersonic jets also destroy ozone.

Ozone is also destroyed due to natural causes. Natural gases like hydrogen oxides (derived from water vapour), methane, hydrogen gas and nitrogen oxides are effective destroyers of stratospheric ozone. Following table reveals the chemicals that destroy ozone and their uses.

TABLE 40.3. Important Ozone–Depleting Chemicals and Uses

Compound name	*O_3 depleting potential*	*uses*
1. CFC-11 ($CFCl_3$)	1.0	Refrigeration, foam
2. CFC-12 (CF_2Cl_2)	0.9–1.9	Refrigeration, foam, food freezing heat detectors, warning systems cosmetic, pressurized blowers
3. CFC-13 (CCl_3CF_3)	0.8–0.9	Cosmetics, solvents
4. Halon 1301 ($CBr.F_3$)	10.0–13.2	Firefighting
5. Halon 1211 ($CCl\ Br\ F_2$)	2.2–3.0	Firefighting
6. HCFC-22 ($CHCl\ F_2$)	0.05	Foam, refrigeration
7. Carbon tetrachloride (CCl_4)	1.2	Solvent

Box 40.2 **Ozone Hole**

Effect of man's activities on the environment has resulted depletion of stratospheric ozone over Antarctica (and recently over Arctic also) only within a short span of 15 years. The conditions necessary for the destruction of and ozone loss are:

(1) Polar winter leads to the formation of the polar vortex, which isolates the air within it.

(2) Cold temperatures formed inside the vortex is cold enough for the formation of Polar Stratospheric Clouds (PSCs). As the vortex air is isolated, the cold temperatures and PSCs persit.

(3) Once PSCs form, heterogenous reactions take place and convert the inactive chlorine and bromine reservoirs to more active forms of chlorine and bromine.

(4) No ozone loss occurs until sunlight returns to the air inside the polar vortex and allows the production of active chlorine and initiates the catalytic ozone destruction cycles, thus causing rapid ozone loss.

The Arctic ozone hole currently covers a geographic region a little bigger than Antarctica and extends nearly 10 km in altitude in the lower stratosphere.

Causes of Depletion of Ozone Layer

When CFCs rise into the stratosphere, the solar UV-radiations decompose them releasing free chlorine radicals. Each of these chlorine atoms reacts with one molecule of ozone and converts ozone into oxygen molecule. The reaction process can be summarised as follows

UV-rays of upper atmosphere break off chlorine atom from a CFC molecule

$$\left[\begin{array}{l} CCl_3F \xrightarrow{UV} CCl_2F + Cl \\ CCl_2F_2 \xrightarrow{UV} CClF_2 + Cl \end{array}\right]$$

↓

The free chlorine atom reacts with ozone molecule (O_3) which splits into oxygen molecule (O_2) and ozone atom (O).

$$Cl + O_3 \rightarrow ClO + O_2$$

The result is that free chlorine combines with oxygen atom and destroy ozone-

$$\underset{\text{Free chlorine}}{Cl_2} + \underset{\text{Ozone}}{2O_3} \longrightarrow \underset{\text{chlorine monoxide}}{2ClO} + \underset{\text{Oxygen molecule}}{2O_2}$$

↓

Chlorine monoxide mutually reacts to produce Cl_2O_2 or Chlorine monoxide is attacked by free oxygen atom forming a free chlorine atom and a free oxygen molecule

$$2ClO \rightarrow Cl_2O_2 \xrightarrow{UV} Cl_2 + O_2$$

↓

Free chlorine atom repeats the process of destroying ozone molecule. The whole process is repeated again and again.

Dr F Sherwood Rowland predicted that each chlorine atom can destroy 1,00,000 ozone molecules by such cyclic process.

Effects of Ozone Layer Depletion

The most important function of the ozone layer present in stratosphere is to protect the troposphere and the earth's surface from most of the ultra violet rays. Thus, ozone layer serves as a shield against the UV-solar radiation and saves the earth's surface from becoming too hot. The various effects of ozone depletion may be summarized as follows-

(1) **Effects on climate**. With the depletion of ozone layer in the stratosphere, the UV-rays of solar radiation will directly reach the earth's surface making it too hot. The substantial increase in the surface temperature on the earth would cause global warming and

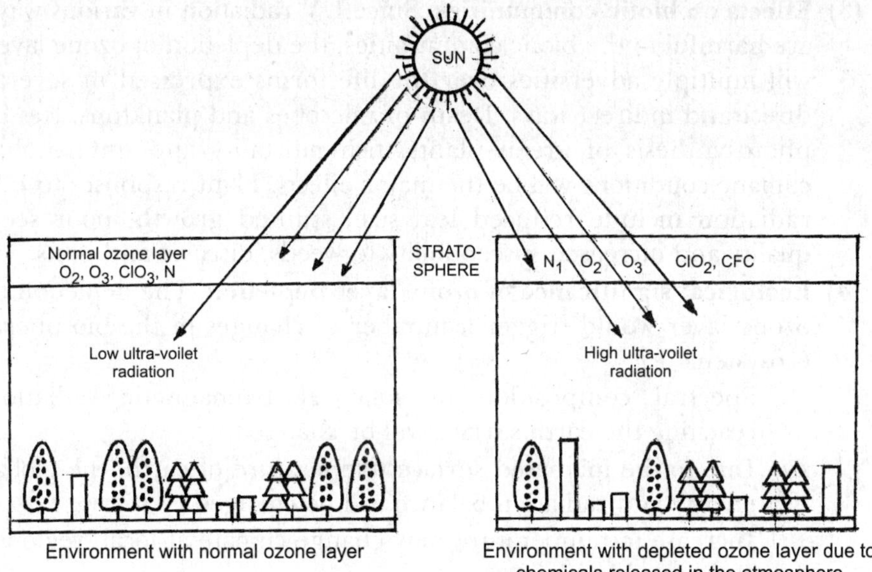

Fig. 40.4. Ozone layer depletion and its effects.

climate changes at regional and global levels. The global warming would cause melting of continental glaciers and ice sheets. This would in turn cause rise in sea level and consequent submergence of coastal low lands.

It is also believed that due to ozone depletion, the level of hydrogen peroxide in the troposphere will increase which will ultimately induce **acid rains**.

(2) **Effects on human beings**. Higher incidence of UV-rays on earth's surface due to ozone layer depletion will increase skin cancer, particularly among the white populations. According to an estimate 1 – 2 per cent decrease in ozone would cause skin cancer to 1,20,000 people per year in USA, alone. According to USA Environment Protection Angency (1986), there are about 3 lac to 4 lac new cases of skin cancer each year in USA.

Increased exposure of the human bodies to UV-rays will decrease their immunity system making more prone to the infectious diseases. General physical growth will be checked and brain retardation will appear in the tropical regions of the world.

Increase of UV-rays will promote photochemical reactions and formations of smog. Smog causes lung cancer and damages to respiratory system.

In general, increase of UV radiation on earth's surface will adversely affect all forms of life causing several serious direct and indirect repercusions.

(3) **Effects on biotic communities**. Since UV radiation in various ways are harmful to the biotic communities, the depletion of ozone layer will multiply adversities over the life forms expressed in several direct and indirect ways. Death of microbes and planktons, loss in photosynthesis of green plants, fish mortality and unfavorable climatic conditions will be the major effects. Plant responses to UV radiation include reduced leaf size, stunted growth, poor seed quality and increased susceptibility to weeds, diseases and pests.

(4) **Ecological significance of ozone layer depletion**. The depletion of ozone layer would trigger a number of changes in the biospheric ecosystems, *e.g.*

(i) Spectral composition of solar electromagnetic radiation reaching the earth surface will be changed.

(ii) Due to the increased surface temperature of earth, the global energy and radiation balance will be changed.

(iii) Increase in temperature may change climate at local, regional and global level.

(iv) Change in climate may alter the physiological characteristics of plants, animals and human beings.

(v) Change will induce imbalance in hydrological cycle and increase in incidence of flood and drought.

(vi) Global warming has been reported as the major cause of coral bleaching. A large scale coral bleaching has already been observed in different parts of the world.

(vii) *El Nino* phenomenon may take place quite frequently causing damage to marine life and change in the climate.

It is speculated that depletion of ozone layer in the stratosphere, would upset whole ecological balance of the biospheric ecosystems and the effects would be catastrophic.

Protection and Maintenance of Ozone Layer

The depletion of stratospheric ozone layer and consequent danger to biological communities in general and human beings in particular has become a matter of serious environmental concern to governments, scientists and general public at local, regional and global levels. **Montreal Protocol** (1987), **London Conference** (1989) and various efforts of United National Environment Programmes (UNEP) reveal global concern and awareness towards it. The international efforts towards remedial measures to control ozone depletion are being taken up at two levels:

(1) To promote reduction in the production and consumption of emissions of ozone depleting chemicals and

(2) To make serious efforts to produce and propagate the use of alternative chemicals which do not deplete ozone.

(1) **Reduction in the production and consumption of ozone depleting chemicals**. Ozone depleting chemicals are mainly CFCs and halons and, therefore, the first step before the world community, is to stop or markedly reduce the production and consumption of ozone depleting chemicals (ODCs). The first substantial step in this regard was the agreement signed by 35 countries in Montreal, Canada (September, 1987) which was called the **Montreal Protocol on Substances that Deplete the Ozone**. Montreal Protocol could be succeeded due to the efforts of UNEP. It has reached the following agreements–

(i) To freeze the production of CFCs at 1986 level by 1999.

(ii) To decrease the production of ozone depleting chemicals by 20 per cent by the end of 1993.

(iii) To allow further 20 per cent cut in their production by 1998.

(iv) To freeze the production of halons at 1986 level starting from 1992.

Thus, the total production of ODCs (mainly CFCs and halons) according to the Montreal Protocol would be reduced by 50 per cent by the beginning of 1999. Now, it is felt that this reduction was not sufficient and it is expected to increase enormously in the 21st century. The US Congress Office of Technology Assessment (OTA) stated that reduction and consumption of ODCs would decrease only 15 to 30 per cent unless Montreal Protocol is accepted by eight major developing countries (China, India, Brazil, Indonesia, Iran, Mexico, Saudi Arabia and South Korea). According to the report of the survey conducted by the US Environment Protection Agency (EPA), the total concentration of chlorine in the atmosphere would increase threefold even if all nations of world sign the Montreal Protocol. So, adequate global efforts are still at large.

(2) **Search of alternative technology and substitute chemicals**. Efforts to reduce global production and consumption of ODCs are not sufficient to cope with the depleting ozone layer. Until substitutes to CFCs and halons are discovered, global attentions would remain centred on improving the use and maintenance of CFCs by developing new efficient equipments.

Some chlorine free chemicals, as substitute to CFCs, are under consideration. For example, US based petroleum company has developed Bioact EC-7, a biodegradable, non-toxic and non-corrosive chemical. With trade name HFC-134 which can be used in place of Freon–12 in air conditioners and refrigerators. Attempts are also being made to search for chlorine free substances that do not have ozone depletion potential and can be used in refrigerators.

Indian Efforts for Ozone Layer Protection

The Ministry of Environment and Forests, has set up a **Ozone Cell** as a national unit to look after and to render necessary services to implement Montreal Protocol and its Ozone Depleting Substances (ODS) phase-out programme in India.

The Ozone Cell and UNEP, has jointly launched a new global initiative to raise awareness on the Ozone Layer Protection and *Remembering O₃ur Future: Commemorating Closure of ODS Production sites* under the Montreal Protocol on 8th March, 2005.

As a part of non-investment activity under the Foam and Commercial Refrigeration Programme, the Ministry of Environment and Forests had organised three technical workshops in Banglore (February, 2005), Mumbai (June, 2005) and Ahmedabad (June, 2005).

The International Ozone Day was celebrated in India on 16 September, 2005. The theme of the year was-

"**Act Ozone Friendly and stay sun safe**".

India was selected as a member of the Executive Committee of Montreal Protocol for the year 2006. India has so far met the following compliance dates as per the control schedule of the Montreal Protocol

(i) Freeze of CFC production and consumption in July, 1999 at 22588 ODP tons and 6681 ODP tons, respectively and

(ii) Freeze of Halon production on 1 January, 2002

NUCLEAR ACCIDENTS AND HOLOCAUST

Nuclear energy was earlier considered to be the cheapest source of power but because of nuclear accidents and disposal problems it is proving not only problematic but also hazardous to all. In general, governments, administrators and scientists assure that nuclear materials are stored and disposed by fool proof methods. But, accidents occurring in nuclear establishments all over the world, are alarming and focus a different picture. Nuclear power stations require much more efficient safety measures to control accidents. As such there is no absolutely safe method of storing high level nuclear wastes which are extremely dangerous and require thousands of years to lose their hazardous effects (some details are given in chapter on Nuclear Hazards). Moreover, tonnes of such nuclear wastes are piling up all over the world. According to Miller (2004), in Yucca of US alone, today there are 45,000 tonnes which would be around

1lac tonnes by 2035 as the deadly stuff to be stored and secured for about 10,000 years. The problem is highly acute and dangerous.

Nuclear bombs have been used only once so far. These were dropped by United States on Hiroshima and Nagasaki, Japan in 1945 (see case studies). However, innumerable nuclear tests conducted by various countries and of and on leakage accidents are certainly of more concern. Earlier before 1963, nuclear tests were conducted above ground exposing radiations to dangerous level. Now, these are conducted underground, but question is that how long and how much safe they are Besides, low level radiation used in hospitals and research institutes, are also subject to exposure in use and disposals, if used carelessly or by untrained technicians.

Nuclear accidents have occurred in various countries but none is absolutely safe. At times small accidents occur but these remain largely unknown. Some serious nuclear accidents are as follows-

(1) In 1957, due to disastrous fire in a major nuclear complex of Sellafield, England, there was a heavy leakage of radioactivity. It was caused due to human error. Unfortunately, no warning was issued and as a result, the accident took the life of 1000 people. Several accidents in Sellafield have occurred in last 40 years causing unsafe exposure to workers leading to various problems including cancer.

(2) Several accidents have taken place in Hanford Nuclear Power Complex of United States. Moreover, in 1973, Hanford released 4,50,000 litters of radio active liquid in the Columbia river, possibly due to lack of storage tanks. Higher than normal rates of cancer have been reported from the area.

(3) Among the most serious nuclear accidents, that occurred in United States, was that of March 29, 1979 at the Three Mile Island Power Plant in Pennsylvania. A human error caused failure of a cooling system that led to a 50 per cent melt down of the reactor core. A huge amount of radioactivity escaped in the atmosphere. Luckily, the area was evacuated and there were no immediate casualities. But accident shook the confidence of the people and there were no further new projects in the area. It took about 12 years to reopen the plant with an expense of US$ 1.2 billion dollars.

(4) The biggest nuclear accident in the world occurred in the early hours of 26 April 1986. The Reacter No. 4 at the nuclear power complex at Chernobyl exploded with terrible and long lasting consequences. More than 30 fires were set off in the complex and it took 10 days for hundreds of firemen to control the fires. Over 1,35,000 people were evacuated and about 60,000 buildings had to be decontaminated (for details see case studies).

(5) In September, 1987, two men found a lead capsule with 3500 cu. m. of radioactive waste in an abandoned radiotherapy unit of Brazilian city of Goiania. Whosoever visited and exchanged with each other these deadly but beautiful crystals, became sick. By the time, the actual cause of sickness was known, the contamination has spread over an area of 300 sq km. Some sick died, more than 200 people were evacuated and several houses were demolished. According to World Health Organisation (WHO), this disaster was second in severity only to the Chernobyl disaster.

In India, there have been several cases of missing of articles containing radioactive substances and some have been stolen from laboratories and hospitals. No one knows their details and when will they unleash their hazardous powers.

Nuclear accidents can be reduced by the safe disposition of nuclear wastes and minimizing the nuclear tests.

Case Studies

CHERNOBYL NUCLEAR ACCIDENT (1986)

Chernobyl nuclear accident was the biggest nuclear accident in the world. The accident took place in the early hours of April 26, 1986 when Reacter No. 4 at the nuclear power complex exploded terribly. It was Ukraine's (then USSR) most efficient nuclear power plant. While conducting some experiments, a scientist switched off the automatic shut down mechanism. Unfortunately it went wrong and massive steam exploded and blew off the roof blowing the radioactive gases and debris high into the air. More than 30 fires were set off in the complex which took hundreds of firemen and 10 days to control the fire. Helicopters were used to drop more than 5000 tonnes of lead, boron and other materials on the core to smother the radioactive wastes.

Some 1,35,000 people were evacuated and about 60,000 buildings spread in an area of 1000 sq. km were decontaminated. This operation continued for 10 days.

The radioactive clouds speared across Europe. High radioactive fall out was found on Poland, Sweden, Norway, Finland, Germany, Italy and France and was recorded upto Siberia, Saudi Arabia and North America. Actual damage could not be known due to the lack of monitoring and the long time spans of radio active decay. It is possible that several parts of Europe may remain contaminated well into the twenty second century or even longer. But all it was very unfortunate, scaring people world over.

According to Green Peace Forces, some 93 thousand people died due to this accident.

Chapter Summary

Constant global pollution has caused a change in climate at an alarming level. Climate changes are basically due to global warming, acid rains, ozone layer depletion, nuclear accidents etc. **Global warming** is resulting an increase in temperature of earth and its environment. The rise in temperature is due to increase of greenhouse gases in the atmosphere which absorb the reflected long wave radiations and cause warming. The global temperature has risen by only 4°C over the last 18,000 years but because of rapid industrialization during recent past, it has increased by about 0.5-1°C since the last century. The rate of global warming in the past quarter century was greater than any previous period since 1880.

Important greenhouse gases are CO_2, CH_4, N_2O, CFCs etc. The approximate contribution of percentage of greenhouse gases to global warming is 61 by CO_2, 15 by CH_4, 11 by CFCs, 4 by N_2O and 9 by others. Several million tonnes of CO_2 per year is emitted in the world, of which USA and China top the list in their contribution of CO_2 emission. An increase in global warming will rise the earth's temperature, change the rainfall pattern, rise in sea water level, cause melting of glaciers, reduce the snow falls, create flood, result submersion of low lying coastal countries, change the agriculture pattern etc. Indian glaciers are rapidly receding, of which small glaciers have receded by 38 per cent and larger ones by 12 per cent. The only remedial measure to reduce the global warming is to decrease the production of greenhouse gases all over the world. Despite international efforts on these lines, the results are not very promising.

Acid rains simply means fallout of acids caused by SO_2, and NO_x with rainfall. Such gases are emitted by industries, factories and autovehicles using fossil fuels as source of energy. These gases react with atmospheric water and produce H_2SO_3, H_2SO_4 and HNO_3. The acid rains are not confined to the source area rather spread over larger areas through wind and clouds. The acid rain damages the flora, fauna and historical buildings and reduce the soil fertility. In plants, acid rains retard the rate of photosynthesis, growth of roots and nutrients absorption from the soil. Reduction of pollutant discharges in the atmosphere is the only remedial measure to prevent acid rains.

Ozone layer depletion is now a major concern to human welfare activists. Ozone layer present in the stratosphere is considered as a protective shield which prevents ultra violet solar radiation reaching the earth surface and protects the biome of the biosphere. Ozone gas is unstable and the creation and destruction of ozone gas is a natural process. In 1947, scientists discovered that chlorofluorocarbons (CFCs) used in refrigerator and air conditioning and aerosols are continuously destroying the ozone layer. Two common CFCs are CFC-11 and CFC-12. The free chlorine atoms of CFCs react with ozone molecules and split them into oxygen molecules and oxygen atoms. The various effects of ozone depletion are changes in the climate and increase in skin cancer and decrease of immunity system in human beings. Plant responses to UV radiation include reduced leaf size, stunted growth, poor seed quality etc. Global efforts are now on to save ozone layer with two different approaches- *(i)* reduction in production of ozone depleting chemicals and *(ii)* production and propagation of use of alternative chemicals which do not deplete ozone. The international theme of 2005 for ozone drive was "Act ozone friendly and stay sun safe."

Nuclear accidents are becoming cause of holocausts. Such accidents are becoming increasingly common. Nuclear tests and leakage accidents by different nations are the prime causes of concern to the human life. Several leakage accidents that occurred in various parts of the world have resulted death or disabilities of lakhs of people. Of such cases, Chernobyl Nuclear Accident was the biggest nuclear accident that occurred in 1986. Nuclear Accidents can be reduced by the safe storage and disposition of nuclear wastes and by minimising the nuclear tests.

Study Questions

1. Enumerate the effect of pollution on climate changes.
2. Describe the various effects of increase of carbon dioxide in the atmosphere.
3. Explain how global warming is affecting our planet?
4. What is global warming? Describe causes, effects and methods to reduce it.
5. What are acid rains and what are the causes for them? Describe various measures to minimise the acid rains.
6. Write a brief note on nuclear accidents and holocausts.
7. What is ozone layer depletion? Describe the causes, effects and mitigation measures to prevent it.
8. Write short notes on:
   (i) Global warming and its consequences
   (ii) Causes and effects of acid rains
   (iii) Causes and effects of ozone layer depletion
   (iv) Nuclear accidents
   (v) Climate change

Objective Questions. *Select correct answers*

1. The biggest nuclear accident occurred in:
   (1) New York (2) Chernobyl
   (3) Bhopal (4) Beijing

2. The major greenhouse gas is:
   (1) CFC (2) CO
   (3) CO_2 (4) Freon

3. Depletion of ozone layer in the stratosphere will cause:
   (1) CO_2 increase (2) Skin cancer
   (3) Cholera (4) All of these

4. The main cause of acid rain is:
   (1) O_2 (2) CO_2
   (3) H_2O (4) SO_2

5. Fluorocarbons are released in the atmosphere by:
   (1) Acid rains (2) Automobiles
   (3) Jet airplanes (4) Strong wind

Answers

1. (2) *2.* (3) *3.* (2) *4.* (4) *5.* (3)

41

CHAPTER

Wasteland Reclamation

LEARNING OBJECTIVES
Introduction • Definitions of Wastelands • Classification of Wasteland • Distribution of Wasteland in India • Wasteland Reclamation Practices • Objectives of National Wasteland Development Board • India

Introduction

Wastelands are the degraded lands, which are currently underutilized and the lands which are deteriorating for lack of appropriate water and soil management or on account of natural causes but which can be brought under vegetation cover with reasonable efforts. These cover derelict, devastated, destroyed, poor or neglected lands. Technical Task Group Report of National Wasteland Development Board, India, in 1985, had defined wastelands as *the land which is presently lying unutilized due to different constraints*. It comprises both culturable and unculturable wastelands. Nearly half of the country's land is considered wasteland. The task for wasteland reclamation is not new. The issue of formation of ravine lands has been with us since nineteenth century. The need to formulate a policy for rational land use management for the degraded land of our country was felt in 1970s. In our country National Wasteland Development Board (NWBD), established in 1985, looks after wastelands of all kinds such as degraded forest lands, strip lands, mine spoils and abannded quarries, waterlogged land, marsh, saline land, barren rock areas, steep slopes, eroded valleys, shifting cultivation areas etc. Formation of wastelands is mainly due to human activities like indiscriminate or over utilisation of forest resources, overgrazing, shifting cultivation, side effects of development projects and unscientific land management practices.

Definitions of Wastelands

A wasteland has been defined variously by different authors/ organisations. National Remote Sensing Agency, Hyderabad has defined wasteland as land that is at present lying unused or land which is not being used to its optimum potential due to various constraints, or land which can not be used. Some other definitions for a wasteland are:

1. Land which is neither under forest cover or agricultural cover nor is assigned for any specific purpose such as national park or wildlife sanctuary or hydel projects.
2. Land which is ecologically unstable, badly eroded and degraded.
3. Land which is not capable of producing materials or services of value.
4. Land which has at less than 20 per cent of economic potential.
5. Land which is incapable of sustaining greenery.

Classification of Wasteland

Wastelands can be broadly classified into two types:

(1) **Culturable Wasteland.** The Culturable wasteland is that which is capable of or has the potential for development for agricultural or pastoral purposes or can be afforested. Presently it is not in use due to constraints such as lack of water, in being saline or alkaline, soil erosion, waterlogging, unfavourable physiographic status, or human neglect. The wastelands included in this category are ravinous or gullied hands, surface waterlogged and marshy lands, saline and alkaline lands, lands with laterite soil, shifting cultivation areas, degraded forest lands, strip lands, degraded pastures, mining and industrial wastelands etc.

(2) **Unculturable Wastelands.** The unculturable or non-culturable wasteland is that which is a barren land and can not be put to any productive use, either for agricultural or pastoral purposes or for afforestation. The wastelands included in this category are barren rocky areas, steep slopes, snow covered or glacial areas.

Distribution of Wastelands of India

Wastelands occur in every state and region of the country under different climatic conditions. Their area may be small or large but it is estimated that in India, the total area under wastelands is about 1295.2 lakh hectare. The area may be classified as saline and alkaline lands, wind eroded area, water eroded area and degraded forest area which occurs around the forest but are not in use (fig . 41.1). Madhya Pradesh and Rajasthan have an area of about 200 lakh hectare of the land as wasteland.

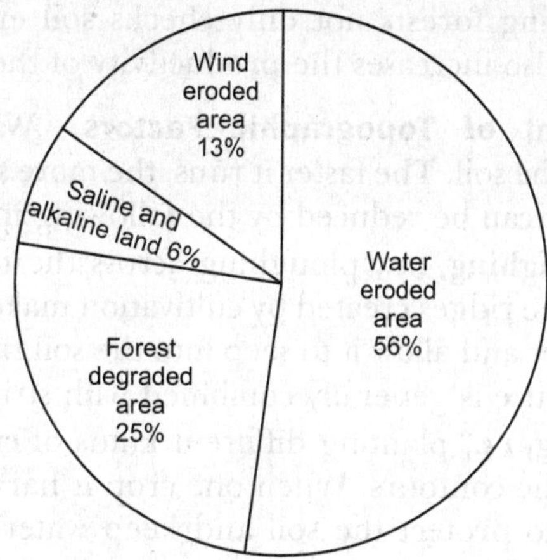

Fig. 41.1. Distribution pattern of Indian wastelands.

Wasteland Reclamation Practices

The development and reclamation of wasteland can provide more land for human activities including agriculture. There is an urgent need to develop a strategy to cure land damage and save the land from the future damage. This can be planned by preparing accurate land use data through remote sensing, a time bound nation wide survey programme of micro-level land, reviewing all existing legislation and updating them, and preparing management plans for land reclamation.

Since reclamation of wasteland is a time taking process and involves high expenditure, every care should be taken to give priority adoption to that wasteland which is reclamable within the financial means and involves known techniques. Thus, usually culturable wastelands are undertaken for reclamation programmes so as to increase the availability of land for productivity. It is estimated that India has the immediate prospect of reclamation of 4 million hectares (mh) of ravines and gullies, 16.73 mh of culturable wastelands, 9.82 mh of current fallows, 5.5 mh of coastal sandy area, and 2.5 mh of alkaline land. An area of 1.5 million hectare was taken up for reclamation scheme under the Seventh Five Year Plan

Various practices for checking land degradation and wasteland management are:

(1) Afforestation and Reforestation. Afforestation refers to growing forest over culturable wastelands where they have never been due to various adverse factors prevailing. Reforestation refers to replanting forest over areas where they were destroyed/degraded by overgrazing, shifting cultivation, excessive felling, forest fires etc. in the culturable wastelands. Restoring forests not only checks soil erosion, floods and water logging but also increases the productivity of the wastelands.

(2) Management of Topographic Factors. Water, on running downhill's, erodes the soil. The faster it runs, the more soil it carries off the fields. Water runoff can be reduced by the following methods:

(i) **Contour Ploughing,** *i.e.,* ploughing across the hill rather than up and down. The ridges created by cultivation make small check dams that trap water and allow it to seep into the soil rather than running off. This practice is generally combined with strip farming.

(ii) **Strip Farming,** *i.e.,* planting different kinds of crops in alternating strips along the contours. When one crop is harvested, the other is still present to protect the soil and keep water from running off straight downhill.

(iii) **Tied Ridges** are often useful in heavy rainfall areas. The method involves making of a series of ridges running at right angles to each other, so that water runoff is blocked from all directions and is encouraged to soak in the soil.

(iv) **Terracing** involves shaping the land in such a way to create level shelves of earth to hold water and soil. The edges of the terrace are planted with soil anchoring plant species. The procedure, though expensive and requiring either much hand labour or expensive machinery, is most effective.

Local people also control water runoff by leaving grass strips in waterways or arranging stones to form a water check wall.

(3) Plantation of Perennial Species. Plantation of perennial species, the plants that grow for more than two years, is the only suitable method for some specific land and soil types. For establishment of forest, grassland or crops such as tea, coffee or some other crops that do not have to be cultivated every year, it is necessary to protect certain unstable soil on sloping sites or water courses, *i.e.,* the low areas where water runs off after rain.

(4) Providing Ground Cover. The easiest way to provide cover that protects soil from erosion is to leave crop residue on the land after harvest of the crop. This not only covers the surface to break the erosive effects of wind and water, but also reduces evaporation and soil temperature during hot weather and protects ground and burrowing organisms that help aerate and rebuild soil. Where crop residues are not adequate to protect the soil or are in appropriate for subsequent crops or farming plans, 'cover crops' such as alfalfa, to be used as green manure, can be planted soon after the harvest to hold and protect the soil.

(5) Social Forestry Programmes. These are community based programmes, such as social forestry mostly involving strip plantation on road, rail road and canal-sides, or degraded forest lands; farm forestry; wasteland forest development etc.

(6) Gully Control. The formation and widening of gullies can be checked by constructing bunds, dams, drains or creating diversions through which excess run off water is channeled.

(7) River and Stream Bank Protection. The cutting and caving of river banks can be checked by developing vegetation alongside river/ stream banks or by building stones or concrete pitching etc.

(8) Mulching. Shifting sand can be controlled by mulching, *i.e.*, use of artificial protective covering. Stem stocks and basal parts of plants left after harvesting of potato, cotton, maize etc. are used to form a protective layer, the mulch. The mulches not only act as wind barriers, but also reduce evaporation and increase soil moisture by adding organic matter. Apart from shifting sand control, mulching is also effective against water erosion.

(9) Changing Agricultural Practices. Shifting cultivation (Jhoom cultivation) can be replaced by crop rotation, mixed cropping or developing plantation crops that would improve fertility and support a larger population.

(10) Leaching. Salt affected lands, especially the area where ground water level is not very high, can be recovered by leaching them with more water which removes the salts from the soil. In case of flood prone and irrigated lands, salinity can be prevented by diverting drainage towards such lands.

(11) Ecological Succession. Ecological succession, a natural process of establishment or re-establishment of an ecosystem, approach is particularly useful in mining and industrial abandoned wastelands where agricultural oriented methods fail. Instead of fast growing, vulnerable agricultural grasses, slow growing native grasses, known to be adopted to mineral deficient soils or soils with undesirable minerals can be planted. This approach has been successful in the reclamation of degraded lands in many parts of the country.

(12) Drainage. This is required to reclaime waterlogged wastelands. The excess water is removed by artificial drainage. Surface drainage is used in areas where water retains on the fields after heavy rains, by providing ditches to allow runoff the excess water. Usually 30-45 cm deep ditches, lying parallel to each other at 20-60 m distance, are able to remove 5 cm of water within 24 hours. For sub-surface drainage, horizontal sub–surface drainage is provided in the form of perforated corrugated PVC pipes or open jointed pipes with an envelope of gravel 2-3 cm below the land surface. This makes more chances of evaporation of water and checking of accumulation of salts. For reducing soil salinity by sub-surface drainage system, World Bank has funded a project to village Sampla, Rohtak (Haryana).

(13) Changing Irrigation Practices. Surface irrigation with efficient hydraulic design with precise land leveling, high frequency irrigation with controlled amount of water and thin and more frequent irrigation have been found more useful to reduce waterlogging and salinity when irrigation water is saline.

(14) Gypsum Amendment. Amendment of sodic soils with gypsum has been recommended for reducing soil sodicity, as calcium abounded with gypsum salt, replaces sodium from exchangeable sites.

(15) Use of Green Manures, Fertilizer and Biofertilizers. Application of farm yard manure or nitrogen fertilizers have been found to reclaime saline soils. Green manuring with *dhaicha,* sunhemp or *guar* improves salt-affected soils. Promoting growth of blue green algae (cyanobacteria) has been found to be quite promising in reclamation of *usar* lands.

(16) Selection of Tolerant Crops and Crop Rotation. It has been found that tolerance of crops for salts ranges from sensitive, semi-tolerant, tolerant to highly tolerant. Sugar beet and date palm are found to tolerate high salinity; wheat, sorghum, pearl millet, soybean, mustard and coconut are salt tolerant; and rice, maize, pulses, sunflower, sugarcane and many vegetables such as bottle gourd, brinjal etc. are semi-tolerant. The care should be taken to select a crop based on the extent of salinity of the land.

Objectives of National Wasteland Development Board, India

Because of increasing populations, the demand for the land for various human activities such as agriculture, industry and settlement is increasing. On the other hand good and usable land is shrinking due to the degradation. Thus it has become necessary to develop and reclaim the cultural wastelands to increase the availability of land for productivity. Government of India, in the year 1985, constituted an apex body, the National Wasteland Development Board (NWDB) to formulate action plans to arrest degradation of land and deforestation. Till 1989, NWDB's programme was focussed on tree planting. But with an indepth review in 1989-90, the programme was suitably diverted to check land degradation and to bring about wastelands into sustainable use and improve biomass availability and restore ecological balance. Some of the objectives of the Board are:

(1) To improve the physical structure and quality of marginal lands.
(2) To improve the availability of good quality of water for facilitating irrigation of these lands.
(3) To prevent soil erosion, flood and landslides.
(4) To conserve biological resources of land for sustainable use.
(5) To fulfill the growing need of the land for rapidly increasing population of the country.

A Centrally Sponsored Scheme of Reclamation of Alkali Soil was taken up in Punjab, Haryana and Uttar Pradesh during Seventh Five Year Plan which was extended in Ninth-Five Year Plan to all other states of India. The major components of the Scheme include help for assured irrigation water for farm development works like land levelling, bunding and ploughing; community drainage systems; application of soil amendment; and organic manure etc. An area of 0.64 mh, out of 3.5 mh, of alkali land has been reclaimed till the end of 2004-05 in the country.

Chapter Summary

Wastelands are the degraded lands which are currently unutilised and the lands which are deteriorating for lack of appropriate water and soil management or on account of natural causes but which can be brought under vegetation cover with reasonable efforts. These cover derelict, devested, destroyed, poor or neglected lands comprising both culturable and unculturable wastelands. Wastelands have been defined variously but they are either culturable or unculturable. Nearly 50 per cent of Indian land is considered as wasteland. They do occur in every state as small or large area but Madhya Pradesh and Rajasthan are the states with larger areas of wastelands. Importance of land increased with the increased population and caused Government of India to form a body-National Wasteland Development Board (NWDB), in 1985, to formulate action plan to arrest degradation of land and deforestation. The Board is actively engaged in checking degradation of land and bring about wastelands into sustainable use to improve biomass availability and restore ecological balance. There is an urgent need to develop a strategy to cure past land damage and save the land from future damage. The various practices to check land degradation and wastelands management include afforestation and reforestation, techniques involved in soil conservation, social forestry, providing groundwater to wastelands, proper drainage, new irrigation practices, selection of tolerant crops for the area etc.

Study Questions

1. Define wastelands. How to classify them? Describe.
2. What are the various wasteland relamation practices? Explain.
3. Discuss distribution of wastelands in India.
4. Discuss about National Wasteland Development Board, India.

5. Discuss Afforestation and Reforestation in Wasteland Management.

Objective Questions. *Select correct answers*

1. Which of the following does not fall under the category of wasteland?
   (1) Degraded Land (2) Divastated Land
   (3) Watershed (4) Poor Land

2. The states with most of their lands as wastelands are:
   (1) Utttar Pradesh and Bihar
   (2) Andhra Pradesh and Tamil Nadu
   (3) Rajasthan and Madhya Pradesh
   (4) Assam and Tripura

3. The state of India with maximum of its land as wasteland is:
   (1) Rajasthan (2) Gujarat
   (3) Madhya Pradesh (4) Uttarakhand

4. More than 50 per cent Indian wastelands are:
   (1) Water eroded (2) Forest degraded
   (3) Wind eroded (4) Saline and alkaline land

5. Replanting forests over areas where they were destroyed or degraded refers to:
   (1) Forestation (2) Aforestation
   (3) Reforestation (4) Deforestation

Answers

1. (3) *2.* (3) *3.* (3) *4.* (1) *5.* (3)

42

CHAPTER

Consumerism and Waste Products

LEARNING OBJECTIVES
Introduction • World Consumerism Patterns • Environment Protection Based Consumerism • Eco-labelling, Environment-Friendly Products • Prohibiting Polybags

Introduction

One of the prime reasons for our destructive impacts on the global environment is our consumption pattern and disposal of wastes. 'Fordism' of 1920s to 1960s emphasised mass production, mass consumption, corporate control and resource exploitation. But since 1960s, economists, environmentalists and various thinkers have raised the question on growing consumerism, *i.e.*, excessive consumption stimulated through marketing. Earlier, societies used to consume much less resources but with the dawn of industrial era, consumerism has shown an exponential rise. The modernisation in the life styles of the people has further increased the demand of consumer products and their consumption. The Indian society is fast adopting to western societies where packed foods and other goods are a common practice. This leads to the generation of solid wastes. Solid wastes are resulted mainly from industries, agriculture, urban localities and domestic activities. The waste products may be sewage, industrial wastes, agricultural wastes, hazardous wastes, packing refuse such as polybags etc.

World Consumerism Patterns

World consumerism patterns vary in more developed countries and less developed countries. In more developed countries, population size is smaller while resources are in abundance and due to luxurious life styles, the people living in these countries consume much more resources than the people living in less developed countries. More the consumption of resources, more is waste generation and greater is the degradation of the environment. In less developed countries, the condition is just reverse but the excessive population living in these countres causes more consumption of limited resources. Since their over all consumption is high, they generate more waste products. The sum total of environmental impact of these two types of consumerism may be the same or may be greater than of more developed countries. Although consumption patterns vary from country to country, but USA is known for the highest rate of consumerism. The throw away attitude and luxurious life styles of the people of western countries not only increase their consumption rate but also the rate of waste generation and environmental degradation. There is a vast difference in consumerism and waste generation pattern between India and USA (table 42.1).

TABLE 42.1. Consumerism Pattern in USA and India

	Per Cent Global Value	
Parameter	USA	India
Population	4.7	16
Production of Goods	21	1
Energy Use	25	3
Pollutants/ Wastes	25	3
CFCs Production	22	0.7

It is evident from the table 42.1 that although population of India is 3.4 times more than that of USA, its over all energy use and waste generation are less than $1/8^{th}$ that of USA. Thus more consumption leads to more waste generation. On an average a US citizen consumes 50 times as much as an Indian. A US borne baby, due to high consumption pattern, will damage the earth's environment 20 to 100 times more in its life time than a baby borne in a poor family of a less developed country. A Japanese with similar life style, as that of an American, causes half the impact on environment due to improved eco-friendly technology, adoption of energy efficiency, and following 3R principle (*Reduce*, *Reuse* and *Recycle*), they have minimised waste generation.

Strategies for Environmental Protection Based Consumerism

Efforts to reduce environmental degradation and health hazards due to industrialisation and to stimulate the development of clean technologies, are on. Rapid unplanned urbanisation and changing consumption patterns in the race to achieve better living standards, is causing tremendous stress on the environment. The time has come for consumers to take the lead in prompting manufactures to adopt clean and eco-friendly technologies and environmentally safe disposal of refuse with preventive and mitigative approaches.

The question arises-How will people, the consumers and the businessmen-the suppliers to consumers, will shift to something more supportive of environmental goals? The time has come when the consumers, the manufactures, the suppliers and the government will arrive on a common platform where each will play its own role. Some of the efforts made in this direction are:

(1) **Industrial Ecology**. Application of the ecosystem concept to industry, *i.e.*, to link the 'metabolism' of one company or body to others, is not new. Many groupings of companies have already adopted this practice. Industrial ecology works on the principle that waste and pollution are uneconomical and harmful and seeks to 'dovetail' them with raw material, *i.e.*, whenever possible industry should use by products and go beyond the minimization of wastes to make use of what remains by the producers or other bodies. For example, in Sweden and Denmark, agricultural waste, sewage and house hold refuse disposal are generally integrated with district heating and electricity generation. Similarly, in Kalundburg (Denmark), many industries are interlinked and the by product of one forms the raw material for the other.

(2) **Green Marketing**. By early 1980s, some companies and public bodies had recognised that a satisfactory green image could improve public relations and provide a marketing niche. And there are manufactures that have gained from this. These companies offer genuinely improved products in various respects. For example, in 1980s, world famed Mc Donald commissioned an environmental audit and shifted from plastic packing foamed with CFCs to environment friendly card board. Similarly, companies started refrigerators that consume less electricity, do not leak CFCs or use alternatives and companies that manufacture equipments for monitoring and managing environmental quality.

(3) **Consumer Protection Bodies**. Paralled to the growth in green marketing, arose the concept of green consumerism. The consumer protection bodies have been active since 1960s. They have not been restricted to only developed nations. For example, the Consumer's Association, Penang, Malaysian body, has been active in its own country and also worked for the consumer's rights elsewhere.

(4) **Pigouvian Taxes**. Making manufacturers responsible for some or all costs of recycling or waste disposal is one way of encouraging waste reduction and industrial ecology. Pigouvian taxes, named after the 1920's UK economist Arthur Pigou, aims to ensure that a

manufacturer pays all costs from raw material and energy provision to final collection and recycling, that inhibits the use of cheap packaging materials.

Box 42.1 Kodaikanal Mercury Case

Large and powerful business organisations are vastly affecting the environment for their profit motive such as in the case of Hindustan Lever Ltd., a subsidiary of the Anglo Dutch multinational Unilever which moved its mercury thermometer factory to Kodaikanal, Tamil Nadu in 1983 from Western Town, New York which dumped tons of mercury waste from broken thermometer on a dirt lot near the factory. In normal case this waste would have slowly contaminated the soil and caused many problems for the years. But the efforts of local people, in collaboration with Tamil Nadu Pollution Control Board (TNPCB), got the mercury waste sent back to United States for recycling on May 2003. Keeping in view the high mercury levels in the Kodaikanal, in September 2004, the Hazardous Waste Monitoring Committee, appointed by Supreme Court, visited the site and ordered the clean up of the affected area as well as medical aid for the affected people. The case study throws light on as to how developed nations are causing damage to the environment of developing countries.

Eco-labelling (Eco-Mark)

Eco-labelling is the marketing of the goods just to indicate that they are environment friendly. The objectives of eco-labelling are influencing the consumer's behaviour, helping them identify the environmental impacts of the products and encouraging manufacturers to produce goods that reduce the effects of their products on the environment. In most cases the product is judged, without formal eco-auditing against similar goods by an independent agency to establish whether it has less environmental impact.

Germany is one of the first countries to introduce eco-labelling of "Blue Angel" relied on a jury of experts by Federal Environment Ministry. Now many countries such as Canada, Japan, Sweden, USA and India have adopted eco-labelling as a mark of display on packaging and advertisement of goods.

To increase consumer awareness, ECO-MARK scheme was launched by the Ministry of Environment and Forests, Government of India in 1991 for labelling of environment friendly consumer products which meet certain environmental criteria along with quality requirements of the Bureau of Indian Standards (BIS). The Ministry continued to encourage activities for building awareness for the scheme. During the period, a project sponsored by the Ministry for comparative testing of five products namely copier paper, exhaust fans, fluorescent lamps, dry cell batteries (zinc chloride and alkaline) and edible oils was completed. Considering the need for review of the ECO-MARK scheme to make it more effective, a workshop was organised by the Central Pollution Control Board on "Future of Eco-labelling" in November 2005 in which representatives of concerned Ministries, Industry Associations, Consumer Groups, Manufactures and other stakeholders actively participated. The scheme is being revived with a view to introduce necessary changes required for its promotion as also to bring it in the line with the procedures in other countries. The eco-mark is "Earthen Pot" for easy identification of environment friendly products for household or other consumer products which fulfill Indian environmental standard requirements for that product.

Eco-labelling regulations encourage manufacturers to reduce the environmental impact of their products and inform consumers about the environmental performance of their products. It also helps products to obtain an advantage over non-labelled competitors. Eco-mark helps consumers identify environment-friendly products because such a product while it is made, used, recycled or disposed significantly reduces the harm it would otherwise had caused to the environment. The avenue of eco-labelling or Eco-Mark will hopefully be pursued for attracting customers.

Environment-Friendly Products

Now environmental awareness is catching on. People are opting to buy the products that are eco-friendly (environment-friendly). The issue of environmental protection has brought consumers, producers, distributors and government to a common plateform. Any consumer item which is made, used, recycled or disposed in such way that it significantly reduces the harm to the environment is called an eco-friendly product. The criteria for an eco-friendly product are based on *cradle-to-grave* approach, *i.e.*, from raw material to manufacturing, use and to disposal. The basic principles cover broad environmental levels and aspects specific at product level. The products are examined in terms of the following environmental aspects-

(i) That they have substantially less potential for pollution than other comparable products on being manufactured, used and disposed.

(ii) That they are recyclable, made from recycled or bio-degradable products, whereas comparable products are not.

(iii) That they make significant contribution to saving non-renewable natural resources including energy resources (coal, oil etc.) in comparison to comparable products.

(iv) That they must contribute to a reduction in the adverse primary criteria, *i.e.*, the environmental impact associated with the use of the product and its disposal and which will be set for each of the product category.

A scheme that labels environment friendly products would surely help the consumers know which product is eco-friendly and has less environmental impacts as each consumer product that is available in the market has some or the other impact on the environment. The scheme will enable the consumer help in preserving the environment and saving the planet.

Environment-friendly products will step out the process of consumerism and preserve one's individual freedom and extend a choice of exercise as to how one wants to live. The problem of disposal of variety of products of modern technologies such as batteries, electronic good's refuse, polybags etc. has became headache even for industrially developed and technologically advanced countries.

Prohibiting Polybags

Growing public awareness about the environmental impact of plastics has resulted a growing lobby of anti-plastic activities worldwide. The more we use polybags, the more we dump them into environment something that would not go away for a long long time. Polybags are difficult to dispose as they remain in the environment for a very long time. This is due to their non-biodegradable nature. Despite, many superiorities than any of their competitors, they are harmful to the environment as they do not disintegrate like other natural organic materials. This non-biodegradable nature of polybags makes them unfit for use. They keep on accumulating in the environment for a very long time and are difficult to dispose.

Today, polybags have become the main cause of blockages in the municipal sewerage and drinking water pipelines and pose serious problems in city garbage disposal. They are damaging many water bodies such as lakes and rivers. Many tourist places and even nature reserves have large accumulations of these polybags. Their use as carrybags has become a fashion in modern society.

Polybags pose a serious threat to the environment by way of remaining in the soil for considerable period and causing problems at all stages of their disposal such as melting, burning, recycling or dumping, during which they cause pollution to the environment.

After much deliberations, Ministry of Environment and Forests, Government of India, eventually banned the use of recycled plastic carrybags and polycotainers for packing and supply of food items. The notification issued under the Environment (Protection) Act, 1986, prohibits use of carrybags or containers made of recycled plastic for storing carrying or packaging of food stuffs. The notification also lays down specification such as minimum thickness of carrybags made of virgin plastic not less than 20 microns in natural colour or white and use of colourants or pigments, if necessary, as per norms of BIS (Bureau of Indian Standards).

Chapter Summary

One of the prime reasons for the destructive impacts on global environment is our consumption pattern and disposal of wastes. Earlier, societies used to consume much less resources. Rapid development and adoption of western patterns have resulted exponential rise in consumerism. More developed countries with smaller population size consume much more resources than less developed countries and generate more wastes. Environment protection based consumerism can be achieved through industrial ecology, green marketing, adopting clean and eco-friendly technologies and producing environmentally safe products which are easy to dispose and cause no harm to the environment. Eco–labelling is the marketing of goods just to indicate that they are environment friendly. To increase consumer awareness, India has launched an eco-mark programme in 1991 for specific products. Indian eco-mark is 'earthen pot' for packaging and advertisement purposes. Polybags are the most dangerous wastes generated by consumer products. These are non bio-degradable and remain in the environment for very long time and pose a serious threat to our planet.

Study Questions

1. What are world consumerism patterns and stategies for environmental protection based consumerism? Discuss.
2. Discuss briefly eco-labelling and eco-mark with reference to environment-friendly products.
3. Give a short account of:

(i) Industrial Ecology
(ii) Pigouvian Taxes
(iii) Green Marketing

Objective Questions. *Select the correct answers*

1. The eco-mark of consumer products in India is:

(1) Blue Angel (2) Earthen Pot
(3) Lotus Flower (4) Tiger

2. The use of polybags was banned in India through a notification by Government of India in the year:

(1) 1970 (2) 1973
(3) 1975 (4) 1986

Answers

1. (2) *2.* (4)

43

CHAPTER

Environment Protection Acts

LEARNING OBJECTIVES

Introduction • Water (Prevention and Control of Pollution) Act, 1974 • Central Pollution Control Board (CPCB) • State Pollution Control Board (SPCB) • Air (Prevention and Control of Pollution) Act, 1981 • The Environment (Protection) Act, 1981 • The Environment (Protection) Act, 1986 • National Environment Tribunal Bill, 1992 • Environmental Impact Assessment, 1994

Introduction

Environmental laws play a crucial role in providing a framework for regulatory use of the environment and its management. It regulates resource use, protects environment and biodiversity; mediates conflicts, resolution and conciliation; and formulates stable unambiguous undertakings and agreements. Environmental laws encourage satisfactory performance and enable authorities to punish those who infringe environmental legislation or confiscate equipment that is misused or is faulty or close a company. They also provide opportunities for employees, bystanders, producers or service users to sue for damages, if they are harmed. Environmental law concerns regulation (protection or prohibition) of activities affecting the environment. Environment is a global issue. The environmental law has unique feature among all subject matters of law. No matter as to how peculiar is an environmental issue to an area or a nation, it always has an international/global repercussions.

Some of the countries such as Sweden, the Netherlands, USA, UK, Canada, Australia and New Zealand have been quite active in framing and implementing environmental laws. Some environmental laws are quite old, such as UK had local pollution control laws as early as twelveth century and nationally enforced pollution control legislation like the Alkali Act (1863), about one and half century old. Ancient Indian rulers promulgated control on hunting and forest felling centuries ago. India is the first country in the world to have made provisions in its constitution for the protection and conservation of the environment. Several Acts were passed from 1972 to 1986 and various amendments thereafter.

Environmental Legislation in India

India is the first nation in the world to have made provisions for the protection and conservation of environment in its constitution. Environment was first discussed as an agenda at an international platform on 5 June, 1972 in UN Conference on Human Environment in Stockholm. Thereafter, 5 June is celebrated as *World Environment Day* every year all over the world. Soon after the conference, the country took legislative steps for the protection of the environment and passed a series of Acts such as Wildlife (Protection) Act, 1972; Water (Prevention and Control of Pollution) Act, 1974; Forest (Conservation) Act, 1980; Air (Prevention and Control of Pollution) Act, 1981 and subsequently Environment (Protection) Act, 1986.

The provisions for environmental protection in the constitution were made in 1976 within four years after the Stockholm Conference 1972, through the fourty second Amendment imposing environmental protection and conservation as the fundamental duty of all citizens of the country. Article 48-A of the constitution provides: "It shall be the duty of every citizen of India to protect and improve the natural environment including forests, lakes, rivers and wildlife and to have compassion for living creatures."

WATER (PREVENTION AND CONTROL OF POLLUTION) ACT, 1974

Keeping in view the critical problems on quality and quantity of water, India enacted *Water (Prevention and Control of Pollution) Act, 1974* and corresponding *Water (Prevention and Control of Pollution) Rules, 1975*, mainly for maintaining and restoring wholesomeness of water by preventing and controlling its pollution. For effective implementation, the various provisions of this Act have been amended such as in 1977, 1988 and so on.

Water pollution is defined as "such contamination of water, or such alteration of the physical, chemical or biological properties of water, or such discharge as is likely to cause a nuisance, or render the water harmful or any injurious effect to public health and safety or harmful for any other use or to aquatic plants or other organisms or animal life." The definition covers entire probable agents in water that may cause any harm or have a potential to harm any kind of life or property in any way such as for the purpose of domestic use or use in agriculture, industries or otherwise.

Objectives of the Act

The principal objectives of the Act are:

(i) Prevention and control of all types of water pollution.

(ii) Maintenance and restoration of wholesomeness of waters of all types such as surface water and ground water.

(iii) Establishing Central and State Boards for water pollution control.

Functions and Powers of Boards

The Act is applicable to both Central Pollution Control Board (CPCB) and State Pollution Central Boards (SPCBs), which coordinate with CPCB. The CPCB has a full-time Chairman (nominated by the Central Govt.) whereas SPCB may have full or part-time Chairman (nominated by the State Govt.). There can also be a joint board between any two or more State Governments. A board may constitute as many committees as are necessary for effective implementation of various provisions of the Act. Under Section 8, the Act directs the Boards to have atleast one meeting in every three months.

Cental Pollution Control Board (CPCB)

The main functions of CPCB (under section 16-A) are:

(1) Advising Central Government in matters related to prevention and control of water pollution.

(2) Coordinating the activities of State Pollution Control Boards, providing technical assistance and guidelines and resolving disputes among them, as and when arising.

(3) Organising training programmes for persons engaged in prevention and control of water pollution.

(4) Organising comprehensive programmes on water pollution-related issues involving mass media.

(5) Collection, compilation and publishing of technical and statistical data related to water pollution.

(6) Preparation of manuals or codes or guidelines for treatment and disposal of sewage and trade effluents.

(7) Prescribing standards for water quality parameters.

(8) Planning nation-wide programmes for prevention control and abatement of water pollution.

(9) Establishing and recognising laboratories for analysis of water samples and samples from any stream, well, trade effluent and sewage effluent.

State Pollution Control Board (SPCB)

The functions of SPCB (under section 7-B) are:

(1) Advising the respective state government regarding location of any industry that might pollute the nearby water bodies.

(2) Prescribing standards for sewage and trade effluents to be discharged into any adjacent water body such as river, stream, canal etc.

(3) Planning a comprehensive programme for prevention, control and abatement of pollution of rivers, streams, wells etc.

(4) Collaborating with CPCB for organising educational programmes and disseminating information for water pollution control through mass media.

(5) Promoting and conducting research on issues for evolving economical and effective measures of treatment of sewage and trade effluents and their disposal.

(6) Inspecting trade effluents and waste water treatment plants.

(7) Establishing or recognising laboratories for analysis of water sample and samples from river, streams, wells etc.

(8) To perform such functions as may be assigned or entrusted by CPCB or state government.

(9) Collecting legal samples of trade effluents in presence of the occupier or his representative in accordance with the procedure laid down in the trade, *i.e.*, sample divided into two parts, sealed and signed by both the parties and sent for analysis to some recognised laboratory. If the sample does not conform to the prescribed water quality, *i.e.*, it crosses the maximum permissible limits, then the consent is refused to the trade unit.

(10) Every industry or the trade has to secure Board's consent, fixed for a specific period, on a prescribed proforma for furnishing all technical details along with a prescribed fee following which analysis is carried out.

(11) The Board suggests efficient methods for utilisation, treatment and disposal of trade/industry effluent.

(12) The Act is empowered to restrict expansion and new outlets in the trade units.

Powers of the State Government

The principle powers of state government are:

(*i*) Under section 20.2, SPCB many carry out surveys like measurements or seek information for the purpose of performing functions under the Act.

(*ii*) Under section 21 (1) A, the state government has the power to collect samples of water of any neighbouring stream or well or any effluent being discharged into such river, stream or well for analysis.

(*iii*) Under section 22.4, the state government has the power to obtain a report of the results of the analysis by a recognised laboratory.

(*iv*) Under section 23.5, the SPCB is empowered by the state government to enter any place for the purpose of performing any of the functions entrusted to it

(*v*) Under section 24.6:

  (*a*) No person shall knowingly allow the entry of any matter into any stream, which may impede the proper flow of water resulting in substantial aggravation of water pollution.

  (*b*) No person shall knowingly allow the entry of any toxic, noxious or polluting matter directly or indirectly into any stream, well, river or on land.

(*vi*) Under section 25.7:

  (*a*) No person shall establish any industry, operation or any treatment disposed system, which is likely to discharge any effluent or sewage into river stream, well or on the land.

  (*b*) No person shall begin to make any new discharge of sewage.

  (*c*) No person shall use any new outlet for discharge of sewage.

(*vii*) The board can also levy and collect a cess on water consumed by industries and authorities for increasing Board's resources under the Water (Prevention and Control of Pollution) Cess Act, 1977.

Penalties for Violating the Provisions of the Act

Various penalties for violating the provisions of the Act are:

(*a*) In case of failure to provide information by a person discharging effluents into streams or well or regarding construction or establishment of an effluent treatment and disposal system, the penalty is imprisonment upto three months or fine upto Rs. 10,000 or both. If omission continues, the offender can be penalised with an additional fine upto Rs. 5000 per day.

(*b*) In case of violation of order for prohibiting discharge of any polluting matter into stream, well or on land, Board can order for closure of industry, or stoppage of water, electricity supply etc. The penalty is imprisonment for one and half years to six years or fine or both. If violation continues, the penalty is an additional Rs. 5000 per day, and if continues beyond one year, then the penalty is imprisonment of seven years. On conviction, the name of the offender is published in the newspapers at offender's expense.

(c) In case of destroying or damaging Board's property, obstructing Board's functions, failure to provide information about accidents under section 31, giving wrong information or making false statements to get Board's consent, the penalty is imprisonment upto three months or fine upto Rs. 10,000 or both.

AIR (PREVENTION AND CONTROL OF POLLUTION) ACT, 1981

It seems that the air pollution has come to the notice of the legislature in India slightly later, *i.e.*, seven years after Water Act. Interestingly, control of water pollution was country's own initiative, whereas the measure on air pollution came only in response to Stockholm Conference, 1972 when it was urged to all participating countries in the Conference to take appropriate steps for the prevention of natural resources of the earth, including the prevention of the quality of air and control of air pollution. The salient features of the Act are:

Objectives of the Act

The objectives of the Act are:

(i) Prevention, control and abatement of air pollution.

(ii) Maintenance of the quality of the air.

(iii) Establishment of Boards for prevention and control of air pollution.

Salient Features of the Act

In order to have an integrated approach for tackling the problems related to air pollution, Air (Prevention and Control of Pollution) Act 1981 provides that the Central and State Boards, established under the Water (Prevention and Control of Pollution) Act, 1974, shall also act as Air Pollution Control Boards under this Act. But in those states where State Boards for Water Pollution and Control are not established, separate State Boards for Air Pollution and Control shall be established. The functions of Central and State Pollution Control Boards for the control of air pollution are also identical to the functions of Central and State Pollution Control Boards for the control of water pollution, except that their aim will be to control air pollution instead of water pollution. The provisions regarding the Board's composition, essential qualifications of the members, meetings, committees, funds, accounts, audits, penalties and procedures are like those of Water (Prevention and Control of Pollution) Act, 1974.

Powers of the Boards

The CPCB and SPCB established under the Water (Prevention and Control of Pollution) Act, 1974 exercise the powers of prevention, control and abatement of air pollution. Noise pollution has also been included in the Act, amended in 1987. The powers of the Boards are as follows:

(*i*) Under sections 17 regarding the discharges of emission of any air pollutant, the Boards have to check whether or not the industry strictly follows the norms or standards laid down by the Board. Based upon analysis report, consent is granted or refused to the industry concerned.

(*ii*) Under section 19.1, the State Government, after consultation with SPCB, may:

(*a*) Declare any area or areas as Air Pollution Control Area/Areas by notification in the official gazette.

(*b*) Prohibit the use of any type of fuel or kind of appliance causing or likely to cause air pollution in the Air Pollution Control Area.

(*c*) Prohibit burning of any material causing or likely to cause air pollution in Air Pollution Control Area.

No person shall, without the permission of SPCB, operate or establish any industrial unit in the Air Pollution Control Area.

(*iii*) Under section 20.2, the Boards have the power to ensure that the laid down emission standards of air pollutants from automobiles are complied with. Based upon this, the state government is empowered to issue instructions to the authority incharge of registration of motor vehicles, under Motor Vehicle Act, 1939, that is bound to comply with such instructions.

(*iv*) Under section 21.3, no person shall, without the pervious consent of the SPCB, establish or operate any industrial plant in the Air Pollution Control Area.

(*v*) Under section 22.4, no person, operating any industrial plant in any Air Pollution Control Area, shall discharge or cause or be permitted to discharge the emission of any air pollutant in excess of the laid down standards by the SPCB.

(*vi*) Under section 24.5, any person empowered by SPCB shall have the right to enter any place for-

(*a*) Performing any of the functions of the Board entrusted to him.

(*b*) Examining or testing any industrial plant, record, register, document, control equipment etc. and

(*c*) Seizing any such control equipment, if he has reasons to believe that it may furnish evidence of an offence punishable under Act.

(*vii*) Under section 26, any person empowered by the Board shall have the power to take samples of air or emission from any chimney, duct or flue etc. for analysis purpose and send the same to the established or recognised laboratory by the SPCB for analysis.

(*viii*) Under section 28 of Water Act, 1974 and 31 of Air Act, 1981, a provision has been made for appeals before the Appellate Authority consisting of a single person or three persons appointed by State Governor within 30 days of passing order.

Penalties for Violating Provisions of the Act

The Penalties for violation or defaults under the Act are the same as under Water (Prevention and Control of Pollution) Act, 1974, except that there is no provision for publication of name/names under this Act.

Drawbacks of the Act

Some of the drawbacks of the Act are:

(1) The main drawbacks of this Act lies in giving the offender or defaulter 60 days notice before taking him to the court and thus by the time the case is filed in the court, the offender or defaulter may destroy the evidence, in that case it may be difficult to prove the offence.

(2) No consent or permission is required to be taken from SPCB for establishing an industry outside the Air Pollution Control Area, even though its emission might be reaching the Air Pollution Control Area.

(3) The emissions of air pollutants by an aircraft or ship are excluded from the purview of the Act.

THE ENVIRONMENT (PROTECTION) ACT, 1986

The Environment (Protection) Act, 1986, introduced on 19 November the birth anniversary of our Late Prime Minister Smt. Indira Gandhi, in wake of the Bhopal disaster, 1984, is the pioneer Environment Protection Act in India. Corresponding to Environment (Protection) Act, 1986, were framed Environment (Protection) Rules, 1986, empowering the Central Government to prevent, control and abate environmental pollution and to State Governments to co ordinate the actions or the effective implementation of the Act. Its various provisions have been amended from time to time. The Act extends to whole of India.

Important Terms Used in the Act

Some important terms related to environment and used in the Act are as:

(*i*) **Environment**: Includes air, water and land and the interrelationship that exists among and between them and human beings, all other living organisms and the property.

(*ii*) **Environmental Pollutant**: Includes any solid, liquid or gaseous substance present in such concentrations as may be injurious to environment.

(*iii*) **Environmental Pollution**: The presence of any environmental pollutant in the environment.

(*iv*) **Hazardous Substance**: Any substance or preparation which is liable to cause harm by its physico-chemical properties or handling to human beings, other living creatures, property or environment.

Objectives of the Act

Following are the objectives of the Act:

(*i*) Protection and improvement of the environment.

(*ii*) Prevention of hazards to all living creatures (plants, animals and humans) and property.

(*iii*) Maintenance of harmonious relationship between humans and their environment.

Some Details of the Act

The Act has given power to the Central Government to take measures to protect and improve environment while State Governments shall coordinate the actions. Most important functions of the Central Government, for the protection and improvement of environment, for the prevention, control and abatement of environmental pollution and prevention of hazards to humans, plants, animals and property under section 3 of this Act include:

(1) To instruct every state to set up 'Green Bench' courts to attend to Public Interest Litigation (PIL) cases concerning environmental hazards affecting the quality of life of the citizens. The 'Green Bench' courts have been empowered to settle the case quickly and provide legal redress to the citizens.

(2) To plan and execute nationwide programmes for the prevention, control and abatement of environmental pollution.

(3) To coordinate activities of State Government Officers and other authorities under this Act and any other related law.

(4) To lay down standards provided in the schedule to the Environment (Protection) Rules, 1986 for quality of environment in its various aspects and for emission or discharge of environmental pollutants from different sources.

(5) To demark and segregate industrial areas from non-industrial areas.
(6) To lay down procedures and safeguards for the handling of hazardous substances for prevention of accidents causing environmental pollution and remedial measures.
(7) To examine the manufacturing processes, materials and substances likely to cause environmental pollution.
(8) To inspect any premises, plant, manufacturing process, equipment or machinery and give direction to prevent, control and abate environmental pollution.
(9) To establish and recognise environmental laboratories and institutes to evolve standardised methods for sampling and analysis of various types of environmental pollutants, to analyse samples, to carry out investigations, to lay down standards for quality of environment and discharge of environmental pollutants, to monitor and enforce the standards set, and to report periodically to the Central Government.
(10) To carry out and sponsor investigation and research in environmental pollution problems.
(11) To prepare manuals, codes or guides to disseminate collected information in matter relating to environmental pollution and its prevention, control and abatement.
(12) To take up any matter, if necessary, for the purpose of performing such functions and powers of Central Government under section 4.2 of this Act.

Further, to regulate environmental pollution, the Central Government may, by notification in official gazette, make rules in respect to call or any of the matters referred in section *3a* to *3l* under section 6.3 of the Act.

Provision for the Prevention, Control and Abatement of Environmental Pollution

For the prevention, control and abatement of environmental pollution, the Act has following important provisions:

(a) Under section 7.1, no person carrying on any industry, operation and process shall be permitted to discharge any environmental pollutant beyond the permissible limits.
(b) Under section 8.2, no person shall handle hazardous substances without complying with the prescribed procedural safeguards.
(c) Under section 9.3, a person is responsible to mitigate environmental pollution and to intimate any occurrence relating to environmental pollution to the concerned authorities and also assist those authorities in preventing or mitigating environmental pollution.

(*d*) Under section 10, the Central Government and its officers have the power to take samples of air, water, soil or substances from the industry or a place for analysis according to the procedures laid down in the Act.

(*e*) Under section 12, the Central Government has the power to establish environmental laboratories or recognise any laboratory or institute as an environmental laboratory.

(*f*) Under section 13, the Central Government may appoint or recognise government analyst for purpose of analysis of samples of air, water, soil or any other substance.

(*g*) Under section 14, the report signed by a government analyst may be used as an evidence of the facts stated therein in any proceeding under the litigation.

The Act also made it clear that the expenses incurred with respect to remedial measures shall be recoverable from the person responsible. Under this Act, the Central Government also made The Hazardous Waste (Management and Handling) Rules, 1989. An Amendment was made in 1994 in Environment (Protection) Rules for Environmental Impact Assessment (EIA) of various development projects.

Penalties for Violation of the Act

Various penalties for violation of the provisions of the Act are:

(1) Under section 15.5, any person violating any of the provisions of the Act shall be punishable with imprisonment for a term which may extend up to five years or with a fine which may extend up to one lakh repees or both. In case the violation continues beyond a period of one year after the date of conviction, the offender shall be punishable with imprisonment for a term of seven years.

(2) Under section 16, the criminal liability is also fixed on the company's directors and principal officers in case of an offence being committed by a company.

(3) Under section 17, the criminal liability is also fixed on the heads of the department of the government where an offence is committed by the concerned department and the head of the department is unable to prove that the offence was committed without his knowledge or that he exercised all diligence to prevent such offence.

National Environment Tribunal Bill, 1992

The bill, to provide strict liability damage arising out of any accident occurring while handling any hazardous substance and for the establishment of National Environment Tribunal for effective and expeditious disposal of case arising from such accident, with a view of giving relief and compensation for damage to persons, property and the environment and for matters connected therewith or incidental thereto, was enacted in 1992. The decisions were taken at Rio Conference, 1992 in which India participated calling upon govenments to develop national laws regarding liability and compensation for the victims of pollution and other environmental damage. The Bill was passed by the Parliament of India in 1992.

ENVIRONMENTAL IMPACT ASSESSMENT, 1994

Analysis of any possible change in the environmental quality, adverse or beneficial, caused by a development project of government or private sector is known as Environmental Impact Assessment (EIA). The programme of EIA was taken up by the Ministry of Environment and Forests, Govt. of India in 1977 in view of Planning Commission's suggestions that all development projects should evaluate themselves from environmental point of view. In 1992, the Ministry, in consultation with all other Ministries issued a Draft Notification making environmental impact assessment statutory for selected activities. This was finalised in 1994.

As a matter of government policy, it is compulsory for any enterprise, whether it is government or private, to include EIA in the planning stage of any development project and submit it to the Central Government for clearance. All major and minor irrigation projects and all highly polluting industries were subjected to EIA for their initiation.

After the issue of the notification in 1994, an amendment was made in 1997 delegating the powers to State Governments in respect to certain categories of thermal power plants.

Environmental Impact Assesement (EIA) is aimed at sustainable development using optimal natural resources without causing damage to the environment. Environmental considerations are required to be integrated in planning, designing and implementation of development projects. It is one of the proven management tools for incorporating environmental concerns in development process and also in improved decision making.

During 2005-06, new project relating to construction of townships, industrial townships, settlement colonies, commercial complexes, hotel complexes, hospitals, office complexes for 1,000 persons and above, a discharging sewage of 50,000 liters/day and above/or with the investment of 50 crores and above, and new industrial estates having an area of 50 hectares and above and the industrial estate irrespective of area, if their pollution potential is high, were brought under the preview of EIA notification published on 7 June, 2004.

Chapter Summary

Environmental law plays a crucial role in providing a framework for regulating the use of environment and its management. In India, several acts were passed from 1972 to 1986 and various amedments thereafter to protect the environment, regulate resource use, protect bio diversity, pollution control, conserve forests etc. India is the first nation in the world to have made provisions for the protection and conservation of environment in its constitution in 1976 within four years of the Stockholm Conference, 1972. Water (Prevention and Control of Pollution) Act was enacted in 1974 for maintaining and restoring wholesomeness of water by preventing and controlling its pollution through Central and State Pollution Control Boards. Air (Prevention and Control of Pollution) Act was enacted in 1981 for prevention, control and abatement of air pollution through Central and State Pollution Control Boards. Persons violating provisions of Water and Air Acts are imposed penalties. The Environment Protection Act, 1986 was introduced in the wake of Bhopal disaster 1984, for the protection and improvement of the environment, prevention of hazards and maintenance of harmonious relationship between humans and their environment. In 1992, soon after Rio Conference, National Environment Tribunal Bill was passed and enacted to develop national laws regarding liabilities and compensation for the victims of pollution and other environmental damages. Environmental Impact Assessment, a Draft Notification, to evaluate all development projects from environment point of view, was initiated by the Ministry of Environment and Forests in 1992 and was finalised in 1994.

Study Questions

1. Discuss the state of environmental legislation in India.
2. What are various powers of Central and State Pollution Control Boards of India and penalties for violating the provisions of the Pollution Control Acts? Explain.
3. Discuss Air (Prevention and Control of Pollution) Act, 1981.
4. Discuss Water (Prevention and Control of Pollution) Act, 1976.
5. Discuss the Environment (Protection) Act, 1986

Objective Questions. *Select the correct answers*

1. National Environmental Tribunal Bill was enacted in the year:

(1) 1985 (2) 1990
(3) 1992 (4) 1995

2. Environmental Impact Assessment was finalised by the Ministry of Environment and Forests, Government of India in the year:

(1) 1970 (2) 1994
(3) 1990 (4) 1995

Answers

1. (3) *2.* (2)

44

CHAPTER

Forest Conservation and Wildlife Protection Acts

LEARNING OBJECTIVES

Introduction • FOREST (CONSERVATION) ACT, 1980: National Forest Policy • Powers of State Government • Penalties for Violating the Provisions of the Act, 1988 and 1992 Amendments • WILDLIFE (PROTECTION) ACT, 1972 • Declaration of Sanctuary and National Parks • Penalties for Violating the Provisions of the Act, Wildlife (Protection) Amendment Bill, 2002

Introduction

India is a unique country with reverence to nature and respect to environment inherent in its cultural ethos. *Isha Upanishad* states "The whole universe together with its creatures including plants and animals belongs to the Lord." The pre-historic Aryans were worshipers of nature and preferred sylvan surroundings. The epic *Vedas* and *Upanishads* were composed by our sages singing "May the God, the Water, the Plants and the Forest Trees accept our prayer ..... May the blessings of our simul (tree) protect us forever". The *Puran* said "One tree is equal to ten sons" and Gautam Buddha has preached not to harm trees. Founder of *Jainism*, Lord Mahavir has told not to harm to creatures. Kautilya's famous treatise *Arthashatra* describes what may be considered as the World's first forest Conservation and Wildlife Management programme. *Ahimsa Parmadharma* is ingrained in our culture. Mauryan Kings maintained forests for different purposes. Mughal emperors promoted tree plantation and development of gardens.

India had a Forest Act in 1927 classifying forests as reserve forests, protected forests, village forests and forests restricted for hunting and authorised establishments or sanctuaries. In 1976, 42nd amendment to the constitution of India transferred forests from state list to the concurrent list, *i.e.*, all India basis. The action plan for wildlife objectives, as an outcome of Wildlife (Protection) Act, 1972, emerged at par with Stockholm Conference, 1972, but Wildlife Protection Rules came into force in 1973.

Forest related legislation in India came quite late as Forest (Conservation) Act 1980 and Forest (Conservation) Rules as legal binding or forest conservation in 1981, thirty three years after the independence. The legal binding for wildlife protection, although initiated quite early as a part of Forest Act, 1927, with the provision of establishment of sanctuaries, marched ahead with the onset of the first National Park, the Corbett National Park in 1936.

FOREST (CONSERVATION) ACT, 1980: NATIONAL FOREST POLICY

This important Act has received very little public attention and is focussed more on the proposed Forest Bill amending the Forest Act, 1927 which stipulates that no forest land or any portion thereof may be used for non-forest purposes without the permission of Central Government. The Forest (Conservation) Act, 1980 and the Forest (Conservation) Rules, 1981 provide for the protection and conservation of forests giving power to State Governments to regulate or prohibit, in any forest, the clearing of the land for cultivation, pasturing of cattles or the clearing of vegetation for multiple use.

The Act is adopted all over India except in Jammu and Kashmir. It covers all types of forests including reserve forests, protected forests or any forest land, irrespective of its ownership. Under this Act, State Governments have been empowered to use the forests only for forestry purposes. If the State wishes to use the forest for some other purpose, it has to take prior approval of the Central Government, after then only it can pass the orders for declaring some part of the reserve forest for non-forest purposes such as quarrying or for clearing some naturally growing trees and replacing them with economically important trees, *i.e.*, reforestation.

India is one of the few countries which has forest policy since 1894. It was revised in 1952 and then in 1980. The **aims** of Forest (Conservation) Act 1980 are:

(*i*) Maintenance of environmental stability through preservation and restoration of ecological balance,

(*ii*) Conservation of natural heritage,

(*iii*) Check on soil erosion and denudation in catchment areas of rivers, lakes and reservoirs, *(iv)* Check on extension of sand dunes in desert areas of Rajasthan and along coastal tracts,

(v) Substantial increase in forest tree cover through massive afforestation and social forestry programmes,

(vi) Steps to meet requirements of fuel wood, fodder, minor forest produce and timber for rural and tribal populations,

(vii) Increase in productivity of forests to meet the national need,

(viii) Encouragement to efficient utilisation of forest produce and optimum substitution of wood and

(ix) Steps to create massive people's movement with involvement of women to achieve the objectives and minimise pressure on existing forests.

Objectives of the Act

The objectives of the Forest (Conservation) Act, 1980 are:

(1) Protection and conservation of forests

(2) To ensure judicious use of forest produce

(3) To check diversion of forest land to non-forest purposes

Important Terms Used in the Act

Forest. A biotic community composed predominantly of trees, shrubs and woody climbers (lianas).

Forest produce. Timber, wood, bark, charcoal, oil, resin, catechu, natural varnish, lac, mahua, seeds etc., whether brought from or found in a forest or not.

Notification of Reserve Forests

Under section 3 of the Act, a forest land or a waste land may be constituted as a Reserve Forest by the State Government. after issuing a notification in the official gazette under section 4 of this Act:

(1) Declaring that it has been decided to constitute such a land as a Reserve Forest,

(2) Specifying the situation and limits of such a land and

(3) Appointing an officer not holding any other forest office except that of Forest Settlement Officer.

According to section 4 of this Act, the State Government may appoint not more than three officers, only one of whom shall be holding forest office except as aforesaid to perform the duties of a Forest Settlement Officer

The proclamation under section 6 of the Act, shall be such as to:

(i) Specify the situation and limits of the proposed forest,

(ii) Explain the consequences that will ensure the reservation of such a forest and

(iii) Fix a period of not less than six months from the date of such proclamation for claiming any right over any land or over forest produce to the Forest Settlement Officer in a written notice specifying the nature of such right and amount and particulars of the claimed compensation

Powers of the State Government

(a) Under section 5, the Forest Officers, with previous sanctions of the State Government, may stop any public or private way or water-courses in the reserved forest, provided that an alternate way or water course already exists or has been provided or constructed.

(b) Under section 32, the State Government is empowered to make rules for Protected Forests to regulate:
   - (i) Protection or management of any portion of a forest.
   - (ii) The cutting, sawing, conversion or removal of trees and timber and removal of forest produce from such protected forests.
   - (iii) The cutting of grass and pasturing of cattle in such forests.
   - (iv) Hunting, shooting, fishing, setting traps and poisoning water or shares in such protected forests.
   - (v) Granting of licence to inhabitants living in the vicinity of protected forests to take trees, timber and forest produce for their own use.
   - (vi) Inspection of forest produce passing out from these protected forests.
   - (vii) Clearing and breaking-up of land for cultivation or other purposes in such protected forests.
   - (viii) Protection from fire of timber lying in these protected forests.

(c) Under section 30, the State Government, by notification in the official gazette, may:
   - (i) declare any tree or trees as reserved in protected forests and
   - (ii) prohibit the quarrying of stone or the burning of lime or charcoal or collection or removal of any forest produce in the protected forests.

(d) Under section 35, the State Government, by notification in official gazette, may prohibit the following in any forest or wasteland:
   - (i) pasturing of cattle
   - (ii) clearing of the vegetation
   - (iii) breaking-up or clearing of land for cultivation

Some non-forest activities such as preservation of soil; maintenance of water supply in rivers, springs and tanks; protection against storms, winds, floods and landslides; protection of roads, railways, bridges and lines of power and telecommunication; preservation of public health; fencing; making water holes, trench, pipelines, check posts etc. are exempted.

Penalties for Violating the Provisions of the Act

Under section 33 of this Act, an imprisonment of six months or more or fine of Rs. 500 or more or both, is the penalty for any person who violates any of the provisions of this Act and commits any of the following offences:

(i) Falls, taps, lops or burns or damages any tree reserved under section 30.

(ii) Clears land in any protected forest for cultivation etc. prohibited under section 30.

(iii) Sets fire to such forests or kindles fire without taking all necessary precautions to prevent its spreading to any tree reserved under section 30 whether standing, fallen or felled.

(iv) Carries out quarrying of any stone or burns lime or charcoal or collects or removes any forest produce prohibited under section 30.

(v) Permits cattle to damage any reserved tree.

(vi) Violates any of the rules under section 32.

1988 Amendment in the Forest Act

The Forest (Conservation) Act, 1980 was amended in 1988 with major objective to incorporate welfare of forest dwellers. The amendment prohibits lease of forest land to any body other than government and removal of reserved trees through reforestation without the permission of the Central Government. It encourages conservation, planting and increase of forest cover to an average of 30 per cent across the country. The scope of existing non-forest purposes has been extended to other areas of cultivation of other crops such as tea, coffee, spices, rubber, palms, medicinal plants etc.

1992 Amendment in the Forest Act

Amendment made in the Forest Act in 1992 has extended the area of 'non-forest activities', without cutting of trees or limited cutting such as for setting of transmission lines, seismic surveys, exploration and drilling with permission of Central Government. For raising hydroelectric project involving large scale destruction, prior approval of the Central

Government is essential. Wildlife Sanctuaries and National Parks etc. are totally prohibited for any exploration or survey, even if no tree felling is required, without the prior permission of the Central Government.

Non-forest activities are not allowed in reserve forests. Cultivation of native species in the forest included in non-forest activity are exempted from prior permission. *Tuser* (a type of silk yielded by insect) cultivation in forest areas by tribals, as a means of their livelihood is treated as a forest activity as long as it does not involve some specific host tree like *Arjun*, in order to discourage monoculture practices. Plantation of mulberry, for rearing silkworm, is considered a non-forest activity because it involves monoculture practice. Mining is a non-forest activity and as such approval of Central Government is mandatory. Removal of stones, *bajari*, boulder etc. from riverbeds located within the forest area is a non-forest activity. Any proposal, for non-forest activity, send to Central Government must have a cost benefit analysis and Environmental Impact Assessment (EIA) of the proposed activity with reference to its ecological and socio-economic impacts.

In 2003, Ministry of Environment and Forests, Government of India, constituted the National Forest Commission for two years to review the working of Forests and Wildlife Sector and a National Forestry Action Programme (NFAP) as a comprehensive long-term plan for next 20 year to bring one third area of the country under tree/forest cover and to arrest deforestation.

WILDLIFE (PROTECTION) ACT, 1972

Wildlife protection in India is as old as its culture. Attempts have been made in the past to protect wildlife such as Madras Elephant Preservation Act, 1879, The Wild Birds and Animal Protection Act, 1912, Indian Forest Act, 1927 (restriction on hunting) and Wildlife National Park Act (establishment of sanctuary – Tiger Reserve the Corbett National Park, 1936). In India the Wildlife (Protection) Act came out as late as in 1972 and Wildlife (Protection) Rules in 1973, imposing ban on trade in rare and endangered species and the products obtained from them. The Act is adopted by all states of the country except Jammu and Kashmir (which has its own Act), and governs the wildlife conservation and protection of endangered species both inside and outside the forest areas.

India is a signatory to Convention on International Trade in Endangered Species of Wild Fauna and Flora (CITES), which came into force in 1975. Under the treaty, import and export of endangered animal and plant species are subject to strict control and commercial exploitation of such species in prohibited. The Wildlife (Protection) Act, 1972 was mainly to prohibit hunting of about 50 species of animals, 43 species of birds and several species of reptiles, amphibians, insects etc., specified in schedule of the Act. Wildlife was transferred from state list to concurrent list in 1976, thus giving power to Central Government to enact the regulation. With several amendments, the Act provides for the constituion of a Wildlife Advisory Board for regulation of hunting of animals, trade in wild animals including birds and establishment of santuaries and national parks and provide advice to the government about the wildlife affairs. Prime Minister is the Chairperson of the Board. Later on State Advisory Boards have also been constituted.

Objectives of the Act

The main objectives of the Act according to section 1 are:

(i) To maintain essential ecological processes and life supporting systems

(ii) To preserve biodiversity, and

(iii) To ensure a continuous use of species, *i.e.*, protection and conservation of wildlife.

Important Terms Used in the Act

Wildlife. 'any animal, bees, butterflies, crustacean, fish, moths, and aquatic and land vegetation which forms part of any habitat.'

Habitat. 'land, water or vegetation which is the natural home of any wild animal'.

Hunting.

(a) 'to capture, kill, poison, share and trap any wild animal or trying to do so'.

(b) 'to enquire, destroy, or take away any part of the body of such animal or damaging or disturbing the eggs or nests of wild birds and reptiles.

Animal Article. 'any article made from any part of a captive or wild animal.'

Wildlife Advisory Board

Under section 6 of the Act, the State Government or the Union Territory Administration may constitute a Wildlife Advisory Board, consisting of the following members-

(1) Minister incharge of the forests in the State or Union territory as Chairman of the Board. In case there is no such Minister, the Chairman will be represented by the Secretary to the State Government or the Union Territory.

(2) Two members of the State Legislature.

(3) Secretary Incharge of forests to the State Government.
(4) Forest Officer Incharge of the State Forest Department, as an ex-officio member.
(5) An Officer to be nominated by the Director of Wildlife Preservation.
(6) Chief Wildlife Warden, as an ex-officio member.
(7) Not more than five Officers of the State Government, including more than three representative tribals, who in the opinion of State Government are interested in the protection of wildlife.

Declaration of Sanctuary and Restriction of Entry

Under section 18 of the Act, if the State Government considers any area is of ecological, floral, faunal, geomorphological, natural or zoological significance, then it may declare such an area as a *Sanctuary* by notification for the purpose of protecting, propagating or developing wildlife therein or its environment.

According to section 27 of this Act, no person other than-
(a) a public servant on duty.
(b) a person permitted by Chief Wildlife Warden or the authorized officer.
(c) a person passing through the sanctuary along a public highway.
(d) a person who has any right over immovable property within the limits of the sanctuary, and
(e) the dependent of the person referred in the above clauses shall enter or reside in the sanctuary

Declaration of National Parks and Closed Areas

Under section 35, if the State Government feels any area is of adequate ecological, floral, faunal, geomorphological or zoological importance, then it may, by notification, declare such an area as *National Park* for the purpose of protecting, propagating, or developing wildlife therein or its environment.

Under section 37 of this Act, the State Government, by notification, can declare any area as closed to hunting for a specified period. No hunting of any wild animal shall be permitted in such a closed area.

Prohibition of Hunting

Under section 9 of this Act, no person shall hunt any wild animal except as provided under section 11 and 12 of this Act.

Under section 11 of this Act, hunting of wild animals is permitted in following cases:

(a) If the Chief Wildlife Warden is satisfied that any wild animal has become dangerous to human life or is disabled or diseased beyond recovery, then he may, by an order in writing and stating the reason thereof, can permit any person to hunt such animal.

(b) The killing or wounding in good faith of any wild animal in self defense or defense of any other person shall not be an offence.

(c) Any wild animal killed or wounded in defense of any person shall be the government property.

Under section 12 of this Act, the Chief Wildlife Warden by an order in writing stating the reason thereof and on the payment of the prescribed fee, may grant a permit to any person allowing him to hunt any wild animal specified in such a permit for the purpose of education, scientific research and management, *i.e.*, translocation of any wild animal to an alternative suitable habitat or population management without killing or poisoning any wild animal, collection of specimens for zoological gardens, museums or similar institutions or derivation, collection or preparation of snake venom for the manufacture of life saving drugs.

Penalties for Violating the Provisions of the Act

Following are penalties for violating the Act:

(1) A person violating any of the provisions of the Act shall be punishable with imprisonment for three years or a fine of Rs. 25,000 or both.

(2) In case a person is convicted of an offence against the Act, the court may order that any captive animal, wild animal article, trophy, weapon, vehicle, trap etc. be forfeited to the State Government and that any permit or license held by such a person be cancelled in addition to the other penalties imposed in such an offence.

(3) In case of cancellation of license, the court may order that such a person shall not be eligible for a license under the Arms Act, 1959 for a period of five years from the date of conviction.

Constitution and Function of Central Zoo Authority

Under section 38-A of the Act, the Central Government shall constitute the Central Zoo Authority consisting of following members:

(1) a Chairperson

(2) not more than ten members and

(3) a member-secretary

Under section 38-C of the Act, the Central Zoo Authority shall perform the following functions:

(a) Specification of the minimum standards for housing upkeep and veterinary care of the animals kept in zoo.
(b) Evaluation and assessment of the functioning of zoo.
(c) Recognition and de-recognition of zoos.
(d) Identification of endangered species for the purpose of captive, *i.e.*, *ex-situ* breeding, and
(e) Coordination of acquisition, exchange and loading of animals for breeding purposes.

Several Conservation Projects for individual endangered species such as Lion (1972), Tiger (1973), Crocodile (1974) and brown-antlered Deer (1981) were started under this Act. Recognition of Zoo Rules for evaluation and recognition of zoos was made in 1992.

Some of the drawbacks of the Act include illegal trading of wildlife trade in Jammu and Kashmir, mild penalty to offenders, personal ownership certificate for animal articles like tiger and leopard skin, no coverage of foreign endangered wildlife, pitiable condition of wildlife in mobile zoos and little emphasis on protection of animal genetic resources.

Wildlife (Protection) Amendment Bill 2002

The bill proposes:

(1) To highlight the ecological and environmental objective in the long title of the Wildlife Act in order to extend its area.
(2) To add more definitions in view of amendments proposed in Wildlife Act.
(3) To give statutory status to the National Board for Wildlife and re-structuring of Wildlife Advisory Board by providing representation of all concerned.
(4) To provide certain safeguards to stop killing of animals on pretext of being dangerous to human life, agriculture, livestock and property.
(5) Rationalise and expedite the process of final notification of Sanctuaries and National Parks and safeguard the decline of bio-diversity during the intervening period between the first and final notification.
(6) To provide that any alteration in the boundaries of National Parks and Sanctuaries shall be made only on the basis of recommendations of the National Board for Wildlife.
(7) To ban commercial sale of forest produce removed from National Parks and Sanctuaries for better management of Wildlife.

(8) To provide that no construction of commercial tourist lodges, hotels zoos and safari parks shall be allotted inside National Parks and Sanctuaries except with the prior approval of the National Board for Wildlife

(9) To empower the officers to evict encroachment from the National Parks and Sanctuaries.

(10) To provide for the creation and management of community reserves as well as conservation reserves.

(11) That zoos shall not acquire or dispose of any wild or captive animal to any organisation other than recognised zoos.

(12) To provide that captive animals and wild animals included in schedule I and part II of schedule II of Wildlife Act and their parts or products can be acquired only by way of inheritance.

(13) To enhance and rationalise penalties prescribed under the Act including the making of suitable provisions on the line of provisions of Chapter VA of Narcotic Drugs and Psychotropic Substances Act, 1985 in case of offences pertaining to wild animals included in schedule I and part II of scheduled II of the Act.

(14) To enhance the amount the rewards payable to persons rendering assistance in detecting the offences and apprehension of offenders.

(15) To increase the amount that can be realised as compensation upto Rs. 25,000 and

(16) To provide that the vehicles, weapons and tools etc. used in committing compoundable offences are not to be refunded to the offenders

The National Wildlife Action Plan provides the framework of strategy as well as programme for conservation of wildlife. The first National Wildlife Action Plan (NWAP) of 1983 has been revised and the new Wildlife Action Plan (2002-16) has been adopted. The Indian Board of Wildlife headed by Prime Minister is the apex advisory body overseeing and guiding the implementation of various schemes for Wildlife Conservation. The Animal Welfare Division became a part of the Ministry of Environment and Forests in July 2002.

Chapter Summary

India has a long history of conserving the forests and protecting wildlife. The efforts in the direction conservation of forests and protection of wildlife, although initiated during pre-independence came in full form during 1972 when the Wildlife (Protection) Act and during 1980 during which Forest (Conservation) Act was formed. Forest Conservation act, 1980, also known as forest policy of India, worked as an umbrella for conservation of forests and to bring India at par with national goal of having 33 per cent of forest area. The aims of the Forest (Protection) Act, 1980 are manifold but the main aim centres round substantial increase in forest area of the country. Various efforts made in this direction include notification of reserve forests, delegation of powers to state governments and provisions of penalties for violating the provisions of the Act. In 1992, amendments were made in the Forest Conservation Act to extend the area of 'non-forested' activities. The wildlife protection in India is as old as its culture. But the full fledged efforts in this area were made in 1972 when India made its Wildlife (Protection) Act to protect wildlife and in 1973 imposed a ban on trade in rare and endangered species or the products obtained from them. India is a signatory to Convention on International Trade in Endangered Species of Wild Fauna and Flora (CITES), which came enforce in 1975. Emphasis is given on imposing penalties for violating the provisions of the Act. Later a Central Zoo Authority was formed to look after minimum housing standards and veterinary care of animals kept in the zoo. Several conservation projects, such as Lion (1972), Tiger (1973), Crocodile (1974) etc. for protection of individual endangered species were started. In 2002, Wildlife (Protection) Amendment Bill was passed to cover more areas under wildlife protection.

Study Questions

1. Briefly discuss the history of forest conservation and wildlife protection in India.
2. What are salient features of Forest (Conservation) Act, 1980? What are the penalties for violating the provisions of the Act? Explain.
3. Write salient features of Wildlife (Protection) Act, 1972.
4. Discuss Central Zoo Authority.
5. Discuss CITES.
6. Discuss declaration of Sanctuary and National Park.
7. Discuss amendment in the Forest Act, 1980.

Objective Questions. *Select the correct answers.*

1. In India, Project Tiger was initiated in the year:

(1) 1970 (2) 1971

(3) 1972 (4) 1973

2. The first National park established in India in 1936 is:

(1) Dudhwa National Park (2) Jim Corbett National Park

(3) Rajaji National Park (4) Kaziranga National Park

Answers

1. (4) *2.* (2)

Issues Involved in Enforcement of Environmental Legislation and Public Awareness

LEARNING OBJECTIVES

Introduction • Environmental Legislation • Drawbacks of Air and Water Pollution (Prevention and Control) Acts • Drawbacks of the Forest (Conservation) Act, 1980 • Drawbacks of the Wildlife (Protection) Act, 1972 • Problems Faced in the Enforcement of Environmental legislation • Environmental legislation and Public Awareness.

Introduction

Most legislation evolved in response to problems. As such there is often delay between the need and the establishment of satisfactory law. Without effective legislation, environmental protection initiatives are likely to fall into chaos and conflict. But more important is enforcement of laws. If laws and international agreements are made and not enforced, they become of no use. Despite there are a number of laws to safeguard our environment, we are not able to achieve our target of environmental protection. It could not become possible to achieve the target of bringing 33 per cent of the land under forest cover. We are loosing our wildlife rapidly. Our rivers in many places are turning into open sewers and air is becoming abusive. The present status of our environment shows that there are many drawbacks in the environmental legislation and problems arising in enforcement of environmental laws and Acts (Box 45.2)

Environmental Legislation

The Ministry of Environment and Forests has undertaken the formulation of a comprehensive Environment Policy to harmonise the demands of development and environment in response to the need to weave environmental considerations into the fabric of development process and national life. Although the process was initiated in the Fourth Five Year Plan, the Department of Environment was created in 1980 and the Ministry of Environment and Forests in September 1985.

The Ministry in consultation with experts has prepared a draft on National Environment Policy, 2005 to harmonise the demands of development and environment.

The Laws and constitutional basis of environmental protection are good as well and are listed in Concurrent List of the Central Government but their drawbacks make them difficult to enforce.

Drawbacks of Air and Water Pollution (Prevention and Control) Acts

The main drawbacks in Pollution (Prevention and Control) Acts of Water in 1974 and Air in 1981 or any such other Acts may be accounted as follows:

Box 45.1 Forms of Legislation

Principle. Broadly, a step towards establishing law. Once established, tested and working, it can be incorporated into law.

Standard. Level of pollution, energy efficiency etc. that are desired or required. They provide a benchmark so that different individuals, bodies, countries, are as far as possible dealing with same values. A treaty may incorporate standards.

Guideline. Suggestions as to how to proceed, usually without real force of law.

Directive. Documents that set out a desired outcome, but to some extent leave the ways of reaching it to companies, states or countries.

License. A right granted to a body, which agrees to terms or pays which requires adherence to strict practice and does not give any guarantee of permanent ownership or usufruct.

Law. Laws and statutes that require certain actions or standards, and may punish failure to achieve them.

Treaty. A solemn binding agreement between international entities especially nations. Treaties can lay down rules or treaty constraints. Treaties can be difficult to enforce-often enforcement is attempted by an international organisation:, *e.g.* the International Whaling Commission. Treaties should bind the nations that sign and ratify them to accept terms as customary law, but in practice they do not always get transformed into law, and some are largely ignored.

Declaration. A general statement of intent or drafting of guidelines to follow 'Softer' than the obligation of a treaty.

Convention. Multilateral instrument signed by many nations or international institutions.

Protocol. Less formal agreement often subsidiary or ancillary to a convention.

Contingency Agreement. A good way of dealing with uncertainty surrounding many global environmental management issues. Agreement of what to do if something happens.

(1) The power and authority has been bestowed upon the Central Government with limited delegation of powers to the State Governments. The centralisation very often hinders efficient execution of provisions of the Acts in the States. Illegal mining taking place in many parts of the country such as in Rajasthan poses difficulty because it is to be checked by Central Government.

(2) Provisions of penalties in the Acts are insignificant as compared to damages caused by heavy industries due to pollution. Often penalty is much less than the cost of the treatment/pollution control equipments. This renders a loose rope to the industries.

(3) 'Right of information' for citizens has not been included in the Acts. This greatly restricts the involvement or participation of public in general.

(4) Environment (Protection) Act, 1986, encompassing earlier two Acts and regarded as an umbrella Act, often seems superfluous due to overlapping areas of jurisdiction. For instance, section 24 (2) of Environment (Protection) Act, 1986 has made a provision that if the offender is punishable under other Acts like Water Act, 1974 or Air Act, 1981 also, then he may be considered under the older two Acts in which case punishment is too less than the provision of penalty in the Environment (Protection) Act, 1986, thus offender easily gets away with lighter punishment.

(5) Under section 19, a person can not directly file a petition in the court on a question of environment and he has to give of minimum 60 days to the Central Government. In case no action is being taken by the latter, then only a person can file a petition which certainly delays the remedial action.

(6) Legislation, especially related to the environment, is very expensive, tedious and difficult as it involves expert testimony and technical knowledge of the issues. It also requires understanding of terminology and technical know how of the unit process and takes lengthy time of imposing prosecution.

(7) State Boards very often do not have sufficient funds and expertise to pursue their objections.

(8) Tendency of exercising pressure on polluter and out of court settlements, usually hinder the implementation of legal measures.

(9) Due to paucity of funds or heavy expenditure involved, small units often find it difficult to install Effluent Treatment Plant (ETP) or air pollution control device and something they do not have any other option except to close the unit. The Act needs to have some provisions for providing subsidies for installing pollution treatment plants/ devices for such small units.

(10) Pollution control laws are not backed by sound policy pronouncements or guiding principles.

(11) Since the position of Chairman of the Board is usually occupied by a political appointee, it becomes difficult to keep political interference at bay.

(12) The policy statement of the Ministry of Environment and Forests of involving public in decision making and facilitating public monitoring of the environmental issues has often remained only on paper.

It is needed that environmental policies and laws should be aimed at democratic decentralisation of power, community stake partnerships, administrative transparency, accountability and more stringent penalties and punishments to the offender. There is also a need for capacity building for managers of environmental issues and environmental law educators.

Drawbacks of the Forest (Conservation) Act, 1980

Some important drawbacks are:

(1) The Act inherits the exploitative and consumeristic elements of forest laws of British Period.

(2) In order to decide conversion of forest lands to non-forest areas, the Act has centralised the power by transferring the power of States to the Centre.

(3) For decision making process regarding the nature of use of forest area for the local communities has been kept out.

(4) There are no clear-cut provisions for tribal sustenance on the forests. Quite often they retaliate when they are stopped from taking any resources from the forests and start criminal activities upon stopping them to do so.

(5) The Act has failed to attract public support because if has infringed upon human rights of the poor native people who argue that the law is concerned about protecting trees, animals and birds but is treating poor people at margin. Very poor community's participation in the Act remains at bay and is one of the major drawbacks which affects its poor execution.

(6) Despite a known fact that forest inhabiting tribals have a rich knowledge of the forest resources including their importance and conservation, their role and contribution is neither acknowledged nor honoured.

Attempts are being made to make the laws perfect by filling up gaps through introducing the principles of human right protection.

Drawbacks of the Wildlife (Protection) Act, 1972

Some major drawbacks of the Act are:

(1) It seems that the Act has been enacted just as a fall out of Stockholm Conference, 1972.

(2) The Act does not have the provision of any locally evolved conservation measures.

(3) The ownership certificates for animal articles such as skin of tiger, leopard etc. are permissible, which very often become the cause for illegal trading.

(4) In Jammu and Kashmir, which has its own Wildlife Act, wildlife traders easily get illegal furs and skins for making caps, belts etc. from other states. These articles are sold in the country or smuggled to other countries because the State does not follow the Central Wildlife (Protection) Act. And thus hunting and trading of several endangered species, prohibited in the other states, are permissible in this State, thereby opening avenues for illegal trading in such animals or the articles.

(5) Act offender is not subjected to very harsh penalties as it is just three years imprisonment or a fine of Rs. 25,000 or both, sometimes too less than the market value of animal or the article.

Problems Faced in Enforcement of Environmental Legislation

Following diffculties commonly arise in enforcement of environmental legislation:

(1) The Precautionary Principle. Precautionary principle for environmental legislation has evolved to deal with the risks and uncertainties faced by environmental management. The principle is based on the proverb *an ounce of prevention is worth a pound of cure*. It does not prevent the problems but may reduce their occurrence and helps ensure that contingency plans are made. The application of this principle requires either cautious progress until a development can be judged 'innocent' or avoiding development until research clearly indicates exactly what the risks are, and then proceeding to minimise them. Once a threat is identified, action should be taken to prevent or control the damage even without being certain about the reality of threat. If delay is made, some of the environmental problems become costlier to control or rather impossible, therefore, waiting for research and legal proof is not costlier. Some people are of the opinion that the precautionary principle should be applied under the situation where both the probability and cost of impacts are unknown. Application of this principle was stressed in many of the decisions reached at the World Earth Summit, 1992 held at Rio. The principle was endorsed in article 15 of the 1992 Rio Declaration on Environment and Development.

(2) **The Polluter-pays Principle**. In additional to obvious, the principle of 'the polluter pays for the damage caused by a development' – this principle also implies that a polluter pays for monitoring and making the policy. This approach sometimes imposes so much of fines that causes the small business to become bankrupt. Sometimes fine may be low enough for a large company to write them off as an occasional overhead, which does little for pollution control. Thus, there is a debate as to whether the principle should be retrospective. If polluter pays then how long back does the liability stretch? Developing countries pay more for carbon dioxide and other emission's control arguing that they polluted the global environment during the industrial revolution, yet enjoy the benefits of investment from that era. Developed countries are of the view of 'pollute and perish' and pay for damages caused to the environment. Infact, this principle is more a way of allocating the costs of polluter than a legal binding.

(3) **Freedom of Information**. Environmental management and planning gets hindered if public, NGOs and GOs are unable to get information as to what is happening to the environment. Release of more information has now begun in several countries such as USA which has now Freedom of Information Act and the European Union is moving ahead in this direction. But still there are many governments and multinational corporations which fear that in doing so their industrial secrets will leak to competitors, if there is so much disclosure. The situation may also arise where authorities declare 'strategic' needs and suspend disclosure.

Environmental laws are strategic in nature and prohibit acts that bring immediate gains which, in long run, may prove harmful to the

Box 45.2 One Man Enviro-Legal Brigade

MC Mehta, known as one man enviro-legal brigade, came into limelight when environmental legislation emerged as a new concept in India. He is hero of public interest environmental cases. He has filed several public interest litigation including Mathura refinery-Taj and other-more than 50 cases against industries polluting river Ganges from Haridwar till its end. In 1993, he was given UN Environment Programme's 'Global Award'. Mehta won the Goldman Environmental Prize in 1996. He has also been awarded Magsaysay Award for Public Service in 1997 for his staunch efforts in protection of environment. He is also credited for getting implemented the mandatory course on Environmental Studies as an outcome of a Public Interest Litigation filed against Union Government in 1988.

life to the planet. In context of International Laws, even if nations may agree to the needs for laws they would prefer that everyone except them should follow them. A classical example of how to break a law is that of many countries have agreed to the Montreal Protocol to limit the emissions of greenhouse gases through various sources to the atmosphere but when it comes to penalise those countries exceeding their agreed limits, then the question arises about possible trading in limits, *i.e.*, buy the upper limits against payment from those countries that emit less, thus making it difficult to enforce laws even among the law abiding nations. Such an attitude is also reflected in enacting local or national laws for the purpose of seriousness with which they are implemented/ enforced within the domestic limits.

Environmental Legislation and Public Awareness

This is quite likely that, in India, the trend towards 'judicial activism' may not continue for as long as there are large pending cases and judges are aware that judicial intervention alone cannot bring about systematic change in such a large and complex country. Despite the fact that environmental legislation is one of the many tools, public awareness by educating them about the environment and impacts of its degradation is one of the most effective methods in many cases. In some cases, political action by organising citizens may be far more effective. In long run, there must be political will to conserve the environment. Allocation of budgetary funds for enforcing environmental laws will also prove effective for long term care of the environment. The court can only supplement the mechanism of administrative enforcement.

For many pressing environmental issues concerned with resource management, which are largely institutional or social issues, participation of all relevant social sectors in environmental management is the cornerstone at the basis of each and every environmental policy and programme. Building on the positive links between economic development and environment, requires the involvement of the people in making decisions as to how resource would be utilised? Neither government nor aid agencies are equipped to make judgement about how people should value their environment. Without public participation and support, environmental programmes/policies can not survive or bring about the permanent change in the people's attitude required. Public participation helps planners in better understanding of local values in increasing knowledge and experience, facilitates in winning community backing and help for the objectives of the project/ policy, and provide more chances of resolving conflict over resource use. If not many, an individual can adopt at least one of the environmental issues that has the most personal meaning and participate directly in the solution and add himself to a local organisation or NGOs working on that problem. Many such individuals will prove boon in solving the environmental problems at local, national, regional and global levels.

Chapter Summary

Without effective legislation and enforcement of laws, environmental protection initiatives are likely to fall into chaos and conflict. If laws and international agreements are made and not enforced, they become of no use. In India, Ministry of Environment and Forests in consultation with expert has prepared a draft on National Environment Policy, 2005 to harmonise the demands of development and environment. The laws and constitutional basis of environmental protection are good as well and are listed in Concurrent List of the Central Government but their drawbacks make them difficult to enforcement. The main drawbacks of Pollution (Prevention and Control) Acts are centralisation of power with Union Government; lack of 'Right of Information' for citizens; insignificant penalties; legislations being expensive, tedious and difficult; lack of involvement of public during policy framing by the Ministry etc. There are many drawbacks in Acts of water and air pollution prevention and control and wildlife protection. Proceeding for environmental legislation requires for adopting precautionary principle, *i.e.*, prevention is better than cure. Despite the fact that the environmental legislation is one of the many tools, public awareness by educating people about the environment and impacts of its degradation, is one of the most effective methods in many cases Without public participation and support, environmental programmes/policies can not survive or bring about the permanent change in the people's attitude required.

Study Questions

1. What are drawbacks of pollution (Prevention and Control) acts? Explain.
2. Discuss the drawbacks of forest and wildlife protection acts.
3. "Environment legislation can be made effective only by making people aware of the importance of environmental laws." Justify the statement.
4. Define environmental legislation.
5. Define a treaty.
6. Define right of Information.
7. Define legislation and tribals.

Objective Questions. *Select the correct answers*

1. National Environment Policy of India aims to bring forest cover:

(1) 25% (2) 32%
(3) 33% (4) 34%

2. Multilateral instrument singed by many nations or international institutions is called:

(1) Treaty (2) Protocol
(3) Convention (4) Declaration

Answers

1. (3) *2.* (3)

Human Population and the Environment

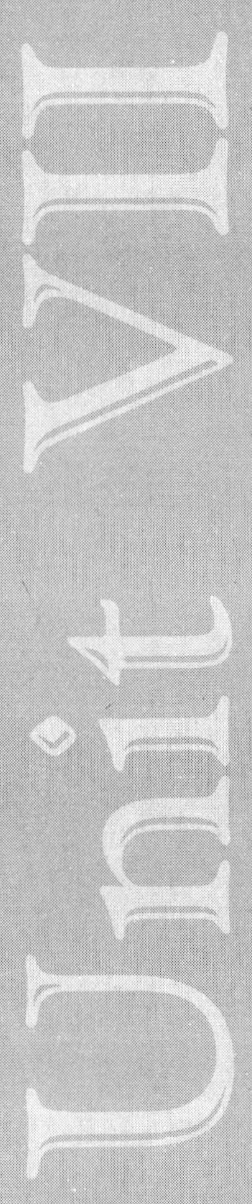

- *Population Growth –Variation Among Nations*
- *Population Explosion-Family Welfare Programme*
- *Environment and Human Health*
- *Human Rights*
- *Value Education*
- *HIV/AIDS*
- *Women and Child Welfare*
- *Role of Information Technology in Environment and Human Health*

46

CHAPTER

Population Growth-Variation Among Nations

LEARNING OBJECTIVES
World Population • Variation Among Nations • Population in India • Population Growth • Growth Curve • Environmental Resistance • Maintenance of Population • Growth Variation Among Nations

World Population

Human history is only about 50,000 years old but most explosive increase in human population has occurred during last two centuries. The steep rise of population in last century or so, has exploited the lithosphere, biosphere as well as atmosphere to the level of great concern (fig 46.1).

The world population in the year 2005 reached to about 6396 million which was only about 810 million in 1800 and 1160 million in 1900. The doubling time of population between 1800 to 1900, (nineteenth century) was 100 years and that of 1900 to 1960 was 60 years and 1960 to 1981 was 25 years. At present, the growth rate of world population is about 2 per cent. As such, these three are considered as important population explosions of the last two centuries, *i.e., (i)* in between 1800-1900, *(ii)* 1900 to 1960 and *(iii)* 1960 and onwards. Now, by developing awareness towards consequences of population explosions, the growth rate of population is regularly decreasing. Some countries have even zero per cent growth rate, *e.g.* Canada.

A rapid and dramatic rise (population explosion) in world population has occurred over the last few hundred years. Between 1959 and 2000, it increased from 2.5 billion to 6.1 billion people. According to UN estimates, the world population will be between 7.9 billion and 10.9 billion by 2050.

Population growth is currently continuing in the developing countries of Asia, Africa and Latin America. It is because of combination of a continued high birth rate and a low death rate. Such duo factors are creating a rapid population increase.

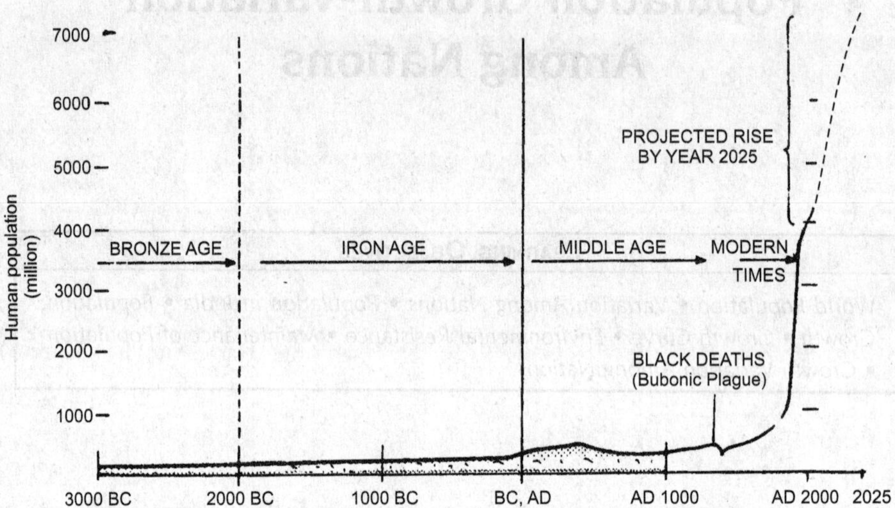

Fig. 46.1. A graph showing population growth of human beings.

Variation Among Nations

The World Population is distributed unevenly in different continents of the world (as shown in table 46.1).

TABLE 46.1. World Population 2005

Continents	*Population*
1. Asia	3,879,000,000
2. Africa	877,500,000
3. Europe	727,000,000
4. North America	501,500,000
5. South America	379,500,000
6. Australia	32,000,000
Total	6,396,500,000

As per the Encyclopedia Britannica Almanac, 2004 the population of world religions is as follows (table 46.2).

TABLE 46.2. World Population (Religion-wise)

Religion	*Population*	*Religion*	*Population*
Christians	2,069,883,000	Ethnic religionists	238,096,000
Roman Catholics	1,092,853,000	New religionists	105,106,100

Protestants	364,530,000	Sikhs	24,295,200
Orthodox	271,030,000	Jews	14,551,000
Anglicans	79,988,000	Spiritists	12,732,600
Muslims	1,254,222,000	Bahais	7,503,000
Hindus	837,262,000	Confucians	6,425,300
Chinese folk religionists	398,106,300	Jains	4,413,700
		Zoroastrians	2,733,900
Buddhists	372,974,000	Shintosists	2,680,300
		Other religionists	1,118,00
		Non-religious	784,269,000
		Atheists	148,660,000

Some least populated countries of the world are (table 46.3).

TABLE 46.3. Some Most and Least Populated Countries of the World (2005)

Most populated countries		*Least populated countries*	
Country	*Population (million)*	*Country*	*Population*
1. China	1306.3	1. Vetican City	900
2. India	1080.2	2. Tuvalu	11,636
3. USA	295.7	3. Nauru	13,048
4. Indonesia	241.9	4. Palau	20,303
5. Brazil	186.1	5. San Marino	28,880
6. Pakistan	162.4	6. Monaco	32,405
7. Bangladesh	144.3	7. Liechtenstein	33,171

Distribution of human population in the world is not balanced as evidenced by Asia where more than half of the global population, (around 3.8 billion) live in one fifth of the land area of the world. Similarly, America (both North and South) has more than a quarter of the total land surface with one fifth of the global population, (about 837 million) and Africa has one fourth of the land surface with about 877 million population. On the other had Europe is about one twenty fifth of the world land but the population is about 727 million, *i.e.,* one nineth of global population.

Variation within the continent is also uneven. In Asia, China alone, with about 1306 million people, accounts for one-third Asian and one fifth of the world population. In INDIAN SUBCONSTINENT, population variation is well evident and it is about 1080 million in India, 162.4 million in Pakistan, 144.3 million in Bangladesh, 27.6 million in Nepal, 20 million in Sri Lanka, 2.2 million in Bhutan and 0.34 million in Maldives.

Population in India

Populationwise, India ranks second in the world with its number about 1080 million (1027 million in March 2001). It shares about 16.7 per cent of world population living in a meager 2.4 per cent of the earth's surface area. It is estimated that at the present rate of population growth (1.93 per cent during 1991-2001). India will overtake China by 2050. The ten most populated states of the country have following percentage composition.

TABLE 46.4. Ten Most Populated States of India

States	*Per cent of total population of India (2001)*	*States*	*Per cent of total population of India*
1. Uttar Pradesh	16.17	6. Tamil Nadu	6.05
2. Maharashtra	9.42	7. Madhya Pradesh	5.88
3. Bihar	8.07	8. Rajasthan	5.5
4. West Bengal	7.81	9. Karnataka	5.14
5. Andhra Pradesh	7.37	10. Gujarat	4.93

The ten most populated cities of the country are Greater Mumbai (16.3m), Kolkata (13.2m), Delhi (12.7m), Chennai (6.4m), Banglore (5.68m), Hyderabad (5.5m), Ahmedabad (4.5 m), Pune (3.7m), Surat (2.8 m) and Kanpur (2.69m), (m=million).

POPULATION GROWTH

The number of individuals living in a particular area at a particular time is referred to as human population of that area. Each such population has a characteristic pattern of increase (growth) called growth form or growth curve.

Growth curve

Normally a growth curve of human population is S-shaped or sigmoidal. However, if growth stops abruptly a J-shaped growth curve is obtained.

The sigmoidal growth curve drawn in terms of time and the number of individuals in the population appears to be S-shaped and comprises four distinctive phases (fig. 46.2).

(i) **Positive acceleration phase.** During this phase population growth increases as reproduction gets underway. It starts slowly because initially there is a shortage of reproducing individuals which may be widely dispersed.

(ii) **Exponential** or **logarithmic phase**. During this phase, a maximum or numerical growth rate is achieved under the optimal environmental conditions. There is no environmental resistance and the birth rate exceeds the death rate.

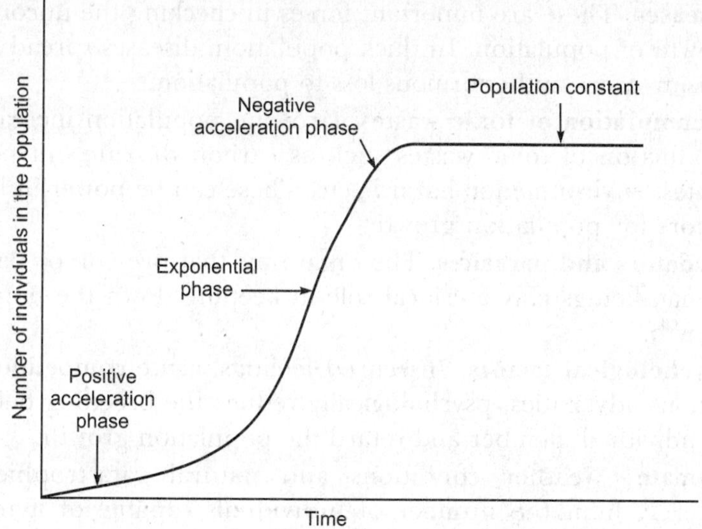

Fig. 46.2 S-shaped growth curve of human population.

(iii) Negative acceleration phase. It is characterised by decelerating population growth as the environmental resistance sets in and there is increasing death rate and/ or decreasing birth rate.

(iv) Upper asymptote or **population constant**. Finally the population becomes constant and there is no further growth as birth rate and death rate exactly balance each other resulting in equilibrium.

Environmental Resistance

Environmental resistance, *i.e.,* factors limiting population growth, depend on the species in question. For human beings, the general factors that provide resistance to the growing population, are:

(1) Shortage of food, water or **oxygen**. These are the basic requirements of human lives. With the growth of population, these commodities become limited and organisms begin to compete for them. As such, greater the population, the more severe is the competition because without them existence of human beings is not possible.

(2) Lack of light. Light is essential for plants and plants are essential for human lives. In absence or scarcity of light, food and oxygen supply will become limited leading to the limiting of population growth.

(3) Lack of shelter. Shelter from adverse physical environment and predators is an important defensive factor. A population without proper shelter is always insecure and vulnerable.

(4) **Diseases**. These are important forces in checking the uncontrolled growth of population. In thick population, diseases spread rapidly causing quick and enormous loss to population.

(5) **Accumulation of toxic wastes**. Growing population increases the production of toxic wastes such as carbon dioxide, nitrogenous wastes, environmental hazards etc. These can be powerful limiting factors for population growth.

(6) **Predators and parasites**. The organisms that prey on or parasitise human beings play a critical role in keeping down the population growth.

(7) **Psychological factors**. Insecured feelings, acute competitions and various adversities, psychologically reduce the breeding behaviour of individual member and retard the population growth.

(8) **Climate**. Weather conditions and natural catastrophies may severely limit the number of individuals capable of living in a particular area.

Maintenance of Population

At population constant phase, the population does not remain absolutely constant but fluctuates because of variations in the environmental resistance. In normal course of events, it fluctuates on either side of the normal point, but regulatory processes prevent fluctuations being excessive. Thus, **homeostatic** control of population is observed. It can be explained as follows:

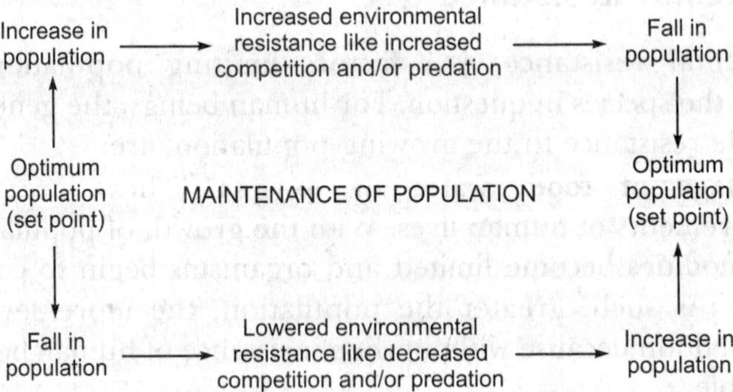

However, changes in the environment can bring alterations in the population.

Growth Variation among Nations

The rate of population growth shows variation among different nations of the world. According to recent studies, the growth rate is one or less than one per cent or zero among developed countries and two to three per cent in developing and underdeveloped countries. The world has been classified into four divisions on the basis of population growth rate pattern.

(1) Very high growth rate (3 per cent). Such growth rate is found in several countries of Africa and Western Asia. Important countries included in this category are – Nigeria, Ghana, Liberia, Mali, Tago, Ethiopia, Uganda, Kenya, Tanzania, Zambia, Congo, Angola, Libia, Western Sahara, Botswana, Israel, Gazastrip, Syria, Lebanon, Zorden, Yamen, Oman, Afghanistan etc. The reason for, so high growth rate in these countries are high birth rate (46 per 1000) and declining death rate (16 per 1000). The population doubling time is about 23 years.

(2) High growth rate (2 to 3 per cent). A high growth rate is met in Africa, Asia and Central America. Important such countries are – all counties of Central America (except Nicaragua), Ecuador, Peraguye, Venezeula, Egypt, Algeria, Moracco, Sudan, Tunesia, South Africa, Namibia, Iran, Iraq, Saudi Arabia, Turkey, UAE, Bahrain, India, Pakistan, Bangladesh, Nepal, Tazakistan, Uzbekistan, Myanmar, Malaysia, Vietnam, Combodia, Laos, Philippines, Brunai etc. Many of these countries are now controlling birth rate.

(3) Medium growth rate (1 to 2 per cent). Such countries have effectively controlled the birth rate to reduce the growth rate of the population. Important such countries are – China, North Korea, South Korea, Mangolia, Indonesia, Thailand, Singapore, Sri Lanka, Brazil, Columbia, Peru, Surinam, Guyana, Argentina, Chili etc. The population doubling time is about 35 years.

(4) Minimum growth rate (1 per cent or less). It is found among developed nations. In some countries like France, Sweden and Canada, the growth rate is zero and the population is nearly constant. Important such counties are USA, Canada, Great Britain, Denmark, Norway, Sweden, Poland, Czekoslovakia, Hungary, Ukrain, France, Switzerland, Holland, Belgium, Italy, Yugoslavia, Greece, Spain, Portugal, Russia, Japan etc. In these countries, birth rate is 12 to 15 per thousand and death rate is 9 to 11 per thousand. In Germany, growth rate is gone below zero as the population is regularly decreasing. The population doubling time is about 70 years.

The variation and changing trend of population can be determined by finding the ratio of age and sex. In population, individuals can be broadly classified into three phases, *i.e.*,

(i) Per-reproductive phase (0 to 14 years)

(ii) Reproductive phase (15 to 44 years)

(iii) Post-reproductive phase (45 and above)

Population growth rate trend of a nation can be predicted by histograms (as follows).

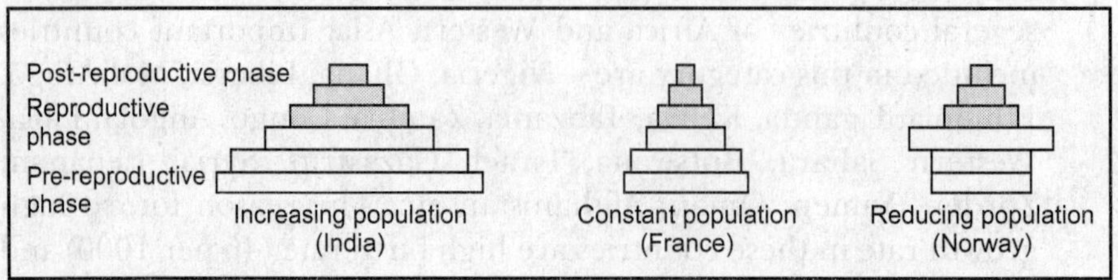

Fig. 46.3. Histograms showing different population patterns.

In simple way, higher the rate of births with decreased rate of deaths, higher will be the increasing rate of population. Higher number of children in a nation is an indication of fastly increasing population, as in India.

Chapter Summary

In the last two centuries, the world population has grown from about 810 million (1800) to 6396 million (2005) with multiple rate comprising three population explosions. Due to awareness, the population growth in developed countries has been significantly reduced but it is taking place in developing countries of Africa, Asia and Latin America. China is the world's most populated country followed by India. The distribution of human population in the world is not balanced, particularly with resources.

As per Census 2001, the population in India is about 1027 million (1080 million in 2005), *i.e.,* 16.7% of world population. Ten most populated states of the country are UP, Maharashtra, Bihar, West Bengal, Andhra Pradesh, Tamil Nadu, MP, Rajasthan, Karnataka and Gujarat, respectively.

The growth curve of human population is S-shaped and comprises, *(i)* Positive acceleration phase, *(ii)* Exponential or logarithmic phase, *(iii)* Negative acceleration phase and *(iv)* Upper asymptote or Population constant phase (fig . 46.2). Environmental factors that resist the growth of human population are *(1)* Shortage of food, water or oxygen, *(2)* Lack of light, *(3)* Lack of shelter, *(4)* Diseases, *(5)* Accumulation of toxic wastes, *(6)* Predators and Parasites, *(7)* Psychological factors, *(8)* Climate etc. These factors keep homeostatic control of population.

On the basis of growth pattern, various countries are classified as *(1)* Very high growth rate (3%) (*e.g.* Nigeria, Congo, Israel etc.), *(2)* High growth rate (2-3%) (*e.g.* South Africa, Iran, India, Namibia etc.), *(3)* Medium growth rate (1-2 %) (*e.g.* China, Indonesia, Brazil etc.) and *(4)* Minimum growth rate (0-1%) (*e.g.* Canada, Norway, Germany etc.). A country with higher number of children has a growing population.

Study Questions

1. Population growth is a major problem of the world. Describe taking suitable examples of various nations.

2. Write a note on world population with variations among nations.
3. Describe the various aspects of population growth in India.
4. What is population growth curve? Describe environmental resistance and maintenance of population.
5. What is population explosion? Describe growth variation among nations.
6. Write short notes on:
   (i) Population growth
   (ii) Environmental resistance against population growth
   (iii) Maintenance of population
   (iv) Growth variations among nations
   (v) Population distribution in India
   (vi) Population growth in developed and developing countries.

Objective Questions. *Select the correct answers.*

1. Number of population explosions occurred in human beings is:

(1) 1	(2) 2
(3) 3	(4) Indefinite

2. Population growth is rapid in:

(1) Syria	(2) France
(3) Germany	(4) Canada

3. Zero per cent population growth is found in:

(1) Lebanon	(2) Kenya
(3) India	(4) Canada

4. A rapid growth of population will take place in a country with high numbers of:

(1) Veterans	(2) Reproducible youth
(3) Children	(4) Male

5. A country with highest human population is:

(1) India	(2) USA
(3) China	(4) Korea

Answers

1. (3) *2.* (1) *3.* (4) *4.* (3) *5.* (3)

47

CHAPTER

Population Explosion - Family Welfare Programme

LEARNING OBJECTIVES
POPULATION EXPLOSION: Population explosion • Causes of Population Explosion, Population Explosion and Rising Concerns • Impact of Population Explosion on Environment • Population Explosion in India *FAMILY WELFARE PROGRAMME: National Population Policy • Family Welfare Programme • Family Planning Methods*

POPULATION EXPLOSION

An enormous growth in number of human beings is called population explosion.

Human history is only about 50,000 year old, and during this period, the social evolution reveals the occurrence of three such population explosions which have heavily affected the global environment. Scientists consider three important events of such explosions –

(1) The first population explosion occurred some 20,000 years back when human beings discovered, developed and improved weapons for hunting and procuring food.

(2) The second population explosion occurred some 6,000 years back when they learned and improved methods of farming (agriculture).

(3) The third population explosion occurred about 300 years back when they learned improved methods of *(i)* food production, *(ii)* industries establishments and *(iii)* medicine production.

These factors led to reduce the doubling time of the world population which was about 100 years between 1800 and 1900 (AD), to 60 years between 1900 and 1960 (AD) and to 25 years between 1960 and 1981 (AD) (as shown in fig. 47.1).

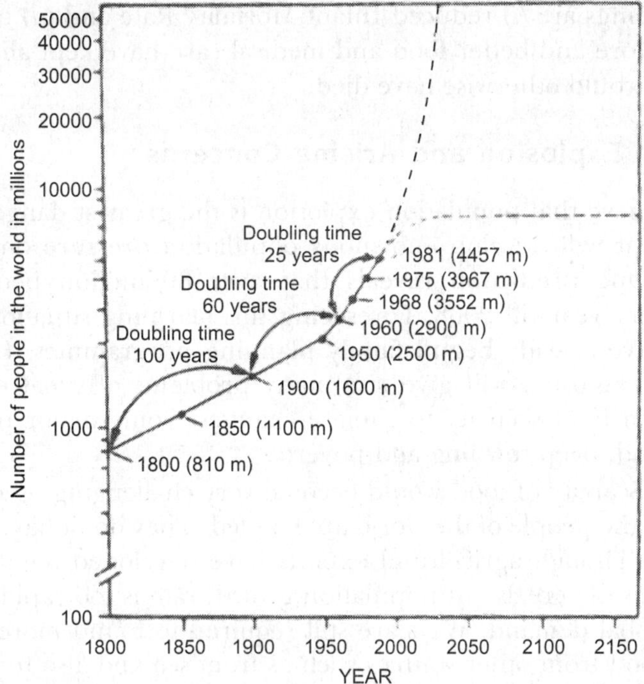

Fig. 47.1. Three population explosions of human beings in last two centuries.

Causes of Population Explosions

During the industrial evolution (a period of history in Europe and North America), there were great advances in science and technology which indirectly proliferated population growth mainly by reducing death rates. The factors for such a contribution were:

(*i*) Increase in food production and distribution,

(*ii*) Improvement in public health and

(*iii*) Medical technology (vaccines, antibiotics, new researches) followed by improved living standard and literacy.

Gradually, over a period of time, these discoveries and inventions spread throughout the world, lowering death rates and improving the life spans.

Malthus proposed following causes of increase in human population-

(1) Increased food production due to scientific and technical advancements.

(2) Reduction in death rates due to famines and epidemic diseases.

(3) Reduction in the Infant Mortality Rate (IMR) due to improved health services.

(4) Eradication of several fatal diseases by community service programmes.

It becomes imperative to conclude that causes of population explosion of human beings are *(i)* reduced Infant Mortality Rate and *(ii)* increased life spans. More and better food and medical care have kept alive many people who would otherwise have died.

Population Explosion and Arising Concerns

Scientists believe that population explosion is the greatest danger facing mankind as it will develop enormous population pressure on natural resources. One prediction reveals that over 80 million people will increase every year till 2008. Foreseeing this alarming situation, many countries have already begun family planning programmes. However, population explosion will give rise many problems of great concerns ranging from food security to climate changes. Some major problems would be food, overcrowding and poverty.

(1) **Food**. Scarcity of food would become very challenging. Even today half of the people of the world are half fed. They don't have enough to eat. Though agricultural experts have developed high yielding varieties of cereals but population growth rate is too rapid to meet the global demand. Ways are still required to found more protein rich food from other sources such as from sea and also to improve the pattern of distribution of food globally. If mass starvation or famines are to be avoided, the countries with excessive food will have to make sacrifices to feed the underfed countries.

(2) **Overcrowding**. Competition for space will develop enormous problems, seen and unseen.

Scientists are predicting numerous problems and tensions that are bound to arise from overcrowding. They feel that the population should not grow beyond the capacity of earth to house people with basic and necessary comforts. They also believe that today's global population is close to the limit that the earth can support and even today the problems are unmanageable.

(3) **Poverty**. Poverty is a major cause of concern and mother of several problems.

Poverty in itself is a curse and problems are more aggravated with population growth. Rapid growth of population tends to create difficulties for the government in providing proper education, health, housing and employment to the needed people. It is also disgraceful if government fails in providing social facilities and fair and equal development. Both these problems are in fact the two sides of the same coin.

About 1.2 billion desperately poor people live in less developed countries and about 33 million people (1/8th population) in USA live below the official poverty line.

Impacts of Population Explosion on Environment

Population growth has manifold consequences affecting environment.

(1) *Global warming and greenhouse effects*

Population growth appears to have a vital factor in the increase of carbon dioxide emissions (a major greenhouse gas). It has been substantiated by the doubling of fossil fuel emissions in developed countries between 1960 and 1988. Due to increased population, nearly half of the tripling of CO_2 emissions in developing counties was due to the same cause. CO_2 emissions have significantly contributed to a potential change in raising the temperature of the earth through the process known as the *Greenhouse effect*. The threat to climate change and global warming is heavily influenced by population growth. China, with its increased population and rapid industrialisation, is expected to become the leading source of global warming gas emission by 2050.

Global warming causes profound environmental effects such as melting of the polar icecaps, raising of sea levels, change of regional and global climate, change of natural vegetation and agricultural systems etc. Such changes will exercise an enormous impact on human civilisation.

(2) *Deforestation*

Population pressure contributes direct effect on loss of a forest cover because of increased land use for various purposes. Deforestation causes several far reaching damages to whole of the living system. Food and Agricultural Organisation (FAO) has begun to predict extent of deforestation on the basis of local population density, because the link between the two is so strong. The land use estimates calculated by UN FAO, revealed the fact that in the past three decades, nearly 59 per cent of the forests have been cleared in the developing countries for human settlements, roads and other non-agricultural developments that are entirely related to population growth. The data from FAO also revealed that a further 20 per cent forest cover might have been lost for farming purposes because of population pressure.

(3) *Biodiversity destruction*

Deforestation of biodiversity enriched regions like loss of tropical rain forests have caused destruction of wildlife and threat to the biodiversity of life on earth. It has been estimated that the current extinction rate is in between 50 and 100 species a day, with millions of species facing extinction in the present decade. Much of this is attributed to the population explosion.

(4) *Water scarcity*

Demand of water is closely related to the population growth. According to the United Nation's estimate, some 2.7 billion to 3.5 billion people lived in water scarce countries in last 30 years time. This scarcity is likely to affect some 4.4 billion people by 2050 but figure may grow to 7.7 billion, out of the 10 billion people of the world, if the population growth is projected in the current rate. UN has forecasted that a third world war may happen due to water scarcity.

(5) *Land loss*

An increased population clearly indicates that more land will be needed to raise food crops, build roads and to construct houses. An increased number of factories, industries and vehicles will further cause land loss. Thus, the population growth will have a widespread effect on the environment.

(6) *Stress on natural resources*

An increased population means a greater drain on the world's limited resources. Whereas the mineral resources on the earth are limited, the population pressure will compel unmindful extraction of the same. A term population / resource ratio is now often used to estimate the future threat and environmental repercussions. Problems related with exhaustible resources are among the great challenges of future.

Population explosion has become a gigantic problem and its consequences will be equally alarming. Important of them are:

(*i*) There will be acute shortage of food leading to famine or starvation deaths.

(*ii*) Housing space will be highly reduced.

(*iii*) There will be acute shortage of potable water.

(*iv*) Pollution of air, water and soil will be aggravated and will become hazardous. Crisis in physical environment will develop.

(*v*) Possibility of spreading infectious diseases will increase.

(*vi*) Illiteracy, unemployment and poverty will increase.

(*vii*) Acute shortage of natural resources will develop.

The world's current and projected population growth demands an increase in efforts to meet the needs for food, water, health care, technology and education. In less developed countries, massive efforts are required. A rapid population growth can affect both the overall quality of life and the degree of human sufferings.

Population Explosion in India

Little awareness with the time reveals the fact that India's enormous population explosion neutralises whatever economic progress the country makes. Although about 70 per cent of the workers are employed in agriculture, yet country is facing acute shortage of food. The reasons being that there are about 25 dependents to feed for every 10 workers, and as a result there is no savings and investments for any kind of future crisis. Nowadays, there is always a news of a farmer committing suicide. The country has a very low per capita annual income ($ 73) and it tops the list of malnutrition. This problem may further increase because 40 per cent population is of teenagers, which poses a potential threat to the social security of the nation if their energies are not properly used and jobs are not provided to them.

The common reasons for high population growth in India are –

(1) Large size of population in the reproductive group.
(2) Higher fertility rate due to failure of family planning.
(3) High demand of fertility due to high Infant Mortality Rate (IMR).
(4) The marriage of girls below the age of 18 years resulting in *too early*, *too frequent* and *too many*. It increases the reproducing period.

According to the report of National Family Health Survey (2007), even now 45 per cent of girls in India are married before the age of 18 years, being highest in Jharkhand (61 per cent) and then in Rajasthan (57 per cent).

(5) Predominating religious thoughts or superstitious nature.
(6) Joint family and importance of male child.
(7) Illiteracy and rural society.
(8) Lack of proper entertainment facilities.
(9) Social insecurity.
(10) Poverty and backwardness.
(11) Lack of sex education etc.

FAMILY WELFARE PROGRAMME (INDIA)

National Population Policy

Increasing problems due to population explosion have forced to think about family welfare programmes to keep control over population growth. Government of India has taken up this programme at priority level and effects are now clear.

The Census of India–2001 indicates the population of India on 1 March 2001 as 1027 million (531 million male and 496 million female). A look at the census figures of last four decades indicates a perceptible decline in recent years.

TABLE 47.1. Growth of Population in India

Census Year	*Decadal growth (per cent)*	*Average exponential growth*
1971	24.8	2.2
1981	24.66	2.22
1991	23.86	2.14
2001	21.34	1.93

Control on population growth could have been more effective, if family welfare programmes were equally adopted by all religious people. The 1991 census showed that of the total Indian population during 1981 to 1991, Hindu percentage decreased by 0.68 and Christians by 0.13, but the percentage of Jains and Sikhs remain constant whereas Muslim community increased by 32.76 per cent.

In 1952, India was the first country in the world to launch a national programme, emphasising fertility regulation to the extent necessary for reducing birth rate "to stabilise the population at a level consistent with the requirement of national economy."

Half a century after formulating the National Family Welfare Programme (NFWP) India's demographic indicators are –

(i) Reduced Crude Birth Rate (CBR) from 40.8 (1951) to 25.0 (SRS, 2002).

(ii) Reduced Infant Mortality Rate (IMR) from 146 (1951) to 63 (SRS-2002).

(iii) Reduced Crude Death Rate (CDR) from 25.1 (1951) to 8.1 (SRS-2002)

(iv) Reduced Total Fertility Rate (TFR) (per woman on an average) from 6.0 (1951) to 3.2 (SRS-1999).

(v) Increased Couple Protection Rate (CPR –per cent) from 10.4 (1971) to 48.2 (1998-99).

(vi) Increased Life Expectancy (ILE) at birth years (male) from 37.2 (1951) to 63.87 (2001-06)*.

(vii) Increased ILE at birth years (female) from 36.2 (1951) to 66.91 (2001-06)*.

(SRS= Simple Registration System of office of the Registrar General India, *=Projected).

As per TFR (SRS-1999), U.P. (undivided), Bihar (undivided) and Rajasthan with their respective values 4.7, 4.5 and 4.2, top the country list.

The National Population Policy, 2000 (NPP 2000) affirms the commitment of the Government towards voluntary and informed choice and consent of citizens while availing of reproductive health care services and continuation of the target free approach administering family planning services. It's immediate objective is to address the unmet needs for contraception, health care infrastructure and health personnel and to provide integrated service delivery for basic reproductive and child health care, the medium-term objective is to bring the TFR to replacement levels by 2010 and the long-term objective is to achieve a stable population by 2045, at a level consistent with the requirement of sustainable economic growth, social development and environmental protection.

Family Welfare Programme

The National Family Welfare Programme (NFWP) was launched in India in 1951 with the objective of reducing and stabilizing population to match with National Economy. The country being democratic, the Family Welfare Programme (FWP) was launched through voluntary and free choice of family planning methods, best suited to individual acceptors. The innovative use of mass media and interpersonal communication is made for highlighting the benefits of a small family. India kept FWP as priority area with 100 per cent centrally sponsored programme. The National Welfare Services are provided to the community through a network of Sub-centres, Primary Health Centres and Community Health Centres in rural areas and hospitals and dispensaries in the urban areas. This network set up under Basic Minimum Services (BMS) Programme is also supported by an expanding number of Post Partum Centres at district and sub-district levels.

Control over the population growth is a national priority. FWP committee has selected certain measures for successful execution of the programme, *e.g.*

(1) To encourage for small family.
(2) To increase gap period between the child births (spacing).
(3) To educate for ideal motherhood.
(4) To provide proper treatments for sex disabilities and related problems.
(5) To take proper care during pregnancy.
(6) To mentally prepare for the first child.
(7) To educate mother for taking the balanced nutrition.
(8) To develop awareness for child mother care.

Family Planning Methods

Family planning methods include all those systems which prevent pregnancy. For an ideal process, certain points are considered, *e.g.*

(1) Methods must be effective to prevent pregnancy.
(2) Methods should not be harmful to the health.
(3) Methods should be acceptable to male and females of all communities.
(4) Techniques for using methods must be simple and not requiring medical supervision.
(5) Methods should be easy to give up when a birth of child is required.
(6) Methods must be quite economic and easily available.

Family planning methods are broadly divided into two categories –

(I) Temporary methods
(II) Permanent methods

(I) Temporary Methods

The main aim of such methods is to increase the gap period (space) between the childbirth. Their use can be stopped as and when required, that is why these are called temporary methods. These can be broadly divided into following groups-

(A) Mechanical contraceptives. These include-
  (i) Intra Uterine Devices (IUD) such as copper–T and Lippes loop.
  (ii) Diaphragm
  (iii) Condom

(B) Chemical contraceptives. These include the usage of foam tablets and cream or jellies.

(C) Hormonal methods. These include various birth control tablets, *e.g.* femilon, Loette, Ovral, Mala–D etc.

(D) Natural methods. These include safe period method and coitus interrupts etc.

(II) Permanent Methods

These include methods of permanent birth control such as tubectomy and vasectomy.

Family Planning Acceptance in India

The use of family planning methods is gradually increasing in the country. During the year 2003-04, the total number of family planning acceptors at all India level was higher by 5.2 per cent than in the corresponding period of 2002-03. The performance figures for the year 2003-04 method wise are –

Sterilisation – 48.74 lakh
IUD insertions – 60.79 lakh
Condom users – 249.9 lakh
Oral pills users – 87.54 lakh

However, vasectomy percentage has reduced from 53.01 (2004-05) to 46.44 (2005-06).

The National Family Welfare Programme (FWP) provides the contraceptive services for spacing births. These include condoms, oral contraceptive pills, Intra-Uterine Devices (IUD) and Emergency Contraceptives. Of these, Condoms and Oral Contraceptive pills are provided through free distribution scheme and social marketing scheme and IUD is being provided under free distribution scheme. In Social Marketing Programmes, Condoms and Oral Contraceptive Pills are sold at subsidised rate. In order to promote use of new kind of IUD (380-A), which provides safety for about ten years and its use can avoid sterilisation, a scheme of the FWP is underway to provide these IUDs through social franchising of private clinics. FWP is also promoting No-Scalpel Vasectomy (NSV) and till March 2004, 2.89 lakh men have accepted it.

Govt. of India has enacted **Medical Termination of Pregnancy Act, 1971,** with the objective to reduce maternal morbidity and mortality due to unsafe abortion. The Act has been amended in 2002 to further suit the people.

The Department of Indian Systems of Medicine and Homoeopathy (ISM & H), established in the Ministry of Health and Family Welfare in March 1995, has been renamed as Department of Ayurveda, Yoga and Naturopathy, Unani, Siddha and Homoeopathy (AYUSH) in November 2003 to make it more efficient in the service of Family Welfare and Health.

FWP are approaching people of India to reduce a Net Reproduction Rate of Unity (NRR-1) by the years 2011-16, though it was aimed for the year 2000. Clinical, extension and education approaches are being adopted to make this programme successful. NRR-1 plans to achieve a Birth Rate of 21 per thousand, death rate of 9 per thousand and natural population growth rate of 1.2 per cent.

Chapter Summary

Enormous growth in number of human beings is called population explosion. The world has witnessed three important population explosions, *(i)* some 20,000 years back, *(ii)* some 6000 years back and *(iii)* some 300 years back. The doubling time of population growth between

1800 and 1900 AD was 100 years, between 1900 and 1960 AD was 60 years and between 1960 and 1981 was 25 years (fig. 48.1). Reduction in doubling time indicates a high rate of growth with rapid increase. The industrial revolution has indirectly proliferated the population growth, *e.g.* through increased food production and improved medical technology supporting public health. It has resulted in *(i)* reduced Infant Mortality Rate (IMR) and *(ii)* increased average life spans. Now population explosion is proliferating and concerns about food, overcrowding and poverty.

Population explosion has caused several impacts on the environment such as *(i)* Global warming and greenhouse effects, *(ii)* Deforestation, *(iii)* Biodiversity destruction, *(iv)* Water scarcity, *(v)* Land loss, *(vi)* Stress on natural resources etc. The main factors for population explosion in India are illiteracy, poverty and religious superstitiousness.

Govt. of India has formulated National Population Policy to check population growth to build up an ideal country. Results of the policy are evident by data on decadal per cent growth which was 24.8 in 1971 and 21.34 in 2001. The 50 year efforts of National Family Welfare Programme have resulted in a reduction in Crude Birth Rate, Infant Mortality Rate, Crude Death Rate and Total Fertility Rate and Increase in Couple Protection Rate, Increased Life Expectancy and Increased ILE at birth years. The NFWP was launched in 1951 with the objective of reducing and stabilising population to match the national economy. It executes programmes through, *(i)* to encourage small family, *(ii)* to increase gap period (spacing) between the child births, *(iii)* to educate for ideal motherhood, *(iv)* to provide proper treatment for sex disabilities and related problems, *(v)* to take proper care during pregnancy, *(vi)* to mentally prepare for the child, *(vii)* to educate mother for taking the balanced nutrition and *(viii)* to develop awareness for child mother care.

Various kinds of temporary and permanent family planning methods are now available. Temporary methods include mechanical and chemical contraceptives and hormonal and natural methods whereas permanent methods include tubectomy and vasectomy. The performance figures for the year 2003-04 method wise are *(i)* Sterilisation - 48.74 lakh, *(ii)* IUD insertions – 60.79 lakh, *(iii)* Condom users – 249.9 lakh and *(iv)* Oral pills users - 87.54 lakh. Many of these are available through free distribution scheme or through subsidy. Govt. of India has enacted Medical Termination of Pregnancy Act, 1971 to reduce maternal morbidity and mortality due to unsafe abortions. FWP has developed NRP-1 (National Reproduction Rate –1) programme to achieve a birth rate of 21/1000, death rate of 9/1000 and natural population growth rate of 1.2 % by 2011 to 2016.

Study Questions

1. Explain the causes and concerns of population explosion.
2. What is population explosion? Explain its causes and concerns.
3. Describe the impact of population explosion on environment.
4. What are the causes of population explosion in India. Describe various Family Welfare Programmes to reduce it.
5. Write a brief account of Family Welfare Programme of India.
6. Write short notes on-
   (i) Causes of population explosion
   (ii) Population explosion and rising concerns.

(iii) Impact of population explosion on environment
(iv) Population explosion in India
(v) National Population Policy
(vi) Family planning methods
(vii) Medical Termination of Pregnancy Act, 1971
(viii) AYUSH

Objective questions. *Select Correct Answers*

1. The doubling time of the world population between 1960 to 1981 was:
(1) 25 years (2) 40 years
(3) 60 years (4) 100 years

2. A problem not concerned with population explosion is:
(1) Overcrowding (2) Food
(3) Population growth (4) Poverty

3. Impact of population explosion on environment includes:
(1) Global warming (2) Biodiversity destruction
(3) Deforestation (4) All of these

4. Main cause of population explosion in India is:
(1) Good health (2) Higher percentage of female
(3) Prosperity (4) Poverty

5. National Family Welfare Programme in India was launched in:
(1) 1947 (2) 1951
(3) 1961 (4) 2001

Answers

1. (1) *2.* (3) *3.* (4) *4.* (4) *5.* (2)

48

CHAPTER

Environment and Human Health

LEARNING OBJECTIVES
Effect of Environment on Human Health • Environmental Factors • Overpopulation • Pollution • Urbanisation • Degradation of Natural Resources

Environment and human health are intricately related. A good environment is an indication of healthy human beings and a developed nation. Ruthless exploitation of nature by human beings has disbalanced whole of the ecosystem and global biosphere. Now polluted atmosphere has become a serious threat to the very existence of human species. The consequences are becoming evident and the benefits of development are gradually being deprived. Pollution growth and intensification of man's activities on various fronts like agricultural development, urbanisation and industrialisation all over the world have created a hazardous environment. The environment is under constant interference by human bound activities.

The effects of environment on human health are generally accounted in the following terms-

(i) The extent to which environmental conditions lead to shortening of life.

(ii) The extent to which environment induced disability or impairment is found, and

(iii) The extent to which the biological potential of an individual is reduced.

ENVIRONMENTAL FACTORS

The various environmental factors which affect human health are –

(1) Overpopulation

(2) Pollution

(3) Urbanisation
(4) Degradation of natural resources

(1) Over Population

Population explosion is the most important cause of all environmental problems. It leads to poverty, over exploitation of resources and environmental degradation. (*Population explosion and its concerns* are discussed in the last chapter.)

(2) Pollution

Any change in the environment which directly or indirectly affects the welfare of the human beings is called pollution. But air pollution, water pollution and solid waste pollution are of prime concern. Other kinds of pollution like nuclear hazards also cause problems related to the health.

(i) Air Pollution

Various pollutants present in the air directly affect the different systems of the human body. These enter the human body by inhalation or absorption through skin or eyes. Some specific effects of air pollutants are- (table 48.1).

TABLE 48.1. Specific Health Effects of Air Pollutants

Air Pollutants		*Health effects*
1. Sulphur oxides, nitrogen oxides, ammonia, chlorine, ozone.	–	are pulmonary irritants and coat the respiratory tract.
2. Carbon monoxide and hydrogen sulphide	–	impede oxygen uptake by haemoglobin and cause asphyxiation.
3. Silica, quartz, carbon asbestos, cobalt	–	damage pulmonary interstitial tissues.
4. Beryllium, hair sprays, talcum powder	–	produce granuloma in lungs.
5. Lead, mercury, fluoride cadmium, chlorinated hydrocarbons	–	are toxins to the nerve tissue, brain, bones, teeth, blood vessels, kindney, liver, fat tissue etc.
6. Pollens, fungal spores, house dust	–	develop allergy in skin, lung and respiratory tract and cause dermatitis, bronchial asthma, hay fever etc.
7. Strontium –90, Iodine –121, chromium, arsenic polyvinyl	–	cause cancer in bones, thyroid, lungs, skin etc. (carcinogens).
8. Mercury, lead, arsenic Cadmium	–	behave like mutagens and effect the central nervous system.

(For more details see chapters on *Air pollution* and *Pollution Case Studies* in Unit 5).

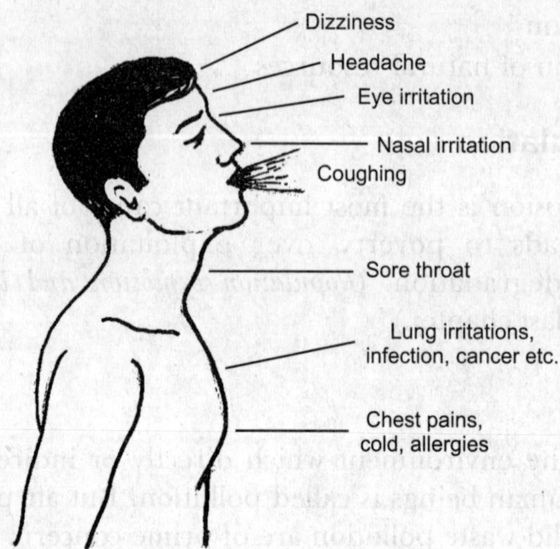

Fig. 48.1. Effects of polluted air on human body.

(ii) Water Pollution

About 70 per cent of the water bodies in India are polluted. Majority of Indian rivers are dangerously polluted. The common pollutants are the industrial effluents, municipal wastes, agro-chemicals, oil spills etc. It is estimated that more than 60 per cent of the diseases are water borne diseases in India, mostly due to pollution of rivers and lakes.

Some important **water borne diseases** are- cholera, typhoid, paratyphoid, bacillary and amoebic dysentery, diarrhoea, vomiting, jaundice (infectious hepatitis), polio etc. (For more details see chapters on *Water Pollution* and *Pollution Case Studies* in Unit-5)

(iii) Solid wastes

Discharge of industrial sludges or dumping of industrial and municipal wastes is the prime cause of land pollution. Such wastes include garbage, rubbish, ashes, hospital refuse, dead animals, agricultural wastes etc. Chemicals like pesticides, weedicides, insecticides, fungicides etc. used in agriculture, food preservation, community health services etc. are toxic substances and further hazardous.

Indiscriminate use of toxic chemicals contaminate ground and surface water, food and food products, cold drinks etc. Most of the pesticides are mutagenic and carcinogenic. Mutagens can cause chromosomal aberrations and various kinds of hereditary abnormalities. DDT (an insecticide) causes hepatocarcinoma, leukemia, aplastic anemia etc., **malathion** causes schizophrenia and depression, **dieldrin** causes insomnia, **BHC** affects central nervous system, **parathion** and **marathion** cause muscular weakness and delayed paralysis, **methoxychlor** damages liver and kidney etc.

Hospital refuses are the source of various infectious diseases and these act as ideal home for disease carriers or vectors like flies, insects, bugs, rodents etc. Heavy metal contaminated industrial effluents cause toxicities and various health problems such as fluorosis, minamata disease etc.

(3) Urbanisation

Increase in urbanisation adversely affects the ecological balance. It not only causes deforestation but also increases uncontrolled discharge of municipal wastes. Most Indian cities have about 40 per cent slum population which leads to deteriorating life quality and poor health services and flourishing of communicable diseases. These area lack basic amenities and are disease prone. As such the down troden areas of the cities are the prime source of infectious diseases. It is the main cause why India is unable to eradicate certain diseases from the country.

Box 48.1 **Polio**

The disease popularly known as Polio is caused by a virus commonly called Polio virus (*Poliomyelitis*) which is basically classified into three types P1, P2 and P3. The disease is mainly found among children below the age of 5 years. It reaches the child body through infected food or water and multiplies in the intestine. Later, the virus enters in blood and damage the nerve cells of the vertebral column resulting into paralysis. The paralytic attack is sudden and rapid and is incurable because the destroyed cells can not be rejuvnated. Thus, a permanent disability is developed. However, paralytic condition appears only in 1 per cent of the infected child and 99 per cent children remain carrier of the virus which is highly dangerous. The virus is passed out with child's faeces and becomes a constant source of further infection.

In 1985, Rotary International took the challenge to make the world polio free and started the ambitious programme, the *Polio plus*. Rotarians of all over world have contributed US$ 61 crore for the programme, run through out the world (upto 2005). Inspite of arduous efforts, India could not be made polio free because of certain superstitious religion oriented population.

The disease can be prevented by providing oral vaccines called OPV (Oral Polio Vaccine) or IPV (Inactivated Polio Vaccine). These vaccines prevent viral multiplication and discharge with faecal matter.

WHO (World Health Organisation) has identified twenty sensitive spots in India based on reports of 2005. These are Gaziabad, Hardwar, Saharanpur, Meerut, Muzzaffarnagar, Moradabad, Bijnaur, Chandausi, Aligarh, Badaun, Hatharas, Rampur, Mau, Lakhimpur-Kheri, Begusarai, Darbhanga, Samastipur, Nalanda, Khageria, Patna (all from North India). People of these area must provide polio vaccine to their children which is made available in civil hospitals round the year and in different NID (National Immunisation Day) camps organised throughout the region. Amitabh Bachchan gave a popular slogan for the vaccine "**Do Boond Jindegi ke**".

(4) Degradation of Natural Resources

Loss of natural resources adversely affects the quality of human life, *e.g.* deforestation results in biodiversity loss because biodeversity is essential for maintaining the basic life supporting process. A number of drugs and medicines procured from various life forms are essential for human health. (*Effects of deforestation on environment* has been discussed in the Unit-2)

As such a clean and green environment is the life of life. A time to take corrective steps is still there, if taken care of. Now a new concept of development is needed that emphasises the relation between human beings and nature. A sustainable development is the only solution to save human life and health.

Chapter Summary

Environment and human health are intricately related. The impact of environment is accounted on the basis of the extent to which *(i)* environmental conditions lead to shortening of life, *(ii)* environment induced disability is found, and *(iii)* the biological potential of an individual is reduced. The various environmental factors which affect human health are *(i)* overpopulation, *(ii)* pollution, *(iii)* urbanisation and *(iv)* degradation of natural resources. All these factors individually harm to the human health and collectively their effects get multiplied.

Study Questions

1. Describe the various effects of environment on human health.
2. Write short notes on the effects of following factors on human health:
   (i) Overpopulation
   (ii) Pollution
   (iii) Urbanisation
   (iv) Degradation of natural resources
   (v) Air pollution

Objective Questions. *Select correct answers*

1. The pulmonary irritant found in air pollution is:
(1) Water vapours (2) Oxygen
(3) Sulphur oxides (4) Nitrogen gas

2. A disease not caused by water pollution is:
(1) Jaundice (2) AIDS
(3) Cholera (4) Dysentery

Answers

1. (3) *2.* (2)

49

CHAPTER

Human Rights

LEARNING OBJECTIVES

Human Rights • Universal Declaration of Human Rights • Human Rights and Environment • Fundamental Rights • Human Rights in India

Human Rights means "the rights relating to life, liberty, equality and dignity of the individual guaranteed under the constitution or embodied in the International Covenants and enforceable by courts in India – Section 2, Protection of Human Rights Act, 1993". *International Covenant* means the International Covenant on Civil and Political Rights and the International Covenant on Economic, Social and Cultural Rights adopted by the General Assembly of the United Nations on the 16 December, 1966 (*Man*. 2006).

Human rights are inherent rights which can be enjoyed by them simply because they are humans. Every human being is entitled to enjoy these rights without any distinction of race, colour, sex, language, religion, political or other opinion, national or social origin, property, birth or any other status. The characteristic of human rights include the:

Right to:

(1) Life, liberty and security of persons.
(2) Adequate standard of living.
(3) Seek and enjoy asylum from persecution in other countries.
(4) Freedom of opinion and expression.
(5) Education, freedom of thought, conscience and religion.
(6) Freedom from torture and degrading treatment.

World is full of disparity wherein one third population is enjoying two third amenities while two third population has to contend with less than one third of global resources. In order to have respect for individual human dignity, the thought of human right was seriously considered by United Nations Organisations long ago.

Universal Declaration of Human Rights

The Universal Declaration of Human Rights was prepared after about three years of continuous efforts by the Human Rights Commission, and it was accepted by UN General Assembly on 10 December 1948.

The Declaration comprises following **30 articles**.

Art. 1. All human beings are free by birth and equal in rights and respect. They possess reason and intelligence. Therefore, everyone should have fraternal behaviour towards others.

Art. 2. Without distinction of race, colour, sex, language, religion and social origin, birth or other distinctions, each human being is worthy of enjoying the rights and freedoms mentioned in this declaration.

No distinction will be made on the basis of political level of any place or country of which one is the citizen, irrespective of the fact whether that country is free or protected, sovereign or limited in any respect.

Art. 3. Every one has the right to life, freedom and security.

Art. 4. No one can be kept under slavery. Slavery system will be prohibited everywhere.

Art. 5. No one will be given inhuman punishment nor shall be subject to cruel or insulting behaviour.

Art. 6. Every one shall have a right to be considered subject to law everywhere.

Art. 7. All shall be equal before law and shall have the right to legal protection without any distinction.

Art. 8. Every one shall have a right to have protection for national courts against violation of fundamental rights granted by the constitution or law.

Art. 9. No one will be illegally arrested, imprisoned or exterminated.

Art. 10. Everyone shall have a right of proper and open hearing in a free and impartial court against the charges leveled against his rights and duties.

Art. 11.

(i) Any person charged of a punishable crime shall be considered innocent till he is proved guilty in an open trial. He will be given full opportunity to prove himself non-guilty.

(ii) No one will be considered guilty of committing a crime which was not considered punishable by any national or international level at the time of its commitment. He can not be given more punishment than the punishment prescribed for it while the crime was committed.

Art. 12. There shall be no arbitrary interference in the confidential nature of any family or familial correspondence and its respect and dignity will not be hurt.

Art. 13.

(*i*) Everyone shall have a right to move and stay within the boundary of his state.

(*ii*) Everyone shall have a right to leave and again return to any country, including his own.

Art. 14. In order to avoid misery and insult, every one has a right to seek asylum in a country and live there comfortably.

Art. 15.

(*i*) Everyone shall have a right to nationality.

(*ii*) No one can be arbitrary deprived of his nationality nor shall he be denied the right to change nationality at will.

Art. 16.

(*i*) Every adult male and female shall have a right to marry and establish family without limitation of race, nationality or relation.

(*ii*) Marriage will be solemnised only among consenting male and female with their complete freedom and sanction.

(*iii*) Family is the natural and original unit of society and has the right of protection from the society and state.

Art. 17.

(*i*) Every one has a right to keep property himself and others.

(*ii*) No one can be forcefully dispossessed of his property.

Art. 18. Every one has freedom of expression of opinion and thought.

Art. 19. Every one has right of freedom of thought, feeling and religion.

Art. 20.

(*i*) Everyone has freedom to peaceful assembly.

(*ii*) No one can be forced to join any institution.

Art. 21.

(*i*) Every one has right to participate in the government of his country through elected representatives.

(*ii*) Everyone has a right to position in government services of his country according to his merit.

(*iii*) Public opinion shall be the basis of right of administration and the public opinion will be decided by ballot.

Art. 22. Everyone shall have a right to social security.

Art. 23.

(*i*) Everyone has a right to work, choose a vocation for earning his likelihood, attain proper conditions of work and defend against unemployment according to his instincts.

(*ii*) Everyone has a right to equal wages for equal work without any distinction.

(*iii*) Everyone has a right to get suitable remuneration for his work.

(*iv*) Everyone has a right to organise and join labour union to safeguard his interests.

Art. 24. Everyone has a right to leisure and vacation.

Art. 25.

(*i*) Everyone has a right to maintain a suitable standard of life.

(*ii*) Every mother has a right to maternity, special care of the young and attainment of maternity aid.

Art. 26. Everyone has a right to education. The aim of education shall be to fully develop human personality and to increase the respect of basic freedoms.

Art. 27. Everyone has a right to culture. Everyone can participate in cultural activity without distinction and profit by his genius.

Art. 28. Everyone has a right to such national and international system by which he may completely attain the rights and freedoms mentioned in this declaration.

Art. 29.

(*i*) Individual has some duties towards the society by fulfillment of which his personality may attain free and full development.

(*ii*) Everyone shall ramain within the boundaries fixed by law to use his rights and liberties.

(*iii*) Enjoyment of any right and freedom against the objectives and principles of UNO is not admissible.

Art. 30. No interpretation should be made of any order given in this proclamation by which any group of states or individual may have a right to be attached to work or do it whose objective may be destruction of any right or freedom.

Human Rights and Environment

During World Earth Summit, Rio de Janeiro (1992), sustainable development and the protection of natural resources, wildlife and environment was a prime topic of discussion. As a result in the conference at Geneva, 1994, the first ever **'Declaration of Human Rights and Environment'** was drafted which affirms-

(1) The human rights to get healthy, safe and secure environment free from degradation and pollution.

(2) The human rights to enjoyment of natural ecosystems and their rich biodiversity in just and equitable manner.

(3) The human rights to lead a dignified life and fulfill their needs.

(4) The human rights to environmental information, awareness, education and participation in environmental decision making.

(5) The right of future generations to get healthy environment and to fulfill their own needs.

It also affirms the duty of every one including individuals, governments, international organisations and multinational cooperations for the protection and preservation of the environment and for the prevention of environmental damages.

Fundamental Rights

All democratic states constitutionally provide some rights to every citizen. Citizens have legal supports behind them and these rights can not be deprived by any government. Such rights are called fundamental rights as human development is not possible without them.

Significance of fundamental rights are–

(i) These rights are necessary for the development of human life, and they assure physical, mental and moral development.

(ii) These rights are necessary because without them no one can make life happy and prosperous.

(iii) These rights have been guaranteed by the Constitution of India. If any government tries to snatch them away, one can go the court to get justice.

Some important fundamental rights are:

(1) Right of Equality (Articles 14 to 18). All citizens of the country have equal rights to get a government job. All other title and untouchability have been abolished except the educational degrees.

(2) Right of Freedom (Articles 19 to 22). Every Indian citizen has following fundamental freedoms-

(i) Freedom of speech and expression.

(ii) Right to assemble peacefully without arms.

(iii) Right to form associations.

(iv) Freedom of movements.

(v) Freedom to reside in any part of India.

(vi) Right to buy, keep and dispose off property.

(vii) Freedom of Profession.

(3) Right against Exploitation (Art. 23 and 24). No one can sell or purchase any man or woman. The children below 14 years can not be employed in a factory or any other dangerous work.

(4) Right to Freedom of Religion (Art. 25 to 28). India has been declared as a secular country. Any citizen can follow, practice and preach any religion he likes and has faith in the same. Freedom to religion has been granted to every individual.

(5) **Cultural and Educational Rights** (Art. 29 and 30). Every citizen of the country has right to follow and protect his culture and right to get admission to any government or aided educational institution.

(6) **Right to Constitutional Remedies** (Art. 32 to 35). Every citizen has right to protect their fundamental rights and can move to the supreme court or any other high court of India.

Human Rights in India

India is one of the member states of the UN General Assembly and a signatory of Human Rights. In October, 1993, Indian Govt. has set up a National Human Rights Commission (NHRC). In its Section 2 of the 'Protection of Human Rights Acts, 1993', Human Rights have been elucidated as-

"Human Rights means – the rights relating to life, liberty, equality and dignity of the individual guaranteed by the constitution or embodied in the International Covenant on Economic, Social and Cultural Rights."

It was accepted by President of India to enforce it from January 8, 1994.

Article 3 provides provision for constituting a Commission of Human Rights comprising:

(i) Chairman –any former Supreme Court Justice of India.

(ii) Member –any former High Court Justice.

(iii) Member –any former or present High Court Judge.

(iv) Two members –experts and experienced.

Its central office is in Delhi. The Commission will take Chairmen of National Minority Commission and National Schedule Castes and Scheduled Tribes Commission of India as additional members. Indian Courts of different levels have been authorised to take decision on cases of Human Rights and can punish to the defaulters.

Human Rights Commission can review the safeguards provided by our Constitution or Law for the protection of rights and recommend measures for their effective implementation. The Commission, can also utilise the services of any officer or investigating agency of Central Government or any State Government, as and when required.

Chapter Summary

Human Rights means the rights relating to life, liberty, equality and dignity of the individual guaranteed under the constitution or embodied in the International Covenants and enforceable by Court in India-Section 2, **Protection of Human Rights Act, 1993.** It was adopted by UN General Assembly on 16 December 1966. The declaration comprises 30

Articles. Declaration of Human Rights and Environment was drafted in the Conference at Geneva, 1994. Fundamental Rights have legal supports and these can not be deprived by any government. These rights are necessary for the development of human life and they assure physical, mental and moral development. These are must for happy and prosperous life and provide Right of Equality and Freedom. Govt. of India has established National Human Rights Commission (NHRC) to look after the interest of Human Rights.

Study Questions

1. What are Human Rights? Discuss them in legitimate brevity.
2. Describe various Human Rights.
3. Write short notes on:
   (i) Human Rights
   (ii) Protection of Human Rights Act, 1993
   (iii) Characteristics of Human Rights
   (iv) Universal Declaration of Human Rights
   (v) Human Rights and Environment
   (vi) Fundamental Rights
   (vii) Human Rights in India
   (viii) National Human Rights Commission

Objective Questions. *Select correct answers*

1. Protection of Human Rights Act in India was enacted in:
   (1) 1950 (2) 1981
   (3) 1993 (4) 2004

2. National Human Rights Commission in India became enforced in:
   (1) 1991 (2) 1994
   (3) 2001 (4) 2004

Answers

1. (3) *2.* (2)

50

CHAPTER

Value Education

LEARNING OBJECTIVES

Definition of Value Education • What is Education • Distorted Aims of Education • Demolition of Value Education • Value Education and Vedanta • Education and Value Education • International • National and Individual Values • Cultural values • Value Education in Education System

Defining Value Education

Monica Thapar (2001) defined that *value education is education in values and education towards the inculcation of values*. Such a definition, provides conviction that value education is a universal phenomenon intrinsic to all learning and education, whether at home or in an institution. But it is incorrect in totality. At present through our habits, we accept most things handed over to us by the media, the government and the polity. This leads inadequate development of individual capabilities and potentialities due to incomplete imbue of self confidence. Personal individual power to make a difference remains largely unidentified. Moreover our education system is of little help to teach life values that are the basis of creative thinking. Neither it is made to learn that personal values such as honesty, charity, personal hygiene and tolerance make man an ideal member of the society, and as a part of human community, time tested values and morals give rise to harmonious living and peace in society. Integrity towards ones' own country, a spirit of brotherhood, religious tolerance and respect for the freedom of others are now essential values that firmly bind our social fabric together. Then a question arises what is education?

What is Education

Education in its proper sense is neither literacy nor information but a systematic attempt towards human learnings, subjective and self related.

Educational activity starts with the individual – Who am I? Where am I going? What exactly is my social niche? Such education will develop understanding of the self to understand our relationships with the others and the environment. It will make us to learn personal, social, national and universal values to provide harmonious living, peace in society and sustainable development. Integrity towards own country, a spirit of brotherhood, religious tolerance and respect for freedom of others, should be emphasised for modern successful social set up. Education must provide knowledge of individual reality and relevance of an individual.

Distorted Aims of Education

Swami Dr. Parthasarathy (2006) stated that-

"Human excellence through education is an effort to stimulate our thinking in that direction. We study books, hear class lectures, take part in sports, sit for and pass examinations. All these are certainly part and parcel of education, but the most important part of education is shaping our mind and character, the training of our mind to fit it to shape ourselves and our environment so as to fulfill our life's purposes. This is the most important work we ought to have today in all our education."

The world is now full of great scientific achievements with vast knowledge of macrocosmic and microcosmic universe. We have plenty of comforts and speedy means of travel and communication of ideas. We have gone to the moon, and one is now racing towards the outskirts of the solar systems and into outer space. But something vital is missing in ourselves. That missing is revealed in its bitter fruit in the worldwide phenomenon of tension, anxiety and unhappiness within the man, and the psychic and social distortions resulting from them. Certainly these can not be aim of education.

Demolition of Value Education

We can fairly evaluate the significance of value education and disastrous effects of its demolition from a statement made by Lord McCauley on February 2, 1835 in the British Parliament. In his words-

> *I have travelled across the length and breadth of India and I have not seen one person who is a beggar, who is a thief. Such wealth I have seen in the country, such high moral values, people of such high caliber, that I do not think we would ever conquer this country unless we break the very backbone of this nation, which is her spiritual and cultural heritage and therefore, I propose that we replace her old and ancient education system, her culture, for if the Indians think that all that is foreign and English is good and greater than their own, they will lose their self esteem, their native self culture and they will become what we want them, a truly dominated nation.*

This quotation exposes the fact that what we were and now what we are. It was McCauley who introduced chalk and talk culture of education in India and the result was our slavery. Think, the disastrous effects of demolition of value education. A famous saying is that "should you want to peacefully destroy the nation, devalue its education and its religion".

Value Education and Vedanta

Developed in India ages ago, the science and philosophy of *Vedanta* revealed a universal, comprehensive, rational and practical philosophy of life possessing high intellectual and spiritual prestige. It has attracted the attention of modern scientists.

In a critical spirit of truth seeking, British astronomer Arthur Eddington said–

> *You can neither understand the spirit of true science nor of true religion, unless you keep seeking in the forefront.*

The philosophy of *Vedanta* breaths with such scientific temper and spirit. It is the product of a sustained rational questioning and investigation of experience, of inquiry into the true nature of man and his destiny conducted by brilliant minds of ancient India thousands of years ago. They knew value education and preferred it, only.

Education and Value Education

Mundaka Upanishad (1.5) revealed that knowledge is not only all positivistic knowledge, such as physical sciences, economics, politics, sociology, but also all the world's ethnical religions and their holy scriptures. Further, it explains that *wisdom* is matured knowledge and value education is a part of wisdom. A simple example neatly explains to a man that the lady is beloved of your father is *knowledge* but she is called mother is *wisdom*. Another example, when you introduce yourself as Mr. so and so, with qualification you reveal knowledge but it leaves out many physical and mental possibilities, *i.e.*, wisdom, or a part of value education which you have. Certainly your essence of personality lies in value education or in your wisdom.

International, National and Individual Values

Pt. Jawahar Lal Nehru, the first Prime Minister of India, advocated the **Panchsheel** for **value based international relations** between nations. The Panchsheel refers to –

(1) Mutual respect for each other's territorial integrity and sovereignty.
(2) Non-aggression.
(3) Non-interference in each other's internal affairs.

(4) Equality for mutual benefits.
(5) Peaceful coexistence.

Similarly, at **national level** value based education (or knowledge) should be given to obey following **fundamental duties** to realise the true spirit of secularism and democracy.

1. To cherish and follow the noble ideals which inspired our national struggle for freedom.
2. To protect the sovereignty, unity and integrity of India.
3. To defend our country by national service.
4. To promote the spirit of common brotherhood amongst all the people of India.
5. To preserve the rich heritage of our composite culture.
6. To abide by the constitution and respect the national anthem.
7. To protect and improve the eco-environment of the country.
8. To develop scientific attitude and progressive spirit.
9. To safeguard public property and socio-economic life.
10. To strive towards excellence in all spheres of individual and collective activity.

At **individual level,** each one of us must identify the values we want to live. Once we are clear about values we shall be better able to sift and control information of the natural world, make choices and be creative in our mental process. Confucius once said '*If a man carefully cultivates values in his conduct, he may still err a little but he won't be far from the standard of truth.*' It is time to clarify those values which are much talked about. It is up to each one of us to determine the society we will create by deciding upon the values we will emphasise today.

Values can be experienced as life, joy, brotherhood, love, compassion, service, bliss, truth and eternity. However, individual values are reflected in individual goals, vows, relationships, commitments and personal preference, *e.g.* a family may be one's liability or a pleasure.

Cultural Values

Cultural values are social values of the day. They are specific to time and place and can be used as much as misused. These are meant to maintain social order as these values are concerned with good and bad, right or wrong, mutual customs and behaviours. These may serve a function in a particular situation or circumstance but in no way should be taken as the only or the best way of doing things. When you elevate cultural values to the status of universal values, there is the risk of intolerance, oppression, demagoguery, brutality and aggression. Human history substantiates these facts. The study of other cultures provides a wider frame of reference. These should be studied through sacred traditions and not only through studying history.

Cultural values are reflected in language, ethics, social hierarchy, aesthetics, education, law, economics, philosophy and social institutions of every kind. These must be respected and appreciated.

Value Education in Education System

At **infant** stage, the first nursery of values of child is their parent in schools, called homes. At this stage only parents teach family traditions, values and customs which have been followed by them for generations.

In **schools**, the children are exposed to more secular and democratic environment where they have to interact with various communities and religions. Such multicultural environment provides a set of social and cultural values that help them to participate openly and whole heartedly in the process of education. Teachers appear in role model and they should give value education enriched with values of simplicity, punctuality, truthfulness, commitment to purpose, self confidence and courage and teach through personal examples.

At **higher level**, the value education is to be made the core of personality development with respectful awareness of social, cultural and environmental values. Such education should be embedded in the curriculum, as-

(1) Literature can emphasise on human values expressed through prose, poetry or writings, as such patterns bring out positive emotional responses.

(2) Social studies emphasise the study of human beings and their interactions.

(3) Historical studies reveal perspective on ancient, medieval and modern events that shaped the destinies of man.

(4) History also traces the track of progressive civilization and their evolutionary patterns which increases worldly wisdom of students.

(5) By history, students come to know the destructive effects of wars which cultivate in them the need for national integration and international cooperation.

(6) Understanding cultural values promote tolerance among students.

(7) Geographical studies reveal knowledge of different places, climate and culture. These imbue international peace and harmony and understanding of life in all its diversity.

(8) Sports and physical education develop universal and wholesome value among students. These breed healthy competitive feelings.

(9) Field trips and educational excursions reveal man's exploitative use and abuse of nature and cultivate among students a love for nature and thoughts of its conservation.

(10) Environmental studies make students to realise the value of nature and adverse effects of deforestation and pollution on human health and quality of life. Such studies make them environment-friendly, especially for the betterment of future.

No responsible man, society, community or nation can ignore value education as it shapes their future. Our ancient India has great giving in value education and one can not forget the contribution of Gurukul System of education and wisdom of Bhagwad Gita. We develop respect when we utter following lines of a sanskrit hymn:

Sarve bhavantu sukhinah
Sarve santu nirmayah
Sarve bhadrani pashyantu
Ma Kascit dukha-bhak bhavet

"May all (beings) be happy; may all (beings) be free from ailments (physical or mental); may all (beings) experience what is good and auspicious; and may none by subject to sorrow and suffering."

To a well groomed value educated person, the whole world is a family, *i.e.*,

Udar charitanam vashudhaiv kutumbakam

Chapter Summary

Value education is education in value and education towards the inculcation of values. Education is neither literacy nor information but a systematic attempt towards human learnings. Education without value education is golden but not gold. The world is now full of great scientific achievements but is enriched with tension, anxiety and unhappiness. Certainly, it can not be aim of education. Demolition of value education resulted slavery in India.

Education must inculcate values and wisdom. Wisdom is matured knowledge and value education is its integrity. Education must be made to learn and obey the International, National and Individual values. It will increase international relations, sovereignty, unity and integrity of India and peaceful, respectful and healthy society and life.

Cultural values are social values that teach us to maintain social order. It develops tolerance and worthyness.

The need of the hour is to introduce value education in educational curricula of the country to imbue awareness of social, cultural and environmental values for healthy growth of civilisation, humanity and prosperity. Studies on literature, sociology, history, geography and environment are essential parts of value education that provide insight in one or other way in developing the valued human life. In ancient India, value education was most preferred, as substantiated by Gurukul system of teaching and wisdom of Bhagwad Gita.

Study Questions

1. What is value education? Evaluate its merits.
2. Defining value education, explain why is it essential to introduce it in various educational curricula of the country.
3. Write short notes on:
   - *(i)* Value education
   - *(ii)* Education and value education
   - *(iii)* Effects of demolition of value education
   - *(iv)* Value education and Vedanta
   - *(v)* International values
   - *(vi)* National values
   - *(vii)* Individual values
   - *(viii)* Cultural values
   - *(ix)* Value education in education

Objective Questions. *Select Correct answers*

1. Panchsheel concept of value education was proposed by:

(1) W. Churchill (2) Mahatma Gandhi

(3) J.L. Nehru (4) Ho Chi Minh

2. Vedanta studies are full of:

(1) History (2) Religion

(3) Value education (4) Astrology

Answers

1. (3) *2.* (3)

51

CHAPTER

HIV/AIDS

Learning Objectives

Introduction • HIV (Human Immuno-deficiency Virus) • Structure of HIV • Genetics of HIV • AIDS (Acquired Immuno-Deficiency Syndrome)- Symptoms • Mode of Transmission • Causes • Prevention • Global Magnitude of AIDS • World AIDS Day and Campaign • AIDS in India and its Mitigation • National AIDS Control Organization (NACO) • Treatments

Acquired Immuno-Deficiency Syndrome (AIDS) is a slow but fatal disease caused by *Human Immuno-deficiency Virus* (HIV). Now no other word engenders as much fear, revulsion, despair and utter helplessness as AIDS.

The first infected sample of HIV was discovered in 1959 which was a blood sample obtained at Leopoldville (now Kinshasa), Belgian Congo, and it was the first known death due to AIDS. It is believed that HIV originally affected chimpanzees from where it passed on to human beings in 1950 through some accident or a bite. All early cases of AIDS originated in the Central African States of Congo, Rwanda or Burundi. The monkey harbor SIV (Simian Immuno-deficiency Virus) is thought to be the ancestor of HIV. Dr. Albert Sabin and Louis Pascal reported that polio vaccines obtained from chopped up chimpanzee kidneys may have carried this virus.

The first cases of AIDS were reported in 1981 among male homosexuals in Los Angeles and New York (USA). It spread globally in last two decades and up to 50 million people might have been infected at the rate of 15000 new infections occurring daily. In 2001, UN called for global fund to fight against AIDS and in 2003, $15 billion US plan for AIDS relief was announced.

HIV (Human Immuno-deficiency Virus)

HIV is a retrovirus that provokes an infection and destroys the body's immune system. AIDS is the advanced stage of this disease when the immune system becomes irreparably damaged, engendering multiple infections and cancers. A person is considered HIV positive when he tests positive for any of 26 diseases covered in Kaposi's sarcoma, lymphoma, pulmonary tuberculosis, wasting syndrome, recurrent pneumonia etc.

Structure of HIV. It is a retrovirus consisting of RNA surrounded by lipo-protein envelope. HIV is a retrovirus because its genetic material is RNA which replicates in the presence of enzyme *reverse transcriptase*, *i.e.*, the RNA first synthesises DNA and this DNA transcribes to form viral RNA.

HIV has two major categories: HIV-1 and HIV-2. HIV–1 has about 10 subtypes and is most common worldwide. HIV-2 is less virulent and is currently confined to West Africa, though spreading steadily.

Genetics of HIV. Since HIV can not multiply at its own, it uses host machinery to produce new viral particles. The main cellular target of it is a special group of white blood cells (WBC) known as helper T-lymphocytes. Once the virus has infected the T-cells, it begins to multiply resulting the loss of most of helper T-cells.

During replication, the virus copies its RNA into a double stranded DNA copy with the help of viral enzyme *reverse transcriptase*, the process is called reverse transcription, *i.e.*,

$$\text{RNA} \xrightarrow[\textit{transcriptase}]{\textit{Reverse}} \text{DNA} \xrightarrow{\textit{Transcription}} \text{RNA}$$

Since the viral enzyme lacks proof reading mechanism, several mutations may arise due to errors during replication. The genetic variants

$$\text{RNA} \xrightarrow[\text{Reverse transcription}]{\textit{Reverse transcriptase}} \text{DNA}$$

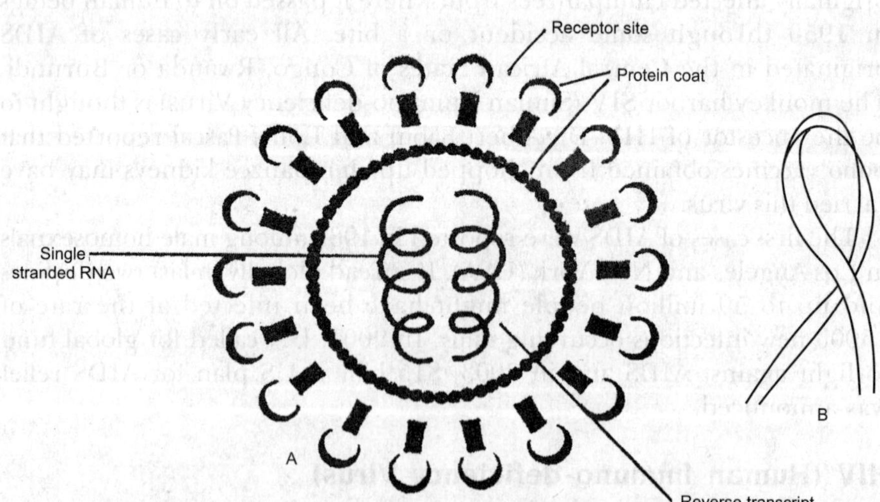

Fig. 51.1. A-structure of HIV and B-international sign (logo) of AIDS.

of HIV have been categorised into several major sub-types which may have different geographical distribution. Such a mechanism leads to rapidly evolve new variants regularly, which allow the virus to escape from antiretroviral drugs and antiviral immune responses. Also due to rapidly changing surface proteins, virus can avoid recognition by human antibodies.

The viral genome copied on DNA, transcribes to produce several new viral RNAs, some of these are incorporated into new RNA particles while others are used as messenger RNA for the production of new viral proteins. Still some others manage to get integrated into host cell DNA and result in the permanent acquisition of viral genes by the host cell.

Most infected T-cells die quickly in about one day. Eventually number of T-cells begins to decline. This decline is measured by CD-4 count method which provides a good indication of the status of the immune system in the human body. The CD-4 molecules present on the surface of T-cells are the sites of parasitization of HIV (or AIDS Virus). National HIV testing strategies involve ELISA (Enzyme-Linked Immunosorbent Assay) test which is performed on serum using laboratory based equipments.

Recently, the first ever rapid home test kit for HIV, Ora Quick has been approved in USA. It analyses saliva and result is available in just 20 minutes. The two blue lines indicate HIV +ve. The kit is very simple.

AIDS (Acquired Immuno Deficiency Syndrome)

AIDS has been defined as any sign or symptom of acquired cellular immunodeficiency in a person having confirmatory positive HIV tests.

Symptoms of AIDS. The term syndrome is used to refer a jumble of symptoms. Symptoms of the disease (AIDS) include loss of apetite and weight, constant fever, prolonged fatigue, diarrhoea, constipation, changing bowel patterns, swollen glands, chills with excessive sweating, lesions in the mouth, sore throat, persistent cough, shortness of breadth, skin rashes, tumours, headache, memory lapses, swelling in joints, body pain, vision problems, regular feeling of lethargy, ill health etc.

HIV positive persons are susceptible to various types of cancers including KS (Kaposi's Sarcoma) which occurs in the mucous membranes of the eyes, nose and mouth.

Mode of transmission of AIDS. HIV is transmitted mainly by unprotected anal, vaginal or oral sex with an infected partner; by sharing contaminated injection needles (used in infected person for drugs or blood transfusion); or cuts due to shared blades and razors or from infected mother to child through pregnancy or breast feeding. Unprotected sex with multiple partners, infected semens, infected vaginal fluids, infected blood and blood products and infected parents are the primary sources of HIV transmission.

Studies reveal that needle sharing and receptive anal sex appear to have the highest risk, with transmission rates estimated to be between 1 per cent and 3 per cent per exposure. Occupational needle stick injuries, receptive vaginal sex, insertive and/or vaginal sex and receptive oral sex have estimated transmission risks <1 per cent per exposure. Most efficient is infected blood transfusion, while most transmissions happen through sexual contacts due to the frequency of acts.

HIV particles are not transmitted by water, air, mosquito bites or superficial contacts with diseased persons. The viral particles are fragile and short-lived outside the host.

AIDS is not an infectious disease rather it is a type of contagious in nature and that is why it can not be transmitted through air (by sneezing, coughing or breathing), water (sharing swimming pools or bath tubs), casual physical contact (touching, hugging or kissing), toilets (sharing), or through mosquitoes.

Course of infection. The course of infection involves three stages:

(i) *The primary HIV infection*. During first stage (lasting for one or two weeks), the HIV particles replicate rapidly in the host cells. Some persons may experience a variety of symptoms such as flue-like illness, fever, sore throat, enlarged lymph nodes, rashes, malaise, muscle and joint pains etc. Standard HIV tests normally remain negative.

(ii) *The asymptomatic phase*. It is the second phase of HIV infection which lasts for 8-12 years. Viruses without producing symptoms continue to replicate and reduce the number of helper T-cells. However, irregular symptoms related with defective immune system may appear.

(iii) *AIDS, the final stage of HIV infection*. Due to loss of immune system in the body, infections that may appear cause pneumonia,

Box 51.1 Summary of HIV Transmission Routes/ Mechanisms

1. Sexual transmission (vaginal, anal or oral)
   (i) Heterosexual
   (ii) Homosexual
2. Blood contact
   (i) Blood transmission
   (ii) Intravenous drug use
   (iii) Occupational exposure
   (Needle stick, cuts, splashes)
3. Mother to Child
   (i) In-utero
   (ii) During delivery
   (iii) Breast feeding

tuberculosis, herpes, toxoplasmosis etc. Patients can develop dementia and certain cancers including KS and lymphomas. Death ultimately results, as the body fails to fight against pathogens or malignancies.

Prevention. Prevention is the only solution of AIDS because it is a high risk disease. It can be prevented with proper precautions but it requires a greater awareness. Some preventive measures are:

(1) High risk sexual behavior must be avoided by all means. These include – oral sex, oral anal sex, vaginal sex without latex condoms, sharing of sex toys, using saliva for lubrication, fisting or manual anal intercourse and blood contact of any kind during performance. If woman is infected, avoid sex during menses.

(2) Latex condoms for male and female protect both partners. Only water based lubricates should be used as oil lubricants tear latex condoms.

(3) Use of spermicidal foams and jellies in addition to condoms provide protection. Only spermicidal jellies may not be effective.

(4) Monogamy is safe but polygamy (having multiple partners) is quite risky.

(5) Alcohol or drugs should be avoided as it may lose control of senses and one gets engaged in unsafe sex.

(6) Disposable syringes and needles must be used for transfusions. Only HIV free (tested) blood should be used for transfusions. Preference should be given to the blood of close family members.

(7) The presence of sexually transmitted diseases increase the risk of transmission.

(8) HIV positive mother should not breast feed her baby. The infant should be given AZT for the first several weeks to reduce the infection risk.

(9) One should not get shaved by shared blade or razor at barber shop.

Global Magnitude of AIDS

High risk disease AIDS is moving with high speed globally. The UN AIDS, *epidemic update of December 2004* reveals that every day some 15,000 new member join AIDS group. The number of people living with HIV has been rising almost in every region of the world with steepest increase in East Asia, Eastern Europe and Central Asia in the last two years. The rise is about 50 per cent in East Asia and 40 per cent in Eastern Europe and Central Asia between 2002 and 2004, but it is 64 per cent in Sub-Saharan Africa. As per UN Report (May, 2006), some 38.6 million people world wide are living with HIV.

The report reveals that in 2004, 19.6 million men, 17.6 million women and 2.2 million children under 15 years live with AIDS. Some 3.1 million people died due to AIDS and about 4.9 million people were newly infected in 2004. About 25.4 million people live with HIV in Sub-Saharan Africa.

AIDS is spreading fast among women and girls. The report reveals that nearly 50 per cent of some 37.2 million adults living with HIV in 2004 were women, which were just 35 per cent in 1985. AIDS is the leading cause of death for African-American women between the ages of 25 to 34 years. Five to six times higher victims are adolescent girls. According to UNICEF: *State of World's Children 2005* report, the number of AIDS orphans has grown from 11.5 million in 2001 to 15 million in 2003. It is estimated that the number of adults and children newly infected globally (2007) with HIV reached a total of 2.5 million, of which South and South-East Asia alone contributed 3,40,000. Estimated adult and child deaths from AIDS during 2007 would be a total of 2.1 million, of which South and South-East Asia alone will contribute 2,70,000. Millions of them live in households with sick and dying family members. Many of them are forced to drop out of school in order to work or care for their families.

UNAIDS *epidemic update* of May, 2006 revealed that by 2025, AIDS would kill 31 million in India, 18 million in China and 100 million in Africa. It would prove worst and deadliest epidemic that humankind has ever experienced.

Box 51.2 **Some Facts About AIDS**

(1) Some 12 million AIDS orphans (children who lost their guardians due to AIDS) are found in South Africa (Rotarian, Nov. 2006), many of them can't afford school.

(2) The World's highest epicentre of the AIDS pandemic is KwaZulu -Natal province of South Africa.

(3) According to UNAIDS report, in 2005, South Africa with 5.5 million HIV cases was topped by India, with 5.7 million, for the first time.

(4) UNAIDS report predicts that by 2010, there will be 40 million AIDS orphans in 9 countries in sub-Saharan Africa. Before the first AIDS case was reported 25 years ago, about 2 per cent of South African children had lost one or both parents. Today, that figure stands at 10 per cent.

(5) There are many reasons behind the epidemic's harsh grip on South Africa: *(i)* the legacy of apartheid, *(ii)* limited access to antiretroviral drugs, *(iii)* wide spread denial, *(iv)* a tradition of polygamy, *(v)* a high rate of child rape and child brides, *(vi)* government apathy, *(vii)* prevalence of diseases like tuberculosis and malaria etc. Above all, there is poverty. Some backward superstitious people of South Africa believe that the AIDS can be cured by raping a minor girl (below the age of 13).

World AIDS Day and Campaign

The world Summit of Ministers of Health held at London in January 1988 on prevention of AIDS proposed to celebrate 1st December as **World AIDS Day** which has been supported by UN World Health Organisation. Now it is celebrated by various governments, communities and individuals around the world. Such activities create awareness about AIDS among the people. Each year some slogans are selected to focus the theme, some selected themes are *Women and Aids* (1990), *Children Living in a World with AIDS* (1997), *Men Make a Difference* (2000), *Women, Girls and AIDS* (2004) etc.

The **World AIDS Campaign** (2004) developed a slogan "*Have you heard me today*?" This Campaign culminated on World AIDS Day, December 1, 2004. The campaign explored how gender inequality fuels the AIDS epidemic and was conceived to help accelerate the global response to HIV and AIDS by encouraging people to address female vulnerability to HIV. It explained that all over the world women do not enjoy the same rights and access to employment, property and education as men. Women and girls are also more likely to face sexual violence, which can accelerate the spread of HIV. Now more than half of all people living with HIV in the world are female. The campaign posters were developed to give a voice to women facing the challenges of gender inequality in the face of HIV and AIDS.

AIDS in India and its Mitigation

Number of HIV–infected people in India has grown considerably from 2 lakh in 1990 to 5.7 million in December, 2005. Union Minister of Health, Hon. A. Ramdas informed (6th July, 2007) that incidences of AIDS in India have been greatly reduced in the year 2006, and there are only about 2.7 million cases (Reports of NACO, UNAIDS and WHO, 2006). India has an estimated 2.5 m people living with HIV/AIDS in 2007 and is third worst affected country after South Africa (5.5 m) and Nigeria (2.9 m). India's HIV prevalence in the general population is 36 per cent. An estimated 1.7 lakh people died of HIV in 2006 as against 4 lakh estimated earlier. No state is free from this virus. India ranks as the second country in the world, next only to South Africa. However, it is listed among Low Prevalence Countries of the world because only 0.91 per cent (less than 1 per cent) among adult population have HIV infections. The prevalence rate of HIV infections in South Africa at present is 21.5 per cent.

As per distribution of HIV infections in the country, Six states are categorised as **High Prevalence States** (Tamil Nadu, Maharashtra, Karnataka, Andhra Predesh, Manipur and Nagaland) and three as

Moderate Prevalence States (Goa, Gujarat and Pondicherry). However, Assam, Bihar, Delhi, Himachal Pradesh, Kerala, Madhya Pradesh, Punjab, Rajasthan, Uttar Pradesh, West Bengal, Chhattisgarh, Jharkhand, Orissa and Uttarakhand are categorised as **Highly vulnerable** and Arunachal Pradesh, Haryana, Jammu and Kashmir, Meghalaya, Mizoram, Sikkim, Tripura, Andaman & Nicobar Islands, Chandigarh, Dadar & Nagar Haveli, Daman and Diu and Lakhsdweep are categorised as **Vulnerable**.

For monitoring the trends of HIV/AIDS epidemic in the country, the National AIDS Control Organisation (NACO) conducts the survey every year.

National AIDS Control Organisation (NACO)

The Government of India launched a comprehensive National AIDS Control Programme (NACP) from April 1992. It's first phase programme (NACP-1) was conducted from 1992 to 1999 and NACP-2 is currently operative.

NACP-2 has two objectives-

(i) to reduce the spread of HIV infections, and

(ii) to strengthen the capacity of Central and State Governments to respond to HIV/AIDS on a long term basis.

It proposes to achieve these objectives through the following key components-

Component–1. Targeted interventions for communities at higher risk
Component–2. Provision of HIV transmission among general population.
Component–3. Provision of low cost care
Component–4. Strengthening institutional capacities
Component–5. Intersectional collaboration

Under this programme, each State and Union Territory has registered a State AIDS Control Society (SACS) responsible for implementing the programmes. Cities like Mumbai, Chennai and Ahmedabad have formed Municipal AIDS Control Societies to implement NAC Programmes.

NACO considers that Information, Education and Communication (IEC) is one of the most important strategies in the fight against HIV/ AIDS, particularly in the absence of a vaccine or a cure. IEC programme will not only raise awareness levels in general population on prevention of the infection but also bring about behavioural change to adopt and maintain healthy life style. NACO implements its strategies at Nation and State levels and formulates IEC guidelines to provide a framework for its focused and cost effective activities from time to time.

Govt. of India is also providing treatment for opportunistic infections. Provisions have been made for free Antriretrovira Therapy (ART) to HIV positive people. Free ART services were launched from 1 April 2004 and till 30 June 2005, a total of 10250 patients have been placed in 25 ART Centers. Govt. plans to establish a total of 100 medical institutions by the end of current financial year (2006-07).

At present, the government also provides prophylaxis to the HIV positive mothers and their new born babies with single dose of Nevirapine tablets/syrup through PPTCT (Prevention of Parent to Child Transmission) centres established in all Govt. Medical Colleges of High Prevalence States of the country.

NACO also deals with the control of Sexually Transmitted Diseases (STDs) and has established 775 STD clinics till the end of financial year 2004-2005, through which free supply of STD drugs and condoms is offered. But even after all these efforts, cases of HIV infections are steadily increasing, calling more adequate government attention.

Treatment of AIDS

In **treatment**, antiretroviral (ARV) agents are used as drugs against HIV. These drugs inhibit the enzymes- *nucleoside reverse transcriptase* and *protease*, used in viral replication. The three classes of ARV's are –

(i) Nucleoside Reverse Transcriptase Inhibitors (NRTI), *e.g.* Zidovudine (AZT), Stavudine (d4T), Lamivudine (3TC).

(ii) Non-Nucleoside Reverse Transcriptase Inhibitors (NNRTI), *e.g.* Nevirapine (NVP), Efavirenz (EFV).

(iii) Protease Inhibitors (PI), *e.g.* Saquinavir (SQV), Ritonavir (RTV), Indinavir (IDV).

It is important not to spread news information like-

- HIV can be cured –myth
- HIV means AIDS –False
- HIV can be spread by casual contact –myth
- HIV can not happen in good families –false
- HIV can be spread through insect bites –myth
- People with HIV are bad people –false

HIV+ persons can lead healthy life by adopting following schedules-

(1) Prevention of serious infections

(2) Vaccination (Hepatitis A and B, Polyvalent Pneumococcal vaccines).

(3) Nutritious food which includes-

(i) Energy giving foods (carbohydrates and fats) (*e.g.* rice, wheat, maize, cereals, potatoes, bread, bananas, sugar, vegetable, oils, animal fats)

(ii) Body building foods (Proteins) (*e.g.* eggs, nuts, beans, peas, soya, milk, fish, meat)

(*iii*) Health-giving foods (protective against infection and providing vitamins) (*e.g.* all vegetables and fruits, green vegetables, orange or red coloured vegetables).

All they need is a regular careful life and encouraging social patronage, coupled with adequate treatments.

Ms Sujatha Rao, Director General, NACO (2006) has requested-

Let us work together towards a world where every person living with HIV is treated with compassion, dignity and high quality of care and no one is denied access for want of ability to pay.

Bill (the richest man of the world) and Malinda Gates Foundation has announced (Oct. 25, 2006) to donate 2.3 crore dollars to NACO for controlling AIDS in India. The amount is to be used in three years.

There is a great hope that current AIDS drugs might prevent high -risk people from becoming infected. One of these, *tenofovir*, is being tested in several countries. The second drug *emtricitabine* or FTC is another promising drug (HT, June 5, 2006).

Chapter Summary

AIDS is a fatal disease caused by HIV (Human Immuno-deficiency Virus). The first cases of HIV were reported in 1981 and uptil now upto 50 million people might have been infected. HIV is a retrovirus consisting of RNA surrounded by lipo-protein envelope. It replicates in the presence of *reverse transciptase* enzyme. The main cellular target of virus is a special group of WBC known as helper T-lymphocytes (T-cells) only where it multiplies by the process called reverse transcription. Now several genetic variants of HIV are found. Most infected T-cells die quickly in about one day. The decline in number of T-cells is measured by CD-4 count method or by ELISA (Enzyme-linked Immunosorbent Assay) test.

AIDS has been defined as any sign or symptom of acquired cellular immuno-deficiency in a person with positive HIV test. Symptoms of the disease include weight loss, constant fever, swollen glands, lesions in mouth etc. Its causal organism HIV is mainly transmitted by unprotected sex, sharing of contaminated injection needles, blood transfusions, breast feeding etc. HIV particles are not transmitted by water, air, superficial contacts, mosquito bites etc. AIDS is a final stage of infection causing total failure of body's immune systems. Death is the ultimate result. The disease can be prevented by avoiding high risk sexual behaviors, using condoms and spermicidal foams, using disposable syringes during injections and blood transfusion, preventing breast feeding of HIV positive mothers etc.

UNAIDS epidemic update of December 2004, reveals that everyday some 15,000 new members join AIDS group. The report reveals that in 2004, 19.6 million men, 17.6 million women and 2.2 million children under 15 years of age live with AIDS. AIDS is spreading fast among children and it has grown from 11.5 million in 2001 to 15 million in 2003. December 1, is celebrated as World AIDS Day to create awareness against AIDS. In India NACO (National AIDS Control Organisation) is carrying out comprehensive programme to control AIDS with the support of Govt. of India. Some treatment methods have been developed which inhibit the enzymes – Nucleoside reverse transcriptase and protease, *e.g.* Zidovudine, Stavudine, Nevirapine, Sesquinavir etc.

Study Questions

1. Describe the structure and multiplication of HIV.
2. What is AIDS? Describe its symptoms, mode of transmission, causes and Prevention.
3. Write short notes on-
   (i) World AIDS Day and Compaign
   (ii) Genetics of HIV
   (iii) Symptoms of AIDS
   (iv) AIDS in India and its mitigation
   (v) NACO
   (vi) Treatments of AIDS

Objective Questions. *Select correct answers*

1. HIV infects:

(1) RBC (2) Helper T-cells
(3) Nucleic acid (4) Brain

2. World AIDS Day is celebrated on:

(1) 1ST January (2) 5th June
(3) 28th February (4) 1st December

3. ELISA is for testing:

(1) Anaemia (2) T.B.
(3) AIDS (4) None of these

4. Govt. Organisation to control AIDS in India is:

(1) UNO (2) UNICEF
(3) NACO (4) NFWP

Answers

1. (2) *2.* (4) *3.* (3) *4.* (3)

52

CHAPTER

Women and Child Welfare

Learning Objectives

Women and Child Development/Welfare • Child Development/Welfare • Cooperation of UNICEF • National Commission for Children • Awards Related with Child Welfare • Women's Welfare • Awards • Autonomous Organisations • Women's Welfare in High–tech India.

Children are the future of a nation and women nurture them. As such these two are essential for the development of a nation. In developing countries seventy per cent of the population comprises these two.

In India, this population is about 62 per cent, in which 22 per cent are women (age 15 to 44) and 40 per cent are children. These two belong to *special risk group*, being vulnerable in nature. The risks are maternity problems in women and growth, development and survival in the infants and children. In developing nations, the maternity death rates are in between 8-1100 per 1,00,000 live births which is very high as compared to developed countries (2-140/1,00,000). The main problems of the two in our country are: *(i)* Illiteracy, *(ii)* Malnutrition, *(iii)* Infections (health) and *(iv)* Employment and poverty. Government of India has taken several measures towards these lines. Year 2001 was celebrated as *Women Empowerment year*.

WOMEN AND CHILD DEVELOPMENT/WELFARE

In 1985, Government of India has set up the Department of *Women and Child Development* as a part of the Ministry of Human Resource Development. It's objective is to provide the much needed impetus to the holistic development of women and children. The Department formulates plans, policies and programmes, enacts and amends legislations affecting women and children and guides and coordinates the efforts of both governmental and non-governmental organisations working in the field of women and child development. The ultimate objective of these programmes is to make *women independent* and to ensure that *children grow and live in a healthy and secure environment.*

The Government has under its aegis one statutory body and three autonomous organisations.

(a) **Statutory body** National Commission for Women

(b) **Autonomous Organisations**

(i) National Institute of Public Cooperation and Child Development (NIPCCD).

(ii) Central Social Welfare Board (CSWB).

(iii) Rashtriya Mahila Kosh (RMK).

In 1993, Food and Nutrition Board has been transferred to the Department of Women and Child Development, and in 2003, a National Nutrition Mission, has been set up. International Organisations like UNICEF and UNIFEM have formulated five Acts to be administrated by the Department,

(i) Immoral Traffic (Prevention) Act, 1956 (amended upto 1986)

(ii) The Indecent Representation of Women (Prevention) Act, 1986.

(iii) The Dowry Prohibition Act, 1961 (amended upto 1986)

(iv) The Commission of Sati (Prevention) Act, 1987

(v) Infant Milk Substitutes, Feeding Bottles and Infant Food (Regulation of Production, Supply and Distribution) Act, 1992.

Child Development/Welfare

As per 2001 Census of India, the country has 157.87 million children (about 15.42%) below the age of six years. Govt. of India has following plan programmes for their proper welfare:

(1) Integrated Child Development Services (ICDS). Started in 1975, this Centrally Sponsored Scheme has following objectives

(i) To improve the nutritional and health status of children (below 6 yrs.), Pregnant and Lactating mothers.

(ii) To lay the foundation for the proper psychological, physical and social development of the child.

(iii) To reduce the incidents of mortality, morbidity, malnutrition and school drop outs.

(iv) To achieve effective coordination of policy and implementation among various departments to promote child development.

(v) To enhance the capability of the mother to look after the health and nutritional needs of the child through proper health and nutrition education.

The scheme covers a package of *(a)* Supplementary nutrition, *(b)* Immunization, *(c)* Health check up, *(d)* Referral Services, *(e)* Pre-school non-formal education and *(f)* Nutrition and Health Education. Many of these schemes are executed through *Aganwadi*. By December 2004, about 371.11 lakh children and 75.24 lakh women have been benefitted.

(2) Creches/Day Care Centres for Children of Working and Ailing Mothers. The scheme was launched in 1975 under National Policy of Children adopted in 1974. The scheme aims to provide day care services to children (0-5 years) of parents whose monthly income is below Rs. 1800=00 per month and to incapacitated mothers. The services provided under the scheme include Sleeping and Day care facilities, Supplementary nutrition, Immunisation, Medicines and Reactions. Central Social Welfare Board takes care of these schemes supporting 12470 creches covering 3.11 lakh children.

(3) Balika Samridhi Yojana. It started in 1997 with a specific sole objective to encourage the enrolment and retention of girl child in the schools. Also it provides financial assistance to poor family for educating girls.

UNICEF Cooperation

UNICEF programmes related with these problems are being implemented in India through Master Plan of Operations (MPO). MPO of 2003-2007 is currently in operation. The MPO has following objectives:

(i) to empower families and communities with appropriate knowledge and skills to care and protection of children,

(ii) to expand partnerships as a way to leverage resources for children and scale up interventions and

(iii) to strengthen the evaluation and knowledge base of best practices on children.

Box 52.1 **Ban on Child Labour**

India has over 1.2 crore child labours (below the age of 14) and it is globally listed as worst exploiters. India has enacted Child Labour Protection Act, 1986 from October 10, 2006. The new provision in the Act makes it mandatory for the government to rescue children from homes and restaurants and produce them before the Child Welfare Committee before they are sent to shelters. However, Labour Department has stated directly that "We are not likely to go into an over drive and rescue more children unless we can accommodate them safely".

According to Child Labour Protection Act, 1986, the offenders will be penalised. The punishment is imprisonment from three months to one year or a fine of upto Rs. 20,000 or both. Now, every year 12th June is celebrated as **Anti child-Labour Day**.

UNICEF Contributes towards-

(a) Reduction in infant and matron mortality
(b) Improvement in the level of child nutrition
(c) Ensuring universal elementary education
(d) Enchanting child protection
(e) Protection of children and adolescents from HIV/AIDS.

The UNICEF has allocated an amount of US$ 400 million for India for the period from 2003-07.

National Commission for Children

The commission is a statutory body with a bill titled "Commissions for Protection of Child Right Bill, 2005". Under the scheme, assistance is given to voluntary organizations working in the field of Child and Women Welfare. India observes **Universal Children's Day** on 14th November every year (a birth day of Late Pt. J.L. Nehru, the first P.M. of India). The National Child Awards for Exceptional Achievement is given every year (instituted in 1996).

Awards Related with Child Welfare

Several awards have been established by the Government of India related with child welfare, *e.g.*

(1) National Child Award for Exceptional Achievement. (established in 1996, includes 1 Gold Medal & 35 Silver Medals)

(2) National Award for Child Welfare. (instituted in 1979, cash prize of Rs. 3 lakh and a certificate, given to 5 institutions and 3 individuals)

(3) Rajiv Gandhi Manav Seva Award. (instituted in 1994, cash prize of Rs. 1 lakh, given to individual on birth anniversary of Rajiv Gandhi, *i.e.*, 20 August. Award of 2003 was given to Ms Aneena Joseph of Kerala)

WOMEN'S WELFARE

Government of India is making arduous efforts for the uplift of women in the country because it believes that no progress is possible after suppressing their welfare. Some of the major objectives considered by the government are-

(1) To educate women to combat their problems
(2) To provide legal and financial assistance for their education
(3) To make women self reliant
(4) To create confidence and awareness regarding women's status, health, nutrition, education, sanitation and hygiene, legal rights, economic uplift

(5) To involve women in local level planning
(6) To provide facilities and assistance to adopt family welfare programmes
(7) To support women for welfare, health and education of their children
(8) To support and rehabilitation of women in difficult circumstances
(9) To train them about environmental protection
(10) To provide training and skills to facilitate them to obtain employment or self employment on a sustainable basis
(11) To take care of women from weaker sections of the society
(12) To educate them to learn and fight for their rights
(13) To train them to participate equally with men towards the development of the country etc.

According to 2001 Census, 498.7 million (about 48.2%) women live in India. For their social and economic upliftment and empowerment, Government of India carries out following programmes:

(1) **Support to Training and Employment Programme (STEP)** for Women started from 1986-87, and during the year 2004-05 benefit to 13,000 women has been given.
(2) **Swawalamban** programme was launched in 1982-83. The objective is to provide training and skills to develop self employment on a sustainable basis.
(3) **Swaymasidha** is an integrated project based on formation of women into Self Help Groups (SHG). By December 2004, some 59,940 SHGs have been formed in the various parts of the country.
(4) **Swa-Shakti** project (also called Rural Women's Development and Empowerment Project) is supported by World Bank and IFAD (International Fund for Agricultural Development) and is run in nine states of the country.
(5) **Working Women's Hostels** are being set up in the country to provide assistance to NGOs, Cooperative Bodies and Other Agencies. During the year 2004, an amount of Rs. 4.82 crore was sanctioned under this scheme. Some 850 such hostels have been planned to develop by the end of 2006-07.
(6) **Swadhar** scheme was launched in the year 2001-02 for the benefit of women in difficult circumstances. An amount of Rs. 1.24 crore was sanctioned for the year 2004-05 to benefit 1250 women.

Some important programmes and activities run by Govt. of India for the welfare of women in the country are:

(1) Operational Literacy Programme
(2) Occupational Training and Education for Adult Women
(3) Rural Women Board

(4) Women's Hostel
(5) Women Training Centres
(6) Establishment of National Women Fund
(7) Women Environment Scheme
(8) Reservation in Panchayati Raj and Municipalities
(9) National Maternity Benefit Scheme
(10) Indira Women Scheme
(11) Women Empowerment Scheme
(12) Self Dependence Scheme
(13) Self Power Project
(14) Girl Enrichment Scheme
(15) Women Enrichment Scheme
(16) Condensed Course for Women Education
(17) Family Counselling Centres
(18) Primary Health Centres
(19) Girl Empowerment Scheme
(20) Short Stay Homes
(21) Initiates to combat Trafficking of Women and Children
(22) Grant in aid for Research, Publication and Monitoring

Awards. The Government has established some prestigious awards to appreciate the outstanding contributions of women, *e.g.*

(a) **Sree Shakti Puraskar** (given on the name of daughters of India-Kannagi, Mata Jijabai, Devi Ahilya Bai Holkar, Rani Laxmibai and Rani Gaindiliu) is given for outstanding work on women welfare and presented annually on the International Women's Day (January, 8).

(b) **Dr. Durgabai Deshmukh Award on Women's Development** carries a cash prize of Rs. 5 lakh and shawl and given annually to any organisation for outstanding contribution towards women welfare.

Autonomous Organisations

Government of India has established several Autonomous Organisations to accelerate women's welfare, such as:

(i) National Commission for women
(ii) Rashtriya Mahila Kosh
(iii) National Institute of Public Cooperation and Child Development (NIPCCD)
(iv) Central Social Welfare Board
(v) Pension and Pensioner's Welfare etc.

Women's Welfare in Hi-Tech India

Governmental efforts of uplifting and empowering women have proven inadequate due to vastness of country and growing population. However, modern India is shaping in different way due to privatisation and industrialisation with new emerging trend of economic growth.

Women are now stepping out silently but surely for building their independent careers at par with male bastion. Economic liberalisation and the new freedom *mantra* has thrown open a new world of market opportunities for women. Today, women in India can do any job they set their minds to.

In India, some ten years ago, 80 per cent of jobs were in the area of manufacturing, while today the majority are in the service sector. Now government jobs are no more alluring due to availability of wide choice of other exciting options, even job security is no more an attraction.

Several hot jobs for women are available in the country, particularly in the field of:

(i) Information Technology
(ii) Telecom Sectors
(iii) Biotechnology
(iv) Insurance
(v) Health Care
(vi) Personal care and fitness
(vii) Retail sectors
(viii) Entertainment
(ix) Designing and Interior Decoration
(x) Education
(xi) Journalism and Media etc.

The present scenario of the country reveals that in these fields women are not only more successful but earning respect and appreciation so much so that they are preferred over men. Kiran Majumdar Shaw started her career in bio-technology only about ten years back and today, she is the richest and most successful woman of India. Now there are several examples of women's success in the country, and their future appears to be very promising.

Chapter Summary

For a better future of the country Women and Child Welfare is inevitable. Govt. of India has kept them in *special risk group* being vulnerable in nature. The main problems of the two are *(i)* illiteracy, *(ii)* malnutrition, *(iii)* infection (health) and *(iv)* employment and poverty. Year 2001 was celebrated as *Women Empowerment year.*

In 1985, Govt. of India has set up the Department of Women and Child Development under the Ministry of Human Resource Development with an objective for their holistic development. The Govt. has constituted one statutory body (National Commission for Women) and three autonomous organisations-NIPCCD, CSWB and RMK. In 1975, the government sponsored scheme "Integrated Child Development Services" started with several welfare objectives. The scheme covers a package of *(i)* Supplementary nutrition, *(ii)* Immunisation, *(iii)* Health, *(iv)* Referral service, *(v)* Pre-school non–formal education and *(vi)* Nutrition and health education. UNICEF is also cooperating in execution of Master Plan of Operations (MOP) and has allocated US $ 400 million for the period from 2003-07. National Commission for Children is a new statutory body which has enacted Child Rights Bill, 2005. Several awards for outstanding contribution towards Child Welfare have been established by the government.

Govt. of India has maintained manifold objectives for the welfare of the women and several programmes are being executed for their social and economic upliftment and empowerment, *e.g.* Swa-Shakti Programme, Swawalamban programme etc. Govt. has established some prestigious awards to appreciate the outstanding contributions in the field, *e.g.* Sree Shakti Puraskar and Dr. Durgabai Deshmukh Award on Women's Development. Several autonomous organisations to accelerate women's welfare have been established by the government, *e.g.* National Commission for Women, NIPCCD, Central Social Welfare Board etc.

With new emerging trend of economic growth in high-tech India, job opportunities for women are multiplying day-by-day. Today, women in India can do any job they set their minds to.

Study Questions

1. Describe various efforts of Govt. of India for the welfare of children in the country.
2. Discuss the role of Govt. of India in upliftment and empowerment of women in India.
3. Briefly describe the women and child welfare programmes executed in India.
4. Write short notes on:
   (i) Special risk group of the country.
   (ii) Department of Women and Child Development.
   (iii) ICDS.
   (iv) Aganwadi.
   (v) Balika Samridhi Yojana.
   (vi) National Commission for Children.
   (vii) Govt.'s objectives for women welfare.
   (viii) Government's programmes for women welfare.
   (ix) STEP and SHG.
   (x) Autonomous organisations for women Welfare.

Objective Questions. *Select correct answers*

1. Women Empowerment year was celebrated in:

(1) 1919 (2) 1984
(3) 2001 (4) 2005

2. Rajiv Gandhai Manav Seva Award is given for the outstanding contribution in the field of:

   (1) Science (2) Peace

   (3) Child Welfare (4) Economics

3. SHG (Self Help Groups) programme runs under the programme entitled:

   (1) Swadhar (2) Swayamsidha

   (3) Swawalamban (4) Swa-Shakti

4. Department of Women and Child Development of Govt. of India was established in the year:

   (1) 1985 (2) 1984

   (3) 1950 (4) 1972

Answers

1. (3) *2.* (3) *3.* (2) *4.* (1)

53

CHAPTER

Role of Information Technology in Environment and Human Health

LEARNING OBJECTIVES

Age of Information Technology • INSAT System • IRS Satellites • Telemedicine • Ocean Development • NCAOR • OOIS, Marine Resources Research SSTWS • ENVIS

Global scenario has rapidly so much changed due to scientific advancement that the present period is popularly referred to as an **Age of Information Technology** (IT-age). People sitting at any distant place are now able to assess detailed global information of their interest in almost no time, and it has led to develop the whole world as one family.

IT (Information Technology) has made unbelievable progress in mass communication making the world up-to-date on various problems. This technology has given a new concept in business, demand and supply. And now IT is being used in Environment and Human health which are problems of great concern to human beings.

IT is now playing a major role to deal with various problems of environment, such as Climatic Changes, Global Warming, Cyclones, Floods, Tsunami, Earthquakes etc. (already discussed separately). IT is a recent tool for communication and dissemination of information for the protection of environment and sustainable development.

IT is now making available details of environmental problems and solutions. Such informations are used in improving the quality of Environmental Education. Since experts of various fields of the world are involved in the preparation of information, one is now able to obtain authoritative, up-to-date and exhaustive details of the subject. These details are available on different web sites and internet to schools, institutions and even to individuals. Such systems appear to rejuvenate the concept of *Vasudhaiv Kutumbakam*.

Role of Information Technology in education and human health is briefly discussed in reference to the India.

Indian National Satellite (INSAT) System

It is one of the largest domestic communication satellite systems in the Asia-Pacific region and has revolutionized the country's communication sector. It is now also providing meteorological services through very High Resolution Radiometer and CCD cameras. This part of satellite carries cyclone monitoring through meteorological imaging and issues warning on impending cyclones through disaster warning receivers. For this, 350 receivers have been installed along the east and west coasts (December 2005). INSAT-3A spacecraft provides distress alert and position location service.

Indian Remote Sensing (IRS) Satellites

Indian Space Research Organisation (ISRO) has launched IRS satellites with the state-of-the-art cameras which take pictures of Earth in several spectral bands. Some important such satellites are- IRS-1C, IRS-1D, OCEANSAT-1, RESOURCESAT-1 etc. Imagery sent by these satellites is being used for ground and surface water harvesting, monitoring of reservoirs and irrigation command areas to optimise water use. Forest survey and management and wasteland indentification and recovery are other uses in his regard. IRS data is being used for urban planning, flood prone area identification and the consequent suggestions for mitigation measures. The data has also helped to evolve concept of *Integrated Mission for Sustainable Development, i.e.,* space craft imagery data integrated with the socio-economic data obtained from conventional sources to achieve sustainable development.

India is a signatory to the International Charter on Disaster Management and is providing remote sensing data for the same.

Telemedicine

IT is a major potential contributor in a large number of economic and social sectors. Of these, telemedicine is one such area which utilises this technology for affecting specialised consultations for diagnosis and treatment of diseases at a distance. Telemedicine helps patients in rural and distant areas to avail timely consultations of specialist doctors without going through the ordeal of travelling long distances. Such IT services will benefit 70 per cent of our population that lives in rural areas. Govt. of India is initiating pilot schemes of telemedicine in the country. Teleconferencing has made it possible to get expert opinion and help from the specialists of the subject world over. Several recent and updated details about various health problems remain regularly available in different web sites, accessible to every one.

Ocean Development

IT is proving enormously helpful to the Department of Ocean Development (DOD) which was started in 1981. The technology is being used by DOD for sustainable and environment-friendly exploration and utilisation of marine living and non-living resources. The government has recently set up a centre of Marine Living Resources and Ecology at Kochi. Considering the devastating effects of Tsunami (2004) in Andamans and East coast, Govt. has taken initiative to set IT based Tsunami Warning System in Indian Ocean (to be completed by 2015).

National Centre for Antarctic and Ocean Research (NCAOR)

It is an autonomous institution of its kind being set up at Goa under Antarctic Research Programme. At Antarctica, it carries out scientific experiments in the field of Measurement of Greenhouse gases, Tele Seismic studies, Study of Crack propagation on Ice Sheet at the Indian station, **Maitri**.

Ocean Observation and Information Services (OOIS)

Climatic variability in the recent past has caused considerable impact on the weather pattern which has resulted in droughts, floods and extreme heat events in the country and also in the various parts of the world. These cause a great deal of impact on country's economy. Recognising the importance of information and knowledge of the sea, the Government of India has began to implement an Integrated Programme on Ocean Observation and Information Services (OOIS) which comprises four major units: *(i)* Ocean observing system, *(ii)* Ocean Information services, *(iii)* Ocean Modelling and Dynamics (INDOMOD) and *(iv)* Satellite Coastal Oceanographic Research (SATCORE). It's National Institute of Ocean Technology (NIOT) supplies data on environmental monitoring and climate studies. The Ocean Information Services provide data on Sea Surface Temperature, Potential Fishing Zone (PFZ), Upwelling zones, eddies, chlorophyll, suspended sediment load etc.

PFZ Advisories are disseminated thrice a week to over 225 nodes located in Gujarat, Maharashtra, Karnataka, Goa, Kerala, Tamil Nadu, Andhra Pradesh, Orissa, West Bengal, Lakshadweep and Andaman & Nicobar Islands.

Under the project of INDOMOD, several models have been developed with primary applications about prediction of monsoon variability, storm surges with cyclones, waves, biological productivity and coastal processes.

Marine Resource Research

IT is providing services through Fisheries and Oceanographic Research vessel *Sagar Sampada* run under Centre for Marine Resources and Ecology, Kochi in studies pertaining to the assessment of environmental parameters, primary and secondary productivity, upwelling, investigation on toxic algal bloom, secondary production studies on biomass, distribution pattern of species at different depths in Exclusive Economic Zones (EEZ) of ocean etc. The study also envisages concentration of heavy metals and pesticide residues to mitigate environmental pollution of assimilative capacities of coastal marine areas for contaminant introduction and to enable the Pollution Control Boards to evolve a mechanism for mitigating the adverse effect of pollution on marine environment.

The Integrated Coastal and Marine Area Management (ICMAM) using satellite based studies plan to amend shoreline management with various programmes scheduled for tenth five-year plan period, which includes R&D activities along with Marine Eco-toxicology.

Establishment of Storm Surge and Tsunami Warning System (SSTWS)

Govt. of India, using IT technology, plans to establish Storm Surge and Tsunami Warning System to prevent further disasters of coastal lives. One can not forget the devastation caused by Indian Ocean Tsunami of 26 December 2004. The SSTWS is proposed to be completed by September 2007, and it will strengthen the existing seismological network to indicate near real time occurrence of Tusnamigenic earthquake, as the sensor will be set at the ocean bottom at appropriate locates in the Indian Ocean. Hope in future IT will be able to provide information to warn to reduce the effect of devastations caused by oceanic earthquakes.

Environmental Information System (ENVIS)

Considering the immense value of Environmental Information, Govt. of India, in December, 1982, established ENVIS as a plan programme under the Ministry of Environment and Forests. It provides environmental information to decision makers, policy planners, scientists, engineers, research workers etc. ENVIS network has 78 subject-specific and state related centres (upto March, 2006). Its immediate objective is to:

(i) provide national environmental information services relevant to present needs and capable of developments to meet the future needs of the users, originators, processors and disseminators of information.

(*ii*) build up storage, retrieval and dissemination capabilities, with the ultimate objective of disseminating information speedily to the users.

(*iii*) promote national and international cooperation and liaison for exchange of environment related information.

Environment information plays a vital role not only in formulation of environmental management policies but also for decision making processes.

The information of ENVIS is available on Website of the Ministry (url: http://envfor.nic.in). Information has also been arranged in various relevant heads which include current events, clearances, legislation, Parliament matters, funding schemes etc.

Chapter Summary

Rapid advancement in Information Technology and total dependence of revolutionary society to this has presented the modern phase of world as if it is in an age of IT. Unbelievable progress in mass communication is typically changing the global concept and world is now appearing as one unit or one family. Like other fields, It has also played a great role in environment and human health.

INSAT system is providing meteorological services. Their advanced appliances carry cyclone monitoring and issue warning on impeding cyclones. INSAT–3A spacecraft provides distress alert. IRS satellites are being used for ground and surface water harvesting and monitoring of reservoirs. Such data help to evolve concept of Integrated Mission for Sustainable Development. Telemedicine provides specialized consultations for diagnosis and treatment of diseases. Teleconferencing permits consultation and discussion with experts sitting at any distance. DOD uses IT for sustainable and environment friendly explorations. NCAOR carries out Measurement of Greenhouse gases, Tele-seismic studies and studies on Crack propagation on Ice Sheet at Maitri, Antarctica. OOIS provide data on sea surface temperature, potential fishing zone, upwelling zones, eddies, chlorophyll, suspended sediment load etc. INDOMOD project concentrates on prediction of mansoon variability, storm surges, cyclones, biological productivity and coastal processes. ICMAM uses satellites for shore-line management and marine Eco-toxicology. SSTWS is in process of establishment to use IT to prevent disasters of coastal lives. One can not forget Tsunami 2004 devastation in several countries including India. ENVIS provides environmental information to decision makers, policy planners, scientists, engineers, research workers etc. It plays a vital role in the formulation of environmental management policies .The website of ENVIS is url:http://envfor.nic.in.

Study Questions

1. Discuss the role of information technology in environment and human health.
2. Write short notes on the following with reference to environment and human health-

   (*i*) INSAT system (Indian National satellite system)

(ii) IRS satellites (Indian Remote Sensing satellites)
(iii) Telemedicine
(iv) NCAOR (National Centres for Antarctic Ocean Research)
(v) OOIS (Ocean Observation and Information Services)
(vi) Marine resources
(vii) SSTWS (Storm Surge and Tsunami Warning System)
(viii) ENVIS (Environment Information System)

Objective questions. *Select Correct answers*

1. Govt. of India established ENVIS in:
   (1) 1952 (2) 1982
   (3) 1992 (4) 2002
2. Indian Research station Maitri is located at:
   (1) Bombay (2) Chennai
   (3) Kochi (4) Antarctic

Answers

1. (2) *2.* (4)

Unit VIII

Field Work

Field Work

Field work provides an opportunity to understand micro-level information of wide variety of environmental problems India is facing. Project work, based on field study, individually or as a group provides a chance of being a part of problem and to become an explorer of solution. Environmental studies related field work and project preparation offers the opportunity to understand the real nature and complexity of the environment.

For field work on general environmental studies, a project should be prepared by visiting different places of various environments. Guidance should be taken from the teacher concerns to record various factors significant to that place. Such activities encourage students to know directly about the environment. The detailed contact with the places make them to learn problems and needs of that environment, and cultivate personal attention and responsibilities towards those places. Some outline helps are given as follows:

PROJECT-1

Visit to local area to decumbent environment assets-River/Forest/Grassland/Hill/Mountain

While making a visit to any place like River, Forest, Grasslands, Hills or Mountains, following are common features to be recorded-

(1) Place of visit
(2) Date and period of visit
(3) Guide teacher and accompanying students
(4) Climatic conditions of visiting duration
(5) Size and location of area undertaken for studies
(6) Observations

Observations should be recorded on two different aspects

(i) Abiotic components. Nature and conditions of non-living components should be recorded on some of following lines such as-
Light Condition and availability of light

Perceptions Average annual rainfall (to be recorded from meteorological information centres)

Temperature Range of annual temperature (from meteorological department) and temperature of the time.

Soil conditions Such as sandy, clayey, stoney, dry or water logged, status of litters, fertility of soil (as observed by type of vegetation present there), nature and level of soil pollution etc.

Water conditions such as clarity of water, pH, temperature, pollution nature, vegetation conditions etc.

Air conditions such as cold, dry, windy, hot, polluted or fresh (if polluted then source and nature of pollutants) etc.

(ii) Biotic components. Occurrence of plants and animals should be separately recorded under different categories of biotic components. Observations on following lines can be carried out-

(1) *Producers*

Types of plants should be recorded with different aspects herbs, shrubs, trees, climbers, creeping, prostrate, procumbent, annuals, perennials, height variations etc. For examples, common or scientific names can be used. In aquatic systems phytoplanktons, free floating, suspended, submerged rooted, emerged rooted, dominance of types of plants etc. should be recorded.

Names of common producers found in different ecosystems are given in Chapter 13 of Unit 3.

Efforts should be made to evaluate impact of climatic conditions on occurrence, distribution, forms and habits of the plants. Figures of normal and affected plants can be drawn to emphasize the environmental conditions and their effects on vegetation.

(2) *Primary Consumers*

These are herbivorous animals such as small fish in aquatic systems and rat, rabbit, squirrels, cow, buffalo, deer etc. in terrestrial ecosystems. Common examples of these organisms should be noted and a tentative speculated and proportionate occurrence may be specified. Figures of common primary consumers should be drawn specifying their preference of nutrition. Life forms and fidelity may be accounted. Special emphasis on birds should be given.

Common examples of primary consumers of different ecosystems are listed in chapter 13 of unit 3.

(3) *Secondary consumers*

These are primary carnivores feeding on herbivorous animals and should be located carefully as these can harm the students. Describe their shape, feeding habit, dwelling names or scientific names. Draw their figures in the project reports.

> Common examples of secondary consumers of different ecosystems are given in chapter 13 of unit 3.

(4) *Top consumers*

These are top or secondary carnivores and strict predators and obtain their food by the killing of primary carnivores. Describe their names (common or scientific), shape and feeding habits. Diagrams should drawn explaining methods of killing their targets. As such, these may not be available to observe, but list them according to specified top consumers of various ecosystems.

> A list of top consumers, as found in different ecosystems, has been given in Chapter 13 of unit 3.

Inference

After describing all these components a summary of overall impact of that place should be outlined throwing light towards merits and demerits of the place and on effects of biological interference on flora and fauna. Personal opinion and suggestive mitigating measures should also be discussed and concluded. Special mention about receiving inspiration from such field studies should be finally included.

PROJECT –2

Visit to a local polluted site – Urban/Rural/Industrial/Agriculture

For such studies, polluted sites are selected such as – in cities these sites may be an area around thermal power station or thickly populated market, – in industrial studies- survey of air pollution or site of discharge of effluents and in agriculture, field supplied with polluted water, or heavy pesticide sprays or heavily fertilized part may be selected.

Urban populated site/ Pollution factors of cities

The main aim and objective of such studies is to make students aware of pollution factors and their effects on various aspects. In urban area the main factor for such studies is to take an account of either air pollution, noise pollution or sewage pollution. Their sources, intensity and effects should be taken up in details.

An account of air pollution should include the discharge and presence of air pollutants and types of their predominance. Their effects should be described taking support from teachers and literature available. Now in several cities, particularly in metropolis, CPCB has installed self analysing pollution indicator machines with expressive boards. Such boards indicate information on various pollutants such as SPM, RSPM, level of SO_2 and oxides of nitrogen etc. These data are quite helpful in preparation of the project and a report on monthly or any duration basis can be nicely prepared.

In cities about 85 per cent noise pollution is caused by automobiles which can be measured by different types of sound meters. The data can be expressed in the form of percentage contribution of various types of automobiles. Effect of noise pollution can be surveyed by peacefully observing the behavior of local people. A description of effect of noise pollution on human health may be accounted in the project. A special observation should be made in places of no sound pollution such as hospitals and institutions. Emphasis on uncultured behavior of different classes of the society may also be considered.

At the end, inference should be drawn logically. Some figures of sources of air pollution and noise pollution be expressed with graphic representations. A photograph of CPCB boards may be taken along with the site. The effects of air pollution on plants and animals and on human beings should be separately elucidated.

Rural Polluted Sites

Important polluted sites in rural areas can be different places of garbage and sewage discharges. These cause soil pollution, water pollution and also air pollution. Details of these spots with photographs should be noted and methodically expressed in the project. Effects of these factors on surrounding plants and animals should be specially recorded. Photographs of sick plants and animals should be taken to express the observations. An account of various pollutants and their effects on human beings should be emphasized. Measures to make rural residents aware about such pollution, should be suggested and elucidated with simple pictorial illustrations.

Inference and feelings developed while visiting to such area may be expressed honestly.

Industrial Polluted Site

A comprehensive field study should be carried out by visiting the industrial areas of the city. Special records should be mantained on various aspects of pollution, such as source of **air pollution** – (including names, location and kind of industries), nature of pollutants, hazardous effects of such pollutants, quality of air, approximate amount of fly ash, effects of air pollution on plants, animals and human beings of such area. A special note should be recorded about health hazards of human beings including workers, and **water pollution**- including discharge of industrial effluents, their common chemical composition and the impact of these discharges on plants and animals of that area. Studies may be extended to record biotic and abiotic components of water bodies of the area.

These observations may be concentrated on feature like-

I. *Water*

(i) Physical properties –colour, smell, temperature, cleanliness, transparency etc.

(ii) Chemical properties –pH (acidity/alkalinity), suspended solid particles, free CO_2, soluble O_2, hardness, salts, BOD and COD.

(iii) Biological properties –*Plants* – aquatic plants (free floating, submerged, emergent etc.), terrestrial plants (herbs, shrubs, climbers, procumbent, trailors, trees etc.), *Animals* - aquatic animals and terrestrial animals etc.

II. *Air*

(i) SO_2, NO_X, CO, HC etc.

(ii) SPM, RSPM etc.

III. *Plants*

(i) General appearance of plants

(ii) Habits of plants

(iii) Effects on leaves, flowers and fruits

(iv) Quantitative abundance etc

IV. *Animals and Human beings*

(i) Types of animals

(ii) General health of animals

(iii) Common diseases found in animals

(iv) General health and diseases of local people etc.

Finally inference and opinion should be expressed.

PROJECT –3

Study of Common Plants, Birds and Insects

Appearance of environment depends upon plants and animals of that area which clearly indicate about the climate. These should be studied in detail, seeking the support of teachers and described in certain systematic manner. Such studies should be made periodically to know the proper details of the area. Names, figures and functional roles of these members should be accounted. Such activities cultivate interest towards environment and have long lasting impacts.

The common members found in plain regions of the country are briefly as follows-

Plants

Herbs – *Ageratum, Rumex, Polygonum, Chenopodium, Boerhaavia, Phyllanthus, Argemone, Achyranthus, Elicpta* etc.

Shrubs – *Cestrum, Tabernaemontana, Calotropis* (aak or madar*), Lantana, Convolvulus, Capparis, Bougainvillea* etc.

Tress – *Mangifera* (mango tree), *Syzigium* (Jamun tree), *Zizyphus* (ber), *Citrus* (neebu), *Azadirachta* (neem), *Acacia* (babool), *Eucalyptus, Salmalia* (red silk tree), *Dalbergia* (sissoo), *Caesalpinia, Cassia, Tamarindus, Emblica* etc.

Insects – Butterfly, Honey bee, Locust, Aphids, Ants, Termites, Mosquitoes, Dragon fly etc.

Birds – Crow, Eagles, King fisher, Parrot, Peacock, Pigeon, Stork, Indian darter, Hornbill, Pheasants, Partridge, dove, Heron, Spoon bills etc.

PROJECT–4

Study of simple ecosystem – ponds, river, hill slopes etc.

Details of various ecosystems are given in Unit 3 and chapters 11-16 and accordingly such projects can be prepared. Special mention to figure and details of forms may be taken up. Pyramids can also be prepared.

Appendices

APPENDIX 1

Web sites of Some Important Policies, Conventions and Related Documents.

(1) Convention on Biological Diversity (CBD), Rio, 1992. www.biodiv.org/doc/legal/cbd-en.pdf

(2) UN Framework on Convention on Climate Change (UNFCCC), New York, 1992.
www.unfcc int/resource/conv/conv.html

(3) Convention on Persistent Organic Pollutant (The POP Stockholm Convention), Stockholm, 2001.
www.pops.int/documents/convtext/convtext_en.pdf

(4) Convention on Access to Environmental Information, Public Participation in Environmental Decision Making and Assess to Justice (The Aarhus Convention), Aarhus, 1998.
www.unece.org/env/pp/documents/cep43e.pdf

(5) Convention to Combat Desertification in Countries Experiencing Serious Drought and /or Desertification, 1994, 1998. www.unced.int/convention/text/convention/php.

(6) International Tropical Timber Agreement (ITTA), Geneva, 1994, 1996, 1997.
www.itto.org.jp

(7) Convention on Transboundary Prevention of Preparedness for the response to Industrial Accidents-Effects of Industrial Accidents (UN/ECE) (The Accident Convention, Helsinki, 1992).
www.unece.org

(8) Convention on the Prior Informed Consent Procedure for Certain Hazardous Chemicals and Pesticides in International Trade (UNEP/FAO) (The PIC Rotterdam Convention), Rotterdam, 1998. www.pic.int/en/View Page.asp?id =104#preamable

(9) World Commission on Forest and Sustainable Development. www.iisd.org/wcfsd/

(10) Citizens Guide to the World Commission on Dams. www.irn.org/wcd/wedguide.pdf

(11) Report of the World Commission on Dams. www.dams.org

(12) UN Water Development Report : Water for People, Water for life. www.unesdo.unesco.org/images/0012/001295/1295556e.pdf

(13) World Summit on Sustainable Development (WSSD Johannesburg) Reports. www.uneptie.org/outreach/wssd/joburgreport.htm

(14) A Compendium of Fisheries Agreements. www.oceanlaw.net/texts/index.htm

(15) Independent World Commission on the Oceans. www.waterland.net/iwco/

(16) Living Planet Report, October 2004. www.panda.org/downloads/general/lpr2004.pdf

(17) Global Environment Outlook 3. www.unep.org./geo/geo3/english/pdf.htm

(18) Millennium Development Goals. www.undp.org/mdg

(19) World Charter for Nature. www.un.org/documents/ga/res/37/a37r007.htm

(20) World Scientist's Warning to Humanity, 1992. www.ucsusa.org/ucs/about/page.cfm?page1D=1009

(21) Scientists Call for Action. www.ucsusa.org/global_environment/archive/page.cfm?page ID=530

(22) Convention on the Protection and Use of Transboundary Watercourse and International Lakes (The UN/ECE Water Convention). Helsinki, 1992. www.unce.org/env/water/text/water-convention/text11toc.htm

(23) Convention on the Control of Transboundary Movements of Hazardous Wastes and their Disposal, Basal, 1989. www.basel.int/text/con-e.htm

(24) Convention for Protection of Ozone Layer (UNEP) (Vienna Convention) Vienna, 1985. www.unep.org/ozone/vienna.shtml

(25) Protocol to the Convention for the Protection of the Ozone Layer on Substances that Deplete the Ozone Layer (MONTREAL protocol) as amended, Montreal, 1987.

www.unep.org/ozone/montreal.shtml

(26) Convention on Long-range Tranboundary Air Pollution (CLRTAP) (UN-ECE), Geneva, 1979.
www.unece.org/env/lrtap-hl.htm

(27) Protocol to the Convention on Long –range Transboundary Air Pollution on Persistent Organic Pollutant (POPs), Aarhus, 1998.
www.unece.org/env/irtap/pops-hl.htm

(28) Convention on International Trade in Endangered Species of Wild Fauna and Flora (CITES), 1973.
www.cites.org/eng/dise/text.shtml

(29) Convention on Physical Protection of Nuclear Material, 1980.
www.unodc.org/unodc/terrorism-convention-nuclear-material.html

(30) Convention on the Conservation of Migratory Species of Wild Animals (CMS) (UNEP), Bonn, 1979.
www.wcmc.org.uk/ems

(31) Convention for the Protection of the Mediterranean Sea against Pollution (Barcelona Convention), Barcelona, 1976.
www.unep.ch/seas/main /med/mediconvi.html

(32) UN Conference on Environment and Development, 1992.
www.un.org/genifo/bp/enviro.html

(33) Rio Declaration on Environment and Development, Principles of Forest Management.
www.iisd.org/educate/learn/rio-decl.doc

(34) Agend 21: Global Programme of Action for Sustainable Development.
www.unep.org/Documents/Default.asp? Documed ID=52 & Article ID=49

(35) United Nations Convention on the Law of the Sea (UNCLOS) of 10 December, 1982.
www.un.org/Depts/los/convention-agreements/texts/unclos/UNCLOS-TOC.htm

(36) The FAO Code of Conduct for Responsible Fisheries.
www.fao.org/fi/agreem/codecond/ecodecon.asp

(37) UN Agreement on Straddling Fish Stocks and High Migratory Fish Socts (UNFA).
www.un.org/Depts/los/convention-agreements/convention-overview-fish-stocks.htm

APPENDIX 2

Timeline Calendar of Important Environment Mishappenings

1973 Drought in Sahel, Africa – kills millions of people.
– First Major Oil Crisis.

1976 Major Earthquake in Tangshan in China and Guatemala.

1977 Toxic Chemical Leaks into homes in Love Canal Niagara Fall, USA.

1978 Floods in West Bengal – Kills 1300 people and destroy 1.3 million homes.

1979 Major accident at the Three Mile Island nuclear power plant, USA.
– Six hundred fourty km oil slick in the Gulf of Mexico from a blow-out.

1983 Monsoon storms in Thailand claim 10,000 victims

1984 Famine in Ethiopia caused by exceptional long lasting drought.
– Chemical accident in Bhopal, India due to leakage of methyl isocyanide in Union Carbide – Kills about 8,000 people and maims many more.
– Typhoon in Philippines – Kills 1300 people and leaves 1.2 million homeless.

1986 World's worst nuclear disaster at Chernobyl, Russia, spreading radioactive fall out over large areas of Europe.
– Fire in Basal, Switzerland, releases toxic chemicals into Rhine, killing fish as far north as to Netherlands.

1988 Hurrican Gilbert kills 320, leaves 750,000 homeless and causes damage worth US$ 10 billion in the Caribbean, Mexico and USA.

1989 Exxon Valdez releases 50 million litres of crude oil into Prince William Sound, Alaska.

1990 Millions of litres of crude oil spilled and burned during the Gulf way.
– Chemical explosion from an over-turned tanker in Dahanu, Maharashtra, India-Kills more than 100 people.

1994 Burst pipeline spills thousands of tones of crude oil on the Tundra on the Kori Peninsula, Russia.

1998 Warmest year of the Millenium
– Extensive forest fires in Amazonia and Indonesia.

2000 Size of Ozone Hole reaches record high, affecting tip of South America.

2003 Serious floods in Mexico and Venezuela
- Heatwaves in Europe
- Size of the Ozone Hole over the Antarctica matches with 2000's size.
- Sustained drought in Argentina, Bolivia, Poraguay and Euruguay.
- Break up of the Arctic's largest ice-shelf: the World Hunt
- Big forest fire in California, USA
- Catastrophic earthquake in the city of Ban in SE Iran

2004 Flood in Haiti and Dominican Republic
- Tsunami, triggered by massive undersea earthquake off Sumatra in Indonesia, leaves over 300,000 dead and many more homeless in India, Sri Lanka and South East Asia.

Glossary

Abatement. To minimise pollution through waste treatments or reuse.

Abiotic. Non-living.

Abyssal zone. A zone of no effective light penetration or at water depth greater than 2000 m.

Acid rain. Fall of atmospheric acidic components (such as H_2SO_4 and HNO_3) along with rains, mist or snow. Such components are formed due to atmospheric (air) pollution.

Activated sludge. Bottom sediments with microbes of the oxidation ponds to be used for aerobic decomposition of sewage.

Adaptation. Adjustment of an organism to its environment.

Aeolin. A sediment deposit carried by wind.

Aerosol. Fine liquid (mist) or solid (smoke) particles floating in air.

AIDS (Acquired Immuno-Deficiency Syndrome). A disease caused by Human Immunodeficiency Virus (HIV).

Albumenuria. Presence of albumen in human urine.

Algal bloom. Abundant growth of planktonic algae imparting colour to the water.

Allelopathy. Inhibition of growth of one organism due to the inhibitory chemical exudation by another organisms, sometimes referred to as allelochemical interaction and antibiosis.

Allergy. Hyper sensitivity to certain chemicals or organisms.

Alpine. A mountain zone covered with snow and above tree line.

Ambient. Surrounding on all sides.

Anemia. Deficiency of blood or haemoglobin.

Anoxia. Oxygen deficiency to tissues.

Aphytic zone. Area of water body lacking plants due to depth factor.

Aquaculture. Soilless culture or hydroponics.

Arboreal. Pertaining to trees.

Arid zone. A desert zone with very low rainfall.

Ash. The solid residue left after thorough burning of combustible material.

Atmosphere. The gaseous envelope surrounding the earth.

Atomic wastes. Radio-active material discarded as waste (after use).

Automobile exhausts. Gases emitted by automobiles on the burning of the fuel. These contain CO, unburnt hydrocarbons, oxides of N and S and soot.

Background pollution. Pollution away from pollutant source.

Benthos. Bottom dwelling in a water body.

Bio-chemical Oxygen Demands (BOD). Microbial requirement of oxygen to decompose the organic matter. It is an indicator of water pollution.

Biocides. Agents that kill living organisms.

Biocoenosis. Ecosystem (living components).

Bioconcentration. Accumulation of specific organic or inorganic compounds within a specific tissue of an organism (usually applied for pesticides, heavy metals etc.).

Biodegradation. Oxidative breakdown of natural or synthetic substances by microbial activity.

Biodiversity. Varietal forms of living organisms.

Bio-geo-chemical cycles. Pathways of circulation of elements within an ecosystem, *e.g.* water and carbon cycles.

Biological amplification (or **magnification**). Successive increase in concentration of persistent chemical by organisms at various trophic levels in a food chain, *e.g.* DDT from algae to fish to fish eating birds and to man.

Biological extinction. The complete disappearance of a species.

Biome. A large community of plants and animals, characterized by its particular type of dominant vegetation and its associated animals.

Biomedical waste. Wastes that originate from hospitals and clinics, *e.g.* blood, diseased organs, medicines etc.

Biosphere. That part of earth and atmosphere in which organisms live.

Biotechnology. Use of biological process to produce useful materials.

Bronchitis. Inflammation of bronchiole tube.

Carbon Cycle. Cyclic movement of carbon from environment to organisms and back to the environment.

Carcinogen. Cancer causing agent.

Carcinoma. A malignant tumour of epithelial origin.

CEP. Commission on Environmental Planning (a unit of IUCN)

Chlorofluro-carbon (CFC). A chemical used as refrigerant or aerosol propellant. It causes ozone depletion in stratosphere.

Climax community. The final stable community developed in the process of succession.

Coastal zone. The area that extends from the high tide mark on land to the edge of the continental shelf, which is the submerged part of the continent.

Community. A complex interacting network of plants, animals and micro-organisms existing in a geographical area.

Consumer. An organism that feeds on producers or other organisms, ***i.e.*** a heterotroph.

Continental shelf. The submerged part of the continent at the edge of the coastal zone.

Coral reef. Stoney substances secreted by corals (reef: chain of rocks).

Crude Birth Rate (CBR). The number of live births per 1000 people in a population in a given year.

Crude Death Rate (CDR). The number of deaths per 1000 people in a population in a given year.

Cyclone. A tropical storm, or a system of winds that rotates about a center of low atmospheric pressure, often brings abundant rain.

DDT (Dichloro-diphenyl-trichlorethane) ($C_{14}H_9Cl_{15}$). A common insecticide (contact poison), now abandoned due to its persistence.

Decibel (db). A logarithmic scale for measuring the intensity of sound.

Decomposers. Microbes obtaining their nourishment from dead organic material.

Deforestation. Destruction of forest cover and the undergrowth.

Demography. Study of population, commonly used for human population.

Denudation. Wearing away of the land surface.

Dermatitis. Inflammation of skin.

Desertification. Process of desert formation.

Detritus. Debris of decaying organic matter.

Detrivore. A consumer that feeds on detritus.

Dichloro-difluoro-methane (Freon –12, a CFC). A colorless, odourless and tasteless gas used in refrigeration and is a strong destroyer of ozone.

Dieldrin. A persistent insecticide used to kill mites, now banned.

Earthwatch. An association of UNEP to monitor pollution of the earth.

Ecological extinction. A species left with few members and no longer playing normal ecological role in the community.

Ecological niche. Physical place occupied with its role in the environment.

Ecological pyramid. Graphic representation of biotic components of different trophic levels. These are drawn on the basis of either number of individual or biomass or energy levels.

Ecological succession. The orderly process of transition from one biotic community to another one in a particular area.

Ecology. The study of relationship between organisms and environment or study of organisms in relation to environment.

Ecosystem. The complete ecological system of an area including living and non-living components.

Effluents. Waste liquid discharges usually from industries that often pollute the environment.

Ecotone. The transitional zone between adjoining communities.

Endangered species. The species threatened or at the verge of extinction due to survival of only few members.

Endemism. Restricted distribution of species to an isolated region.

Energy flow. The flow of energy in a food chain, from one organism to the other.

EPA. Environmental Protection Agency of USA.

Erosion. Surface destruction and loss of soil fertility, mainly by water or wind.

Euphotic zone. Upper zone of ocean with enough light for photosynthesis.

Eutrophication. The enrichment of water by nutrients.

Exponential growth. Multiple growth of a quantity with time.

***Ex-situ* conservation**. Conservation to protect the species away from their natural habitat.

Extractive reserve. Protected forests with restricted permission of use without harming them (forests).

Fall out. The descending radio active fission products from atmosphere to the surface of earth.

FAO. Food and Agriculture Organisation of the UNO.

Flue gas. Mixture of gases emitted from the chimney of the factory or industry.

Food chain. A series of organisms arranged in a linear manner with repeated eating and being eaten.

Food web. A complex network of interconnected food chains.

Fossil fuels. Fuels derived from fossil plants and animals that lived 200-500 million years ago, such as coal, petroleum and natural gases.

Fresh water. Potable water with less than 0.2 per cent salt content.

Genetic diversity. Genetic variabilities among individuals within a species.

Global warming. Rise in the temperature of earth due to abnormal increase in greenhouse gases.

Green economics. Economics that seeks to conserve the environment and increases employment and income.

Greenhouse effect. Hot and humid environment due to absorption of heat by greenhouse gases like CO_2.

Green revolution. A rapid growth of food production through the use of high yielding varieties.

Habitat. A place of living in the environment.

HIV. The Human Immuno-deficiency Virus that causes AIDS.

Homeostasis. The capacity of ecosystem to maintain a balanced state.

Humification. Formation of humus.

Humus. More or less decomposed finely divided amorphous organic matter in the soil.

Hurricane. Violent storm with very strong winds (found in Western Atlantic Ocean).

Hydrosphere. The part of the earth composed of water (*e.g.* ocean, sea, ice cap, lake etc.)

IAP. Index of Atmospheric Purity.

Indicator species. Species very sensitive to any specific environment that give early warning of problem.

***In-situ* conservation**. Conservation in protected natural habitat.

Intertidal zone. The area of shore line between the low and high tides, *i.e.* transition zone between the land and the ocean.

Ionosphere. The layer of atmosphere above the stratosphere where ionization occurs.

Itai-itai. A disease of bone atrophy.

IUCN. International Union for the Conservation of Nature and Natural Resources, with headquarter at Gland, Switzerland.

Kaposi Sarcoma. A kind of lethal cancer of tissues affected with AIDS.

Keystone species. Species that play roles affecting many other organisms in an ecosystem. Such species determine the ability of a large number of other species to survive.

Landfill. An area on which municipal waste is dumped.

Lithosphere. The solid rock crust of earth.

Littoral. Shallow zone of water body.

London smog. Air pollution with reducing nature due to oxides of sulphur, soot and smog. It has been related with the disastrous air pollution accident of London.

Los Angeles smog. Air pollution of oxidizing nature (presence of nitrogen oxides and ozone) forming a photochemical smog. It has been related with air pollution accidents of Los Angeles, USA.

Mangrove Vegetation. Plants with salt tolerant trees characterised by stilt roots and shallow marine sediments.

Minamata tragedy. A disastrous tragedy occurred in Minamata Bay of Japan due to methyl mercury pollution. The disease known as Minamata disease causes paralysis and disabilities.

National Park. A well-demarcated and protected area of natural landscape.

Natural Reserve. A well protected area to conserve its natural fauna and flora.

Oil spills. Large scale leakage of oil affecting the plant and animal life.

Organic farming. Farming supported with traditional fertilizers like mannures and not using chemical fertilizers and pesticides.

Ozone Depletion Potential (ODP). Refers to amount of ozone depletion caused by a substance.

Ozone layer. Refers to layer of ozone found in the stratosphere of the atmosphere (some 30-50 km. above the earth).

PAN. Peroxyacetyl nitrate, a secondary pollutant found in smog and is oxidizing in nature.

PCBs. Polychlorinated biphenyls, a highly persistent toxic chemical compound. They can pass through food chains.

Pelagic. Free from bottom and shore effects (marine communities).

POPs. Persistent organic pollutants, toxic chemicals organic in nature.

Photochemical smog. A form of air pollution formed by chemical reactions between sunlight, unburnt hydrocarbons, ozone and other pollutants.

Photosynthesis. A process found in green plants for the manufacture of food using sunlight, water and CO_2.

Phytoplanktons. Free floating micro-producers found in aquatic ecosystems.

Pollution. Undesirable change in the environment harmful directly or indirectly to the human welfare.

Potable water. Clean water suitable for drinking.

Primary air pollutants. A pollutant released directly from a source into the atmosphere.

Producers. Autotrophic green plants.

Putrefaction. Anaerobic decomposition of proteins by microbes.

Rainforest. A type of forest found in hot and humid region near equator.

Remote sensing. Procuring information from far away recording device such as aerial photography, radars and UV and IR detectors.

Reverse osmosis. Purification of water by forcing it through a semi-permeable membrane.

Secondary air pollutant. A pollutant produced from chemical reactions involving primary pollutants.

Smog. A combination of smoke, fog and chemical pollutants, commonly found in air polluted cities.

Species diversity. Number of plant and animal species found in a community or ecosystem.

Stratosphere. A zone of atmosphere found above troposphere.

Sustainable development. Development that meets the needs of today without harming the future generations to meet their own needs.

Swamp. A wetland dominated by trees and shrubs.

Temperate forest. A forest found in region with freezing winter and warm and humid summer.

Threatened species. A species with declining numbers and is likely to become endangered.

Trophic. Related with food and feeding habits.

Trophic level. The specific feeding stage of an organism in the ecosystem.

Troposphere. Lower zone of atmosphere extending upto 15-18 km above the earth surface.

Tsunami. Large waves in the ocean, often formed due to the earthquake.

Typhoon. Violent storm with strong winds, found in Western Atlantic Ocean.

Water cycle. Cycle of water that includes precipitation, evaporation and flow to the ocean.

Water harvesting. The process of catching the rain water.

Water shed. The area drained by a stream or a river.

Zero population growth. Constant population without increase.

Selected Readings

Alam, A. (ed.), 2003. *Human Rights in India: Issues and challenges.* Raj. Pub., Delhi.

Atchia, M. and S. Tropp (Eds), 1995. *Environment Management: Issues and Solutions*. John Wiley & sons.

Brasseur, G. 1987. The endangered ozone layer: New theories on ozone depletion. *Environment*, **29 (1)**:2

Burton, I., R.W. Kates and G.F. White, 1978. *Environment as hazard.* Oxford Univ. Press, N.Y.

Caldwell, L.K., 1991. *International Environment Policy*: *Emergency and Dimensions*. (2nd ed.). Affiliated E.W.P., N.D.

Chary, S.N. and V. Vyasulu (eds.), 2001. *Environmental Management, an Indian Perspective*. Macmillan India Ltd.

Clark, W. C. and R. E. Munn (eds.) 1986. *Sustainable development of the biosphere*. Cambridge Univ. Press, Cambridge.

Cunningham, W.P. and M.A. Counningham, 2003. *Principles of Environmental Science: Inquiry and Applications*. Tata McGraw Hill Pub., N.D.

Dix. J.H., 1989. *Environmental Pollution: Atmosphere, Land, Water and Noise*. Wiley Chichester.

Douglas, I., 1983. *The Urban Environment*. Edward Arnold Pub., Baltimore.

Eckholm, E.P., 1977. *The Picture of Health: Environmental Source of Diseases*. W.W. Norton, N.Y.

Eckholm, E.P., 1991. Biological diversity and economic development. (In *Down to Earth*–Environment and Human Health). Aff. E.W. Press, N.D.

Ehrich, C., 1980. The next flood knocking at the door: a foreigner's view on the ecological crisis in India. (In *Studies on Himalayan Ecology and Development Strategies*, Ed. Singh, T.V.), English Book Store, N.D.

FAO-UNEP Studies, 1981. *Forestry Resources of Tropical Asia, Africa and America*. FAO, Rome.

Firor, J. and J.E. Jacobson, 2003. *Crowded Greenhouse*. Univ. Press, N.D.

Frank, F. (Ed.), 1973. *Water–A Comprehensive Treatise*. Plenum Press, N.Y.

Frankel, O.H. and M.E. Soule, 1982. *Conservation and Evolution*. Cambridge Univ. Press, Cambridge.

Frankena, W.K., 1979. Ethics and Environment. (In *Ethics and Problems of 21st Century*, Eds. Goodpaster, K.E. and K.M. Sayre). Univ. of Notre Dame Press, Notre Dame, Indiana.

Frekla, T. 1981. Long term prospects of world population growth. *Pop. Dev. Rev.* **7(3)**: 489.

French, H.F., 1990. *Clearing the Air: A Global Agenda*. World Watch Inst., Washington D.C.

Gadgil, M., 1994. Bio-diversity; Reckoning with life. In *The Hindu: Survey of Environ.*–1994. Madras.

Gandhi, P.R. 1995. *International Human Rights Documents*. London.

Gangrade, K.D., 2004. Erosion of Values. Silver Jubilee Vol. 2004-05. *Citizenship Dev. Soc.*, N.D.

Galley, F.F., 2000. *Primer for Environmental Literacy*. Univ. Press., N.D.

Gupta, V.K. and A. K. Ganguly, 1983. Environmental Impact of Nuclear Industry. In *Water Pollution and Management*, Ed. Varshney, C.K.), Wiley East. Ltd. N.D.

Gururaja, P., 1991. Energy: Non-conventional approach. *The Hindu Survey of Environ.*, 1991. Madras.

Heywood, V.H. and R.T. Watson, 1995. *Global Diversity Assessment*. Cambridge Univ. P., Cambridge.

Hileman, B., 1982. The Greenhouse effect. *Environ. Sci. Tech.* **17**: 237

Hordoy, J.E. and D. Satterthwaite, 1984. Third world cities and the Environment of Poverty. *Geoforum*, **15**:3

Ion, D.C. 1980. *Availability of World Energy Resources*. Graham and Tortman, London.

Ison, S., S. Peake and S. Wall, 2002. *Environmental Issues and Policies*. Prentice-Hall, N.D.

John –Roth, K., 2000. *International Encyclopaedia of Ethics*. S. Chand and Co., N.D.

Kothari, A., 1993. Biodiversity Conservation for Vanishing Species. *The Hindu Survey of Environ.*, Madras

Kumar, H.D., 2001. *Forest Resources: Conservation and Management*. Aff. E.W. Press. N.D.

Lewis, M., 2003. *Inventing Global Ecology: Tracking the bio-diversity ideal in India* (1945-97). Orient Long., N.Y.

Mace, G.M. and R. Lande, 1991. Assessing extinction threats towards revaluation of IUCN threatened species category. *Cons. Biol.* **5**:148.

Maduro, R. and R. Schauerhammer, 1992. *Hole in ozone scare*. 21st Century Sci. Associates, Washington D.C.

Machta, L., 1983. Effect of non-CO_2 greenhouse gases. In *changing climate*, Rep. of CO_2 Assess. Comm., N.A.P., Washington, D.C.

Meher –Homji, V.M., 1991. Climate change: the forest link. *The Hindu Surv. Environ.*, 1991. Madras.

Miller, G.T., 2001. *Environmental Science* (8th ed.). Brooks-Cole Thomas Learning Inc. U.S.A.

Miller, G.T., 2002. *Living in the Environment* (12th ed.). Brooks Cole Thomas learning Inc., U.S.A.

Misra, S.P., 2002. Water Blooms of River Chambal Ravine Wetland. *Res. J.P.L. Environ*. **18**:27-32, India.

Mitra, A.P. 1991. Global Warming: the greenhouse effect. *The Hindu Survey of Environ*. 1991. Madras.

Mayers, N., 1988. Threatened biotas: 'Hot Spots' in tropical forests. *The Environmentalists* **8**:187

Myers, N., 1990. Bio-diversity Challenge: Expanded Hot Spot Analysis. *The Environmentalists* **10**:126.

Nadkarni, M.V., 2006. *Hinduism A Gandhian Perspective*. Ane Books, India, N.D.

Nagor, A.P. 1996. *Biological Diversity and International Environmental Law*. A.P.H. Pub. Corp., N.D.

NCERT, 2003. *Journal of Value Education: Special issues on Environment Values*. **3** (2).

Peter, H.R., 1976. Ethics and Attitudes. In *Conservation of Threatened Plants*. Plenum Press, N.Y.

Punita, J.C., 1993. Flora and Fauna –they perish one by one. *The Hindu Survey of Environment,* 1993. Madras.

Ravindranath, N.H. and P. Sudha, 2004. *Joint Forest Management: Spread, Performance and Impact*. Univ. Press, N.D.

Repetto, R., 1991 (ed.). *The Global Possible: Resource Development and New Century*. Aff. E.W.P., N.D.

Robertson, A.H. and J.G. Merills, 2005. *Human Rights in the World*. Univ. Law Publ. Co., N.D.

Sahu, K.C., 1994. Power plant pollution: the cost of coal combustion. *The Hindu Survey of Environ*. 1994, Madras.

Sareiya, K.P., 1994. *Forest and Forestry* (5th ed.). N.B.T., N.D.

Schneider, S.H., 1989. Greenhouse effect: science and policy. *Science* **243**:771

Sharma, H.S. and T.T. Khan (eds.), 2003. *Environmental Conservation, Depleting Resources and Sustainable Development*. Avishkar Pub., Jaipur.

Sharma, J.N. and K.M. Gupta, 1996. *Value Education*. Citizenship Dev. Soc., N.D.

Shukla, P.P., S.K., Sharma, N.H. Ravindranath, A. Garg and S. Bhattacharya, 2003. *Climatic Change and India: Vulnerability Assessment and Adaptation*. Univ. Press., N.D.

Silver, C.S. and R.S. DeFries, 1991. *One Earth One Future: Our Changing Global Environment*. N.A. Sc., Aff. E.W. Press, N.D.

Subba Rao, S. 2001. *Ethics of Ecology and Environment*. Rajat Pub., N.D.

Swami, Dr. Parthasarthy, 2006. *Human Values Development*. Ane Books India, N.D.

Swaminathan, M.S., 1991. Sustainability: Beyond the economic factor. *The Hindu Surv. Agri.*, 1991, Madras.

Thaper, S.D., 1975. *India's Forest Resources*. Mac Millan and Co., N.D.

Tiwari, T.N., 1993. Ecological impacts of oil spills on land and water. In *Advances in Environmental Sciences*, Tripathi, A.K. *et al.* (eds.). Ashish Pub. House, N.D.

Tripathi, A.K. and S.N. Pandey, 1990. *Water Pollution*. Ashish Pub. House, N.D.

Tripathi, A.K., A.K, Srivastava and S.N. Pandey (eds.) 1993. *Advances in Environmental Sciences*. Ashish Pub. House, N.D.

Vaidyanathan, A., 1994. Water resources: What we do not know. *The Hindu Surv. Environ.* 1994. Madras.

World Commission on Environment and Development, 1987. *Our Common Future*. Oxford Univ. Press, Oxford.

Index

G